CW00591291

Heat Capacities
Liquids, Solutions and Vapours

Heat Capacities
Liquids, Solutions and Vapours

Edited by

Emmerich Wilhelm
Institute of Physical Chemistry, University of Wien (Vienna), Austria

Trevor M. Letcher
Emeritus Professor, University of KwaZulu-Natal, Durban, South Africa

RSCPublishing

ISBN: 978-0-85404-176-3

A catalogue record for this book is available from the British Library

© The Royal Society of Chemistry 2010

All rights reserved

Apart from fair dealing for the purposes of research for non-commercial purposes or for private study, criticism or review, as permitted under the Copyright, Designs and Patents Act 1988 and the Copyright and Related Rights Regulations 2003, this publication may not be reproduced, stored or transmitted, in any form or by any means, without the prior permission in writing of The Royal Society of Chemistry or the copyright owner, or in the case of reproduction in accordance with the terms of licences issued by the Copyright Licensing Agency in the UK, or in accordance with the terms of the licences issued by the appropriate Reproduction Rights Organization outside the UK. Enquiries concerning reproduction outside the terms stated here should be sent to The Royal Society of Chemistry at the address printed on this page.

The RSC is not responsible for individual opinions expressed in this work.

Published by The Royal Society of Chemistry,
Thomas Graham House, Science Park, Milton Road,
Cambridge CB4 0WF, UK

Registered Charity Number 207890

For further information see our website at www.rsc.org

Foreword

The role of energy in our society, whether it is from fossil fuels or an alternative source such as hydrogen, has been the topic of much discussion recently. An example of this is to be found in the volume entitled *Future Energy: Improved, Sustainable and Clean Options for our Planet* (Elsevier, Amsterdam 2008) edited by T. M. Letcher, published under the banner of the International Union of Pure and Applied Chemistry (IUPAC) and the International Association of Chemical Thermodynamics (IACT), which is an Associate Organization of IUPAC. Not surprisingly, there is also resurgence in the broad topic often termed thermochemistry that is covered by the remit of IACT, and was previously the concern of the IUPAC Commission on Thermodynamics. These organizations take an active role in the definition and maintenance of standards in the fields of thermochemistry. This role includes, but is not limited to, the recommendation for calorimetric procedures, the selection and evaluation of reference standards for thermodynamic-measurement techniques of all types, the dissemination of evaluated thermodynamic data of the fluid state, the standardization of nomenclature and symbols in chemical thermodynamics as well as the surveillance of international pressure and temperature scales and promotion of the subject with awards by IACT to young scientists (further information can be found at www.IACTweb.org).

This book, entitled *Heat Capacities: Liquids, Solutions and Vapours,* is also published under the auspices of both IUPAC and IACT. It is another in a lineage that started in 1956 with *Experimental Thermochemistry* Volume I (Interscience Publishers, Inc. New York) edited by F. D. Rossini, who is honoured at each biennial International Conference on Chemical Thermodynamics, sponsored by both IACT and IUPAC, with the delivery of the coveted Rossini Lecture. Volume II, edited by H. A. Skinner was published in 1962 (Interscience-Wiley, New York). An update of the material in these volumes was given in *Combustion Calorimetry*, edited by S. Sunner and M. Månsson (Pergamon Press, Oxford, 1979). There is in addition a series of volumes on "Experimental Thermodynamics", for which there are now seven volumes in print and an eighth in preparation: (a) Volume I, *Calorimetry of Non-reacting Systems*, edited by J. P. McCullough and D. W. Scott

(Butterworths, London, 1968); (b) Volume II, *Experimental Thermodynamics of Non-reacting Systems*, edited by B. LeNeindre and B. Vodar, (Butterworths, London, 1975); (c) Volume III, *Measurement of the Transport Properties of Fluids*, edited by W. A. Wakeham, N. Nagashima, and J. V. Sengers (Blackwell Science Publications, Oxford, 1991); (d) Volume IV, *Solution Calorimetry*, edited by K. N. Marsh and P. A. G. O'Hare (Blackwell Science Publications, Oxford, 1994); (e) Volume V, *Equations of State for Fluids and Fluid Mixtures*, edited by J. V. Sengers, R.F. Kayser, C.J. Peters, and H.J. White Jr. (Elsevier Science, New York, 2000); (f) Volume VI, *Measurement of the Thermodynamic Properties of Single Phases*, edited by A. R. H. Goodwin, K. N. Marsh, and W. A. Wakeham (Elsevier 2003); (g), Volume VII, *Measurement of the Thermodynamic Properties of Multiple Phases*, edited by R. D. Weir and T. W. de Loos (Elsevier 2005); and (h) Volume VIII, *Applied Thermodynamics*, edited by A. R. H. Goodwin, J. V. Sengers, and C. J. Peters (Royal Society of Chemistry, Cambridge, to be published 2010).

The current text has been produced with an international team of distinguished experimentalists who describe the current state of development of heat capacity measurements. It continues the theme started with volume (f) above to provide a framework for both academia and industry. The latter often require measurements fit-for-purpose obtained from instruments with working equations based on the principles of physics, without recourse to maintaining the breath of expertise required to perform these measurements. Naturally, many of the author team are IACT members and to all great appreciation and gratitude are owed for their willing and enthusiastic contributions. International cooperation in this venture fulfils one mission for IUPAC and all should not forget the tireless editors who, faced with the task (common to all editors of co-operative efforts) of constructing a coherent whole from the independent contributions have achieved their goal. The material contained in this volume is of considerable value, perhaps also providing guidance for the development of new and more precise techniques.

Anthony R. H. Goodwin
Chairman
International Association of Chemical Thermodynamics

Preface

Many of the most significant developments in physical chemistry and chemical engineering during the last century have been based on contributions chemical thermodynamics has provided. The continuously increasing number of articles containing experimental data on thermodynamic properties and on phase equilibria, as well as on new experimental techniques and advances in theory and computer simulation, demonstrate the unabated growth of this field. Most noteworthy is the accelerating thrust in biophysical chemistry towards achieving a broader, quantitative thermodynamic basis of the physicochemical phenomena involved in biological processes. Heat capacities belong to the most important thermodynamic/thermophysical properties, playing a central role in the pure sciences as well as in chemical engineering and industrial applications. In this monograph the reader will find 22 contributions dealing with heat capacity, from various angles, of (mostly) liquids and gases/vapours, both pure and mixed. It may thus be regarded as a natural complement to the volumes on *Experimental Thermodynamics* (Volume I, 1968, through Volume VII, 2003) with their quite general coverage of the field, in that the focus is now on a *single property*. Thermodynamics is a theoretical discipline in close contact with experiment, and the 22 chapters are, in fact, testimony to it. However, they do not cover the entire subject. Amongst other topics, chemically reactive systems are not included. We intend to fill the gaps at a later point of time.

This book has its origins in committee meetings of the International Association of Chemical Thermodynamics. It is a project produced under the auspices of the International Union of Pure and Applied Chemistry (IUPAC). In true IUPAC image, the authors, which represent some of the most important names in their respective fields, come from many countries around the world, including: Austria, Belarus, Belgium, Canada, Czech Republic, France, Germany, Israel, Italy, Japan, Poland, South Africa, United Kingdom, and the United States of America.

Heat Capacities: Liquids, Solutions and Vapours
Edited by Emmerich Wilhelm and Trevor M. Letcher
© The Royal Society of Chemistry 2010
Published by the Royal Society of Chemistry, www.rsc.org

Two features are of paramount importance in monographs like this one: the timeliness of the topic and the coverage and critical evaluation of the pertinent publications. In fact, this book highlights the underlying theory, some of the most important experimental techniques, modelling and computer simulation, as well as significant and new results related to heat capacity, thereby underscoring its importance; and as shown by the contributions, the authors have endeavoured to cover the relevant literature up to about 2008. Overwhelmingly, the contributing authors have adhered to the nomenclature/symbols suggested by the Green Book of IUPAC (3rd edition, 2007). The few differences are either due to their desire to present a more concise, unequivocal notation, or to compliance with usage accepted by the scientific communities working in their respective specialised fields. We, as editors, did not interfere if exact definations were supplied, though in a few instances we added explanatory footnotes at the relevant places. Such as approach is in accord with the spirit of the Green Book as expressed so admirably by Martin Quack in his Historical Introduction on p. XII of its 3rd edition: *It is not the aim to present a list of recommendations in form of commandments. Rather we have always followed the principle that this manual should help the user in what may be called "good practice of scientific language".*

One of the objectives of the book is to bring together research from disparate disciplines which have a bearing on HEAT CAPACITIES. Links between these chapters, we believe, could lead to new ways of solving problems and looking at new and also old HEAT CAPACITIES related issues. Underlying this philosophy is our inherent belief that a book is still an important vehicle for the dissemination of knowledge.

This book is meant for researchers in chemical thermodynamics, either from academia or from (applied) chemical engineering, to show the progress recently achieved. Its success rests with the 35 authors. We would like to thank all of them for their cooperation; and we would also like to thank the publishers, the Royal Society of Chemistry, whose representatives were helpful *and* patient in producing this monograph on heat capacity.

Emmerich Wilhelm
Institute of Physical Chemistry, University of Wien
Wien (Vienna), Austria

Trevor M. Letcher
University of KwaZulu-Natal
Durban, South Africa

Contents

Heat Capacities: Liquids, Solutions and Vapours
Edited by Emmerich Wilhelm and Trevor M. Letcher
© The Royal Society of Chemistry 2010
Published by the Royal Society of Chemistry, www.rsc.org

CHAPTER 1

Heat Capacities: Introduction, Concepts and Selected Applications

EMMERICH WILHELM

Institute of Physical Chemistry, University of Wien, Währinger Strasse 42, A-1090, Wien (Vienna), Austria

Once the fabric is woven it may be embellished at will.

Nero Wolfe in *The Golden Spiders*, by Rex Stout, Bantam edition, New York, NY, 1955.

1.1 Introduction

Heat capacities belong to the most important thermophysical properties of matter: they are intimately related to the temperature dependence of fundamental thermodynamic functions; they may be determined in the laboratory with great accuracy; and they are of key importance for linking thermodynamics with microscopic fluid structure and dynamics, as evidenced by the contributions to this book. They are thus indispensable in physical chemistry as well as in chemical engineering. For instance, as a classical example, consider the standard entropies of liquids at $T = 298.15$ K. They are evaluated from experimental heat capacities at constant pressure from low temperatures to 298.15 K and entropies of phase changes in between (assuming applicability of

Heat Capacities: Liquids, Solutions and Vapours
Edited by Emmerich Wilhelm and Trevor M. Letcher
© The Royal Society of Chemistry 2010
Published by the Royal Society of Chemistry, www.rsc.org

the third law of thermodynamics). The measured heat capacity of an organic compound can usually be extrapolated to 0 K by fitting a Debye heat capacity function to the experimental values at, say, 10 K.

The nature and the size of this monograph's topic make it impractical to cover the entire subject in one volume. As indicated in the title, the focus will be on heat capacities of chemically *non-reacting* liquids, solutions and vapours/gases (though polymers and liquid crystals are also covered). The individual specialised chapters have been written by internationally renowned thermodynamicists/ thermophysicists active in the respective fields. Because of their topical diversity, in this introductory chapter I shall try to summarise concisely the major aspects of the thermodynamic formalism relevant to fluid systems, to clarify, perhaps, some points occasionally obscured, to indicate some ramifications into neighbouring disciplines, and to point out a few less familiar yet potentially interesting problems. The omission of any topic is not to be taken as a measure of its importance, but is predominantly a consequence of space limitations.

Calorimetric determinations of heat capacities of liquids have a long tradition, and many distinguished scientists have contributed to this subject. One can only marvel about the careful work of some of the early researchers, such as Eucken and Nernst,[1,2] who developed precursors of modern, adiabatic calorimeters. The *adiabatic* method for heat capacity measurements at low temperatures was pioneered by Cohen and Moesveld,[3] and Lange,[4] and became widely used. Indeed, during the following decades, many alternative designs of increasing sophistication have been devised and successfully used. A selection of adiabatic calorimeters which were described in the literature up to about 1970 is provided by references 5 through 19. For details, the interested reader should consult the classic IUPAC monograph edited by McCullough and Scott,[20] or the more recent ones edited by Marsh and O'Hare,[21] and by Goodwin, Marsh and Wakeham,[22] or the monograph on calorimetry by Hemminger and Höhne.[23]

More specialised reviews have been prepared by Lakshmikumar and Gopal,[24] Wadsö[25] and Gmelin.[26] A monograph focusing on differential scanning calorimetry has been presented by Höhne, Hemminger and Flammersheim.[27]

To date, the most widely used instruments for measuring heat capacities of liquids and liquid mixtures are based on the differential flow calorimeter designed by Picker,[28,29] which was commercialised by Setaram. Because of the absence of a vapour space, differential flow calorimeters are particularly useful. They may be fairly easily modified to be used at elevated temperatures and pressures, including the critical region. The first instrument of this type was constructed by Smith-Magowan and Wood,[30] with improved versions being due to White *et al.*,[31] and Carter and Wood.[32] However, comparison of heat capacities measured by different types of flow calorimeters and differential thermopile conduction calorimeters shows *small* differences in measured heat capacities, which are attributed to conductive and convective heat losses. Conductive heat losses, the principal problem in flow calorimetric heat capacity measurements on liquids, have recently been analysed by Hei and Raal[33] for a five-zone model calorimeter.

Because of the importance of heat capacity data of liquids in chemical thermodynamics and chemical engineering, numerous critical data compilations have been published – starting at the end of the nineteenth century with Berthelot's *Thermochimie*,[34] and including such well-known publications as the *International Critical Tables*,[35] Timmermans' *Physicochemical Constants of Pure Organic Compounds*,[36] *Landolt-Börnstein*,[37] and Daubert and Danner's *Physical and Thermodynamic Properties of Pure Chemicals: Data Compilation, DIPPR® Database*.[38] The most recent and the most comprehensive compilation of critically evaluated heat capacity data of pure liquids is the monograph on *Heat Capacities of Liquids: Volumes I and II. Critical Review and Recommended Values*, authored by Zábranský *et al.* (1996),[39] with *Supplement I* of 2001.[40] This monograph also contains a valuable survey of calorimetric techniques for determining heat capacities of liquids, and useful comments on terminology and criteria for the classification of calorimeters.

As concerns heat capacity data of mixtures, the situation is somewhat less satisfactory. Critically selected excess molar heat capacities at constant pressure of binary liquid organic mixtures have been included in the *International DATA Series, SELECTED DATA ON MIXTURES, Series A*,[41] and the *Dortmund Data Bank (DDB)* contains a large number of data sets on heat capacities of mixtures/excess heat capacities.[42] However, a monograph devoted to a reasonably comprehensive compilation of heat capacity data of liquid mixtures, though highly desirable, is not available.

For more than a century, experimental studies of real-gas behaviour at low or moderate densities, have held a prominent position in physical chemistry. They were motivated, and still are, either by the need to solve practical problems – such as those encountered in the reduction of vapour–liquid equilibrium data – or by their usefulness as valuable sources of information on intermolecular interactions in both pure gases/vapours and gaseous mixtures. In this context, *perfect-gas* (ideal-gas) state heat capacities are of central importance, say, in the calculation of property changes of single-phase, constant-composition fluids for any arbitrary change of state. They may be determined by vapour-flow calorimetry, or by speed-of-sound measurements. The statistical–mechanical calculation of perfect-gas state heat capacities (they are 1-body properties which do not depend on molecular interactions) has reached a high level of sophistication, with obvious great practical advantages. For instance, the calculations readily allow extension of experimental data to temperature ranges currently inaccessible to measurement. Data compilations of heat capacities of pure substances in the perfect-gas (ideal-gas) state may be found in *Selected Values of Physical and Thermodynamic Properties of Hydrocarbons and Related Compounds*,[43] *Landolt-Börnstein*,[37] in Stull, Westrum and Sinke's *The Chemical Thermodynamics of Organic Compounds*,[44] in the *TRC Thermodynamic Tables*,[45] in the book by Frenkel *et al.*,[46] and in the *NIST-JANAF Thermochemical Tables*.[47] One should always keep in mind, however, that only comparison of experimental with calculated values leads to better approximations and/or new concepts.

1.2   Thermodynamics: Fundamentals and Applications

To set the scene for this monograph, a few selected basic thermodynamic relations will be summarised below. For further aspects and details the interested reader should consult a textbook close to his/her taste, perhaps one of those listed in references 48 through 58.

Convenient starting points are the *fundamental property equations* (also called the *Gibbs equations*) of a *single-phase PVT* system, either open or closed, where P denotes the pressure, V is the molar volume and T is the thermodynamic temperature. No electric, magnetic or gravitational fields are considered in such a simple system. For a multicomponent system, where the total amount of substance is given by $n = \sum_i n_i$, with n_i being the amount of substance of component i, the fundamental property equation in the *energy representation* is

$$\mathrm{d}(nU) = T\mathrm{d}(nS) - P\mathrm{d}(nV) + \sum_i \mu_i \mathrm{d}n_i \tag{1}$$

and, equivalently, in the *entropy representation*

$$\mathrm{d}(nS) = \frac{1}{T}\mathrm{d}(nU) + \frac{P}{T}\mathrm{d}(nV) - \sum_i \frac{\mu_i}{T}\mathrm{d}n_i \tag{2}$$

Here, U is the molar internal energy, S is the molar entropy of the fluid. The intensive parameter furnished by the first-order partial derivatives of the internal energy with respect to the amount of substance of component i,

$$\mu_i \equiv \left[\frac{\partial(nU)}{\partial n_i} \right]_{nS,nV,n_j} \tag{3}$$

is called the chemical potential of component i. Its introduction extends the scope to the general case of a single-phase system in which the n_i may vary, either by exchanging matter with its surroundings (open system) or by changes in composition occurring as a result of chemical reactions (reactive system) or both. Corresponding to Equations (1) and (2), the primary functions (or cardinal functions, or Euler equations) are

$$(nU) = T(nS) - P(nV) + \sum_i \mu_i n_i \tag{4}$$

in the energy representation, and

$$(nS) = \frac{1}{T}(nU) + \frac{P}{T}(nV) - \sum_i \frac{\mu_i}{T}n_i \tag{5}$$

in the entropy representation.

In both the energy and entropy representations the extensive quantities are the mathematically independent variables, while the intensive parameters are derived, which situation does not conform to experimental practice. The choice of nS and nV as independent variables in the fundamental property equation in the energy representation is not convenient, and Equation (4) suggests the definition of useful alternative energy-based primary functions. The appropriate method for generating them without loss of information is the Legendre transformation. These additional equivalent primary functions are the molar enthalpy

$$H = U + PV \qquad (6)$$

the molar Helmholtz energy

$$F = U - TS \qquad (7)$$

and the molar Gibbs energy

$$G = H - TS \qquad (8)$$

Substituting for U in Equation (6) from Equation (4) yields the alternative form

$$H = TS + \sum_i x_i \mu_i \qquad (9)$$

where $x_i = n_i/n$ is the mole fraction. Substitution of U in Equation (7) yields

$$F = -PV + \sum_i x_i \mu_i \qquad (10)$$

as alternative grouping, and substitution of U in Equation (8) yields the Euler equation as

$$G = \sum_i x_i \mu_i \qquad (11)$$

The alternative primary functions H, F and G allow the development of alternative energy-based fundamental property equations:

$$d(nH) = Td(nS) + (nV)dP + \sum_i \mu_i dn_i \qquad (12)$$

$$d(nF) = -(nS)dT - Pd(nV) + \sum_i \mu_i dn_i \qquad (13)$$

$$\mathrm{d}(nG) = -(nS)\mathrm{d}T + (nV)\mathrm{d}P + \sum_i \mu_i \mathrm{d}n_i \tag{14}$$

The four fundamental equations presented so far are equivalent; however, each is associated with a different set of canonical variables $\{nS, nV, n_i\}$, $\{nS, P, n_i\}$, $\{T, nV, n_i\}$ and $\{T, P, n_i\}$.

A primary function which arises naturally in statistical mechanics is the grand canonical potential. It is the Legendre transform when simultaneously the entropy is replaced by the temperature and the amount of substance by the chemical potential:

$$J = U - TS - \sum_i x_i \mu_i \tag{15}$$

with the alternative form

$$J = -PV \tag{16}$$

and the corresponding Gibbs equation

$$\mathrm{d}(nJ) = -(nS)\mathrm{d}T - P\mathrm{d}(nV) - \sum_i n_i \mathrm{d}\mu_i \tag{17}$$

with the canonical variables $\{T, nV, \mu_i\}$.

The *complete* Legendre transform vanishes identically for any system. The complete transform of the internal energy replaces all extensive canonical variables by their conjugate intensive variables, thus yielding the null-function

$$0 = U - TS + PV - \sum_i x_i \mu_i \tag{18}$$

as final alternative primary function in the energy representation. This property of the complete Legendre transform gives rise to

$$0 = -(nS)\mathrm{d}T + (nV)\mathrm{d}P - \sum_i n_i \mathrm{d}\mu_i \tag{19}$$

as the corresponding alternative form of the fundamental property equation. It represents an important relation between the intensive parameters T, P and μ_i of the system and shows that they are not independent of each other.

While the extensive parameters of a simple phase are independent of each other, the conjugate intensive parameters are not, as shown above. For a given phase, the number of intensive parameters which may be varied independently is known as the number of thermodynamic degrees of freedom.

Treating the sum $\sum \mu_i n_i$ as a single term, the *total* number of equivalent primary functions and therefore the *total* number of equivalent fundamental property equations for a thermodynamic system is 2^k. Thus for $nU = nU(nS, nV, n)$

there are but eight distinct equivalent primary functions [nU, Equation (4), plus seven alternatives] and eight distinct forms of the fundamental equation [d(nU), Equation (1), plus seven alternatives]. Of the seven Legendre transforms of the internal energy, five have been treated above (including the null-function). The remaining two, $X = U - \sum_i x_i \mu_i$ and $Y = U + PV - \sum_i x_i \mu_i$, with the alternative forms $X = TS - PV$ and $Y = TS$, respectively, have not received separate symbols or names. The corresponding fundamental property equations are $d(nX) = Td(nS) - Pd(nV) - \sum_i n_i d\mu_i$ and $d(nY) = Td(nS) + (nV)dP - \sum_i n_i d\mu_i$.

Since all the fundamental property equations are equivalent, alternative expressions for the chemical potential are possible, of which

$$\mu_i \equiv \left[\frac{\partial(nG)}{\partial n_i} \right]_{T,P,n_j} \tag{20}$$

is the preferred one, because temperature and pressure are by far the most useful experimental parameters. We recognise that the chemical potential of component i is just the partial molar Gibbs energy of i,

$$\mu_i = \overline{G}_i \tag{21}$$

which quantity is of central importance in mixture/solution thermodynamics.

For a *homogeneous fluid of constant composition*, the following four energy-based fundamental property relations apply:

$$dU = TdS - PdV \tag{22}$$

$$dH = TdS + VdP \tag{23}$$

$$dF = -SdT - PdV \tag{24}$$

$$dG = -SdT + VdP \tag{25}$$

It follows that

$$T = (\partial U/\partial S)_V = (\partial H/\partial S)_P \tag{26}$$

$$P = -(\partial U/\partial V)_S = -(\partial F/\partial V)_T \tag{27}$$

$$V = (\partial H/\partial P)_S = (\partial G/\partial P)_T \tag{28}$$

$$S = -(\partial F/\partial T)_V = -(\partial G/\partial T)_P \tag{29}$$

which relations establish the link between the natural independent variables T, P, V, S and the energy-based functions U, H, F, G. In view of the definitions of F and G and Equation (29), the Gibbs–Helmholtz equations

$$U = F - T(\partial F/\partial T)_V \tag{30}$$

$$H = G - T(\partial G/\partial T)_P \tag{31}$$

are obtained.

A Legendre transformation of the primary function in the *entropy repre-sentation*, Equation (5), resulting in the replacement of one or more extensive variables by the conjugate intensive variable(s) $1/T$, P/T and μ_i/T, defines a *Massieu–Planck function*. For instance, *the* molar Massieu function is

$$\Psi = S - \frac{1}{T}U \tag{32}$$

with its alternative form

$$\Psi = \frac{P}{T}V - \sum_i \frac{\mu_i}{T}x_i \tag{33}$$

Its differential form, an entropy-based alternative fundamental property equation, is

$$d(n\Psi) = -(nU)d\left(\frac{1}{T}\right) + \frac{P}{T}d(nV) - \sum_i \frac{\mu_i}{T}dn_i \tag{34}$$

From a second-order Legendre transformation, *the* molar Planck function

$$\Phi = S - \frac{1}{T}U - \frac{P}{T}V \tag{35}$$

is obtained, with its alternative form

$$\Phi = -\sum_i \frac{\mu_i}{T}x_i \tag{36}$$

Its differential form is another alternative entropy-based fundamental property equation:

$$d(n\Phi) = -(nU)d\left(\frac{1}{T}\right) - (nV)d\left(\frac{P}{T}\right) - \sum_i \frac{\mu_i}{T}dn_i \tag{37}$$

Note that

$$\Psi = -\frac{F}{T} \tag{38}$$

and

$$\Phi = -\frac{G}{T} \tag{39}$$

Another second-order transform is the molar Kramer function

$$\Omega = S - \frac{1}{T}U + \sum_i \frac{\mu_i}{T}x_i \tag{40}$$

Its alternative form is

$$\Omega = \frac{P}{T}V \tag{41}$$

whence

$$\Omega = -\frac{J}{T} \tag{42}$$

The corresponding alternative entropy-based fundamental property equation is

$$d(n\Omega) = -(nU)d\frac{1}{T} + \frac{P}{T}d(nV) + \sum_i n_i d\left(\frac{\mu_i}{T}\right) \tag{43}$$

Again, the complete Legendre transform is identical zero, yielding in the entropy representation

$$0 = (nU)d\left(\frac{1}{T}\right) + (nV)d\left(\frac{P}{T}\right) - \sum_i n_i d\left(\frac{\mu_i}{T}\right) \tag{44}$$

Evidently, also the intensive parameters $1/T$, P/T and μ_i/T in the entropy representation are not independent of each other.

Equations (22) through (25) are exact differentials, whence application of the reciprocity relation yields the Maxwell equations for a constant-composition *PVT* system, of which the following two are particularly useful:

$$(\partial S/\partial V)_T = (\partial P/\partial T)_V \tag{45}$$

$$(\partial S/\partial P)_T = -(\partial V/\partial T)_P \tag{46}$$

Two heat capacities are in common use for homogeneous fluids. Both are state functions defined rigorously in relation to other state functions: the *molar heat capacity at constant volume* (or the molar isochoric heat capacity) C_V and the *molar heat capacity at constant pressure* (or the molar isobaric heat capacity) C_P. At constant composition,

$$C_V = (\partial U/\partial T)_V = T(\partial S/\partial T)_V = -T(\partial^2 F/\partial T^2)_V \tag{47}$$

and

$$C_P = (\partial H/\partial T)_P = T(\partial S/\partial T)_P = -T(\partial^2 G/\partial T^2)_P \tag{48}$$

At this juncture it is convenient to introduce, by definition, a few auxiliary quantities commonly known as the *mechanical* and the *isentropic coefficients*. Specifically, these are the *isobaric expansivity*

$$\alpha_P = V^{-1}(\partial V/\partial T)_P = -\rho^{-1}(\partial\rho/\partial T)_P \tag{49}$$

the isothermal compressibility[†]

$$\beta_T = -V^{-1}(\partial V/\partial P)_T = \rho^{-1}(\partial\rho/\partial P)_T \tag{50}$$

the isochoric pressure coefficient

$$\gamma_V = (\partial P/\partial T)_V \tag{51}$$

and the *isentropic compressibility*[†] (often loosely called adiabatic compressibility)

$$\beta_S = -V^{-1}(\partial V/\partial P)_S = \rho^{-1}(\partial\rho/\partial P)_S \tag{52}$$

where $\rho = M/V$ is the density and M is the molar mass.

Note that

$$\gamma_V = \alpha_P/\beta_T \tag{53}$$

The isentropic compressibility is related to the thermodynamic low-frequency speed of ultrasound v_0 (negligible dispersion) by

$$\beta_S = 1/\rho v_0^2 \tag{54}$$

The ratio of the heat capacities and their difference may now be presented in several compact forms, where the most profitable are given below:

$$C_P/C_V = \kappa \tag{55}$$

$$\kappa = 1 + TM\alpha_P^2 v_0^2/C_P \tag{56}$$

$$C_P - C_V = TV\alpha_p^2\Big/\beta_T \tag{57}$$

Since by definition the compression factor is given by $Z \equiv PV/RT$, alternatively

$$C_P - C_V = R\frac{[Z + T(\partial Z/\partial T)_P]^2}{Z - P(\partial Z/\partial P)_T} \tag{58}$$

where R is the gas constant.

At low temperatures, where γ_V of liquids is large, direct calorimetric determination of C_V of liquids is difficult (it becomes more practicable near the

---

[†] In this chapter the isothermal compressibility is represented by the symbol β_T and not by κ_T as was recently recommended by IUPAC. Similarly, the isentropic compressibility is represented by the symbol β_S and not by κ_S

critical point, where γ_V is much smaller). Thus most of the isochoric heat capacity data for liquids reported in the literature have been obtained *indirectly* through the use of Equations (55) and (56), that is to say from experimental molar isobaric heat capacities, isobaric expansivities and ultrasonic speeds. However, see for instance reference 59. Since also

$$\beta_T/\beta_S = \kappa \tag{59}$$

Equations (54), (56) and (59) may be used for the *indirect* determination of isothermal compressibilities from densities, isobaric expansivities, ultrasonic speeds and molar isobaric heat capacities. All these quantities may now be reliably and accurately measured, whence the indirect method for determining the isothermal compressibility of liquids has become an attractive alternative to the *direct* method of applying hydrostatic pressure and measuring the corresponding volume change. For the difference between β_T and β_S one obtains, for instance,

$$\beta_T - \beta_S = TV\alpha_P^2/C_P \tag{60}$$

A convenient way to derive the volume or pressure dependence of the heat capacities is *via* the differentiation of the appropriate Gibbs–Helmholtz equations. Starting from

$$(\partial U/\partial V)_T = -P + T(\partial P/\partial T)_V = -P + T\gamma_V \tag{61}$$

$$(\partial H/\partial P)_T = V - T(\partial V/\partial T)_P = V - TV\alpha_P \tag{62}$$

these equations lead to

$$(\partial C_V/\partial V)_T = T(\partial^2 P/\partial T^2)_V = T(\partial\gamma_V/\partial T)_V \tag{63}$$

$$(\partial C_P/\partial P)_T = -T(\partial^2 V/\partial T^2)_P = -TV\left[\alpha_P^2 + (\partial\alpha_P/\partial T)_p\right] \tag{64}$$

The pressure or volume dependence of the heat capacities may thus be determined from *PVT* data.

The molar thermodynamic properties of homogeneous constant-composition fluids are functions of temperature and pressure, *e.g.*

$$dH = \left(\frac{\partial H}{\partial T}\right)_P dT + \left(\frac{\partial H}{\partial P}\right)_T dP \tag{65}$$

Replacing the partial derivatives through use of Equations (48) and (62) yields

$$dH = C_P dT + V(1 - T\alpha_P)dP \tag{66}$$

Entirely analogous procedures, using Equations (46) and (48), give

$$dS = \frac{C_P}{T}dT - V\alpha_p dP \tag{67}$$

When T and V are selected as independent variables,

$$dU = C_V dT + (T\gamma_V - P)dV \tag{68}$$

and

$$dS = \frac{C_V}{T} dT + \gamma_V dV \tag{69}$$

are obtained. All the coefficients of dT, dP and dV are quantities reasonably accessible by experiment. For some applications it may be convenient to treat S as a function of P and V. Using

$$\left(\frac{\partial S}{\partial P}\right)_V = \left(\frac{\partial S}{\partial T}\right)_V \left(\frac{\partial T}{\partial P}\right)_V \text{ and } \left(\frac{\partial S}{\partial V}\right)_P = \left(\frac{\partial S}{\partial T}\right)_P \left(\frac{\partial T}{\partial V}\right)_P \tag{70}$$

one obtains

$$dS = \frac{C_V}{T\gamma_V} dP + \frac{C_P}{TV\alpha_P} dV \tag{71}$$

Finally we note the useful relations

$$(\partial\alpha_P/\partial P)_T = -(\partial\beta_T/\partial T)_P \tag{72}$$

and

$$\begin{aligned}
\mu_{JT} &\equiv (\partial T/\partial P)_H \\
&= -\left(\frac{\partial H}{\partial P}\right)_T \bigg/ C_P \\
&= \frac{V(T\alpha_P - 1)}{C_P}
\end{aligned} \tag{73}$$

where μ_{JT} is the Joule–Thomson coefficient. All three quantities C_P, $(\partial H/\partial P)_T$ and μ_{JT} may be measured by flow calorimetry.[60,61] $(\partial H/\partial P)_T$ is also known as the *isothermal* Joule–Thomson coefficient, and frequently given the symbol φ. For ideal gases $T\alpha_P = 1$ and thus $\mu_{JT} = 0$. For real gases, the temperature T_i (at the inversion pressure P_i) where $T_i\alpha_P = 1$ is called the inversion temperature. At that point the isenthalpic exhibits a maximum: for initial pressures $P < P_i$, $\mu_{JT} > 0$, and the temperature of the gas always decreases on throttling; for initial pressures $P > P_i$, $\mu_{JT} < 0$, and the temperature of the gas always increases on throttling. The maxima of the enthalpics form a locus known as the inversion curve of the gas. There exists a maximum inversion temperature at $P = 0$. For pressures above the maximum inversion pressure, μ_{JT} is always negative.

Because of Equation (67) one obtains, for instance, for the isentropic compression or expansion of a gas

$$\left(\frac{\partial T}{\partial P}\right)_S = \frac{TV\alpha_P}{C_P} \tag{74}$$

Since α_P of gases is always positive, the temperature *always* increases with isentropic compression and decreases with isentropic expansion.

In principle, the exact methods of classical thermodynamics are the most general and powerful predictive tools for the calculation of property changes of single-phase, constant-composition fluids for any arbitrary change of state, say, from (T_1,P_1) to (T_2,P_2). For a *pure* fluid, the corresponding changes of molar enthalpy $\Delta H \equiv H_2 - H_1$ and molar entropy $\Delta S \equiv S_2 - S_1$ are, respectively,

$$\Delta H = H_2^R - H_1^R + \int_{T_1}^{T_2} C_P^{pg} dT \tag{75}$$

and

$$\Delta S = S_2^R - S_1^R - R\ln(P_2/P_1) + \int_{T_1}^{T_2} \frac{C_P^{pg}}{T} dT \tag{76}$$

where H^R and S^R are the molar residual enthalpy and the molar residual entropy, respectively, in (T,P)-space, and $C_P^{pg} = C_P^{pg}(T)$ is the molar heat capacity at constant pressure of the fluid in the *perfect-gas* (ideal-gas) state. The general definition for such molar *residual* properties is $M^R \equiv M - M^{pg}$, where M is the molar value of any extensive thermodynamic property of the fluid at (T,P), and M^{pg} is the molar value of the property when the fluid is in the perfect-gas state at the *same* T and P. Given any *volume-explicit* equation of state, these residual functions may be calculated from

$$H^R = -RT^2 \int_0^P (\partial Z/\partial T)_P P^{-1} dP, \text{ (constant } T) \tag{77}$$

and

$$S^R = -R \int_0^P [T(\partial Z/\partial T)_P + Z - 1] P^{-1} dP, \text{ (constant } T) \tag{78}$$

respectively. Thus, application of Equations (75) and (76) requires PVT information for the real fluid as well as its isobaric heat capacity in the perfect-gas state. We note that one may also define residual functions in (T,V)-space: $M^r \equiv M - M^{pg}$, where the Ms are now at the *same* T and V. In general $M^R(T,P) \neq M^r(T,V)$ unless the property M^{pg} is *independent* of density at constant temperature, which is the case for C_P^{pg} and $C_V^{pg} = C_P^{pg} - R$. Since the perfect-gas state is a state where molecular interactions are absent, residual quantities characterise molecular interactions alone. They are the most direct measures of intermolecular forces. In statistical mechanics, however, *configurational* quantities are frequently used. The differences between these two sets are the configurational properties of the perfect gas, and for U and C_V they vanish.

In actual practice, this approach would be severely limited by the availability of reliable data for pure fluids and mixtures. The experimental determination of such data is time-consuming and not simple, and does not impart the glamour associated with, say, spectroscopic studies. Fortunately, statistical–mechanical calculations for C_P^{pg} are quite dependable for many substances, and so are group-contribution theories, for instance the techniques based on the work by Benson and co-workers.[62,63]

The search for generalised correlations applicable to residual functions has occupied scientists and engineers for quite some time. The most successful ones are based on versions of *generalised corresponding-states theory*, which is grounded in experiment as well as statistical mechanics. The *three-parameter* corresponding states correlations, pioneered by Kenneth Pitzer and co-workers,[64–67] have been capable to predict satisfactorily the *PVT* behaviour of normal, nonassociating fluids. They showed that the compression factors of normal fluids may be satisfactorily expressed as

$$Z = Z^{(0)}(T_r, P_r) + \omega Z^{(1)}(T_r, P_r) \tag{79}$$

where

$$\omega \equiv -1 - \log_{10}(P_{\sigma,r})_{T_r=0.7} \tag{80}$$

is Pitzer's *acentric factor*, $T_r = T/T_c$ is the reduced temperature, $P_r = P/P_c$ is the reduced pressure, $P_{\sigma,r} = P_\sigma/P_c$ is the reduced vapour pressure, here evaluated at $T_r = 0.7$, P_σ is the vapour pressure of the substance, T_c is the critical temperature of the substance and P_c is its critical pressure. In fact, this method is a thermodynamic perturbation approach where the Taylor series is truncated after the term linear in ω. The generalised $Z^{(0)}$ function is the simple-fluid contribution and applies to spherical molecules like argon and krypton, whose acentric factors are essentially zero. The generalised $Z^{(1)}$ function (deviation function) is determined through analysis of high-precision *PVT* data of selected normal fluids where $\omega \neq 0$. One of the best of the generalised Pitzer-type corresponding-states correlations for $Z^{(0)}$, $Z^{(1)}$ and the derived residual functions is due to Lee and Kesler.[68]

An alternative to the direct experimental route to *high-pressure PVT* data and $C_P(T,P)$ is to measure the thermodynamic speed of ultrasound v_0 as a function of P and T (at constant composition), and to combine these results, in the spirit of Equations (50), (54) and (60) with data at ordinary pressure, say $P_1 = 10^5$ Pa, *i.e.* $\rho(T,P_1)$ and $C_P(T,P_1)$. For a pure liquid, upon integration at constant temperature, one obtains[69,70]

$$\rho(T,P) = \rho(T,P_1) + \int_{P_1}^{P} v_0^{-2}\mathrm{d}P + TM \int_{P_1}^{P} \alpha_P^2 C_P^{-1}\mathrm{d}P \tag{81}$$

The first integral is evaluated directly by fitting the ultrasonic speed data with suitable polynomials, and for the second integral several successive integration algorithms have been devised. The simplicity, rapidity and precision of this method makes it highly attractive for the determination of the density, isobaric expansivity, isothermal compressibility, isobaric heat capacity and isochoric heat capacity of liquids at high pressures. Details may be found in the appropriate chapters of this book, and in the original literature.

From experimentally determined heat capacities of liquids, relatively simple models have been used to extract information on the type of motion executed by molecules in the liquid state. In general, they are based on the separability of contributions due to translation, rotation, vibration and so forth. Though none of them is completely satisfactory, they have provided eminently useful insights and thereby furthered theoretical advances. Following the early work of Eucken,[71] Bernal,[72] Eyring,[73] Stavely,[74] Moelwyn-Hughes,[75] Kohler,[18,76] Bondi[77] and their collaborators, one may resolve the molar heat capacity C_V of simple, nonassociated liquids into the following contributions:[78,79]

$$C_V = C_{tr} + C_{rot} + C_{int} + C_{or} \qquad (82)$$

The translational (tr) contribution arises from the motion of the molecules under the influence of all molecules (translational movement within their respective free volumes), the rotational (rot) contribution arises from rotation or libration of the molecules as a whole, the internal (int) contribution arises from internal degrees of freedom, and the orientational (or) contribution, for dipolar substances, results from the change of the dipole–dipole orientational energy with temperature. C_{int} can be subdivided into a part stemming from vibrations (C_{vib}) which usually are not appreciably influenced by density changes (*i.e.* by changes from the liquid to the perfect-gas state), and another part, C_{conf}, resulting from internal rotations (conformational equilibria), which does depend on density. Preferably, all these contributions to C_V are discussed in terms of residual quantities in (T,V)-space.[78,79] The residual molar isochoric heat capacity of a pure liquid is defined by

$$C_V^r \equiv C_V(T, V) - C_V^{pg}(T) \qquad (83)$$

For liquids composed of fairly rigid molecules, such as tetrachloromethane, benzene or toluene, to an excellent approximation $C_{int}^r \approx 0$, whence

$$C_V^r = C_{tr}^r + C_{rot}^r \qquad (84)$$

where $C_{tr}^r = C_{tr}-3R/2$, and $C_{rot}^r = C_{rot}-3R/2$, for nonlinear molecules, represents the excess over the perfect-gas phase value due to *hindered rotation* in the liquid of the molecules as a whole. Using corresponding states arguments to obtain reasonable estimates for C_{tr}^r, values for the residual molar rotational heat capacity C_{rot}^r may be obtained, which quantity may then be discussed in terms of any suitable model for restricted molecular rotation.[61,78,79]

The resolution of the variation of C_V of pure liquids along the orthobaric curve (subscript σ), *i.e.* for states (T, P_σ), into the contributions due to the increase of volume and to the increase of temperature, respectively, is a highly interesting problem.[78-80] It is important to realise that due to the close packing of molecules in a liquid, even a rather small change of the average volume available for their motion may have a considerable impact on the molecular dynamics: volume effects may become *more* important in influencing molecular motion in the liquid state than temperature changes. Since

$$\left(\frac{\partial C_V}{\partial T}\right)_\sigma = \left(\frac{\partial C_V}{\partial T}\right)_V + \left(\frac{\partial C_V}{\partial V}\right)_T V\alpha_\sigma \tag{85}$$

evaluation of $(\partial C_V/\partial T)_V$ requires knowledge of the *second* term of the right-hand side of Equation (85). At temperatures below the normal boiling point, the saturation expansivity $\alpha_\sigma = V^{-1}(\partial V/\partial T)_\sigma$ is practically equal to α_P of the liquid [see below, Equation (109)]. In principle, the quantity $(\partial C_V/\partial V)_T$ is accessible *via* precise PVT measurements, see Equation (63), but measurements of $(\partial^2 P/\partial T^2)_V$ are not plentiful. Available data[70,81] indicate that it is small and negative for organic liquids, that is to say, C_V decreases with increasing volume. Alternatively, one may use[18,78,79]

$$\left(\frac{\partial C_V}{\partial V}\right)_T = T\beta_T^{-1}\left[\left(\frac{\partial \alpha_P}{\partial T}\right)_P - 2\frac{\alpha_P}{\beta_T}\left(\frac{\partial \beta_T}{\partial T}\right)_P - \left(\frac{\alpha_P}{\beta_T}\right)^2\left(\frac{\partial \beta_T}{\partial P}\right)_T\right] \tag{86}$$

where the last term in parenthesis on the right-hand side can be evaluated by means of a modified Tait equation,[82] that is

$$\left(\frac{\partial \beta_T}{\partial P}\right)_T = -m\beta_T^2 \tag{87}$$

This equation holds remarkably well up to pressures of several hundred bars, and for many liquid nonelectrolytes $m \approx 10$. For liquid tetrachloromethane at 298.15 K,[78] the calculated value of $(\partial C_V/\partial V)_T$ amounts to $-0.48\,\mathrm{J\,K^{-1}\,cm^{-3}}$, for cyclohexane[78] $-0.57\,\mathrm{J\,K^{-1}\,cm^{-3}}$ is obtained, and for 1,2-dichloroethane[83] it is $-0.60\,\mathrm{J\,K^{-1}\,cm^{-3}}$. These results indicate a *substantial* contribution of $(\partial C_V/\partial V)_T V\alpha_\sigma$ to the change of C_V along the orthobaric curve as well as to the corresponding change of C_V^{r}.

Equation (56) is a suitable starting point for a discussion of the temperature dependence of $\kappa \equiv C_P/C_V$ of a liquid along the orthobaric curve:

$$\left(\frac{\partial \kappa}{\partial T}\right)_\sigma = (\kappa - 1)\left[\frac{1}{T} + \frac{2}{\alpha_P}\left(\frac{\partial \alpha_P}{\partial T}\right)_\sigma + \frac{2}{v_0}\left(\frac{\partial v_0}{\partial T}\right)_\sigma - \frac{1}{C_P}\left(\frac{\partial C_P}{\partial T}\right)_\sigma\right] \tag{88}$$

Usually, the second term in parenthesis on the right-hand side of Equation (88) is positive and the third term is negative; the fourth term may contribute positively or negatively. Thus κ may increase or decrease with temperature.

The importance of the heat capacity in the perfect-gas state has been stressed repeatedly. Flow calorimetry is a commonly used method for measuring C_P of gases and vapours,[84] and allows straightforward extrapolation to zero pressure[85] to obtain C_P^{pg}. The virial equation in pressure

$$Z \equiv \frac{PV}{RT} = 1 + B'P + C'P^2 + \ldots \tag{89}$$

where B' is the corresponding second virial coefficient and C' the third virial coefficient, may be used to calculate the residual heat capacity of a pure fluid according to

$$C_P^{\text{R}} = -RT \int_0^P \left[T\left(\frac{\partial^2 Z}{\partial T^2}\right)_P + 2\left(\frac{\partial Z}{\partial T}\right)_P \right] P^{-1} \mathrm{d}P, \text{ (constant } T) \tag{90}$$

Since the second virial coefficient B' of the pressure series is related to the second virial coefficient B of the series in molar density $(1/V)$ by

$$B' = \frac{B}{RT} \tag{91}$$

one obtains from the two-term equation in pressure

$$C_P^{\text{R}} = -T\frac{\mathrm{d}^2 B}{\mathrm{d}T^2} P \tag{92}$$

Thus the pressure derivative of C_P is given by

$$\lim_{P \to 0} \left(\frac{\partial C_P}{\partial P}\right)_T = -T\frac{\mathrm{d}^2 B}{\mathrm{d}T^2} \tag{93}$$

thereby providing an experimental route to the determination of the second temperature derivative of B.

Flow-calorimetric measurements of deviations from perfect-gas behaviour, particularly *via* the isothermal Joule–Thomson coefficient $\varphi \equiv (\partial H/\partial P)_T$, have the advantage over compression experiments that adsorption errors are avoided, and that measurements can be made at lower temperatures and pressures.[86,87] Specifically,

$$\varphi = B - T\frac{\partial B}{\partial T} + \left(C'' - T\frac{\partial C''}{\partial T}\right)\frac{P_1 + P_2}{2} + \ldots \tag{94}$$

where

$$C'' = C'RT = \frac{C - B^2}{RT} \tag{95}$$

Here, C is the third virial coefficient of the series in molar density, $P_2 - P_1$ is the pressure difference maintained across the throttle, and $(P_1 + P_2)/2$ is the mean pressure. The zero-pressure value of the isothermal Joule–Thomson coefficient is thus given by

$$\varphi_0 = B - T\frac{dB}{dT} \tag{96}$$

Integration between a suitable reference temperature T_{ref} and T yields[61]

$$\frac{B(T)}{T} = \frac{B(T_{ref})}{T_{ref}} - \int_{T_{ref}}^{T} \frac{\varphi_0}{T^2}dT \tag{97}$$

This relation is of considerable importance for obtaining virial coefficients (of vapours) in temperature regions where conventional measuring techniques are difficult to apply. The isothermal Joule–Thomson coefficient of *steam*, the most important vapour on earth, was recently reported by McGlashan and Wormald[88] in the temperature range 313 K to 413 K, and values of φ_0 derived from these measurements were compared with results from the 1984 NBS/NRC steam tables,[89] with data reported by Hill and MacMillan,[90] and with values derived from the IAPWS-95 formulation for the thermodynamic properties of water.[91]

The thermodynamic speed of ultrasound (below any dispersion region) is related to the equation of state, and hence to the virial coefficients. For a real gas, v_0^2 may thus be expressed as a virial series in molar density $1/V$,[92] *i.e.*

$$v_0^2 = \frac{\kappa^{pg}RT}{M}\left(1 + B_{ac}V^{-1} + C_{ac}V^{-2} + \cdots\right) \tag{98}$$

where

$$\kappa^{pg} = \frac{C_P^{pg}}{C_V^{pg}} = 1 + \frac{R}{C_V^{pg}} \tag{99}$$

For constant-composition fluids, the acoustic virial coefficients B_{ac}, C_{ac}, $\ldots$ are functions of temperature only. They are, of course, rigorously related to the ordinary (PVT) virial coefficients. For instance,

$$B_{ac} = 2B + 2(\kappa^{pg} - 1)T\frac{dB}{dT} + \frac{(\kappa^{pg} - 1)^2}{\kappa^{pg}}T^2\frac{d^2B}{dT^2} \tag{100}$$

Since pressure is the preferred experimental parameter, one may also write a virial expansion for v_0^2 in powers of the pressure with corresponding virial coefficients B'_{ac}, C'_{ac}, ... The coefficients of the density and pressure expansion are interrelated; for example

$$B'_{ac} = \frac{B_{ac}}{RT} \qquad (101a)$$

$$C'_{ac} = \frac{C_{ac} - BB_{ac}}{(RT)^2} \; etc. \qquad (101b)$$

Thus, measurements of the speed of ultrasound as function of density (or pressure) will yield information on B together with its first and second temperature derivatives, and C_V^{pg} (or κ^{pg}) through extrapolation of v_0^2 to zero density. The principal advantages of the acoustic method are its rapidity and the greater accuracy at temperatures where adsorption effects become important.[93]

All this valuable thermophysical information can then be used to obtain reliable second virial coefficients over large temperature ranges. For a fluid with spherically symmetric pair potential energy $u(r)$,

$$B(T) = -2\pi N_A \int_0^\infty \left[e^{-u(r)/k_B T} - 1 \right] r^2 dr \qquad (102)$$

where N_A is the Avogadro constant and k_B is the Boltzmann constant. Inversion[94] then yields the fundamentally important potential energy function $u(r)$ for a pair of molecules.

While a discussion of experimental acoustical methods is way outside the scope of this introductory chapter, the following comment is indicated. For gases/vapours at low to moderate pressures not too close to saturation, the highest experimental precision, when measuring v_0, is obtained through use of a spherical resonator, a technique which was pioneered by Moldover, Mehl and co-workers.[95,96]

So far, the focus was on *homogeneous* constant-composition fluids, of which pure fluids are special cases. I will now briefly consider the case where a *pure* liquid is in equilibrium with its vapour. Such a situation is encountered, for instance, in adiabatic calorimetry, where the calorimeter vessel is *incompletely* filled with liquid in order to accommodate the thermal expansion of the sample (usually, the vapour space volume is comparatively small). One has now a closed two-phase single-component system. The heat capacity of such a system is closely related to C_σ^L, *i.e.* the molar heat capacity of a liquid in equilibrium with an *infinitesimal* amount of vapour (as before, the saturation condition is indicated by the subscript σ). For a detailed analysis see Hoge,[97] Rowlinson and Swinton,[56] and Wilhelm.[98]

The molar heat capacity at saturation of the substance in the equilibrium phase π (denoting either the liquid, $\pi = $ L, *or* the vapour, $\pi = $ V) is given by $C_\sigma^\pi \equiv T(\partial S^\pi / \partial T)_\sigma$, whence one obtains, for instance,

$$C_\sigma^\pi = C_P^\pi + T\left(\frac{\partial S^\pi}{\partial P}\right)_T \left(\frac{\partial P}{\partial T}\right)_\sigma \tag{103}$$

$$= C_P^\pi - TV^\pi \alpha_P^\pi \gamma_\sigma \tag{104}$$

$$= C_P^\pi - TV^\pi(\alpha_P^\pi - \alpha_\sigma^\pi)\gamma_V^\pi \tag{105}$$

$$= C_V^\pi + T\left(\frac{\partial S^\pi}{\partial V}\right)_T \left(\frac{\partial V^\pi}{\partial T}\right)_\sigma \tag{106}$$

$$= C_V^\pi + TV^\pi \gamma_V^\pi \alpha_\sigma^\pi \tag{107}$$

$$= C_V^\pi + TV^\pi \alpha_P^\pi(\gamma_V^\pi - \gamma_\sigma) \tag{108}$$

Here, $\gamma_\sigma \equiv (\partial P / \partial T)_\sigma$ is the slope of the vapour-pressure curve, and

$$\alpha_\sigma^\pi \equiv (1/V^\pi)(\partial V^\pi / \partial T)_\sigma = \alpha_P^\pi(1 - \gamma_\sigma / \gamma_V^\pi) \tag{109}$$

denotes the expansivity of a pure substance in contact with the other equilibrium phase (*i. e.* along the saturation curve). As already pointed out, below the normal boiling point, the difference $\alpha_P^L - \alpha_\sigma^L$ is usually negligibly small. At the critical point

$$\gamma_V^L = \gamma_V^V = \gamma_\sigma \tag{110}$$

Neither C_P^π nor C_σ^π is equal to the change of enthalpy with temperature along the saturation curve. From Equation (65) one obtains

$$\left(\frac{\partial H^\pi}{\partial T}\right)_\sigma = C_P^\pi + V^\pi(1 - T\alpha_P^\pi)\gamma_\sigma \tag{111}$$

$$= C_\sigma^\pi + V^\pi \gamma_\sigma \tag{112}$$

Since $U = H - PV$,

$$\left(\frac{\partial U^\pi}{\partial T}\right)_\sigma = C_\sigma^\pi - P_\sigma V^\pi \alpha_\sigma^\pi \tag{113}$$

Thus for the saturated *liquid* at $[T, P_\sigma(T)]$ at temperatures where $T\alpha_P^L < 1$, the following sequence is obtained:

$$\left(\frac{\partial H^L}{\partial T}\right)_\sigma > C_P^L > C_\sigma^L > \left(\frac{\partial U^L}{\partial T}\right)_\sigma > C_V^L \tag{114}$$

The differences between the first four quantities are generally *much* smaller than between C_V^L and $(\partial U^L/\partial T)_\sigma$.

While the *general* equations apply also to the saturated vapour ($\pi = V$), the inequality *does not*. Since $\alpha_P^V V^V$ is always large, for saturated vapours the difference $C_\sigma^V - C_P^V$ is always significant [see Equation (104)]. In fact, for vapours of substances with *small* molecules, such as argon, carbon dioxide, ammonia and water (steam), $\alpha_P^V V^V$ may be large enough to make C_σ^V even *negative*. Finally we note that the difference between the saturation heat capacities in the vapour phase and the liquid phase may be expressed as[98]

$$C_\sigma^V - C_\sigma^L = \left(\frac{\partial \Delta_{\text{vap}} H}{\partial T}\right)_\sigma - \frac{\Delta_{\text{vap}} H}{T} \tag{115}$$

and the difference between the isobaric heat capacities in the vapour phase and the liquid phase as

$$C_P^V - C_P^L = \left(\frac{\partial \Delta_{\text{vap}} H}{\partial T}\right)_\sigma - \frac{\Delta_{\text{vap}} H}{T} + \Delta_{\text{vap}} H \left(\frac{\partial \ln\left(\Delta_{\text{vap}} V\right)}{\partial T}\right)_P \tag{116}$$

where $\Delta_{\text{vap}} H$ denotes the molar enthalpy of vaporisation, and $\Delta_{\text{vap}} V \equiv V^V - V^L$ is the volume change on vaporisation. In deriving these equations, use was made of the exact Clapeyron equation

$$\gamma_\sigma = \frac{\Delta_{\text{vap}} H}{T \Delta_{\text{vap}} V} \tag{117}$$

and the exact Planck equation.[99]

There are, of course, many additional details and fascinating topics, in particular when mixtures and solutions are considered, which fact is amply evidenced by the contributions to this monograph. Enjoy!

1.3 Concluding Remarks

Calorimetry and *PVT* measurements are the most fundamental and also the oldest experimental disciplines of physical chemistry. Although simple in principle, enormous effort and ingenuity has gone into designing the vast array of apparatus now at our disposal. In this introductory chapter, I did not cover design of experiments beyond the bare rudiments – the reader is referred to the relevant articles and books quoted, and to the chapters of this book focusing on this aspect. Let it suffice to say that the advances in instrumentation during the last decades have greatly facilitated the high-precision determination of caloric and *PVT* properties of fluids over large ranges of temperature and pressure. At the same time cross-fertilisation with other disciplines, notably with ultrasonics and hypersonics, and with biophysics, is becoming increasingly important, as is the close connection to equation-of-state research and, of course, chemical engineering.[56,61,79,98,100–103] The discussion presented here and in the chapters to follow may perhaps best be characterised by a statement due to Gilbert Newton Lewis (1875–1946) on the practical philosophy of scientific research:

The scientist is a practical man and his are practical aims. He does not seek the ultimate but the proximate. He does not speak of the last analysis but rather of the next approximation. . . . On the whole, he is satisfied with his work, for while science may never be wholly right it certainly is never wholly wrong; and it seems to be improving from decade to decade.

By necessity, this introductory chapter is limited to a few topics, the selection of which was also influenced by my current interests. In conclusion, I hope to have:

- formulated concisely some important aspects of the thermodynamic formalism needed in this area of research;
- discussed and made transparent some key aspects of experiments;
- shown how to apply and to appropriately extend well-known concepts to perhaps less familiar, yet potentially important, problems;
- stimulated some colleagues to enter this fascinating and important field of research.

Success in any of these points would be most rewarding.

References

1. A. Eucken, *Phys. Z.*, 1909, **10**, 586.
2. W. Nernst, *Ann. Phys. (Leipzig)*, 1911, **36**, 395.
3. E. Cohen and A. L. Th. Moesveld, *Z. Phys. Chem.*, 1920, **95**, 305.
4. F. Lange, *Z. Phys. Chem.*, 1924, **110**, 343.
5. J. C. Southard and D. H. Andrews, *J. Franklin Inst.*, 1930, **209**, 349.
6. J. C. Southard and F. G. Brickwedde, *J. Am. Chem. Soc.*, 1933, **55**, 4378.
7. J. G. Aston and M. L. Eidinoff, *J. Am. Chem. Soc.*, 1939, **61**, 1533.
8. N. S. Osborne and D. C. Ginnings, *J. Res. Natl. Bur. Stand.*, 1947, **39**, 453.
9. H. M. Huffman, *Chem. Rev.*, 1947, **40**, 1.
10. H. L. Johnston, J. T. Clarke, E. B. Rifkin and E. C. Kerr, *J. Am. Chem. Soc.*, 1950, **72**, 3933.
11. A. Eucken and M. Eigen, *Z. Elektrochem.*, 1951, **55**, 343.
12. R. W. Hill, *J. Sci. Instrum.*, 1953, **30**, 331.
13. D. R. Stull, *Anal. Chim. Acta*, 1957, **17**, 133.

14. E. D. West and D. C. Ginnings, *J. Res. Natl. Bur. Stand.*, 1958, **60**, 309.

15. L. J. Todd, R. H. Dettre and D. H. Andrews, *Rev. Sci. Instrum.*, 1959, **30**, 463.

16. R. D. Goodwin, *J. Res. Natl. Bur. Stand.*, 1961, **65C**, 309.

17. R. J. L. Andon, J. F. Counsell, E. F. G. Herington and J. F. Martin, *Trans. Faraday Soc.*, 1963, **59**, 850.

18. E. Wilhelm, R. Schano, G. Becker, G. H. Findenegg and F. Kohler, *Trans. Faraday Soc.*, 1969, **65**, 1443.

19. J. C. Van Miltenburg, *J. Chem. Thermodyn.*, 1972, **4**, 773.

20. *Experimental Thermodynamics, Volume I: Calorimetry of Non-reacting Systems*, J. P. McCullough and D. W. Scott, eds., Butterworths/IUPAC, London, 1968.

21. *Solution Calorimetry. Experimental Thermodynamics, Volume IV*, K. N. Marsh and P. A. G. O'Hare, eds., Blackwell Scientific Publications/ IUPAC, Oxford, 1994.

22. *Measurement of the Thermodynamic Properties of Single Phases. Experimental Thermodynamics, Volume VI*, A. R. H. Goodwin, K. N. Marsh and W. A. Wakeham, eds., Elsevier/IUPAC, Amsterdam, 2003.

23. W. Hemminger and G. Höhne, *Calorimetry. Fundamentals and Practice*, Verlag Chemie, Weinheim, 1984.

24. S. T. Lakshmikumar and E. S. R. Gopal, *Int. Rev. Phys. Chem.*, 1982, **2**, 197.

25. I. Wadsö, *Thermochim. Acta*, 1985, **96**, 313.

26. E. Gmelin, *Thermochim. Acta*, 1987, **110**, 183.

27. G. Höhne, W. Hemminger and H.-J. Flammersheim, *Differential Scanning Calorimetry. An Introduction for Practitioners*, Springer, Berlin, 1996.

28. P. Picker, P.-A. Leduc, P. R. Philip and J. E. Desnoyers, *J. Chem. Thermodyn.*, 1971, **3**, 631.

29. J.-P. E. Grolier, G. C. Benson and P. Picker, *J. Chem. Eng. Data*, 1975, **20**, 243.

30. D. Smith-Magowan and R. H. Wood, *J. Chem. Thermodyn.*, 1981, **13**, 1047.

31. D. E. White, R. H. Wood and D. R. Biggerstaff, *J. Chem. Thermodyn.*, 1988, **20**, 159.

32. R. W. Carter and R. H. Wood, *J. Chem. Thermodyn.*, 1991, **23**, 1037.

33. T. K. Hei and J. D. Raal, *AIChE J.*, 2009, **55**, 206.

34. M. P. E. Berthelot, *Thermochimie, Vol. I and II*, Gautier-Villars et Fils, Paris, 1897.

35. *International Critical Tables of Numerical Data, Physics, Chemistry and Technology, Vol. V*, prepared by the National Research Council of the United States of America, E. W. Washburn, editor-in-chief, McGraw-Hill Book Company, New York, 1929, pp. 78–129.

36. J. Timmermans, *Physicochemical Constants of Pure Organic Compounds*, Vol. I, 1950; Vol. II, 1965, Elsevier, Amsterdam.

37. Landolt-Börnstein. *Zahlenwerte und Funktionen aus Physik, Chemie, Astronomie, Geophysik und Technik, 6. Auflage, II. Band, Eigenschaften der Materie in ihren Aggregatzuständen, 4. Teil, Kalorische Zustandsgrößen*, K. Schäfer and E. Lax, eds., Springer-Verlag, Berlin, 1961.

38. T. E. Daubert and R. P. Danner, *Physical and Thermodynamic Properties of Pure Chemicals: Data Compilation, DIPPR® Database*, Hemisphere Publishing Corp., New York, 1989.

39. M. Zábranský, V. Růžièka Jr, V. Majer and E. S. Domalski, *Heat Capacity of Liquids: Volumes I and II. Critical Review and Recommended Values, J. Phys. Chem. Ref. Data, Monograph No. 6*, American Chemical Society and American Institute of Physics, 1996.

40. M. Zábranský, V. Růžièka Jr and E. S. Domalski, *Heat Capacity of Liquids: Critical Review and Recommended Values. Supplement I, J. Phys. Chem. Ref. Data*, 2001, **30**, 1199.

41. *International DATA Series, SELECTED DATA ON MIXTURES, Series A*, published by the Thermodynamics Research Center, Texas A&M University, College Station, TX 77843, USA, from 1973 through 1994.

42. *Dortmund Data Bank Software and Separation Technology*, www. ddbst. de.

43. *Selected Values of Physical and Thermodynamic Properties of Hydrocarbons and Related Compounds*, F. D. Rossini, K. S. Pitzer, R. L. Arnett, R. M. Braun and G. C. Pimentel, eds., published for the American Petroleum Institute by Carnegie Press, Pittsburgh, PA, 1953.

44. D. R. Stull, E. F. Westrum Jr. and G. S. Sinke, *The Chemical Thermodynamics of Organic Compounds*, Wiley, New York, 1969.

45. *TRC Thermodynamic Tables: Hydrocarbons (formerly TRC Hydrocarbon Project) and TRC Thermodynamic Tables: Non-Hydrocarbons (formerly TRC Data Project)*, loose-leaf format, Thermodynamics Research Center, The Texas A&M University System, College Station, TX.

46. M. Frenkel, K. N. Marsh, G. J. Kabo, R. C. Wilhoit and G. N. Roganov, *Thermodynamics of Organic Compounds in the Gas State, Vol. I*, Thermodynamics Research Center, College Station, TX, 1994.

47. M. W. Chase Jr., *NIST-JANAF Thermochemical Tables: Parts I and II, J. Phys. Chem. Ref. Data, Monograph No. 9*, 4th edition, 1998.

48. I. Prigogine and R. Defay, *Chemical Thermodynamics*, translated and revised by D. H. Everett, Longmans, Green and Co., London, 1954.

49. R. Haase, *Thermodynamik der Mischphasen*, Springer-Verlag, Berlin, 1956.

50. E. A. Guggenheim, *Thermodynamics*, 5th edition, North-Holland, Amsterdam, 1967.

51. M. L. McGlashan, *Chemical Thermodynamics*, Academic Press, London, 1979.

52. Joseph Kestin, *A Course in Thermodynamics*, McGraw-Hill, New York, 1979.

53. G. Kortüm and H. Lachmann, *Einführung in die chemische Thermodynamik. Phänomenologische und statistische Behandlung*, 7. Auflage Verlag Chemie, Weinheim, 1981.
54. K. Denbigh, *The Principles of Chemical Equilibrium*, Cambridge University Press, Cambridge, 1981.
55. D. Kondepuni and I. Prigogine, *Modern Thermodynamics. From Heat Engines to Dissipative Structures*, John Wiley and Sons, Chichester, 1998.
56. J. S. Rowlinson and F. L. Swinton, *Liquids and Liquid Mixtures*, Butterworth Scientific, London, 1982.
57. H. B. Callen, *Thermodynamics and an Introduction to Thermostatics*, John Wiley & Sons, New York, 1985.
58. S. E. Wood and R. Battino, *Thermodynamics of Chemical Systems*, Cambridge University Press, Cambridge, 1990.
59. (a) J.W. Magee, *J. Res. Natl. Inst. Stand. Technol*, 1991, **96**, 725; (b) R.A. Perkins and J.W. Magee, *J. Chem. Eng. Data*, 2009, **54**, 2646.
60. T. Miyazaki, A. V. Hejmadi and J. E. Powers, *J. Chem. Thermodyn.*, 1980, **12**, 105.
61. E. Wilhelm, *Thermochim. Acta*, 1983, **69**, 1.
62. S. W. Benson, F. R. Cruickshank, D. M. Golden, G. R. Haugen, H. E. O'Neal, A. S. Rodgers, R. Shaw and R. Walsh, *Chem. Rev.*, 1969, **69**, 279.
63. E. S. Domalski and E. D. Hearing, (a) *J. Phys. Chem. Ref. Data*, 1993, **22**, 805; (b) *J. Phys. Chem. Ref. Data*, 1994, **23**, 157.
64. K. S. Pitzer, *J. Am. Chem. Soc.*, 1955, **77**, 3427.
65. K. S. Pitzer, D. Z. Lippmann, R. F. Curl Jr., C. M. Huggins and D. E. Petersen, *J. Am. Chem. Soc.*, 1955, **77**, 3433.
66. K. S. Pitzer and R. F. Curl Jr., *J. Am. Chem. Soc.*, 1957, **79**, 2369.
67. R. F. Curl Jr and K. S. Pitzer, *Ind. Eng. Chem.*, 1958, **50**, 265.
68. B.-I. Lee and M. G. Kesler, *AIChE J.*, 1975, **21**, 510.
69. L. A. Davies and R. B. Gordon, *J. Chem. Phys.*, 1967, **46**, 2650.
70. M. J. P. Muringer, N. J. Trappeniers and S. N. Biswas, *Phys. Chem. Liq.*, 1985, **14**, 273.
71. (a) E. Bartholomé and A. Eucken, *Trans. Faraday Soc.*, 1937, **33**, 45; (b) A. Eucken, *Z. Elektrochem.*, 1948, **52**, 255.
72. J. D. Bernal, *Trans. Faraday Soc.*, 1937, **33**, 27.
73. J. F. Kincaid and H. Eyring, *J. Chem. Phys.*, 1938, **6**, 620.
74. (a) L. A. K. Staveley, K. R. Hart and W. I. Tupman, *Disc. Faraday. Soc.*, 1953, **15**, 130; (b) L. A. K. Staveley, W. I. Tupman and K. R. Hart, *Trans. Faraday. Soc.*, 1955, **51**, 323.
75. D. Harrison and E. A. Moelwyn-Hughes, *Proc. Roy. Soc. (London)*, 1957, **A 239**, 230.
76. G. H. Findenegg and F. Kohler, *Trans. Faraday Soc.*, 1967, **63**, 870.
77. A. Bondi, *Physical Properties of Molecular Crystals, Liquids and Glasses*, Wiley, New York, 1968.
78. E. Wilhelm, M. Zettler and H. Sackmann, *Ber. Bunsenges. Phys. Chem.*, 1974, **78**, 795.

79. E. Wilhelm, *Pure Appl. Chem.*, 2005, **77**, 1317.
80. F. Kohler, *The Liquid State*, Verlag Chemie, Weinheim/Bergstr., 1972.
81. R. E. Gibson and O. H. Loeffler, *J. Am. Chem. Soc.*, 1941, **63**, 898.
82. E. Wilhelm, *J. Chem. Phys.*, 1975, **63**, 3379. See also E. Wilhelm, *Proc. 14th Intl. Conf. Chem. Thermodyn.*, Montpellier, France, August 26–30, 1975, Vol. II, pp. 87–94.
83. E. Wilhelm, J.-P. E. Grolier and M. H. Karbalai Ghassemi, *Ber. Bunsenges. Phys. Chem.*, 1977, **81**, 925.
84. J. P. McCullough and G. Waddington, in *Experimental Thermodynamics, Volume I: Calorimetry of Non-reacting Systems*, J. P. McCullough and D. W. Scott, eds., Butterworths/IUPAC, London, 1968, pp. 369–394.
85. (a) S. S. Todd, I. A. Hossenlopp and D. W. Scott, *J. Chem. Thermodyn.*, 1978, **10**, 641; (b) I. A. Hossenlopp and D. W. Scott, *J. Chem. Thermodyn.*, 1981, **13**, 405; (c) I. A. Hossenlopp and D. W. Scott, *J. Chem. Thermodyn.*, 1981, **13**, 415.
86. P. G. Francis, M. L. McGlashan and C. J. Wormald, *J. Chem. Thermodyn.*, 1969, **1**, 441.
87. N. Al-Bizreh and C. J. Wormald, *J. Chem. Thermodyn.*, 1977, **9**, 749.
88. M. L. McGlashan and C. J. Wormald, *J. Chem. Thermodyn.*, 2000, **32**, 1489.
89. L. Haar, J. S. Gallagher and G. S. Stell, *NBS/NRC Steam Tables*, Hemisphere Publishing Corporation, New York, 1984.
90. P. G. Hill and R. D. C. MacMillan, *Ind. Eng. Chem. Res.*, 1988, **27**, 874.
91. W. Wagner and A. Pruß, *J. Phys. Chem. Ref. Data*, 2002, **31**, 387.
92. W. Van Dael, in *Experimental Thermodynamics, Volume 2: Experimental Thermodynamics of Non-reacting Fluids*, B. Le Neindre and B. Vodar, eds., Butterworths/IUPAC, London, 1975, pp. 527–577.
93. M. B. Ewing, M. L. McGlashan and J. P. M. Trusler, *Mol. Phys.*, 1987, **60**, 681.
94. G. C. Maitland, M. Rigby, E. B. Smith and W. A. Wakeham, *Intermolecular Forces. Their Origin and Determination*, Clarendon Press, Oxford, 1981.
95. J. B. Mehl and M. R. Moldover, *J. Chem. Phys.*, 1981, **74**, 4062.
96. M. R. Moldover, J. P. M. Trusler, T. J. Edwards, J. B. Mehl and R. S. Davis, *Phys. Rev. Lett.*, 1988, **60**, 249.
97. H. J. Hoge, *J. Res. Natl. Bur. Stand.*, 1946, **36**, 111.
98. E. Wilhelm, in *Les Capacités Calorifiques des Systèmes Condensés*, H. Tachoire, ed., Société Française de Chimie, Marseille, 1987, pp. 138–163.
99. M. Planck, *Ann. Physik*, 1887, **30**, 574.
100. E. Wilhelm, *High Temp.-High Press.*, 1997, **29**, 613.
101. *Experimental Thermodynamics, Volume V: Equations of State for Fluids and Fluid Mixtures*, J. V. Sengers, R. F. Kayser, C. J. Peters and H. J. White Jr., eds., Elsevier/IUPAC, Amsterdam, The Netherlands, 2000.

102. E. Wilhelm, in *Experimental Thermodynamics, Volume VII: Measurement of the Thermodynamic Properties of Multiple Phases,* R. D. Weir and Th. W. de Loos, eds., Elsevier/IUPAC, Amsterdam, The Netherlands, 2005, pp. 137–176.
103. E. Wilhelm, in *Development and Applications in Solubility,* T. M. Letcher, ed., The Royal Society of Chemistry/IUPAC, Cambridge, UK, 2007, pp. 3–18.

CHAPTER 2

Calorimetric Methods for Measuring Heat Capacities of Liquids and Liquid Solutions

LEE D. HANSEN[a] AND DONALD J. RUSSELL[b]

[a] Department of Chemistry and Biochemistry, Brigham Young University, Provo, Utah 84602, USA; [b] TA Instruments, 890 W 410 N, Lindon, Utah 84042, USA

2.1 Introduction

Heat capacity is defined as the amount of heat required to cause a given temperature change, *i.e.* $(\partial Q/\partial T)_x$ and is arguably the most fundamental of thermodynamic properties since it is the most closely related to molecular properties, see reference 1 and Chapters 1, 16 and 19 in this book. Heat capacity is thermodynamically defined at constant pressure, $C_P = (\partial H/\partial T)_P = (\partial Q/\partial T)_P$ or at constant volume, $C_V = (\partial U/\partial T)_V = (\partial Q/\partial T)_V$. The relation between the two heat capacities in terms of mechanical variables is:

$$C_P - C_V = \alpha_P^2 TV/\beta_T \tag{1}$$

where α is the coefficient of thermal expansion, $\beta_T{}^\dagger$ is the isothermal compressibility, and V is the volume. For practical reasons the heat capacity of a liquid is also sometimes defined as C_{sat}, the heat capacity at the saturation

---

$\dagger$ In this chapter the isothermal compressibility is represented by the symbol β_T and not by κ_T as was recently recommended by IUPAC.

Heat Capacities: Liquids, Solutions and Vapours
Edited by Emmerich Wilhelm and Trevor M. Letcher
© The Royal Society of Chemistry 2010
Published by the Royal Society of Chemistry, www.rsc.org

vapor pressure of the liquid. C_{sat} is related to C_P by:

$$C_{sat} = C_P - T(\partial V/\partial T)_p(\partial p/\partial T)_{sat} \qquad (2)$$

The difference between C_{sat} and C_P is usually negligible at vapor pressures below about 1 atmosphere. Complete exposition of the thermodynamic relations involving heat capacity is given in references 2–4.

Specialized calorimeters for measurements of the heat capacities of liquids were reviewed in 1994 in references 2 and 3 and in 1996 in reference 4. Reference 4 also includes an extensive compilation of data on heat capacities of liquids. The purpose of this chapter is to review the application of currently available commercial calorimeters (Table 2.1) to measurement of heat capacities of liquids.

2.2 Methods

Heat capacities of liquids can be measured in any calorimeter capable of simultaneously measuring both the amount of heat input or removed and the accompanying temperature change in the sample. Figure 2.1 illustrates the various ways that heat capacities of liquids can be measured. Design of the calorimeter and the properties of the liquid limit the measurements that can be made accurately in a given calorimeter. Measurement of heat capacities of liquids is complicated by chemical equilibria, vapor–liquid equilibria, and thermal expansion of the liquid. The type of calorimeter used determines how significant each of these complications is.

Chemical and vapor–liquid equilibria may have significant effects on the measurement. Only chemical equilibria can occur in calorimeters with no vapor space, but both chemical and vapor–liquid equilibria can exist in calorimeters with vapor space. In a constant pressure experiment, the effects of equilibria that shift with temperature can be expressed as:

$$(dQ/dT)_{measured} = (C_P)(n_s) + \sum_r (\Delta_r H)(dn_r/dT) \qquad (3)$$

where n_s is the moles of sample, $\Delta_r H$ is the enthalpy change for each reaction in the sample, and n_r is the amount of substance of each reaction in the reaction vessel. By convention a correction for vapor–liquid equilibria is included when appropriate, but correction for chemical equilibria within the liquid phase may or may not be included, depending on the desired result, see Chapter 18 in this book. Heat capacity and reaction enthalpy may be separated by temperature modulated DSC and AC calorimetry if reactions are slow compared with the modulation frequency, otherwise obtaining the heat capacity without reaction enthalpies requires independent measurement of the enthalpy change and equilibrium constant for each reaction.

Another complication in the measurement of liquid heat capacities occurs because temperature changes in liquids can cause a significant change in

Table 2.1 Commercially available calorimeters for measuring heat capacities of liquids.

Vessel type	Measurement	Representative calorimeters	Temperature and pressure ranges	Relative quality of heat capacity measurement
Sealed, constant volume, with vapor space, <0.1 cm³	Approximately C_P (saturated vapor pressure + external pressure) by temperature-scanning/temperature-modulated calorimetry	DSC from various manufacturers (e.g., TA Instruments, Perkin-Elmer, Netzsch, Mettler)	93–2273 K, depending on specific calorimeter; pressure capability depends on ampule	Poor
Sealed, constant volume, with vapor space, 1–10 cm³	Approximately C_P (saturated vapor pressure + external pressure) by temperature-scanning heat-conduction calorimetry	• TA Instruments MC-DSC (1 cm³) • Setaram C80 (10 cm³) • Setaram MicroDSC (1 cm³)	• 233–473 K, atmospheric to 41 MPa • room to 573 K, atmospheric to 100 MPa • 253–393 K, atmospheric to 70 MPa	Good–better
Removable, overflow with back pressure regulator	C_P by temperature-scanning heat-conduction calorimetry	• TA Instruments MC-DSC (1 cm³) • Setaram C80 (10 cm³) • Setaram MicroDSC (1 cm³)	• 233–473 K, atmospheric to 41 MPa • room to 573 K, atmospheric to 100 MPa • 253–393 K, atmospheric to 70 MPa	Good–better
Open to atmosphere, 5 to 140 cm³	C_P by isoperibol temperature-rise calorimetry	• CSC-ISC (25 or 50 cm³ Dewar) • Parr Instrument Co. solution calorimeter (140 cm³) • TA Instruments SolCal (25 or 100 cm³) • Kyoto Electronics SHA-500 (5 cm³)	• 273–353 K • 283–323 K • 288–353 K • 277–358 K	• Better • Good • Better • Good

Sealed, no vapor space, constant volume or constant pressure, 0.5 cm³	C_V or C_P by heat-conduction calorimetry. Pressure, volume and temperature can each be controlled and can be scanned independently.	Transitiometer[a]	220–503 K, atmospheric to 400 MPa	Good
Flow, with back pressure regulator or at atmospheric pressure	C_P by isothermal heat-conduction calorimetry	• TA Instruments TAM	• 288–423 K, atmospheric to 0.3 MPa (higher pressures with custom ampule)	Good–better
		• Setaram C80	• room to 573 K, atmospheric to 100 MPa	
Fixed-in-place, overflow with back pressure regulator, 0.1–0.5 cm³	C_P by temperature-scanning power-compensation calorimetry	• TA Instruments NanoDSC	• 263–433 K, atmospheric to 0.6 MPa	• Best
		• MicroCal DSC	• 263–403 K, atmospheric to 0.5 MPa	• Better
4 × 4 × (0.1–0.3) mm	AC calorimetry	Sinku-Riko ACC-1	70–800 K, vacuum or He atmosphere	Accuracy poor, precision best
small diameter tubing, <3 cm³	Density and speed of sound	Anton Paar DSA5000	273–343 K, atmospheric to 0.3 MPa	Poor, ±5 µg cm³, ±0.5 m s⁻¹

[a]A commercial version has been introduced by Stanislaw Randzio, Polish National Academy of Sciences, Warsaw.

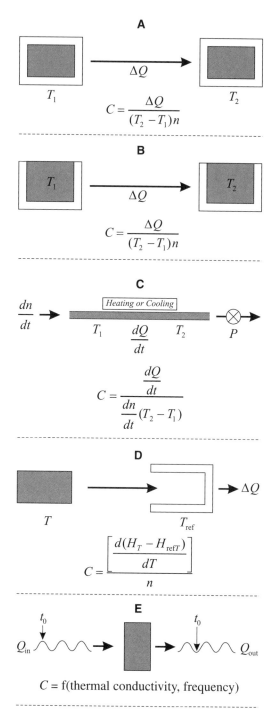

Figure 2.1 Types of calorimeters for measurement of molar heat capacities of liquids. *C* is heat capacity, either C_P, C_V, or C_{sat} depending on how the experiment is done. *Q* is heat, *t* is time, *T* is temperature, and *n* is moles of liquid.

volume. In constant pressure systems with overflow reaction vessels, *i.e.*, the reaction vessel and access tube are completely filled with liquid and the tube is open to a constant pressure reservoir, the mass of liquid in the active volume of the calorimeter vessel changes with temperature. In a constant volume reaction vessel with vapor space, expansion of the liquid changes the volume of the vapor space and this affects both the heat due to vaporization and the heat capacity of the gas phase. The coefficient of thermal expansion for the liquid must be known to correct for these effects.

Heat capacity may be measured either by continuously scanning temperature or by stepwise measurements. In continuous scanning measurements, heat capacity is calculated as the ratio $(dQ/dt)/(dT/dt)$. A correction for instrument response may be necessary under some conditions. If the calorimeter response can be described with a first-order exponential in time, this can be done with the Tian equation:

$$(\partial Q/\partial t)_c = (\partial Q/\partial t)_m + \tau[(\partial Q/\partial t)_m/\partial t] \tag{4}$$

where $(\partial Q/\partial t)_c$ is the corrected heat rate, $(\partial Q/\partial t)_m$ is the measured heat rate, and τ is the time constant for the response of the calorimeter and sample to a temperature change. Stepwise measurements do not require correction for time response, but note that ΔT must be kept small to satisfy the differential definition of heat capacity.

Error propagation analysis and estimates of potential systematic errors allows quantitative comparison of various calorimeters and methods:

$$\delta(\Delta Q/\Delta T) = \{[(\delta\Delta Q)/\Delta T]^2 + [(\delta\Delta T)\Delta Q/(\Delta T)^2]^2\}^{0.5}$$
$$+\delta(\text{correction for equilibria}) + \delta(\text{correction for time constant})$$
$$+\delta(\text{correction for thermal expansion of liquid})$$

$$\tag{5}$$

where δ indicates the estimated error. The first two terms account for random errors in the measurements of heat and temperature and the other three terms are systematic errors. Systematic errors in the measurements of heat and temperature are not included in Equation (5) because these depend on the accuracy of calibration of the calorimeter and thermometer. Estimates of random and systematic error in the measurements of heat ($\delta\Delta Q$), temperature ($\delta\Delta T$), and in the correction for equilibria are needed to estimate the uncertainty of measurements in all calorimeters. Uncertainty in calorimeter response only applies to continuous scanning, AC calorimetry and other dynamic methods. Uncertainty in the correction for thermal expansion of the liquid only applies to measurements in partially filled and overfilled, constant pressure ampules.

Aside from conventional calorimetry, various types of dynamic measurements have been used for determination of heat capacities of liquids, see Chapters 10–12 in this book and references 4–20. AC calorimetry, in which one side of a sample is heated with a sinusoidal power input, either electrically or with a light source, and the resulting temperature signal is measured on the other side of the sample, is a close relative of temperature-modulated DSC.[16] In AC calorimetry the heat capacity is calculated from the phase shift between the input and output signals. Pulse or relaxation methods have been used similarly. The phase shift in AC calorimetry depends on a function of the heat capacity and the thermal conductivity of the sample; the function depends on the thickness and properties of the sample and design of the calorimeter.[5]

Two indirect methods have been used for determination of heat capacities: measurement of the density and speed of sound in the liquid as functions of temperature and pressure; and the measurement of the pVT properties of the liquid. Speed of sound data are fitted to:

$$C_P = T\alpha_P^2/[(\partial\rho/\mathrm{d}p)_T - u^{-2}] \qquad (6)$$

and

$$C_V = C_P/u^2[(\partial\rho/\mathrm{d}p)_T \qquad (7)$$

where ρ is the density and u is the speed of sound.[19,20] This method is generally less accurate than direct calorimetric measurement because of the dependence of the heat capacity on the temperature and pressure derivatives of the primary measurements.[4] The method is however very useful for obtaining compressibility data. The heat capacity is related to pVT properties by:

$$C_P = T(\partial V/\partial T)_P/(\partial T/\partial p)_S \qquad (8)$$

where S is entropy, *i.e.*, $(\partial T/\partial p)_S$ is measured under adiabatic conditions[21] (see also Equation (1)). This indirect method is also of low accuracy because it requires taking derivatives of the measured data.[4]

2.3   Comparison of Methods

Temperature scanning calorimetry (DSC) is probably the most common method currently used for heat capacity determinations (Type A in Figure 2.1). The most commonly used DSCs have sample volumes < 0.1 cm^3, but DSCs with sample volumes ≥ 1 cm^3 and better heat detection limits (nanowatts) are available (see Table 2.1 and Chapter 13 in this book). Larger sample size is advantageous because ΔQ is proportional to sample size, and increasing sample size thus decreases the relative error in the measurement of ΔQ. Two types of ampules are used in DSCs; sealed ampules with some vapor space which measure C_{sat} and overflow ampules which measure C_P. Transitiometry (see Chapter 8 in this book) uses a sealed ampule with no vapor space. Because pressure or volume can be held constant, transitiometry can be used to determine either C_P or C_V.[22] Ampules with vapor space must have a hermetic seal to prevent loss of sample as the temperature increases and total volume of the ampule must be known to make corrections for vapor–liquid equilibria. Because the vapor pressure over the liquid varies with temperature, the measurement gives neither C_P nor C_V directly, but C_{sat}. C_{sat} approximates C_P for small pressure changes. Removable overflow ampules are available for DSCs with larger sample volumes, and are typically a cylindrical ampule with a fill tube extending outside the calorimeter.[23] The fill tube may be at atmospheric pressure or may be fitted with a pressure regulator. Fixed-in-place overflow reaction vessels are used in nanowatt DSCs. These reaction vessels may be a coiled capillary tube, a cylinder, or a lollypop shaped vessel. The last is subject to larger volume changes with pressure changes. Depending on the temperature range, correction for thermal expansion of the ampule may be necessary to obtain high accuracy in DSC measurements of the heat capacity of a liquid sample.[23] Nanowatt DSC has been used to determine the partial molar heat capacities of solutes in dilute solutions. Uncertainty in these measurements can be as good as $\pm(5$–$50)\times10^{-6}$.[24]

Isoperibol temperature-rise calorimetry, type B in Figure 2.1, does not have the temperature range or pressure capability of nanowatt DSC, but is another very accurate method for determining heat capacities of liquids. The large volume, typically ≥ 25 cm^3, and very accurate determination of a small ΔT provide very favorable conditions for determination of C_P that can be as good as $\pm(10$–$100)\times10^{-6}$. The reaction vessel is typically a special Dewar flask, temperature changes are measured with a thermistor bridge with microdegree resolution,[25] and ΔQ is provided by an electrical heater submerged in the liquid. The vapor space in the reaction vessel is typically connected to the atmosphere through a capillary tube to limit the rate of loss of vapor during a measurement. To speed up reaching a steady state before taking measurements on volatile liquids, the capillary tube and head space can be pre-equilibrated with liquid vapor. Because it is difficult to seal the system,[25] the method is limited to liquids and temperatures where the vapor pressure is relatively low, *i.e.* <1 atmosphere.

For the determination of C_P of liquids over a wide range of temperatures and pressures, isothermal flow calorimetry, type C in Figure 2.1, is generally the best choice. Because there is no vapor space, no correction for vapor–liquid equilibria is necessary, and if the flow rate is calibrated in moles per unit time, no correction for thermal expansion of either the liquid or the reaction vessel is necessary. Accurate measurements require thermostatting and careful calibration and maintenance of the pumps. The method is limited to low viscosity liquids; liquids with high viscosity are difficult to pump through the tubing and viscous flow may generate sufficient heat to interfere with the measurement. Two methods can be used, either the temperature change between the input and output flow from a known rate of heat input is measured, or the heat rate required to change the temperature by a known amount is measured. In the first method, the temperature of the input flow is controlled, heat is added to the liquid at a constant known rate, and the temperature of the output flow is measured. In the second method, the temperature of the input flow is controlled

Table 2.2   Properties of calibration and test substances for determination of
heat capacities of liquids. Heat capacities of liquids are recommended values from reference 4. Sapphire data are from reference
26. Units of heat capacities are $J\,K^{-1}\,g^{-1}$. Data for corrections for
vapor–liquid equilibria and thermal expansion are given in the
footnotes with the original units to avoid errors of translation.

T/K	Sapphire C_P	Hexane[a,b] C_P	C_{sat}	Decane[c] C_P	C_{sat}	1,2-ethanediol[d] C_P	Water[e] C_P	C_{sat}
180	0.429	1.981	1.981					
190	0.466	1.980	1.980					
200	0.501	1.987	1.987					
210	0.536	2.001	2.001					
220	0.568	2.020	2.020					
230	0.600	2.044	2.044					
240	0.629	2.072	2.072					
250	0.658	2.101	2.101	2.091	2.091			
260	0.685	2.133	2.133	2.108	2.108			
270	0.707	2.166	2.166	2.129	2.129	2.262		
273.15	0.718	2.177	2.177	2.137	2.137	2.279		
280	0.734	2.201	2.201	2.155	2.155	2.315	4.201	4.201
290	0.757	2.238	2.238	2.184	2.184	2.367	4.187	4.188
298.15	0.775	2.270	2.270	2.210	2.210	2.409	4.181	4.181
300	0.779	2.278	2.278	2.216	2.216	2.418	4.180	4.180
310	0.799	2.320	2.320	2.250	2.250	2.469	4.177	4.177
320	0.819	2.365	2.364	2.286	2.286	2.520	4.178	4.177
330	0.837	2.413	2.412	2.323	2.323	2.570	4.182	4.181
340	0.855	2.464	2.462	2.361	2.361	2.620	4.189	4.187
350	0.871	2.518	2.515	2.400	2.400	2.669	4.198	4.196
360	0.887	2.574	2.570	2.441	2.441	2.718	4.208	4.206
370	0.902	2.632	2.626	2.483	2.483	2.766	4.218	4.217
380	0.916	2.691	2.682	2.526	2.526	2.814	4.229	4.229
390	0.930	2.751	2.738	2.570	2.570	2.862	4.238	4.241
400	0.942	2.811	2.792	2.616	2.615	2.909	4.249	4.254
410	0.954	2.871	2.845	2.662	2.661	2.955	4.261	4.268
420	0.966	2.930	2.896	2.709	2.707	3.001	4.277	4.286
430	0.977	2.988	2.943	2.757	2.755	3.047	4.297	4.308
440	0.988	3.045	2.986	2.805	2.803	3.092	4.324	4.334
450	0.998	3.099	3.025	2.855	2.851	3.137	4.359	4.367
460	1.007	3.151	3.058	2.905	2.901	3.182	4.403	4.406
470	1.016					3.226	4.458	4.454
480	1.025					3.269	4.525	4.511
490	1.033					3.312	4.606	4.577
500	1.041						4.701	4.655
510	1.048						4.814	4.745
520	1.056						4.944	4.848
530	1.063						5.094	4.965
540	1.069						5.265	5.097
550	1.076						5.459	5.245
560	1.082						5.676	5.411
570	1.088						5.919	5.594
580	1.093						6.188	5.797
590	1.099						6.486	6.020
600	1.104						6.900	6.300

Table 2.2 (*Continued*)

T/K	Sapphire C_P	Hexane[a,b] C_P	C_{sat}	Decane[c] C_P	C_{sat}	1,2-ethanediol[d] C_P	Water[e] C_P	C_{sat}
610	1.109						7.89	6.85
620	1.114						10.0	7.89
630	1.118						13.8	9.66
640	1.123						25.1	13.7

[a]Values of C_P and C_V are given as functions of pressure (p_{sat} to 700 MPa) and temperature (243–503 K) in reference 27.

[b]n-Hexane:

$\Delta_{vap}H/\text{kJ mol}^{-1} = -1.99775 \times 10^{-4}x^2 + 7.07046 \times 10^{-2}x + 28.0397$

$R^2 = 0.9998$, $x = T/K$, $298 < T < 444$ K (fitted to experimental data in references 28 and 29)

$\log_{10}(p_{vap}/\text{mmHg}) = A + B\left[(t/°C) + C\right]^{-1}$, $A = 6.98950$, $B = 1216.9154$, $C = 227.451$, –95 to 234 °C (reference 30)

$\alpha_p/\text{K}^{-1} = a(p - p_{sat} + b)^{-0.5}$, $a = 7.117 \times 10^{-3} - 2.484456 \times 10^{-6}h + 4.01821 \times 10^{-8}h^2 + (3.207 \times 10^{-3})e^h$, $b = -2.344 \times 10^{-2}h + 1.29388 \times 10^{-3}h^2 + 5.76483 \times 10^{-6}h^3 + 1.97294 \times 10^{-8}h^4 + (2.952 \times 10^2)e^h$, $h = [(T/K) - 507.8)]$ (reference 27)

$\beta_T/\text{MPa}^{-1} = C/(\{B + p\}\{1 - C \ln[(B + p)/(B + p_{sat})]\})$, $C/\text{MPa}^{-1} = -2.39259 \times 10^{-15}x^6 + 5.11378 \times 10^{-12}x^5 - 4.50677 \times 10^{-9}x^4 + 2.09189 \times 10^{-6}x^3 - 5.38157 \times 10^{-4}x^2 + 7.25381 \times 10^{-2}x - 3.89380$, $R^2 = 0.9994$, $B/\text{MPa} = 3.23979 \times 10^{-8}x^4 - 5.47940 \times 10^{-5}x^3 + 3.54646 \times 10^{-2}x^2 - 10.6256x + 1268.61$, $R^2 = 0.9999$, $243 < T < 503$ (reference 27)

$\rho/\text{g cm}^{-3} = \rho(T,p_{ref})/\{1 - C \ln[(B + p)/(B + p_{ref})]\}$, $\rho(T,p_{ref}) = 0.234\{1 + 1.597561(1 - T/507.9)^{1/3} + 1.842657(1 - T/507.9)^{2/3} - 1.726311(1 - T/507.9) + 0.4943082(1 - T/507.9)^{4/3} + 0.6463138(1 - T/507.9)^{5/3}\}$, $p_{ref} = 0.101325$ MPa up to the normal boiling point (341.9 K), $p_{ref} = p_{sat}$ above boiling point, $C = 0.092380 - 0.004522[(T - 298.15)/100]$, $56.0622 - 56.5365[(T - 298.15)/100] + 18.6982[(T - 298.15)/100]^2 - 2.9566[(T - 298.15)/100]^3 + 0.2334[(T - 298.15)/100]^4$, $223 < T/$ K < 498, $0.57 < p/\text{MPa} < 1079$ (reference 31)

[c]n-Decane:

$\Delta_{vap}H/\text{kJ mol}^{-1} = -6.07499 \times 10^{-5}x^2 - 3.90777 \times 10^{-2}x + 68.3950$, $R^2 = 0.9998$, $x = T/K$, $298 < T < 444$ K (fitted to experimental data in reference 28)

$\log_{10}(p_{vap}/\text{mmHg}) = A + B[(t/°C) + C]^{-1}$, $A = 7.21745$, $B = 1693.9274$, $C = 216.459$, –30 to 345 °C (reference 30)

$\rho = \rho(T,p_{ref})/\{1 - C \ln[(B + p)/(B + p_{ref})]\}$, $\rho(T,p_{ref}) = 0.239\{1 + 0.3291388(1 - T/617.61)^{1/3} + 7.364340(1 - T/617.61)^{2/3} - 9.985096(1 - T/617.61) + 5.283608(1 - T/617.61)^{4/3}\}$, $p_{ref} = 0.101325$ MPa up to the normal boiling point (447.3 K), $p_{ref} = p_{sat}$ above boiling point, $C = 0.087992 - 0.000816[(T - 294.35)/100]$, $B = 83.5746 - 61.9418[(T - 294.35)/100] + 21.8935[(T - 294.35)/100]^2 - 6.4316[(T - 294.35)/100]^3 + 1.0545[(T - 294.35)/100]^4$, $248 < T/K < 503$, $0.60 < p/\text{MPa} < 800$ (reference 31)

[d]Ethanediol (ethylene glycol)

$\Delta_{vap}H = 64.2 \pm 1.2 \text{ kJ mol}^{-1}$ at 298.15 K, $\Delta_{vap}H = 65.99 \pm 0.25 \text{ kJ mol}^{-1}$ (reference 32)

$\log(p_{vap}/\text{kPa}) = 8.3726 - 2994.4/(T/K)$ or $= 8.2549 - 2926/(T/K)$ (reference 32)

$\ln(p_{vap}/\text{kPa}) = (291.2/R) - [80744.3/R(T/K)] - (49.5/R)\ln(T/K)/298.15$ (reference 33)

$\log_{10}(p_{vap}/\text{mmHg}) = A + B[(T/°C) + C]^{-1}$, $A = 9.69941$, $B = 3147.0950$, $C = 264.246$, –13 to 372 °C (reference 30)

[e]Water

$\Delta_{vap}H/\text{kJ mol}^{-1} = -5.1176 \times 10^{-13}x^6 + 5.0733 \times 10^{-10}x^5 - 1.8858 \times 10^{-7}x^4 + 3.1881 \times 10^{-5}x^3 - 2.4351 \times 10^{-3}x^2 + 2.3179 \times 10^{-2}x + 44.872$ ($x = T/°C$) $R^2 = 0.9979$, 0 to 374 °C (fitted to data in reference 34)

$\log_{10}(p_{vap}/\text{mmHg}) = A + B[(T/°C) + C]^{-1}$, $A = 8.05573$, $B = 1723.6425$, $C = 233.08$; 0 to 374 °C (reference 30)

$\alpha_p/\text{K}^{-1} = 2.731325 \times 10^{-14}x^5 - 1.890695 \times 10^{-11}x^4 + 4.716929 \times 10^{-9}x^3 - 5.036596 \times 10^{-7}x^2 + 2.752322 \times 10^{-5}x - 1.228750 \times 10^{-4}$, $R^2 = 0.9993$, $0 < T/°C < 374$ (reference 19, p. 46)

$\beta_T/\text{Mbar at } 20 °C = 45.2 \times 10^{-6}$ (0–100 Mbar), 44.1×10^{-6} (100–200 Mbar), 41.8×10^{-6} (200–300 Mbar), 41.1×10^{-6} (300–400 Mbar), 39.4×10^{-6} (400–500 Mbar) (reference 35)

$\beta_T/\text{bar}^{-1} = 2.516826 \times 10^{-17}x^6 - 2.052429 \times 10^{-14}x^5 + 6.269729 \times 10^{-12}x^4 - 8.747548 \times 10^{-10}x^3 + 5.631529 \times 10^{-8}x^2 - 1.433965 \times 10^{-6}x + 5.434808 \times 10^{-5}$, $R^2 = 0.9999$, $0 < T/°C < 374$ (reference 19, p. 46)

at a different temperature to the isothermal calorimeter (either heat-conduction or power-compensation) so the output flow is at the calorimeter temperature. With either method careful attention must be paid to potential heat exchange with the surroundings that is not sensed by the calorimetric measurement.

Drop calorimetry, type D in Figure 2.1, is the simplest method to implement for measuring heat capacities of liquids over a wide range of temperatures. Although rarely used today, the method was widely used in the past.[4] In this method, an ampule containing the liquid is moved from a controlled temperature to a calorimeter at a different temperature. The calorimeter is often as simple as a copper block or water-filled Dewar with a recording thermometer and calibration heater. Care must be taken to avoid heat exchange between the sample and surroundings and between the thermostat and calorimeter during the transfer. Any commercially available isothermal calorimeter which has good access to the measuring chamber may be adapted for this method. Because ΔT is typically large in this method, data should be reported as enthalpy increments between the reference (calorimeter) temperature and the sample temperature, *i.e.*, $(H_T - H_{\mathrm{ref}T})$, and the heat capacity is then given by the derivative of the curve of $(H_T - H_{\mathrm{ref}T})$ versus T, *i.e.* $\mathrm{d}(H_T - H_{\mathrm{ref}T})/\mathrm{d}T$.

AC calorimetry, type E in Figure 2.1, provides very good precision of the measurement of heat capacity, typically $\pm 0.2\,\mu\mathrm{J\,K^{-1}}$, but accuracy is poorer than that achievable with conventional calorimetry, typically several percent uncertainty.[16]

2.4 Test Compounds

Sapphire is an appropriate calibration standard for temperature-scanning calorimeters if the reaction vessel can accept solids.[26] For liquid-only calorimeters, we recommend water, n-hexane, n-decane, and 1,2-hydroxyethane (ethylene glycol) as test substances. The properties of these liquids, presenting a range of properties that can be used to test the capabilities of most calorimeters for heat capacity measurements on liquids, are given in Table 2.2. Because little or no change in enthalpy is associated with the phase transition, the nematic to smectic-A transition in 4,4'-n-octyloxycyanobiphenyl has been suggested as a temperature calibration standard for AC calorimetry and temperature-modulated DSC.[12]

References

1. G. Somsen, Chapter 1. Introduction in *Solution Calorimetry. Experimental Thermodynamics Vol. IV,* ed. K. N. Marsh and P. A. G. O'Hare, Blackwell Scientific, London, 1994, pp. 5–6.
2. J.-P. E. Grolier, Chapter 4. Heat Capacity of Organic Liquids in *Solution Calorimetry. Experimental Thermodynamics Vol. IV*, ed. K. N. Marsh and P. A. G. O'Hare, Blackwell Scientific, London, 1994, pp. 43–75.

3. S. L. Randzio, Chapter 13. Calorimetric Determination of Pressure Effects in *Solution Calorimetry. Experimental Thermodynamics Vol. IV*, ed. K. N. Marsh and P. A. G. O'Hare, Blackwell Scientific, London, 1994, pp. 303–324.

4. M. Zábranský, V. Růžièka Jr., V. Majer and E. S. Domalski, *Heat Capacity of Liquids. Vols. 1 and 2. Critical Review and Recommended Values,* American Institute of Physics, Woodbury, NY, 1996.

5. P. F. Sullivan and G. Seidel, *Phys. Rev.*, 1987, **173**(3), 679–685.

6. N. O. Birge, *Phys. Rev. B*, 1986, **34**(3), 1631–1642.

7. N. O. Birge, *Rev. Sci. Instrum.*, 1987, **58**(8), 1464–1470.

8. J. E. Smaardyk and J. M. Mochel, *Rev. Sci. Instrum.*, 1987, **49**, 988–993.

9. H. Yao and K. Ema, *Rev. Sci. Instrum.*, 1998, **69**(1), 172–178.

10. J. Caerels, C. Glorieux and J. Thoen, *Rev. Sci. Instrum.*, 1998, **69**(6), 2452–2458.

11. H. Yao, K. Ema and I. Hatta, *Jpn. J. Appl. Phys.*, 1999, **38**, 945–950.

12. C. Schick, U. Jonsson, T. Vassiliev, A. Minakov, J. Schawe, R. Scherrenberg and D. Lõrinczy, *Thermochim. Acta*, 2000, **347**, 53–61.

13. H. Yao and K. Ema, *Rev. Sci. Instrum.*, 2003, **74**(9), 4164–4168.

14. A. A. Minakov, S. A. Adamovsky and C. Schick, *Thermochim. Acta*, 2003, **403**, 89–103.

15. J.-L. Garden, E. Château and J. Chaussy, *Appl. Physics Lett.*, 2004, **84**(18), 3597–3599.

16. Y. Kraftmakher and I. A. A. Kraftmakher, *Modulation Calorimetry: Theory and Applications,* Springer, Berlin, 2004.

17. N. J. Chen, J. Morikawa, A. Kishi and T. Hashimoto, *Thermochim. Acta*, 2005, **429**, 73–79.

18. Y. J. Yun, D. H. Jung, I. K. Moon and Y. H. Jeong, *Rev. Sci. Instrum.*, 2006, 77, Art. No. 064901.

19. J. S. Rowlinson and F. L. Swinton, *Liquids and Liquid Mixtures,* Butterworth, London, 1982, pp. 11–58.

20. A. Pal and R. Gaba, *J. Chem. Thermo.*, 2008, **40**, 750–758.

21. J. S. Burlew, *J. Am. Chem. Soc.*, 1940, **62**(4), 681–689.

22. S. L. Randzio, *J. Thermal Analysis Calorimetry*, 2007, **89**(1), 51–59.

23. D. Gonzalez-Salgado, J. L. Valencia, J. Troncosos, E. Carballo, J. Peleteiro, L. Romani and D. Bessieres, *Rev. Sci. Instrum.*, 2007, **78**(5), Art. No. 055103.

24. E. M. Woolley, *J. Chem. Thermo.*, 2007, **39**(9), 1300–1317.

25. L. D. Hansen and R. M. Hart, *Thermochim. Acta*, 2004, **417**(2), 257–273.

26. National Bureau of Standards Certificate, Standard Reference Material 720, Synthetic Sapphire (α-Al_2O_3), U.S. Department of Commerce, Washington, D.C., 1982.

27. S. L. Randzio, J.-P. Grolier, J. R. Quint, D. J. Eatough, E. A. Lewis and L. D. Hansen, *International J. Thermophysics*, 1994, **15**(3), 415–441.

28. V. Majer, V. Svoboda, Enthalpies of Vaporization of Organic Compounds: A Critical Review and Data Compilation. IUPAC Chemical Data Series No. 32. Blackwell, Oxford. 1985.

29. F. Veselý, L. Šváb, R. Provazník and V. Svoboda, *J. Chem. Thermo-dynamics*, 1988, **20**, 981–983.
30. C. L. Yaws, *The Yaws Handbook of Vapor Pressure,* Gulf Publishing Co, Houston, TX, 2007.
31. I. Cibulka and L. Hnědkovský, *J. Chem. Eng. Data*, 1996, **41**, 657–668.
32. P. Umnahanant, S. Kweskin, G. Nichols, M. J. Dunn, H. Smart-Ebinne and J. S. Chickos, *J. Chem. Eng. Data*, 2006, **51**(6), 2246–2254.
33. S. P. Verevkin, *Fluid Phase Equil.*, 2004, **224**(1), 23–29.
34. K. N. Marsh, in *CRC Handbook of Chemistry and Physics,* ed. D. R. Lide, CRC Press, Boca Raton, Florida, pp. 6–10.
35. T. W. Richards and W. N. Stull, *New Method for Determining Compressibility,* Carnegie Institution, 1903.

CHAPTER 3

An Analysis of Conductive Heat Losses in a Flow Calorimeter for Heat Capacity Measurement

J. DAVID RAAL

Thermodynamics Research Group, School of Chemical Engineering, University of K-Z Natal King George V Ave, Durban 4041, South Africa

With them the Seed of Wisdom did I sow,
And with my own hand labour'd it to grow

****

There was a Door to which, I found no Key:
There was a Veil past which I could not see
Omar Khayyám, The Rubáiyát XXVIII, XXXII

3.1 Introduction

Liquid heat capacity data are required for energy balances on processes involving thermal energy transfer and can provide insight into molecular structure. Both the measurement and prediction of heat capacity as a function of temperature have posed considerable challenges to researchers, as will be evident from perusal of several chapters in this volume. Although moderate accuracy

Heat Capacities: Liquids, Solutions and Vapours
Edited by Emmerich Wilhelm and Trevor M. Letcher
© The Royal Society of Chemistry 2010
Published by the Royal Society of Chemistry, www.rsc.org

may suffice for some purposes, others require more precise data for isobaric specific heat capacities, c_P, as a function of temperature. One such example[1] is in the application of the steady-state macroscopic energy balance across the terminals of a Cottrell pump in differential ebulliometry to measure the evaporation ratio, Φ. Design and analysis of refrigeration cycles would benefit from c_P data of improved accuracy, particularly for the newest refrigerants.

Prediction procedures for pure liquids are reviewed by Poling *et al.*[2] and also in Chapter 19 of this volume. Remarkably elegant prediction procedures have recently been developed by Gmehling's Group[3] at the Carl von Ossietzky University, based on a modified Peng-Robinson group contribution equation of state (EoS). With optimized procedures the mean deviation of predicted c_P values from experimental data was an impressive 0.78 % for 33 components over substantial temperature ranges.

Although sophisticated commercial instruments are now available for c_P measurement (some reviewed elsewhere in this volume), they tend to be expensive and still require careful calibration procedures. Flow calorimeters are comparatively simple and exclusion of the vapor phase eliminates thermal effects from evaporative processes. Energy losses by conduction (along the heater lead-in wires) and by convection, and fluid frictional effects must however be accounted for and new procedures to accomplish this are presented here.

3.2   Flow Calorimetry for c_P Measurement

Typically, in a flow calorimeter, electrical energy (Q) supplied to a flowing fluid from a heater element immersed in the fluid is measured together with the fluid flow rate (dm/dt), and temperature rise. In the absence of any heat losses to the environment or fluid frictional effects, the isobaric heat capacity would then be given by.

$$\int_{T_0}^{T} c_P dT = \frac{Q}{(dm/dt)} \tag{1}$$

Although convection heat losses (q_{CV}) from the flow tube to the environment can be minimized if not eliminated (e.g. by vacuum jacketing) conductive heat losses through the heater lead-in wires constitute a more serious problem. The lead-in wires (typically copper or platinum) must have negligible electrical resistance and offer a ready path for axial heat conduction to the environment, as will be seen in the results presented below. Quantification of such losses as a function of flow rate and fluid property is a challenging exercise and remains the principal problem in flow calorimetry for c_P measurement.

In earlier work[4-8] several elegant procedures have been proposed to account for conductive heat losses (q_{HL}) but basically *ad hoc* procedures did not provide insight into the dependence of heat losses on fluid flow rate and physical properties.

3.3 Mathematical Formulation for a Model Flow Calorimeter

To pose realistic boundary conditions, a five-zone model flow calorimeter is proposed, as shown in Figure 3.1. Heat is generated electrically only in the central zone III of the flow tube and there is axial conduction along the lead-in wires (e.g. platinum) of low electrical resistance but unavoidably high thermal conductivity. In zones I and V the lead-in wire temperatures have reached the environment temperature T_C and heat transfer by convection from the wire to the flowing fluid, characterized by the heat transfer coefficient h, equals zero. Energy balances can be performed separately on the heater and lead-in wires, and on the flowing fluid in the annular area of the calorimeter.[9]

For one-dimensional axial conduction in the metal elements, the first energy balance for a differential element dx (e.g. in zone III) gives:

$$\frac{d^2t}{dx^2} + \frac{QPf^2}{ka} - \frac{hPf^2}{ka}(t - T) = 0 \qquad (2)$$

where t refers to the wire temperature, T to the fluid temperature, P the wire circumference (uniform), Q the power input per unit surface area of wire, k the thermal conductivity of the wire and a is the wire cross-sectional area.

To obtain sufficient electrical resistance for a flow-tube of reasonable length, the heater wire is coiled and the *actual* length l is related to x by $f = dl/dx$. In zones other than zone III there is no heat generated and the second term in Equation (2) disappears. For the second differential balance (on the flowing fluid), the first law of thermodynamics for steady horizontal flow in an annular area gives

$$(t - T) = \frac{C_H}{hPf}\frac{dT}{dx} + \frac{U}{hPf}\frac{dA}{dx}(T - T_C) \qquad (3)$$

where U is the overall heat transfer coefficient which characterizes heat losses by convection to the environment by mechanisms *other than the axial*

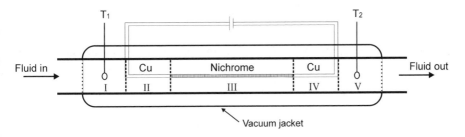

Figure 3.1 Schematic diagram of flow calorimeter. (From: *Heat Capacity Measurement by Flow Calorimetry: an Exact Analysis*, T. K. Hei, J. D. Raal, © AIChE J., 2009, 1, 2006.)

conduction q_{HL}, i.e., the analysis separates the two possible modes of heat loss. In Equation (3), C_H refers to the fluid thermal capacity, A is the interior tube wall area and T is the fluid temperature (assumed radially uniform).

The fluid enthalpy has been assumed independent of pressure, limiting the analysis to pressures remote from the critical pressure.

Equations (2) and (3) are coupled and can be reduced to a single third-order ordinary differential equation (ODE) with constant coefficients by double differentiation of Equation (3) and substitution into Equation (2) to give:[9]

$$\frac{d^3 T}{dx^3} + \left(\frac{hPf}{C_H} + \frac{U}{C_H} \frac{dA}{dx} \right) \frac{d^2 T}{dx^2} - \frac{hPf^2}{ka} \frac{dT}{dx} + \frac{QhP^2f^3}{kaC_H} - \frac{UhPf^2}{kaC_H} \frac{dA}{dx} (T - T_C) = 0$$

$$(4)$$

Mathematical or numerical solution of Equation (4) for appropriate boundary conditions would provide the longitudinal temperature profile in the flowing fluid and through Equation (3) also for the wire. From the latter, the conductive heat losses at the extremities (q_{HL}) are obtained as $q_{HL} = -k_1 a \left(\frac{dt}{dx} \right)$ at $x = 0$ and $x = x_5$.

The solution of Equation (4) for the five zones presents a formidable problem but a reasonable simplification was found[9,10] by neglecting the terms containing the overall convective heat transfer coefficient U (reasonable since the experimental calorimeter was vacuum jacketed).

3.4   Solutions to the Governing Equation

When the terms containing U in Equation (4) are neglected, the equation can be reduced to a second order ODE in the temperature gradient Z ($= dT/dx$) since it does not then contain the dependent variable T explicitly. Solutions in terms of non-dimensional variables ($\bar{T}$ and $\bar{x}$) were found by Hei and Raal,[9] e.g. for the central heated region (zone III, $Q \neq 0$):

$$\bar{T} = C_4 \frac{e^{-q_1' \bar{x}}}{-q_1'} + C_5 \frac{e^{q_2' \bar{x}}}{q_2'} + Q \frac{\Delta x_3 Pf}{C_H T_{\text{ref}}} \bar{x} + C_6 \qquad (5)$$

The reference temperature T_{ref} arises in defining a non-dimensional temperature $\bar{T}$ as:

$$\bar{T} = \frac{T - T_c}{(T - T_c)_{\text{ideal}}}$$
$$= \frac{T - T_c}{(T_{\text{ref}})}$$

where $(T - T_c)_{\text{ideal}}$ represents a situation of no heat loss, i.e. $Q = mc_P(T - T_c)_{\text{ideal}}$.

The boundary conditions required that there be no discontinuities in temperature or heat flux in passing from one zone to another. The q_i in Equation (5) are dimensionless quantities containing the calorimeter characteristics and fluid properties. Application of the boundary conditions to the other four zones gave equations similar to Equation (5), with ten additional integration constants. The large system of linear equations for $\overline{T}$ contained thirteen integration constants, C_1 to C_{13}, and their evaluation for specified fluid and calorimeter properties (by matrix inversion) produces both temperature profiles and conductive heat losses (q_{HL}) at the calorimeter extremities. The work is complex and exacting and has only been briefly sketched here. The heat transfer coefficient, h, for convective heat transfer from the wire to the flowing fluid, is contained in the dimensionless groups q_i and must be evaluated. From a literature search the empirical correlation proposed by Mills[11] was selected. This gives the Nusselt number Nu ($= hD_{eq}/k$) as a function of the Peclet number (Pe), where $Pe = Re \times Pr$ ($=$ Reynolds number $\times$ Prandtl number).

$$Nu_m = 3.66 + \frac{0.065\left(\frac{D_{eq}Pe}{L_t}\right)}{1 + 0.04\left(\frac{D_{eq}Pe}{L_t}\right)^{2/3}} \tag{6}$$

where D_{eq} is the equivalent diameter and L_t the tube length.

3.5 Computational Results

When the dimensions of an experimental calorimeter (such as described below) are entered into the computational program, together with fluid physical properties (heat capacity c_P, viscosity μ, density ρ and thermal conductivity k), the fluid and wire temperatures can be obtained for any specified flow rate for which Equation (6) is valid. An example of such a computation is shown in Figure 3.2 for a fluid with the properties of hexane. The temperature gradient in the wire ($d\overline{t}/d\overline{x}$), which determines the conductive heat loss q_{HL}, is dramatically larger at the calorimeter *exit,* a novel finding. Another significant discovery, seen from Figures 3.3 and 3.4, is that the conductive heat loss increases rapidly with *decrease* in fluid flow rate and also with a *decrease* in fluid heat capacity.

3.6 Experimental Measurements

3.6.1 Correlating Equation

From the computational results and from experimental data for several liquids (see below), it was found that the fractional conductive heat loss (q_{HL}/Q) could be correlated with an equation which separates the fixed calorimeter characteristics from the fluid volumetric flow rate (V) and physical properties,

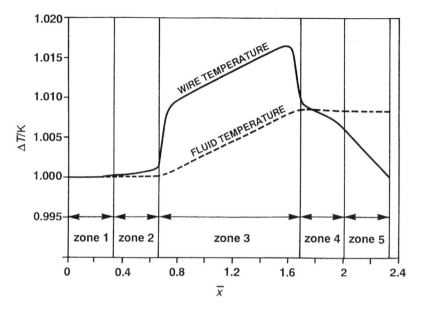

Figure 3.2   Non-dimensional fluid and wire temperatures versus $\bar{x}$ for a fluid with the properties of hexane. Computed from Equation (7) with constants obtained from inversion of the matrix (appendix). (From: *Heat Capacity Measurement by Flow Calorimetry: an Exact Analysis*, T. K. Hei, J. D. Raal, © *AIChE J.*, 2009, **1**, 206.)

a long-sought goal:

$$\frac{q_{HL}}{Q} = \frac{1}{X\sqrt{h}}\left\{ C_8'(V)^{2.2}e^{-X\sqrt{h\cdot\bar{x}_4}} - C_9'\left(\frac{h}{V\rho C_P}\right)^{2.47}e^{X\sqrt{h\cdot\bar{x}_4}} \right\} \tag{7}$$

where $X = \sqrt{\frac{P(\Delta x_3)^2}{k_1 a}}$, P is the circumference of the wire, k_1 is the lead-in wire thermal conductivity, A is the wire cross-sectional area, Δx_3 is the wire heated length and $\bar{x}_4$ is the dimensionless length of wire in zone IV.

3.6.2   Equipment and Procedure for Measuring Conductive and Convective Heat Losses

To test the theoretical approach described above, a glass and Teflon flow calorimeter similar to the model in Figure 3.1 was constructed. The heating element was a coiled nichrome wire and fluid flow rates of up to $0.30\,cm^3\,s^{-1}$ were produced with two Beckman Model (10A) micro-flow pumps operating in parallel. Flow rates were reproducible to $\pm0.002\,cm^3\,s^{-1}$. Inlet and outlet temperatures were sensed with Pt-100 elements with special precautions to

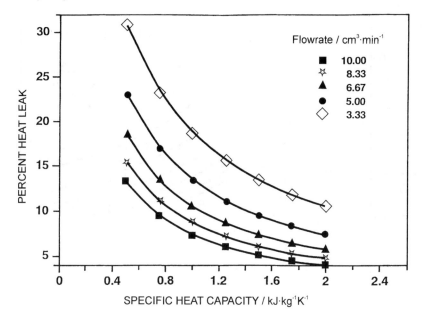

Figure 3.3 Conductive heat loss (q_{HL}/Q) as a function of fluid heat capacity, c_P, for several flow rates, computed from Equation (7). Properties (other than c_P) are those of hexane. Calorimeter characteristics, $X = 3.5$. (From: *Heat Capacity Measurement by Flow Calorimetry: an Exact Analysis*, T. K. Hei, J. D. Raal, © *AIChE J.*, 2009, **1**, 206.)

prevent false readings through heat conduction along the Pt-100 leads. A schematic diagram of the calorimeter is shown in Figure 3.5. Although convective heat loss from the calorimeter tube to the environment, q_{CV} (represented by UdA) was neglected in Equation (4), its influence in the determination of c_P was examined in a separate series of experiments. In these measurements the experimental fluid was brought to a temperature a few degrees above or below that of the calorimeter bath by passage through a coiled tube immersed in another constant temperature bath.[12]

With no electrical energy supplied to the calorimeter, the fluid temperature change across the calorimeter (ΔT) was measured for runs at two different but constant temperatures. The macroscopic energy balance for horizontal flow of an incompressible fluid at a mass flow rate m becomes:

$$\Delta h = q_{cv}$$
$$= c_P \Delta T + \Delta P / \rho = \frac{U_i A_i (\Delta T)_{lm}}{m} \tag{8}$$

where ΔT_{lm} is the logarithmic mean temperature difference between the fluid and the bath temperatures. Thus for measurement at two temperatures but at

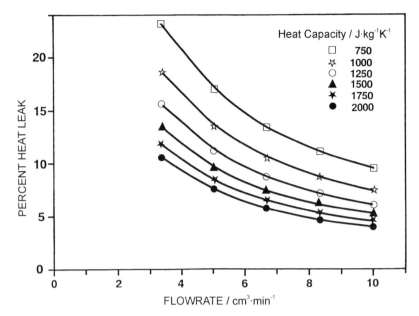

Figure 3.4   Conductive heat loss (q_{HL}/Q) as a function flow rate for various assumed
heat capacities, computed from Equation (7). Calorimeter characteristics,
$X = 3.5$. (From: *Heat Capacity Measurement by Flow Calorimetry: an
Exact Analysis*, T. K. Hei, J. D. Raal, © *AIChE J.*, 2009, **1**, 206.)

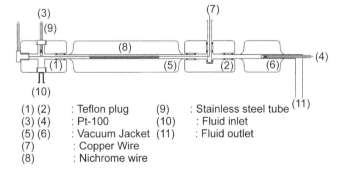

(1) (2)	: Teflon plug	(9)	: Stainless steel tube
(3) (4)	: Pt-100	(10)	: Fluid inlet
(5) (6)	: Vacuum Jacket	(11)	: Fluid outlet
(7)	: Copper Wire		
(8)	: Nichrome wire		

Figure 3.5   Schematic diagram of the flow calorimeter. (From: *Heat Capacity Measurement by Flow Calorimetry: an Exact Analysis*, T. K. Hei, J. D. Raal, ©
AIChE J., 2009, **1**, 206.)

the same volume flow rate V we obtain:

$$U_i A_i = \frac{V\{\rho_{\mathrm{I}} c_{p_{\mathrm{I}}} \Delta T - \rho_{\mathrm{II}} c_{p_{\mathrm{II}}} \Delta T\}}{(\Delta T)_{\mathrm{lm(I)}} - (\Delta T)_{\mathrm{lm(II)}}} \tag{9}$$

Equation (9), with mild assumptions, permits good estimates of the convective heat loss to be obtained from relatively simple terminal temperature measurements. For nearly isothermal operation, the pressure drop term reflects the frictional energy loss, lw_f which can be found from Equation (8) once q_{CV} has been found for the particular flow rate (lw_f was found to be negligible by Hei and Raal).[9] With electrical energy Q supplied to the calorimeter, the conductive heat loss q_{HL} can now be found from an extension of Equation (8):

$$Q - q_{HL} - q_{CV} = mc_P \Delta T + m\Delta P/\rho \qquad (10)$$

Once q_{HL} has been found as a function of mass flow rate m for fluids of known physical properties, the predictive capabilities of the correlating equation, Equation (7) can be tested.

3.7 Discussion of Results

A large program of measurements was conducted by Hei and Raal[9] to test the predictive capabilities of Equation (7) and the trends shown in Figures 3.3 and 3.4. Measurements were made on propanol, butanol, hexane, heptane and water for a range of power inputs and with flow rates from about (0.1 to 0.3) cm^3 s^{-1}. Results for butanol, similar to those found for the other solvents, are shown in Table 3.1 and suggest the following conclusions:

- The convective heat loss to the environment (q_{CV}), though much smaller than the conductive heat loss (q_{HL}), is not negligible and increases with flow rate, as expected.
- The conductive heat losses are much larger. They decrease with increasing flow rate, as found from the computational predictions (Figure 3.4), and range up to more than 20 % of the heat input. The decrease in q_{HL} with

Table 3.1 Measured Conductive and Convective Heat Losses, q_{HL} and q_{CV}, for Varying Power Inputs, Q, for butanol. (From: *Heat Capacity Measurement by Flow Calorimetry: an Exact Analysis*, T. K. Hei, J. D. Raal, in *AIChE J.*, 2009, 1, 206.)

Flow rate, V/cm^3 s^{-1}	Power input, Q/W	$(T_{out}-T_{in})$/K	U_iA_i/W K^{-1}	q_{CV}/W	q_{HL}/W	(q_{HL}/Q)/%
0.2	1.123	2.441	0.201	0.089	0.228	20.33
0.2	0.943	2.047	0.201	0.077	0.190	20.17
0.2	0.776	1.684	0.201	0.066	0.154	19.89
0.2	0.626	1.361	0.201	0.056	0.122	19.42
0.3	1.554	2.430	0.226	0.100	0.251	16.12
0.3	1.337	2.088	0.226	0.088	0.215	16.11
0.3	1.139	1.773	0.226	0.077	0.184	16.15
0.3	0.957	1.487	0.226	0.067	0.154	16.12

Table 3.2 Fractional conductive heat loss (q_{HL}) as a function of power input, Q. (From: *Heat Capacity Measurement by Flow Calorimetry: an Exact Analysis*, T. K. Hei, J. D. Raal, © *AIChE J.*, 2009, **1**, 206.)

Q/W	(q_{HL}/Q)/%	
	Experimental	*Theory*
butanol, V=0.30 cm³ s⁻¹		
1.554	16.12	12.80
1.337	16.11	12.82.
1.139	16.15	12.83
0.957	16.12	12.85
Average	16.13	12.83
hexane, V=0.27 cm³ s⁻¹		
1.152	11.98	12.32
0.969	11.78	12.32
0.802	12.00	12.32
0.648	11.26	12.32
Average	11.76	12.32

increase in fluid heat capacity, at constant flow rate predicted in Figure 3.3, was also confirmed.

In Table 3.2 measured and predicted conductive heat losses are compared for butanol and hexane for varying power inputs. To use the correlating equation for q_{HL}/Q, the calorimeter characterizing parameters X ($= 3.5$ m K$^{0.5}$ W$^{0.5}$) and $\bar{x}_4$ ($=0.255$) were determined from the appropriate physical constants and instrument dimensions, and the constants C'_8 and C'_9 (from Equation (7)) were determined from experiments with n-propanol. The comparisons are satisfactory and very good respectively for the two liquids.

3.7.1 Heat Capacity Measurement

Since Equation (10) contains the initially unknown c_P and q_{HL}, a trial-and error procedure was necessary using Equations (7) and (10) alternately, until converged values were obtained for the two unknowns. The measured c_P values were then compared with literature data, as shown in Table 3.3 for heptane and hexane. The agreement is very good, ranging from less than 1 % to 2.31 %, depending on the choice of literature data.

3.8   Conclusion

From a rigorous analysis of energy flows in a five-zone model flow calorimeter, it was found that longitudinal temperature distributions in both the heating

Table 3.3 Measured fluid heat capacities, using Equation (7) for conductive heat losses and calibration with n-propanol. All c_P values at 299.7 K and atmospheric pressure. (From: *Heat Capacity Measurement by Flow Calorimetry: an Exact Analysis*, T. K. Hei, J. D. Raal, © *AIChE J.*, 2009, **1**, 206.)

Nominal flow rate V/cm^3 s^{-1}	Starting value of c_P/J K^{-1} kg^{-1}	Convergent value of c_P/J K^{-1} kg^{-1}	Literature value [13] of c_P/J K^{-1} kg^{-1}	Difference/%	Literature value [14]/ J K^{-1} kg^{-1}	Difference/%
Experimental fluid: heptane						
0.233	1000	2286.17	2250.51	1.58	2253.70	1.44
0.266	1000	2266.79	2250.45	0.72	2253.62	0.58
Experimental fluid: hexane						
0.266	1000	2250	2275	1.06	2276	1.11
0.283	1000	2246	2275	2.26	2276	2.31

element and flowing fluid were governed by a third-order ordinary differential equation with constant coefficients. Successful solutions of the ODE for realistic boundary conditions produced novel findings with implications for calorimeter design and operation. In particular, conductive heat losses were large, principally at the calorimeter exit, as had been expected and fluid property and flow rate dependent.

Reasonable estimates of q_{HL} can be obtained from the proposed correlating equation for a fluid of known properties (other than c_P) for a given flow rate and energy input. The heat capacity can then be obtained by an iterative procedure, more soundly based than previous *ad hoc* calibration methods. Calorimeter design could be improved by splitting the incoming flow into equal arms (with a single exit) and arranging the heater element so that both heater leads project through an *incoming* stream. Convective heat losses, if not negligible, can be measured by the relatively simple experimental method described above, leading to Equation (9), and subtracted form the energy input as in Equation (10). Measured conductive heat losses could possibly also be correlated directly with the dimensionless groups which emerged from Equation (4) in dimensionless variables. These are $\left[hPf(\Delta x)_3/C_H\right]$, $\left[hPf^2(\Delta x_3)^2/ka\right]$ and $\left[(\Delta x_3)^3 QhP^2f^3/mc_PkaT\right]$.

Improvements in the predictive capabilities of Equation (7) (found unsatisfactory for distilled water) could probably be obtained with an equation more suitable for correlating the heat transfer coefficient (h) for convective energy transfer between a coiled heater element and a flowing fluid.

References

1. J. D. Raal, V. Gadodia, D. Ramjugernath and R. Jalari, *J. Mol. Liquids,* 2006, **125**, 45–57.
2. B. E. Poling, J. M. Prausnitz and J. P. O'Connell, *The Properties of Gases and Liquids,* Boston, McGraw-Hill, 2001.
3. A. Diedrichs and J. Gmehling, *Fluid Phase Equilib.,* 2006, **248**, 56.
4. D. Smith-Magowan and R. H. Wood, *J. Chem. Thermodyn.,* 1981, **13**, 1047.
5. P. S. Z. Rogers and K. S. Pitzer, *J. Phys. Chem.,* 1981, **85**, 2886.
6. P. S. Z. Rogers and C. J. Duffy, *J. Solution Chem.,* 1989, **21**, 595.
7. A. Saitoh, S. Nakagawa, H. Sato and K. Watanabe, *J. Chem. Eng. Data,* 1990, **35**, 107.
8. S. Nakagawa, T. Holi, H. Saito and K. Watanabe, *J. Chem. Eng. Data,* 1993, **38**, 70.
9. T. K. Hei and J. D. Raal, *AIChE J.,* 2009, **1**, 2006.
10. T. K. Hei, MSc. Eng. Thesis, University of Natal.
11. F. Mills, *Heat Transfer,* Irwin, Boston, 1992.

12. J. D. Raal and P. A. Webley, *AIChE J.*, 1987, **33**, 604.
13. J. F. Messerly, G. B. Guthrie, S. S. Todd and H. L. Finke, *J. Chem. Eng. Data*, 1967, **12**, 336.
14. G. V. Reklaitis, *Introduction to Material and Energy Balances,* New York, John Wiley & Sons, 1983.

CHAPTER 4

Heat Capacities and Related Properties of Liquid Mixtures

EMMERICH WILHELM[a] AND JEAN-PIERRE E. GROLIER[b]

[a] Institute of Physical Chemistry, University of Wien, Währinger Strasse 42, A-1090, Wien (Vienna), Austria; [b] Laboratoire de Thermodynamique des Solutions et des Polymères, Université Blaise Pascal, F-63177, Aubière, France

4.1 Introduction

In chemical thermodynamics, experiment, theory and application are so closely interwoven that it is highly subjective and not easy to divide even a subtopic, such as that of this book, into sections suitable for fairly comprehensive, profitable discussion. The topic of this chapter is heat capacities of liquid mixtures, and perhaps a few words are in order to indicate its importance. Quite generally, there are three main reasons for the enormous efforts invested into experimental and theoretical work on thermodynamic properties of liquid mixtures. First, it is hoped that by studying mixture properties we will be able to improve our knowledge of intermolecular interactions in bulk liquid phases. Second, the appearance of new physical phenomena *not* found in the pure liquid components is scientifically exciting and adds a new dimension to research. This is of greatest importance in the real world of applied chemistry/ chemical engineering and biophysical chemistry, since the majority of industrially important processes involve liquid phases, and the practitioner there has always to deal with multicomponent *mixtures*, as does of course, the biophysical chemist. This covers the third major reason for work in this field.

Heat Capacities: Liquids, Solutions and Vapours
Edited by Emmerich Wilhelm and Trevor M. Letcher
© The Royal Society of Chemistry 2010
Published by the Royal Society of Chemistry, www.rsc.org

Real mixtures are non-ideal, and the extent to which they deviate from ideal behaviour is most conveniently expressed through use of thermodynamic *excess* quantities. When relating the excess molar Gibbs energy G^E of a mixture to the activity coefficients $\gamma_i(T, P, \{x_i\})$ of each component i at the same temperature T and pressure P using, say, the *symmetric* convention (the composition is characterised by the set of mole fractions $\{x_i\}$), other excess quantities as well as the corresponding excess partial molar quantities may be obtained by applying the usual formalism. For instance, with:

$$
\begin{aligned}
G^E &= H^E - TS^E \\
&= RT \sum_i x_i \ln \gamma_i
\end{aligned}
\tag{1}
$$

where R denotes the gas constant, the excess molar enthalpy is given by:

$$
\begin{aligned}
H^E &= -T^2 \left[\partial (G^E/T)/\partial T \right]_P \\
&= -RT^2 \sum_i x_i (\partial \ln \gamma_i/\partial T)_{P,x}
\end{aligned}
\tag{2}
$$

the excess molar entropy by:

$$
\begin{aligned}
S^E &= -\left(\partial G^E/\partial T \right)_{P,x} \\
&= -R \sum_i x_i \ln \gamma_i - RT \sum_i x_i (\partial \ln \gamma_i/\partial T)_{P,x}
\end{aligned}
\tag{3}
$$

the excess molar heat capacity at constant pressure by:

$$
\begin{aligned}
C_P^E &= \left(\partial H^E/\partial T \right)_{P,x} = -T \left(\partial^2 G^E/\partial T^2 \right)_{P,x} \\
&= -2RT \sum_i x_i (\partial \ln \gamma_i/\partial T)_{P,x} - RT^2 \sum_i x_i \left(\partial^2 \ln \gamma_i/\partial T^2 \right)_{P,x}
\end{aligned}
\tag{4}
$$

and the excess molar volume by:

$$
\begin{aligned}
V^E &= \left(\partial G^E/\partial T \right)_{T,x} \\
&= RT \sum_i x_i (\partial \ln \gamma_i/\partial P)_{T,x}
\end{aligned}
\tag{5}
$$

During the last fifty years or so, considerable effort was invested in designing and producing increasingly more sophisticated instruments which have provided thousands of high quality excess property data,[1,2] the majority being excess Gibbs energies, or excess enthalpies, or excess volumes. Though the excess molar heat capacity C_P^E at constant pressure is a key thermophysical property, experimental papers dealing with it were quite scarce before the 1970s, and the associated experimental inaccuracies were usually substantial. A dramatic change of the situation was brought about by the introduction of the

Picker calorimeter, which for the first time provided a means for reliably measuring high-precision excess heat capacities of liquid mixtures at hitherto unheard of convenience and speed.[3-5]

Since C_P^E is the first derivative with respect to temperature of the excess molar enthalpy, and $-C_P^E/T$ is the second derivative of the excess molar Gibbs energy, see Equation (4), excess heat capacities are pivotal for the *global* description of thermodynamic behaviour of bulk liquid mixtures. They may be used, together with G^E data (say, from isothermal vapour pressure measurements, VLE[6]), H^E data (say, from flow calorimetry[7]) and V^E data (say, from vibrating-tube densimetry[8]), to construct, by suitable integration, $G^E(T,P,\{x_i\})$ surfaces for mixtures over wide ranges of temperature and pressure, though well below the vapour–liquid critical region. Incorporating also experimental data on the activity coefficients γ_i^∞ at infinite dilution,[9] this is/should be the *method of choice* for global fitting. It goes almost without saying that in this way G^E (or $\ln\gamma_i$) values may be obtained reliably in temperature ranges which are *not* easily accessible to VLE experiments. These data may eventually be used to test and improve theoretical methods for the estimation of excess quantities, and at the same time provide fundamental insight on the microscopic, molecular level through comparison with predictions based on judiciously selected models. However, any attempt to explain mixture properties from information on fundamental, molecular pure component properties, *i.e.* on the respective pair-potential-energy functions, is hampered by our insufficient knowledge of the pair-potential-energy function between unlike pairs (as well as by our meagre knowledge of the contributions due to three-body interactions, *etc.*). If ε_{ij} denotes the energy parameter (the well depth) and σ_{ij} the distance parameter of a two-parameter potential-energy function between molecules of type i and j, and as it is well-known that the classical Lorentz–Berthelot combining rules are rather unsatisfactory, it is common practice to introduce parameters ξ and η to describe deviations from this rule, *i.e.*

$$\varepsilon_{ij} = \xi\left(\varepsilon_{ii}\varepsilon_{jj}\right)^{1/2} \tag{6}$$

$$\sigma_{ij} = \eta\left(\sigma_{ii} + \sigma_{jj}\right)/2 \tag{7}$$

Deviations from the geometric mean rule are sometimes also denoted by $(1-k_{ij})$, where k_{ij} is then known as the binary interaction parameter. A rather successful prescription for ξ and η, which includes as a novel aspect the constraint of additivity of the *temperature-dependent effective* hard-sphere diameters,[10,11] has been suggested by Kohler, Fischer and Wilhelm,[12] and greatly improved predictions for G^E, H^E and V^E for fairly simple mixtures. For more complex mixtures, however, *a priori* estimations of excess quantities have profited greatly by the development of heuristic model theories (for instance those based on Guggenheim's quasi-chemical approximation[13]), in conjunction with group contribution concepts (DISQUAC, UNIFAC).[14-20] An overview of some of these activity coefficient-based methods is given by Sandler[21] and by Prausnitz *et al.*[22] While they work reasonably well for G^E and H^E, predicted results for C_P^E are frequently not satisfactory.

This observation underscores once more the role of the excess heat capacity as an important discriminatory property well suited for testing and improving model theories. Similar comments apply to COSMO-RS and related theories,[23–33] which are based upon unimolecular quantum chemical calculations of the individual species in the system (that is, not of the mixture itself). To conclude this section, we would like to emphasise the mutually stimulating long-time relation to, and cooperation with, H.V. Kehiaian within the frame of his TOM (Thermodynamics of Mixtures) project,[34] where mixtures are systematically investigated in order of increasing complexity of molecular structures and interactions (including, for instance, proximity effects, tautomerism, *etc.*).

Despite the importance of C_P or C_P^E, knowledge of other complementary, thermophysical properties is indispensable for a better understanding on the microscopic, molecular level. The most important are:[35–37] molar heat capacity C_V at constant volume, isobaric expansivity α_P, isothermal compressibility κ_T, and isentropic compressibility κ_S, the latter being related to the thermodynamic speed of sound through $v = (\rho \kappa_S)^{-1/2}$, where $\rho = M/V$ denotes the density, M the molar mass, and V the molar volume. For a single phase at constant composition, a few useful relations between these quantities are given below:

$$\kappa_T - \kappa_S = TV\alpha_P^2/C_P \tag{8}$$

$$C_P - C_V = TV\alpha_P^2/\kappa_T \tag{9}$$

$$C_P/C_V = \kappa_T/\kappa_S = 1 + TM\alpha_P^2 v^2/C_P \tag{10}$$

$$(\partial C_P/\partial P)_T = -T(\partial^2 V/\partial T^2)_P = -TV[\alpha_P^2 + (\partial \alpha_P/\partial T)_P] \tag{11}$$

$$(\partial C_V/\partial V)_T = T(\partial^2 P/\partial T^2)_V = T(\partial \beta_V/\partial T)_V \tag{12}$$

Here, $\beta_V = (\partial P/\partial T)_V$ denotes the isochoric thermal pressure coefficient. In particular, Equations (11) and (12) have proved to be of great value in high-pressure research.

Relatively simple models have been used to extract from accurate molar heat capacities at constant volume information on the type of *motion* executed by the molecules in *pure* liquids. In general, these models are based on assuming separability of contributions due to translation, rotation, vibration and so forth. Early work along this line was reported by Eucken,[38,39] Bernal,[40] Eyring,[41] Staveley,[42,43] Moelwyn-Hughes,[44] Kohler,[45] Bondi[46] and their collaborators. Preferably, all these contributions to C_V are discussed in terms of residual quantities[35,37,47,48] as elaborated by Wilhelm.[37,48] The residual molar heat capacity C_V^r at constant volume is defined by:

$$C_V^r(T, V) = C_V(T, V) - C_V^{pg}(T, V) \tag{13}$$

where the superscript pg refers to the perfect-gas (ideal-gas) state at the same temperature and molar volume. It is the most direct measure of the contributions due to intermolecular interactions at any given state condition.[49] Results so obtained were quite satisfactory and explained successfully, for example, the highly unusual temperature dependence of C_P as well as C_V of tin tetrachloride, SnCl$_4$, between 273.15 K and 323.15 K (at orthobaric conditions):[37,48] both quantities *decrease* with *increasing* temperature. While the associated formalism can be extended to include mixtures, discussions communicated so far remained by and large qualitative and any conclusions drawn have to be regarded as being tentative.[5,42–45,47,48,50,51]

4.2 Experiment

As indicated above, from a practical point of view C_P, or C_P^E, is the quantity of interest, while C_V is the thermophysical property needed in more theoretically oriented work. At temperatures well below the critical temperature, where β_V of liquids is large, the heat capacity at constant pressure can be determined much more readily (say, with batch or flow methods) than the isochoric heat capacity. In fact, rather few *direct* calorimetric measurements of C_V of *dense* liquids and liquid mixtures[52–58] have been reported, though it becomes more practical near the critical point where β_V is much smaller. Thus, the vast majority of researchers in this field use the convenient *indirect* method described by Equation (10); for satisfactory results, high-precision data on isobaric heat capacity, isobaric expansivity, and thermodynamic speed of ultrasound are needed. The last specification requires ultrasound speeds at frequencies well below the dispersion region, yet high enough to provide isentropic conditions. As an additional bonus, isothermal compressibilities are obtained simultaneously. With the availability of fast high-precision calorimetric equipment[3,4,59,60] in conjunction with modern techniques for measuring densities of liquids[8,61] and speeds of ultrasound,[62–66] the *indirect* method for determining κ_T yields results which are comparable with the very best isothermal compressibilities measured directly.[5,35,48,50,67]

Details associated with the direct measurement of heat capacities using calorimetry have been discussed, for instance, by Wilhelm,[35,68] Grolier,[69] Hemminger and Höhne,[70,71] and Lakshmikumar and Gopal,[72] and a classic overview is provided by the IUPAC monograph on calorimetry of non-reacting systems,[73a] which was supplemented recently.[73b] For liquids, the heat capacity at constant pressure may be measured directly, at pressures larger than the vapour pressure, by using a flow calorimeter. Determination of C_P in a batch calorimeter is definitely less convenient, and in practice one measures the heat capacity when the calorimetric vessel is filled under vacuum so that the liquid is in equilibrium with a *small* amount of its vapour. This quantity is closely related to C_σ, the molar heat capacity at saturation,[49,68,69,74] which is the molar heat capacity of a liquid maintained at all temperatures in equilibrium with an infinitesimal amount of vapour. We note that:

$$C_\sigma = C_P - TV\alpha_P\beta_\sigma \tag{14}$$

where $\beta_\sigma = (\partial P/\partial T)_\sigma$ denotes the rate of increase of the vapour pressure with temperature.

For obtaining excess molar heat capacities, two main methods are in use:

(I) excess enthalpies $H^E(T,P,\{x_i\})$ are measured at several temperatures over the whole composition range, and the excess heat capacities at constant pressure are then calculated by:

$$C_P^E(T, P, \{x_i\}) = \left(\partial H^E(T, P, \{x_i\})/\partial T\right)_{P,x} \qquad (15)$$

(II) molar heat capacities of the pure liquid components, $C_{P,i}^*$, $i = 1,2$ for a *binary* mixture, and of a sufficiently large number of mixtures at reasonably spaced compositions are measured, and the excess molar heat capacities are then obtained according to:

$$C_P^E(T, P, x_1) = C_P(T, P, x_1) - \left[x_1 C_{P,1}^*(T, P) + (1 - x_1)C_{P,2}^*(T, P)\right] \quad (16)$$

For the sake of clarity, pure-substance quantities are indicated by a superscript asterisk.

Clearly, method (I) requires high-precision measurements of excess enthalpies,[75] and only relatively few meet the expectations.[76–78] Thus, most of the excess heat capacities are obtained by method (II) either using differential scanning calorimeters (DSC) and carefully avoiding any vapour space within the measuring cell, or, *preferably*, with flow calorimeters[35] with no vapour space at all. Recently, a versatile calorimeter for measuring isobaric heat capacities of liquids at pressures up to 60 MPa was presented by Gonzáles-Salgado *et al.*[79] It is based on the atmospheric-pressure micro-DSC II of Setaram. Besides heat capacities of pure liquids, C_P^E of (hexan-1-ol + hexane) was reported. The instrument could also be useful for measurements in the critical region. A flow calorimetric system consisting of three heat capacity calorimeters (two for the pure components and one for the mixture) and one adiabatic mixing calorimeter has been described by Ernst and co-workers.[80] Excess molar enthalpies and specific heat capacities[81,82] C_P/M of (water + methanol) were measured from 323 K to 513 K at pressures ranging from 2 MPa to 10 MPa. However, to date the most successful instruments for obtaining excess heat capacities are based upon Picker's design[3,4] of the 1970s, which was commercialised by Setaram. It was modified by Smith-Magowan and Wood[83] to be used also at elevated temperatures and pressures including the critical region. Other calorimeters based on the Picker design have been constructed by Rogers and Pitzer,[84] Criss and co-workers,[85,86] Conti *et al.*,[87] and White and Downes.[88] There can be little doubt that *"pickering"* still represents the best way to rapidly measure with good accuracy heat capacities of pure liquids and solutions.

Figure 4.1 shows schematically an ordinary-pressure Picker flow calorimeter of the kind we used for most of our investigations. Full advantage of the high

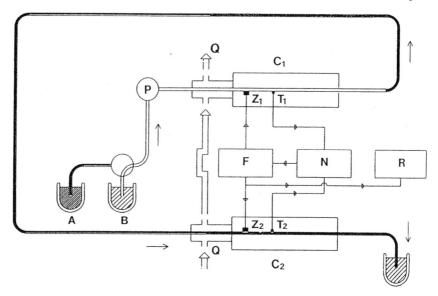

Figure 4.1 Schematic presentation of Picker's flow calorimeter.[3] A, reference liquid; B, liquid/mixture to be investigated; P, constant-flow circulating pump; Z_1 and Z_2, Zener diodes acting as heaters; T_1 and T_2, matched thermistors acting as temperature measuring devices; F, feedback control; N, null detector; R, recorder; Q, thermostat. Here, A is the reference liquid and thus C_2 is the reference cell. When liquid/mixture B circulates through cell C_1, this cell is the working cell.

sensitivity of the instrument was taken by measuring stepwise[4,89] only *small* differences of C_P/V. Specifically, starting with, say, pure component 1, mixtures are studied in the order of increasing mole fraction, with each mixture serving as reference for the subsequent mixture in the series, until the second pure component is reached. Thus, for the *n*th mixture:

$$(C_P/V)_n = \left(C_{P,1}^* / V_1^* \right) \prod_{i=1}^{n} (1 + \Delta w_i/w) \tag{17}$$

Here, V_1^* denotes the molar volume of pure component 1, w is the initial heating power (about 20 mW) applied to both measuring and reference cells, and Δw_i is the change in power supplied by the internal feedback circuit to keep the temperature gradients equal in both cells. One point should be emphasised: any error in the assigned heat capacity $C_{P,1}^*$ of the selected pure reference liquid will be reflected in an equivalent error of the heat capacity of the pure second component. However, for the *excess* molar heat capacity its influence will in general be extremely small and rarely exceed $\pm 0.01 \, \text{J K}^{-1} \, \text{mol}^{-1}$.

Comparison of heat capacities measured by different types of flow calorimeters and static heat-flow type calorimeters (see Figure 4.2) indicate *small* differences in measured heat capacities. These are attributed to conductive and convective heat losses, and quite a few *ad hoc* recipes to account for them have

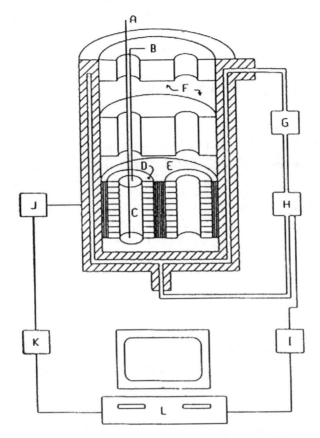

Figure 4.2 Schematic diagram of the micro-DSC of Setaram.[93] A, filling tube; B, removal tube; C, measuring cell; D, thermopile; E, thermostat block; F, thermal guards; G, circulating pump; H, thermostat; I, operation control; J, amplifier and voltmeter; K, interface; L, computer. Only the measuring cell is shown, not the reference cell.

been published.[83,84,88,90,91] Conductive heat losses, the principal problem in flow calorimetric heat capacity measurements on liquids, have recently been analysed by Hei and Raal[92] for a five-zone model calorimeter. Experiments on five pure liquids confirmed the predictions of the model by yielding heat capacities in good agreement with literature data.

4.3   Selected Results and Discussion

In this section we shall present results on heat capacities/excess heat capacities of selected binary systems belonging to one of the following three groups: (I) mixtures of organic liquids, (II) mixtures of water with an organic liquid, and (III) mixtures containing ionic liquids. We have deliberately made no attempt to cover the whole of the field, but rather have chosen to focus on systems we believe to be representative of interesting research trends either in theory or in application.

4.3.1 Mixtures of Organic Liquids

The main reasons for measuring thermodynamic properties of liquid none-lectrolyte mixtures in general, and heat capacities in particular, have already been pointed out, as has our close long-time relation with the TOM project.[34] Work in this direction is continued by Cobos and co-workers in Valladolid, Spain,[93] for instance on mixtures of alkoxyethanols with n-alkanes.[94] The DISQUAC model yields a consistent description of vapour–liquid equilibria including azeotropes, liquid–liquid equilibria, excess molar enthalpies *and* excess molar heat capacities at constant pressure. Indeed, mixtures containing components associating *via* hydrogen bonds, such as alcohols[95–113] or amines,[114] continue to be investigated experimentally as well as theoretically. For instance, we note the systematic work on systems of type (an alkanol + an n-alkane) by Romani and co-workers,[95–99] which complements earlier work by Tanaka *et al.*[104] and Costas and Patterson,[105] to name but a few researchers in this field. Of particular note are the results in the series $\{x_1 C_n H_{2n+1} OH + x_2 n\text{-} C_7 H_{16}\}$ at 298.15 K and atmospheric pressure:[104] the curves C_P^E *versus* x_1 are all skewed towards the heptane side (the methanol system shows phase separation) with maxima between, roughly, $11\,\mathrm{J\,K^{-1}\,mol^{-1}}$ and $14\,\mathrm{J\,K^{-1}\,mol^{-1}}$, yet at very high dilution ($x_1 < 0.0025$) C_P^E becomes negative for the mixtures with the C_3 to C_6 alkanols. The composition dependence of the partial molar heat capacities $\bar{C}_{P,1}$ at constant pressure of the alkan-1-ols is very interesting in that when diluting the alkan-1-ols with heptane, $\bar{C}_{P,1}$ remains nearly constant at the pure substance value $C_{P,1}^*$ down to $x_1 \approx 0.1$. Then it starts to increase very sharply reaching a peak exceeding $540\,\mathrm{J\,K^{-1}\,mol^{-1}}$ at $x_1 \approx 0.008$. Excess molar heat capacities C_P^E of mixtures of 2,2,2-trifluoroethan-1-ol with strongly dipolar liquids (p is the absolute value of the molecular electric dipole moment), such as N,N-dimethylformamide (DMF: $p = 12.9 \times 10^{-30}\,\mathrm{C\,m}$), acetonitrile (AN: $p = 13.1 \times 10^{-30}\,\mathrm{C\,m}$), and dimethylsulfoxide (DMSO: $p = 13.0 \times 10^{-30}\,\mathrm{C\,m}$), as well as isochoric heat capacities and isothermal compressibilities, have been reported by Tamura and Murakami's group.[106–110]

For mixtures, second-order thermodynamic excess quantities, such as C_P, α_P and κ_T may be modelled by a variety of solution models, such as ERAS,[101,115,116] SERAS,[117] KT[116,118,119] or TSAM.[120,121] Since we focus here on the isobaric heat capacity, of course DISQUAC[15,16] may also be added. Of these models, the two-state association model, TSAM, developed by Costas and co-workers,[120,121] has been particularly successful in describing the temperature *and* pressure dependence of the isobaric heat capacity of pure associated liquids as well as of mixtures containing associated components.

Heat capacities of mixtures containing non-hydrogen-bonding (aprotic), strongly polar compounds have been investigated for some time.[42–45,47] In fact, early in our collaboration (in the 1970s) we started with systematic studies of heat

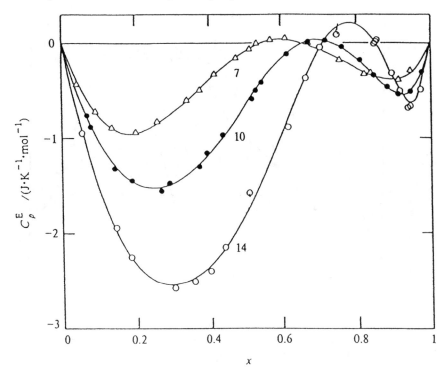

Figure 4.3 Excess molar heat capacities C_P^E at constant pressure of $\{x1,4\text{-}C_4H_8O_2 +$ $(1-x)n\text{-}C_nH_{2n+2}\}$ at 298.15 K and atmospheric pressure: *W-shape*.[123] The triangles and circles (open and filled) represent experimental results obtained for the series of mixtures of 1,4-dioxane with, respectively, heptane ($n=7$), decane ($n=10$), and tetradecane ($n=14$).

capacities of such systems,[5,51] as evidenced by the selected and representative references 122 through 140, occasionally coming up with (then) quite spectacular results, such as the *first* W-shaped curves[123,124] C_P^E *versus* x ever reported in literature. Figure 4.3 shows these seminal W-shaped curves for three mixtures of type (1,4-dioxane + an n-alkane) at 298.15 K, as reported at the *37th Annual Calorimetry Conference* in Snowbird, Utah, USA, in 1982. Our results have been confirmed and augmented by the later work of Roux and co-workers.[141,142] In fact, once the general, at first rather qualitative, ideas had been digested, such mixture behaviour was found fairly often.[36,126,131,135,136,139,143–158] Figure 4.4 shows the W-shaped $C_P^E(x)$ curves we observed in mixtures of pyridine with n-alkanes[131] plus the unusual $C_P^E(x)$ curve for the mixture (pyridine + cyclohexane).[131] Additional interesting work on systems containing polar substances has been reported in references 159 through 167.

In systems of type (a very polar substance + an aromatic substance) highly interesting and *unusual* composition dependences of some excess quantities may be observed, the most spectacular being the *M-shaped* composition dependence of H^E of (benzonitrile + toluene) and (benzonitrile + benzene), as recently reported

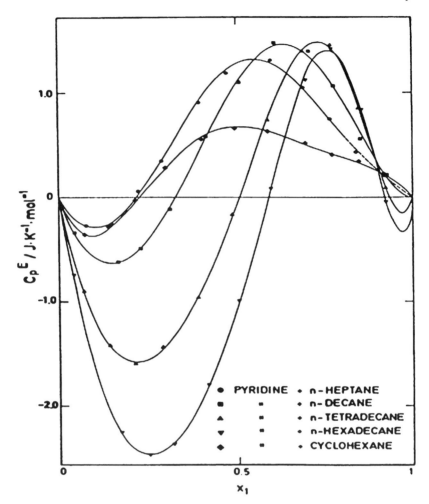

Figure 4.4   Excess molar heat capacities C_P^E at constant pressure at 298.15 K and atmospheric pressure of $\{x_1C_5H_5N + x_2n\text{-}C_nH_{2n+2}\}$ and $\{x_1C_5H_5N + x_2c\text{-}C_6H_{12}\}$.[131] The filled circles, squares and triangles represent experimental results obtained for the series of mixtures of pyridine with, respectively, heptane, decane, tetradecane, and hexadecane, the filled diamonds represent results obtained for the mixture of pyridine with cyclohexane.

by Wilhelm *et al.*,[138] and Horstmann *et al.*,[168] and of (γ-butyrolactone + benzene), as reported by Bjola *et al.*[169] For older reports on M-shaped excess enthalpy curves, see McGlashan *et al.*,[170] Howell and Stubley,[171] Tanaka *et al.*,[172] and Letcher and Naicker.[173] For (benzonitrile + toluene), at 298.15 K C_P^E is positive and somewhat skewed towards toluene (the maximal value is *ca.* 1.2 J K^{-1} mol^{-1} at $x_{BN} \approx 0.34$), while for (benzonitrile + benzene) C_P^E is negative (with a very small positive part near benzene) with a minimal value of *ca.* -0.6 J K^{-1} mol^{-1} at $x_{BN} \approx 0.54$. Of particular note are the *very small* positive deviations of the total vapour pressures from additivity and the corresponding small excess molar Gibbs energies, amounting to, roughly, maximal 230 J mol^{-1} for the former system, and to maximal 180 J mol^{-1} for the latter (both at 298.15 K). All these results have been successfully discussed[138] in terms of the *aromatic effect*.[174]

Directional intermolecular forces contribute significantly to the thermodynamic properties of pure liquids and liquid mixtures. A versatile approximate statistical–mechanical method for estimating such contributions was presented some time ago by Pople.[175] At high densities and low temperatures they lead to local structure, that is to ordering in the *pure* liquid. The presence of dipolar order *lowers* the Helmholtz energy as well as the internal energy and the entropy, and increasingly so at low temperatures, while the contribution to the heat capacity C_V^* of the pure liquid is *positive* and varies as follows:

$$C_{V,\text{or}}^*/R \propto y^{*2} \tag{18}$$

Here, y^* denotes the *effective dipolar interaction parameter* introduced by Wilhelm[36,37] which is defined by

$$y^* = p_r^2/(V_r^* T_r) = N_A p^2/(4\pi\varepsilon_0 V^* k_B T) \tag{19}$$

Here, V^* is the pure-substance molar volume, $V_r^* = V^*/V_c$ is the reduced molar volume, V_c is the critical molar volume, $T_r = T/T_c$ is the reduced temperature, T_c is the critical temperature, N_A is the Avogadro constant, $\varepsilon_0 = 8.854187816 \times 10^{-12}$ C^2 J^{-1} m^{-1} is the permittivity of vacuum, and k_B is the Boltzman constant. The *reduced* dipole moment p_r is expressed as:

$$p_r = [N_A p^2/(4\pi\varepsilon_0 V_c k_B T_c)]^{1/2} \tag{20}$$

Thus, the dipolar contribution to C_V^* varies with $p^4/(V^*T)^2$ and is, of course, superimposed on the contribution due to shape anisotropy.

Extension to mixtures is fairly straightforward. Kalali *et al.*[146] have used an approach similar in spirit involving the corresponding set of thermodynamic quantities in (T,P,x)-space, which are more readily accessible through experiment. In close analogy to the Pople approach they also concluded that the net destruction of order resulting from mixing a dipolar liquid with a non-polar liquid, say an *n*-alkane, results in *positive* contributions to G^E, H^E and S^E, and a *negative* contribution to C_P^E (see also references 176 and 177), which leads to:

$$C_P^E = -2(H^E - G^E)/T \tag{21}$$

This simple relation appears to hold well for mixtures of non-polar or weakly polar substances.[178] However, for mixtures of a strongly polar liquid with a non-polar liquid, the experimental C_P^E is usually *less* negative than demanded by Equation (21). This indicates that some dipole–dipole orientations have considerably *greater* stability than accounted for by the angle-averaging procedure involved. As was pointed out some time ago by Wilhelm,[179] as long as the temperature is sufficiently high so as to make $y \ll 1$, orientational effects are essentially swamped by thermal randomisation.[36,37]

One possible way of treating these non-random effects is through application of Guggenheim's quasi-chemical theory,[13] as suggested by Saint-Victor and Patterson.[145] Denoting the molar cooperative free energy by:

$$W = A + B/T \tag{22}$$

where A and B are constants for a given pair of substances, the excess molar heat capacity $C_P^{\rm E}$ may be separated, to an excellent approximation, into a random (R) and a non-random (NR) contribution[36,37,179] (we note that the original expression of Saint-Victor and Patterson[145] is not quite correct in that in their Equation (7) the factor η^2 is missing from the first term inside the wavy brackets):

$$
\begin{aligned}
C_P^{\rm E}/R &= C_P^{\rm E}({\rm R})/R + C_P^{\rm E}({\rm NR})/R \\
&= -x_1 x_2 T R^{-1} \left({\rm d}^2 W/{\rm d}T^2 \right) \\
&\quad + (x_1 x_2)^2 \left\{ 2[W - T({\rm d}W/{\rm d}T)]^2 \eta^2 \left(z R^2 T^2 \right)^{-1} + \left(\eta^2 - 1 \right) T R^{-1} \left({\rm d}^2 W/{\rm d}T^2 \right) \right\}
\end{aligned}
\tag{23}
$$

where $\eta = \exp(W/zRT)$, and z is the coordination number. This expression is in agreement with the first-order result of the Taylor expansion of the exact quasi-chemical equation as obtained by Cobos.[180] Evidently, the random term is *always negative* with a parabolic composition dependence, as expected for mixtures where dipole–dipole order is being destroyed in the mixing process. In contradistinction, the non-random term is *always positive* and has *zero* slope against the mole fraction axis at both ends of the composition range. Thus, the superposition of the two contributions $C_P^{\rm E}({\rm R})$ and $C_P^{\rm E}({\rm NR})$ accounts qualitatively for the appearance of W-shaped curves $C_P^{\rm E}$ *versus* x. Note that with decreasing temperature, the maximum caused by $C_P^{\rm E}({\rm NR})$ increases, thus making the W-shape more pronounced. For a more detailed treatment, see Cobos.[180] Recently, equation-of-state as well as local composition models (NRTL) have also been used to predict and correlate W-shaped $C_P^{\rm E}(x)$ curves.[181–183]

From a theoretical point of view, perhaps the best way to look at the local composition in a mixture, and thus at non-randomness, is by focussing on the long-wavelength limit of the *partial composition–composition structure factor*[184,185] $S_{\rm cc}(0)$ (unfortunately, this quantity is frequently called the concentration–concentration structure factor) which may be obtained, for instance, from vapour pressure measurements or *via* Rayleigh light scattering measurements. It is related to experimentally accessible thermodynamic quantities:

$$S_{\rm cc}(0) = \frac{RT}{\left(\partial^2 G/\partial x_1^2 \right)_{T,P}} \tag{24}$$

Thus, for a binary ideal, that is random, mixture:

$$S_{cc}^{id}(0) = x_1 x_2 \qquad (25)$$

$S_{cc}(0) > S_{cc}^{id}(0)$ indicates a tendency of the components of the mixture for homocoordination, while $0 < S_{cc}(0) < S_{cc}^{id}(0)$ indicates a tendency for hetero-coordination. There seems to be a good correlation between the value of the maximum of the curve $S_{cc}(0)$ *versus* x_1 of a given system and the appearance of W-shaped C_P^E curves; as threshold value, above which they are observed, Rubio *et al.*,[186] Andreolli-Ball *et al.*,[187] and Lainez *et al.*[136] have suggested $S_{cc}(0)_{max} \approx$ 0.7.

However, it is important to note that many of the mixtures showing W-shaped C_P^E curves are quite close to phase separation with an upper critical solution temperature (UCST), T_{UC}. When a UCST is approached from the homogeneous region at constant pressure and constant critical composition x_c, the heat capacity along this path diverges weakly[146,188,189] according to:

$$C_P^E(x_c) = C_P^E(x_c, \text{ non} - \text{diverging}) + A_c t^{-\alpha} \qquad (26)$$

where A_c is the critical amplitude, $t = |(T - T_{UC})/T_{UC}|$ is the reduced distance, temperature wise, from T_{UC}, and $\alpha = 0.11$ is a universal critical exponent, which may be computed, for instance, for a generalised three-dimensional Ising model.[190,191] In order to correlate data obtained farther away from the UCST, extended scaling has to be used. Evidently, the quasi-chemical approximation underestimates the non-random contribution to C_P^E and becomes qualitatively incorrect for $T \to T_{UC}$.

4.3.2 Mixtures of Organic Liquids with Water

This topic is huge and would certainly merit a monograph of its own, in particular when results obtained at high dilution are to be included. Here, only *very* few papers will be commented on, just to give the reader a general flavour and an indication of current research activities in this field.

In 1981, Grolier and Wilhelm[192] reported precise excess molar heat capacity data of the important system $\{x_1 C_2H_5OH + x_2 H_2O\}$ at 298.15 K. They were obtained with a Picker-type flow calorimeter, and clearly showed the rather complex, highly asymmetric composition dependence characteristic for this kind of mixture. As can be seen in Figure 4.5, the maximum of C_P^E amounts to more than $14 \text{ J K}^{-1} \text{mol}^{-1}$ and is located around an ethanol mole fraction $x_1 = 0.19$. In order to satisfactorily represent the experimental data, we used a generalised Myers-Scott equation:[193]

$$C_P^E = x_1 x_2 \sum_i A_i (x_1 - x_2)^i \Big/ \sum_j B_j (x_1 - x_2)^j \qquad (27)$$

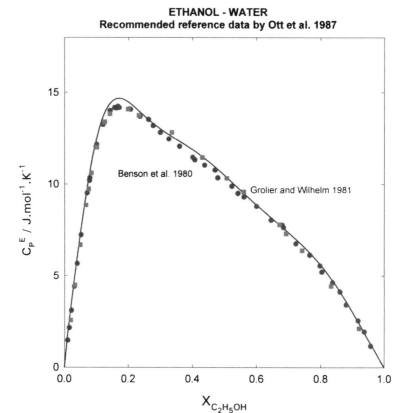

Figure 4.5 Comparison of C_P^E of $\{x_1 C_2 H_5 OH + x_2 H_2 O\}$ at 298.15 K and atmospheric pressure. ■, Grolier and Wilhelm;[192] ●, Benson, D'Arcy and Kiyohara;[194] ——, Ott *et al.*[195] This system was recommended as a reference system for testing mixing calorimeters.

with $B_0 = 1$. The terms of the denominator are constrained in such a way as not to yield a pole within the range $0 \leq x_1 \leq 1$. Agreement with other high-precision results is highly satisfactory. Specifically, this is the case with the directly measured excess heat capacities of Benson *et al.*,[194] and with the excess heat capacities derived from the high-precision excess enthalpies measured by Ott *et al.*[195] at 25 K intervals from 298.15 K to 473.15 K (at 5 MPa and 15 MPa) *via* $C_P^E (x_1) = [\partial H^E(x_1)/\partial T]_{P,x}$. Agreement with the correlating equation of Larkin and Pemberton,[196] though not shown in Figure 4.5, is also good. These results led Ott *et al.* to recommend that (ethanol + water) be considered as a *reference mixture* for testing mixing calorimeters at temperatures as high as 473.15 K. Finally, it is worth to point out the reasonably good accord of Bose's calorimetrically determined heat capacities [197] published in 1907 (!) with ours.

Recent heat capacity measurements on (methanol + water) have been presented by Tanaka *et al.*[198] and Dettmann *et al.*[80] They both cover large temperature and pressure ranges, and special attention was directed by the latter to the critical region of methanol. Alkoxyethanols show also appreciable self-association, and when mixed with water a fairly complicated mixing behaviour is expected. Indeed, as shown by Tamura and co-workers,[199] H^E of $\{x_1\text{ethoxyethanol} + x_2\text{water}\}$, at 298.15 K and ambient pressure, is surprisingly negative with a minimum of about $-1000\,\text{J mol}^{-1}$ at $x_1 \approx 0.2$; the excess molar heat capacity curve is of similar complexity as that of (ethanol + water); a maximum of *ca.* $9.5\,\text{J K}^{-1}\,\text{mol}^{-1}$ is observed around $x_1 = 0.1$. For additional work in this area, that is thermodynamic properties of aqueous solutions of 2-isopropoxyethanol and 2-isobutoxyethanol, see Tamura *et al.*[200–202] Heat capacity measurements on (butoxyethanol + water) and (isobutoxyethanol + water) of critical composition as function of temperature were reported by Würz *et al.*[203] near the *lower* critical point of the systems (both have a *closed* miscibility gap). Analysis of the data showed that both systems belong to the same universality class as "simple" binary mixtures. The value of the critical amplitude ratio of the specific heat capacity, $A^+/A^- = 0.53\pm0.01$, is consistent with theoretical predictions.[191,204] Here, the superscript + refers to the homogeneous and the – to the heterogeneous state.

Though ethylene glycol (1,2-ethanediol, $HOCH_2CH_2OH$) is an industrially important chemical, there are surprisingly few thermodynamic studies published in the open literature; one such paper is by Huot *et al.*[205] Recently, Nan *et al.*[206] reported excess molar heat capacities of $\{xC_2H_4(OH)_2 + (1-x)H_2O\}$ in the temperature range 273.15 K to 373.15 K with 5 K intervals. The excess heat capacity $C_P^E(x)$ changes its sign, *i.e.* it is negative at low temperatures and positive at high temperatures, and in between it shows a very complex composition dependence which should, however, be corroborated by more detailed measurements, preferably with a different type of calorimeter. This is even *more* needed, when their results are compared with the heat capacities recently reported by Yang *et al.*:[207] discrepancies of up to about 10 % in C_P are observed, and the composition dependence of C_P is different.

Discussion of thermodynamic quantities concerning biochemical systems is generally based upon results obtained for relatively simple model compounds dissolved in liquid water. In this context, the partial molar heat capacity of a solute at infinite dilution, $\bar{C}_{P,i}^{\infty}$, is regarded as being one of the most interesting thermodynamic quantities, in particular when combined with concepts taken from group-contribution theory. While considerable effort has been placed upon securing reliable *high-dilution* data for aqueous solutions of alkanoic acids, results covering the whole composition range are rather scarce, whence we determined V^E and C_P^E at 298.15 K of $\{x\text{RCOOH} + (1-x)H_2O\}$ for R = H, CH_3, C_2H_5, and C_3H_7.[208] Since our work was not a high-dilution study, it was highly gratifying to see our extrapolated values for $\bar{C}_{P,\text{RCOOH}}^{\infty}$ in satisfactory accord with the experimental high-dilution results of Konicek and Wadsö.[209] Figure 4.6 shows our results[208] together with the only other directly determined heat capacities of an aqueous alkanoic acid system covering the entire composition range; the results of Neumann's[210] careful measurements, obtained more than 70 years ago, are in excellent agreement with our "modern" results, and confirm the *M-shaped*

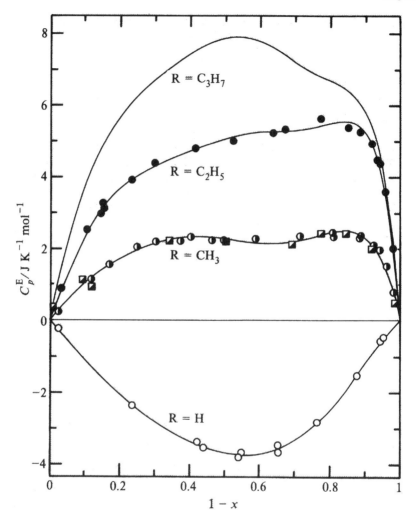

Figure 4.6 Excess molar heat capacities C_P^E at constant pressure of $\{xRCOOH + (1-x)$ $H_2O\}$ at 298.15 K and atmospheric pressure. R = H: ○;[208] R = CH$_3$: ◑;[208] ◪;[210] R = C$_2$H$_5$: ●;[208] R = C$_3$H$_7$: ——, only the fitting curve is shown; the original experimental results were published by E. Wilhelm, C. Casanova and J.-P. E. Grolier, *Int. DATA Ser., Ser. B, Aq. Org. Syst.*, 1977, p. 6. All curves were calculated from a Redlich-Kister type fitting equation.

curve $C_P^E(x)$ for the system $\{xCH_3COOH + (1-x)H_2O\}$ at 298.15 K, which is a *very* rare composition dependence indeed. For a recent paper on heat capacities (and a few other properties) of aqueous solutions of dimethyl sulfoxide, ethylenediamine, polyethylene glycol 200 *etc.*, see Dhondge *et al.*[211]

When dissolved in water, the partial molar heat capacity at infinite dilution of the majority of liquid organic substances is considerably larger than their molar heat capacity in the pure liquid state: $\bar{C}_{P,i}^{\infty} > C_{P,i}^{*}$. These large increments have generally been connected with changes in the structure of water around the solute molecule. This idea is perhaps best illustrated by the classic article of Frank and Evans of 1945,[212] and has continued to challenge theoretical developments. From a practical point of view, the excess partial molar heat capacity at infinite dilution, $\bar{C}_{P,i}^{E\infty} = \bar{C}_{P,i}^{\infty} - C_{P,i}^{*}$, together with the excess partial molar enthalpy at infinite dilution, $\bar{H}_{i}^{E\infty}$, may be used to describe reliably the *temperature dependence* of the activity coefficient at infinite dilution, γ_{i}^{∞}, of solute i. The limiting activity coefficients, in particular of organic solutes dissolved in water, play a central role in separation technology and environmental protection issues. From Equations (1), (2) and (4), the activity coefficient at infinite dilution and its temperature derivatives are connected with the corresponding excess partial molar quantities as follows:

$$\bar{G}_{i}^{E\infty} = \mu_{i}^{E\infty} = RT \ln \gamma_{i}^{\infty} \tag{28}$$

$$\bar{H}_{i}^{E\infty} = -RT^2 \left(\partial \ln \gamma_{i}^{\infty} / \partial T\right)_{P,x} \tag{29}$$

$$\bar{C}_{P,i}^{E\infty} = \left(\partial \bar{H}_{i}^{E\infty} / \partial T\right)_{P,x} = -2RT\left(\partial \ln \gamma_{i}^{\infty} / \partial T\right)_{P,x} - RT^2\left(\partial^2 \ln \gamma_{i}^{\infty} / \partial T^2\right)_{P,x} \tag{30}$$

By combining experimental γ_{i}^{∞}, measured, say, *via* headspace analysis at some convenient temperature, with thermal data ($\bar{H}_{i}^{E\infty}, \bar{C}_{P,i}^{E\infty}$), successive integration of Equations (29) and (30) yields the desired correlating equation, in complete analogy to the method of choice for the global fitting of excess quantities indicated in the Introduction. Usually, a 4-parameter Clarke-Glew type equation[213–215] for $\ln\gamma_{i}^{\infty}(T)$ is selected, which implies the assumption that $\bar{C}_{P,i}^{E\infty}$ is a *linear* function of temperature:

$$\ln \gamma_{i}^{\infty} = A_0 + A_1 \tau^{-1} + A_2 \ln \tau + A_3 \tau \tag{31}$$

where, $\tau = T/T_0$ and, say, $T_0 = 298.15$ K.

Out of the large body of papers devoted to the subject of *heat capacities of dilute aqueous solutions of nonelectrolytes* we quote the pioneering high-temperature, high-pressure work by Inglese and Wood[216] on aqueous solutions of various alcohols, and the work of Hnědkovsky and Wood[217] on aqueous solutions of CH_4, CO_2, H_2S and NH_3. A very recent high-temperature, high-pressure investigation of the heat capacity of dilute aqueous solutions of acetone, diethyl ether, 1,2-dimethoxyethane *etc.*, was presented by Slavik *et al.*[218] The apparent molar heat capacities of dilute solutions of ethanol, propan-1-ol and propan-2-ol were determined by Origlia-Luster and Woolley[219] at temperatures from 278.15 K to 393.15 K and at $P = 0.35$ MPa. For data at atmospheric pressure, we refer to the recent contributions of Hovorka *et al.*,[220,221] Fenclová *et al.*,[222] and Piekarski and Pietrzak.[223] All this work is eminently useful for improving the performance of existing group contribution schemes, such as the method of Cabani *et al.*,[224] or Plyasunov and Shock's approach,[225–227] or modified UNIFAC (Dortmund).[19]

4.3.3 Mixtures Containing Ionic Liquids

Room Temperature Ionic Liquids (RTILs) are liquids composed of ions having melting points, by convention, below 100 °C. Perhaps the first documented RTIL, ethylammonium nitrate, $[C_2H_5NH_3][NO_3]$, with a melting point of 12 °C, was discovered by Walden in 1914,[228] but it took about seven decades before their potential as versatile solvents was fully realised.[229–232] They are good solvents for many organic and inorganic compounds over a wide range of temperature, yet when being immiscible with some organic solvents they may provide non-aqueous, polar alternatives for two-phase extraction systems. They have negligible vapour pressure, and compared to conventional organic solvents they are, in general, less toxic. RTILs are usually composed of a large organic cation with a low degree of symmetry, and an inorganic or organic anion. They have been dubbed *designer solvents*, since their physical properties can be adjusted, for instance by varying the length and the degree of branching of the alkyl groups of the cation, to suit the requirements of a particular application, say, chemical synthesis reactions or novel extraction processes. However, ionic liquids are also finding applications as thermal fluids, lubricants, hydraulic fluids or pharmaceutical ILs. They are indeed multipurpose materials.

Published data on heat capacities of RTILs are not plentiful. A selection of recent publications is provided by references 233 through 242. Of special note is the work on several pure RTILs of Gomes de Azevedo and co-workers,[243,244] who carried out speed-of-sound measurements in the temperature range 283 K through 323 K and in the pressure range 0.1 MPa through 150 MPa. In addition, densities were measured between 298 K and 333 K and between 0.1 MPa and 60 MPa. Combining Equation (11) with a slightly rearranged Equation (8), and remembering that $\alpha_P = -(\partial \ln \rho / \partial T)_P$ and $\kappa_T = \rho^{-1}(\partial \rho / \partial P)_T$, yields:

$$\left(\frac{\partial \rho}{\partial P}\right)_T = \frac{1}{v^2} + \frac{TM}{C_P}\left(\frac{\partial \ln \rho}{\partial T}\right)_P^2 \tag{32}$$

$$\left(\frac{\partial C_P}{\partial P}\right)_T = -\frac{TM}{\rho}\left[\left(\frac{\partial \ln \rho}{\partial T}\right)_P^2 - \left(\frac{\partial^2 \ln \rho}{\partial T^2}\right)_P\right] \tag{33}$$

Using the 0.1 MPa isobars of the density and of the heat capacity, respectively, successive integration over pressure yields (ρ, P, T) and (C_P, P, T) surfaces within the ranges of pressure and temperature of the experimental speed-of-sound data.[245,246] The two papers of Gomes de Azevedo and co-workers[243,244] report the first sound speeds and heat capacities at constant pressure at *high pressures* for RTILs, such as 1-butyl-3-methylimidazolium tetrafluoroborate, [bmim][BF₄], 1-butyl-3-methylimidazolium bis(trifluoromethylsulfonyl)amide, [bmim][NTf₂] *etc.*

So far, only a few research groups have published results on $C_P^E(x)$ of mixtures of ionic liquids with nonelectrolytes. Waliszewski[247] used a differential scanning calorimeter (Micro DSC III, from Setaram) to measure the heat capacity of (methanol + 1-hexyl-3-methylimidazolium tetrafluoroborate) and (methanol + 1-methyl-3-octylimidazolium tetrafluoroborate) over the temperature range 283.15 K to 323.15 K. In both systems, C_P^E is positive and distinctly skewed towards methanol, and the maximum values vary from about $6\,\mathrm{J\,K^{-1}\,mol^{-1}}$ at 283.15 K to about $2.5\,\mathrm{J\,K^{-1}\,mol^{-1}}$ at 323.15 K. For a series of mixtures of ethanol with RTILs, such as [bmim][BF$_4$] or 1-ethyl-3-methylimidazolium ethylsulfate, [emim][EtSO$_4$], molar heat capacities at constant pressure as well as densities were measured between 293.15 K and 318.15 K by García-Miaja et al.[248] In addition, they measured H^E at 303.15 K. The excess enthalpies are all rather strongly endothermic (maximum values range from ca. $1300\,\mathrm{J\,mol^{-1}}$ to $2600\,\mathrm{J\,mol^{-1}}$), and the C_P^E curves show mostly a sigmoidal composition dependence, the exception being C_P^E of (ethanol + [emim][EtSO$_4$]) which is negative throughout with a minimum value of about $-4\,\mathrm{J\,K^{-1}\,mol^{-1}}$ at 298.15 K, and $-7\,\mathrm{J\,K^{-1}\,mol^{-1}}$ at 318.15 K. The Extended Real Associated Solution (ERAS) model[115,249,250] was shown to give good results for V^E and H^E, but was not able to reproduce the experimental excess heat capacities.

Many of the heat capacity studies concern mixtures of RTILs with water, such as those presented in references 251 through 255. For instance, García-Miaja et al.[253] report C_P^E of (water + 1-butyl-3-methylpyridinium tetrafluoroborate, [bpyr][BF$_4$]) in the temperature range 293.15 K to 318.15 K; it shows a strong sigmoidal composition dependence with a maximum value, at 298.15 K, of about $9\,\mathrm{J\,K^{-1}\,mol^{-1}}$ at $x_{water} \approx 0.85$, and a minimum value of about $-1.5\,\mathrm{J\,K^{-1}\,mol^{-1}}$ at $x_{water} \approx 0.15$. Apparently, with increasing temperature, the sigmoidal shape becomes even more pronounced. A somewhat similar curve shape is observed for (water + [emim][EtSO$_4$]),[255] albeit the temperature behaviour is different; while at higher temperatures the maximum of the $C_P^E(x)$ curve, situated in the water-rich region, increases, the minimum at, roughly, $x_{water} \approx 0.4$, becomes less negative. We note that, on the other hand C_P^E of (nitromethane (NM) + [bpyr][BF$_4$]) shows "inverse" composition behaviour;[253] the sigmoidal curve has a *minimum* at $x_{NM} \approx 0.75$ (ca. $-1.4\,\mathrm{J\,K^{-1}\,mol^{-1}}$ at 298.15 K) which becomes more negative with increasing temperature, while the *maximum* at $x_{NM} \approx 0.15$ becomes less positive and has practically disappeared at 318.15 K. For additional recent results on (nitromethane + alkylimidazolium-based IL), see García-Miaja et al.[256] Heat capacities and excess enthalpies were measured for the three binary mixtures (water + 1-ethyl-3-methylimidazolium ethylsulfate, [emim][EtSO$_4$], or + 1-ethyl-3-methylimidazolium trifluorome-thanesulfonate, [emim][OTf], or + 1-ethyl-3-methylimidazolium trifluoroacetate, [emim][TFA]) by Ficke et al.[254] from 283.15 K to 348.15 K. Surprisingly, the values of the C_P^Es of these three aqueous systems are very small (of the same order of magnitude as the experimental uncertainty), so that $C_P^E \approx 0$ has been reported for the entire composition range at all the temperatures investigated.

Of particular note is the detailed, quite comprehensive thermodynamic study of (water + [bmim][BF$_4$]) by Rebelo *et al.*[251] Specifically, they determined densities as a function of temperature (278.15 < T/K < 333.15) and pressure (0.1 < P/MPa < 60) over the whole composition range, heat capacities and excess enthalpies from 278.15 K to 333.15 K and atmospheric pressure, and the liquid–liquid phase diagram at pressures up to 70 MPa, including the impact of isotopic substitution (H$_2$O/D$_2$O). For (H$_2$O + [bmim][BF$_4$]) at 0.1 MPa, Rebelo *et al.* observe an upper critical solution temperature T_{UC} = 277.6 K at $x_{IL,c}$ = 0.07 (or $w_{IL,c}$ = 0.49), and for (D$_2$O + [bmim][BF$_4$]) at 0.1 MPa, T_{UC} = 281.3 K at $x_{IL,c}$ = 0.07 (or $w_{IL,c}$ = 0.46). Evidently, the phase diagram becomes more symmetric when the mole fraction x is replaced by the mass fraction w.

The liquid–liquid phase separation behaviour of binary mixtures is important for improving our understanding of critical phenomena. The mixture (water + choline bis(trifluoromethylsulfonyl)imide, [choline][Tf$_2$N]) was carefully studied by Nockemann *et al.*[257] It exhibits UCST behaviour with T_{UC} = 345.25 K at the critical composition $w_{IL,c}$ = 0.524. We note that this shows again that the terms *hydrophobic* IL and *hydrophilic* IL are only of limited use. Nockemann used high resolution adiabatic slow-scanning calorimetry which is a technique well suited for the analysis of critical behaviour of the heat capacity at constant pressure.[258] An excellent fit for the diverging heat capacity $C_P(T, w_{IL,c})$ *versus* t = |(T − T_{UC})/T_{UC}| was obtained with the 3D-Ising value[188,189] 0.110 for the critical exponent α, *i.e.*

$$C_P(T, w_{IL,c}) = E + FT + A^{\pm} t^{-\alpha} \tag{34}$$

Also the ratio of the experimental critical amplitudes A^+/A^- = 0.54±0.07 is consistent with the theoretical 3D-Ising value of 0.54.[190–191,204]

So far, reports in the literature on liquid–liquid equilibria involving RTILs have been almost exclusively concerned with (RTIL + an alcohol) or (RTIL + water), and all show UCST behaviour.[251,257,259–261] We note that for a given kind of anion, the UCSTs of mixtures with alcohols depend systematically on the chain length n of the alkan-1-ol, $C_nH_{2n+1}OH$, *i.e.* they increase with increasing n. For a comprehensive study of liquid–liquid equilibria involving 1-hexyloxymethyl-3-methylimidazolium-based RTILs mixed with n-alkanes (C$_5$ to C$_8$), cyclohexane and aromatics (benzene, toluene, *etc.*), see Domańska and Marciniak.[262]

Undoubtedly, more experimental work as well as theoretical studies are needed to clarify several interesting questions related to the behaviour in the critical region of mixtures containing RTILs, such as the crossover from Ising-like behaviour generally observed with nonelectrolyte mixtures to mean-field behaviour.[263–266]

4.4   Concluding Remarks

While calorimetry certainly belongs to the oldest well-established experimental disciplines in thermodynamics, it is by no means jaded. Quite the contrary, continuing advances in instrumentation (including automation and miniaturisation) leading to increased precision, accuracy and speed of measurement on the one hand, and the ever widening ranges of application (higher temperatures, higher pressures, smaller concentrations and sample sizes) on the other, provide the impetus for calorimetry to remain an active and developing discipline. This is furthered by cross-linking with other important experimental techniques, such as high-pressure ultrasonics, which has provided high-pressure data for heat capacities, isothermal compressibilities, isobaric expansivities, isochoric thermal pressure coefficients *etc.* of excellent quality which would have been difficult to obtain otherwise. Parallel advances in the statistical–mechanical treatment of liquid mixtures/solutions and increasingly sophisticated computer simulation techniques provide new insights and stimulating connections at a microscopic level. Without doubt, heat capacities belong to the thermodynamic quantities occupying the centre of the stage called scientific endeavour.

References

1. J.-P. E. Grolier, C. J. Wormald, J.-C. Fontaine, K. Sosnkowska-Kehiaian and H. V. Kehiaian, in *Landolt-Börnstein, Group IV: Physical Chemistry, Vol. 10 Heats of Mixing and Solution, Subvol. A*, H.V. Kehiaian, ed. Springer, Berlin, 2004.
2. J. C. Fontaine, J.-P. E. Grolier, H. V. Kehiaian, K. Sosnkowska-Kehiaian and C. J. Wormald, in *Landolt-Börnstein, Group IV: Physical Chemistry, Vol. 10. Heats of Mixing and Solution, Subvol. B*, H.V. Kehiaian, ed. Springer, Berlin, 2005.
3. P. Picker, P.-A. Leduc, P. R. Philip and J. E. Desnoyers, *J. Chem. Thermodyn.*, 1971, **3**, 631.
4. J.-P. E. Grolier, G. C. Benson and P. Picker, *J. Chem. Eng. Data*, 1975, **20**, 243.
5. E. Wilhelm, J.-P. E. Grolier and M. H. Karbalai Ghassemi, *Ber. Bunsenges. Phys. Chem.*, 1977, **81**, 925.
6. *Measurement of the Thermodynamic Properties of Multiple Phases. Experimental Thermodynamics, Vol. VII*, R. D. Weir and Th. W. De Loos, eds., IUPAC Commission on Thermodynamics/Elsevier, Amsterdam, 2005.
7. *Solution Calorimetry. Experimental Thermodynamics, Vol. IV*, K. N. Marsh and P. A. G. O'Hare, eds., IUPAC Commission on Thermodynamics/Blackwell Scientific Publications, Oxford, 1994.
8. E. Wilhelm, J.-P. E. Grolier and M. H. Karbalai Ghassemi, *Monatsh. Chem.*, 1978, **109**, 369.
9. *Developments and Applications in Solubility*, T.M. Letcher, ed., RSC Publishing/IUPAC, Cambridge, 2007.
10. E. Wilhelm and R. Battino, *J. Chem. Phys.*, 1971, **55**, 4012.
11. E. Wilhelm, *J. Chem. Phys.*, 1973, **58**, 3558.

12. F. Kohler, J. Fischer and E. Wilhelm, *J. Mol. Stucture*, 1982, **84**, 245.
13. E. A. Guggenheim, *Mixtures*, Oxford University Press, London, 1952.
14. H. V. Kehiaian, J.-P. E. Grolier and G. C. Benson, *J. Chim. Phys.*, 1978, **75**, 1031.
15. H. V. Kehiaian, *Fluid Phase Equilib.*, 1983, **13**, 243.
16. H. V. Kehiaian, *Pure Appl. Chem.*, 1985, **57**, 15.
17. J. Gmehling, J. Li and M. Schiller, *Ind. Eng. Chem. Res.*, 1993, **32**, 178.
18. J. Gmehling, J. Lohmann, A. Jakob, J. Li and R. Joh, *Ind. Eng. Chem. Res.*, 1998, **37**, 4876.
19. J. Lohmann, R. Joh and J. Gmehling, *Ind. Eng. Chem. Res.*, 2001, **40**, 957.
20. S. Delcros, J. R. Quint, J.-P. E. Grolier and H. V. Kehiaian, *Fluid Phase Equilib.*, 1995, **113**, 1. See also S. Delcros, Ph.D. Dissertation, Université Blaise Pascal, Clermont-Ferrand, France, 1995. Here, DISQUAC calculations are compared with UNIFAC (modified) calculations of GE, HE and C_P^E of ether + 1-alkanol systems..
21. S. I. Sandler, *Chemical and Engineering Thermodynamics*, 3rd Edition, Wiley, New York, 1998.
22. J. M. Prausnitz, R. N. Lichtenthaler and E. Gomez de Azevedo, *Molecular Thermodynamics of Fluid-Phase Equilibria*, 3rd Edition, Prentice Hall, Upper Saddle River, NJ, 1999.
23. A. Klamt and F. Eckert, *Fluid Phase Equilib.*, 2000, **172**, 43.
24. F. Eckert and A. Klamt, *AIChE J.*, 2002, **48**, 369.
25. A. Klamt, G. J. P. Krooshof and R. Taylor, *AIChE J.*, 2002, **48**, 2332.
26. R. Putnam, R. Taylor, A. Klamt, F. Eckert and M. Schiller, *Ind. Eng. Chem. Res.*, 2003, **42**, 3635.
27. S. T. Lin and S. I. Sandler, *Ind. Eng. Chem. Res.*, 2002, **41**, 899.
28. S. T. Lin, J. Chang, S. Wang, W. A. Goddard and S. I. Sandler, *J. Phys. Chem. A*, 2004, **108**, 7429.
29. E. Mullins, R. Oldland, Y. A. Liu, S. Wang, S. I. Sandler, C.-C. Chen, M. Zwolack and K. C. Seavey, *Ind. Eng. Chem. Res.*, 2006, **45**, 4389.
30. S. Wang, S. I. Sandler and C.-C. Chen, *Ind. Eng. Chem. Res.*, 2007, **46**, 7275.
31. H. Grensemann and J. Gmehling, *Ind. Eng. Chem. Res.*, 2005, **44**, 1610.
32. T. Mu, J. Rarey and J. Gmehling, *Ind. Eng. Chem. Res.*, 2007, **46**, 6612.
33. T. Mu, J. Rarey and J. Gmehling, *AIChE J.*, 2007, **53**, 3231.
34. H. V. Kehiaian, *Ber. Bunsenges. Phys. Chem.*, 1977, **81**, 908.
35. E. Wilhelm, *Thermochim. Acta*, 1983, **69**, 1.
36. E. Wilhelm, *High Temp.-High Press.*, 1997, **29**, 613.
37. E. Wilhelm, *Pure Appl. Chem.*, 2005, **77**, 1317.
38. A. Bartholomé and A. Eucken, *Trans. Faraday Soc.*, 1937, **33**, 45.
39. A. Eucken, *Z. Elektrochem.*, 1948, **52**, 255.
40. J. D. Bernal, *Trans. Faraday Soc.*, 1937, **33**, 27.
41. J. F. Kincaid and H. Eyring, *J. Chem. Phys.*, 1938, **6**, 620.
42. L. A. K. Staveley, K. R. Hart and W. I. Tupman, *Disc. Faraday Soc.*, 1953, **15**, 130.

43. L. A. K. Staveley, W. I. Tupman and K. R. Hart, *Trans. Faraday Soc.*, 1955, **51**, 323.

44. D. Harrison and E. A. Moelwyn-Hughes, *Proc. Roy. Soc. (London)*, 1957, **A239**, 230.

45. G. H. Findenegg and F. Kohler, *Trans. Faraday Soc.*, 1967, **63**, 870.

46. A. Bondi, *Physical Properties of Molecular Crystals, Liquids and Glasses*, Wiley, New York, 1968.

47. E. Wilhelm, R. Schano, G. Becker, G. H. Findenegg and F. Kohler, *Trans. Faraday Soc.*, 1969, **65**, 1443.

48. E. Wilhelm, M. Zettler and H. Sackmann, *Ber. Bunsenges. Phys. Chem.*, 1974, **78**, 795.

49. J. S. Rowlinson and F. L. Swinton, *Liquids and Liquid Mixtures*, Butterworth Scientific, London, 1982.

50. J.-P. E. Grolier, E. Wilhelm and M. H. Hamedi, *Ber. Bunsenges. Phys. Chem.*, 1978, **82**, 1282.

51. E. Wilhelm, J.-P. E. Grolier and M. H. Karbalai Ghassemi, *Thermochim. Acta*, 1979, **28**, 59.

52. M. O. Bryant and G. O. Jones, *Proc. Phys. Soc.*, 1953, **B66**, 421.

53. R. D. Goodwin and L. A. Weber, *J. Res. Natl. Bur. Stand. (U.S.)*, 1969, **73A**, 15.

54. J. W. Magee, J. C. Blanco and R. J. Deal, *J. Res. Natl. Inst. Stand. Technol.*, 1998, **103**, 63.

55. J. W. Magee and N. Kagawa, *J. Chem. Eng. Data*, 1998, **43**, 1082.

56. T. Kuroki, N. Kagawa, H. Endo, S. Tsuruno and J. W. Magee, *J. Chem. Eng. Data*, 2001, **46**, 1101.

57. H. Kitajima, N. Kagawa, H. Endo, S. Tsuruno and J. W. Magee, *J. Chem. Eng. Data*, 2003, **48**, 1583.

58. R. A. Perkins and J. W. Magee, *J. Chem. Eng. Data*, 2005, **50**, 1727.

59. J.-L. Fortier, G. C. Benson and P. Picker, *J. Chem. Thermodyn.*, 1976, **8**, 289.

60. J.-L. Fortier and G. C. Benson, *J. Chem. Thermodyn.*, 1976, **8**, 411.

61. O. Kratky, H. Leopold and H. Stabinger, *Z. Angew. Phys.*, 1969, **27**, 237.

62. E. P. Papadakis, *J. Acoust. Soc. Am.*, 1967, **42**, 1045.

63. A. Z. Zak, M. Dzida, M. Zorebski and S. Ernst, *Rev. Sci. Instrum.*, 2000, **71**, 1756.

64. P. J. Kortebeek, M. J. P. Muringer, N. J. Trappeniers and S. N. Biswas, *Rev. Sci. Instrum.*, 1985, **56**, 1269.

65. K. Tamura, K. Ohomuro and S. Murakami, *J. Chem. Thermodyn.*, 1983, **15**, 859.

66. T. Takagi and H. Teranishi, *J. Chem. Thermodyn.*, 1987, **19**, 1299.

67. A. Asenbaum and E. Wilhelm, *Adv. Mol. Relax. Interact. Processes*, 1982, **22**, 187.

68. E. Wilhelm, in *Les Capacités Calorifiques des Systèmes Condensés*, H. Tachoire, ed., Société Française de Chimie, Marseille, 1987, pp. 138–163.

69. J.-P. E. Grolier, in *Solution Calorimetry. Experimental Thermodynamics, Vol. IV*, K. N. Marsh and P. A. G. O'Hare, eds., IUPAC Commission on

Thermodynamics/Blackwell Scientific Publications, Oxford, 1994, pp. 43–75.

70. W. Hemminger and G. Höhne, *Calorimetry. Fundamentals and Practice*, Verlag Chemie, Weinheim, 1984.

71. G. Höhne, W. Hemminger and H.-J. Flammersheim, *Differential Scanning Calorimetry. An Introduction for Practitioners*, Springer, Berlin, 1996.

72. S. T. Lakshmikumar and E. S. R. Gopal, *Int. Rev. Phys. Chem.*, 1982, **2**, 197.

73. (a) *Experimental Thermodynamics, Volume I. Calorimetry of Non-reacting Systems*, J.P. McCullough and D.W. Scott, eds.; prepared under the sponsorship of the International Union of Pure and Applied Chemistry, Butterworths, London, 1968; (b) *Measurement of the Thermodynamic Properties of Single Phases. Experimental Thermodynamics, Vol. VI*, A.R.H. Goodwin, K.N. Marsh and W.A. Wakeham, eds., IUPAC/ Elsevier, Amsterdam, 2003.

74. H. J. Hoge, *J. Res. Natl. Bur. Stand.*, 1946, **36**, 111.

75. J. B. Ott and C. J. Wormald, in *Solution Calorimetry. Experimental Thermodynamics, Vol. IV*, K. N. Marsh and P.A.G. O'Hare, eds., IUPAC Commission on Thermodynamics/Blackwell Scientific Publications, Oxford, 1994, pp. 161–194.

76. R. H. Stokes, K. N. Marsh and R. P. Tomlins, *J. Chem. Thermodyn.*, 1969, **1**, 211.

77. M. B. Ewing, K. N. Marsh, R. H. Stokes and C. W. Tuxford, *J. Chem. Thermodyn.*, 1970, **2**, 751.

78. K. Elliott and C. J. Wormald, *J. Chem. Thermodyn.*, 1976, **8**, 881.

79. D. González-Salgado, J. L. Valencia, J. Troncoso, E. Carballo, J. Peleteiro, L. Romani and D. Bessières, *Rev. Sci. Instrum.*, 2007, **78**, 55.

80. C. Dettmann, G. Ernst and H. Wirbser, *J. Chem. Thermodyn.*, 2006, **38**, 56.

81. G. Ernst, G. Maurer and E. Wiederuth, *J. Chem. Thermodyn.*, 1989, **21**, 53.

82. G. Ernst, J. Gürtner and H. Wirbser, *J. Chem. Thermodyn.*, 1997, **29**, 1113.

83. D. Smith-Magowan and R. H. Wood, *J. Chem. Thermodyn.*, 1981, **13**, 1047.

84. P. S. Z. Rogers and K. S. Pitzer, *J. Phys. Chem.*, 1981, **85**, 2886.

85. J. I. Lankford and C. M. Criss, *J. Solution Chem.*, 1987, **16**, 885.

86. S. Boyette and C. M. Criss, *J. Chem. Eng. Data*, 1988, **33**, 426.

87. G. Conti, P. Gianni, A. Papini and E. Matteoli, *J. Solution Chem.*, 1988, **17**, 481.

88. D. R. White and C. J. Downes, *J. Phys. E: Sci. Instrum.*, 1989, **22**, 79.

89. J.-L. Fortier and G. C. Benson, *J. Chem. Thermodyn.*, 1976, **8**, 411.

90. P. S. Z. Rogers and C. J. Duffy, *J. Solution Chem.*, 1989, **21**, 595.

91. R. W. Carter and R. H. Wood, *J. Chem. Thermodyn.*, 1991, **23**, 1037.

92. T. K. Hei and J. D. Raal, *AIChE J.*, 2009, **55**, 206.

93. J. C. Cobos, I. Garcia, C. Casanova, A. H. Roux, G. Roux-Desgranges and J.-P. E. Grolier, *Fluid Phase Equilib.*, 1991, **69**, 223.

94. J. A. González, J. C. Cobos, F. J. Carmona, I. Garcia De La Fuente, V. R. Bhethanabotla and S. W. Campbell, *Phys. Chem. Chem. Phys.*, 2001, **3**, 2856.

95. J. Peleteiro, D. González-Salgado, C. A. Cerdeiriña, J. L. Valencia and L. Romani, *Fluid Phase Equilib.*, 2001, **191**, 83.

96. J. Peleteiro, D. González-Salgado, C. A. Cerdeiriña and L. Romani, *J. Chem. Thermodyn.*, 2002, **34**, 485.

97. C. A. Cerdeiriña, C. A. Tovar, E. Carballo, L. Romani, M. C. Delgado, L. A. Torres and M. Costas, *J. Phys. Chem. B*, 2002, **106**, 185.

98. J. Peleteiro, J. Troncoso, D. González-Salgado, J. L. Valencia, C. A. Cerdeiriña and L. Romani, *Int. J. Thermophys.*, 2004, **25**, 787.

99. J. Peleteiro, J. Troncoso, D. González-Salgado, J. L. Valencia, M. Souto-Caride and L. Romani, *J. Chem. Thermodyn.*, 2005, **37**, 935.

100. E. Zorebski, M. Chorażewski and M. Tkaczyk, *J. Chem. Thermodyn.*, 2005, **37**, 281.

101. M. Dzida and P. Góralski, *J. Chem. Thermodyn.*, 2006, **38**, 962.

102. E. Zorebski and P. Góralski, *J. Chem. Thermodyn.*, 2007, **39**, 1601.

103. M. Dzida and P. Góralski, *J. Chem. Thermodyn.*, 2009, **41**, 402.

104. R. Tanaka, S. Toyama and S. Murakami, *J. Chem. Thermodyn.*, 1986, **18**, 63.

105. M. Costas and D. Patterson, *J. Chem. Soc., Faraday Trans. I*, 1985, **81**, 635.

106. S. Miyanaga, K. Tamura and S. Murakami, *J. Chem. Thermodyn.*, 1992, **24**, 291.

107. S. Miyanaga, K. Tamura and S. Murakami, *Thermochim. Acta*, 1992, **198**, 237.

108. S. Miyanaga, K. Chubachi, M. Nakamura, K. Tamura and S. Murakami, *J. Chem. Thermodyn.*, 1993, **25**, 331.

109. M. Nakamura, K. Chubachi, K. Tamura and S. Murakami, *J. Chem. Thermodyn.*, 1993, **25**, 525.

110. M. Nishimoto, K. Tamura and S. Murakami, *J. Chem. Thermodyn.*, 1997, **29**, 15.

111. H. Piekarski, A. Pietrzak and D. Waliszewski, *J. Mol. Liq.*, 2005, **121**, 41.

112. F. Comelli, R. Francesconi, A. Bigi and K. Rubini, *J. Chem. Eng. Data*, 2006, **51**, 1711.

113. R. Francesconi, A. Bigi, K. Rubini and F. Comelli, *J. Chem. Eng. Data*, 2005, **50**, 1932.

114. R. F. Checoni and A. Z. Francesconi, *J. Solution Chem.*, 2007, **36**, 913.

115. A. Heintz, *Ber. Bunsenges. Phys. Chem.*, 1985, **89**, 172.

116. S. Figueroa-Gerstenmaier, A. Cabanas and M. Costas, *Phys. Chem. Chem. Phys.*, 1999, **1**, 665.

117. A. Pineiro, *Fluid Phase Equilib.*, 2004, **216**, 245.

118. H. V. Kehiaian and A. J. Treszczanowicz, *Bull. Acad. Chim. France*, 1969, **5**, 1561.

119. C. A. Cerdeiriña, C. A. Tovar, E. Carballo, L. Romani, M. del Carmen Delgado, L. A. Torres and M. Costas, *J. Phys. Chem. B*, 2002, **106**, 185.

120. C. A. Cerdeiriña, D. González-Salgado, L. Romani, M. del Carmen Delgado, L. A. Torres and M. Costas, *J. Chem. Phys.*, 2004, **120**, 6648.

121. C. A. Cerdeiriña, J. Troncoso, D. González-Salgado, G. Garcia-Miaja, G. O. Hernández-Segura, D. Bessières, M. Medeiros, L. Romani and M. Costas, *J. Phys. Chem. B*, 2007, **111**, 1119.

122. J.-P. E. Grolier, A. Inglese, A. H. Roux and E. Wilhelm, *Ber. Bunsenges. Phys. Chem.*, 1981, **85**, 768.

123. (a) A. Inglese, E. Wilhelm and J.-P. E. Grolier, communicated at the *37th Annual Calorimetry Conference*, Snowbird, Utah, U.S.A., 20 to 23 July 1982, Paper No. 54; (b) J.-P. E. Grolier, A. Inglese and E. Wilhelm, *J. Chem. Thermodyn.*, 1984, **16**, 67.

124. A. Inglese, J.-P. E. Grolier and E. Wilhelm, *Fluid Phase Equilib.*, 1984, **15**, 287.

125. A. H. Roux, J.-P. E. Grolier, A. Inglese and E. Wilhelm, *Ber. Bunsenges. Phys. Chem.*, 1984, **88**, 986.

126. A. Lainez, G. Roux-Desgranges, J.-P. E. Grolier and E. Wilhelm, *Fluid Phase Equilib.*, 1985, **20**, 47.

127. A. Lainez, M. Rodrigo, A. H. Roux, J.-P. E. Grolier and E. Wilhelm, *Calorim. Anal. Therm.*, 1985, **16**, 153.

128. E. Wilhelm, A. Lainez, A. H. Roux and J.-P. E. Grolier, *Thermochim. Acta*, 1986, **105**, 101.

129. E. Wilhelm, E. Jimenez, G. Roux-Desgranges and J.-P. E. Grolier, *J. Solution Chem.*, 1991, **20**, 17.

130. J.-P. E. Grolier, G. Roux-Desgranges, M. Berkane and E. Wilhelm, *J. Chem. Thermodyn.*, 1991, **23**, 421.

131. A. Lainez, M. M. Rodrigo, E. Wilhelm and J.-P. E. Grolier, *J. Solution Chem.*, 1992, **21**, 49.

132. J.-P. E. Grolier, G. Roux-Desgranges, M. Berkane, E. Jimenez and E. Wilhelm, *J. Chem. Thermodyn.*, 1993, **25**, 41.

133. J.-P. E. Grolier, G. Roux-Desgranges, M. Berkane and E. Wilhelm, *J. Solution Chem.*, 1994, **23**, 153.

134. A. Lainez, E. Wilhelm and J.-P. E. Grolier, *Monatsh. Chem.*, 1994, **125**, 877.

135. E. Jimenez, L. Romani, E. Wilhelm, G. Roux-Desgranges and J.-P. E. Grolier, *J. Chem. Thermodyn.*, 1994, **26**, 817.

136. A. Lainez, M. R. Lopez, M. Cáceres, J. Nuñez, R. G. Rubio, J.-P. E. Grolier and E. Wilhelm, *J. Chem. Soc., Faraday Trans.*, 1995, **91**, 1941.

137. E. Wilhelm, A. Inglese, A. Lainez, A. H. Roux and J.-P. E. Grolier, *Fluid Phase Equilib.*, 1995, **110**, 299.

138. E. Wilhelm, W. Egger, M. Vencour, A. H. Roux, M. Polednicek and J.-P. E. Grolier, *J. Chem. Thermodyn.*, 1998, **30**, 1509.

139. M. Pintos-Barral, R. Bravo, G. Roux-Desgranges, J.-P. E. Grolier and E. Wilhelm, *J. Chem. Thermodyn.*, 1999, **31**, 1151.

140. A. H. Roux and E. Wilhelm, *Thermochim Acta*, 2002, **391**, 129.

141. E. Calvo, P. Brocos, R. Bravo, M. Pintos, A. Amigo, A. H. Roux and G. Roux-Desgranges, *J. Chem. Eng. Data*, 1998, **43**, 105.

142. P. Brocos, E. Calvo, A. Amigo, R. Bravo, M. Pintos, A. H. Roux and G. Roux-Desgranges, *J. Chem. Eng. Data*, 1998, **43**, 112.

143. A. Lainez, E. Wilhelm, G. Roux-Desgranges and J.-P. E. Grolier, *J. Chem. Thermodyn.*, 1985, **17**, 1153.
144. G. C. Benson, M. K. Kumaran, T. Treszczanowicz, P. J. D'Arcy and C. J. Halpin, *Thermochim. Acta*, 1985, **95**, 59.
145. M. E. Saint-Victor and D. Patterson, *Fluid Phase Equilib.*, 1987, **35**, 237.
146. H. Kalali, F. Kohler and P. Svejda, *Monatsh. Chem.*, 1987, **118**, 1.
147. M. Pintos, R. Bravo, M. C. Baluja, M. I. Paz Andrade, G. Roux-Desgranges and J.-P. E. Grolier, *Can. J. Chem.*, 1988, **66**, 1179.
148. T. Takiwaga, H. Ogawa, M. Nakamura, K. Tamura and S. Murakami, *Fluid Phase Equilib.*, 1995, **110**, 267.
149. C. A. Tovar, E. Carballo, C. A. Cerdeiriña, M. I. Paz Andrade and L. Romani, *Fluid Phase Equilib.*, 1997, **136**, 223.
150. K. Nishikawa, K. Tamura and S. Murakami, *J. Chem. Thermodyn.*, 1998, **30**, 229.
151. J. M. Pardo, C. A. Tovar, C. A. Cerdeiriña, E. Carballo and L. Romani, *J. Chem. Thermodyn.*, 1999, **31**, 787.
152. J. M. Pardo, C. A. Tovar, C. A. Cerdeiriña, E. Carballo and L. Romani, *Fluid Phase Equilib.*, 2001, **179**, 151.
153. J. M. Pardo, C. A. Tovar, D. González, E. Carballo and L. Romani, *J. Chem. Eng. Data*, 2001, **46**, 212.
154. T. Takigawa and K. Tamura, *J. Therm. Anal. Calorim.*, 2002, **69**, 1075.
155. D. G. González-Salgado, J. Peleteiro, J. Troncoso, E. Carballo and L. Romani, *J. Chem. Eng. Data*, 2004, **49**, 333.
156. J. M. Pardo, C. A. Tovar, J. Troncoso, E. Carballo and L. Romani, *Thermochim. Acta*, 2005, **433**, 128.
157. M. Chorażewski and M. Tkaczyk, *J. Chem. Eng. Data*, 2006, **51**, 1825.
158. M. Chorażewski, *J. Chem. Eng. Data*, 2007, **52**, 154.
159. E. Wilhelm, A. H. Roux, G. Roux-Desgranges, M. Rodrigo, A. Lainez and J.-P. E. Grolier, *Calorim. Anal. Therm.*, 1986, **17**, 12.
160. J.-P. E. Grolier and E. Wilhelm, *Pure Appl. Chem.*, 1991, **63**, 1427.
161. S. Miyanaga, K. Tamura and S. Murakami, *J. Chem. Thermodyn.*, 1992, **24**, 1077.
162. M. Nakamura, K. Chubachi, K. Tamura and S. Murakami, *J. Chem. Thermodyn.*, 1993, **25**, 1311.
163. S. Baluja, T. Matsuo and K. Tamura, *J. Chem. Thermodyn.*, 2001, **33**, 1545.
164. (a) P. Brocos, A. Piñeiro, R. Bravo, A. Amigo, A. H. Roux and G. Roux-Desgranges, *J. Chem. Eng. Data*, 2002, **47**, 351; (b) P. Brocos, A. Piñeiro, R. Bravo, A. Amigo, A. H. Roux and G. Roux-Desgranges, *J. Chem. Eng. Data*, 2003, **48**, 712; (c) P. Brocos, A. Piñeiro, R. Bravo, A. Amigo, A. H. Roux and G. Roux-Desgranges, *J. Chem. Eng. Data*, 2003, **48**, 1055; (d) P. Brocos, A. Piñeiro, R. Bravo, A. Amigo, A. H. Roux and G. Roux-Desgranges, *J. Chem. Eng. Data*, 2004, **49**, 647.
165. F. Comelli, R. Francesconi, A. Bigi and K. Rubini, *J. Chem. Eng. Data*, 2006, **51**, 665; Corrections: *J. Chem. Eng. Data*, 2007, **52**, 318. See also K.F. Loughlin, *J. Chem. Eng. Data*, 2007, **52**, 1149.

166. F. Comelli, R. Francesconi, A. Bigi and K. Rubini, *J. Chem. Eng. Data.*, 2007, **52**, 639.

167. E. Calvo-Iglesias, R. Bravo, M. Pintos, A. Amigo, A. H. Roux and G. Roux-Desgranges, *J. Chem. Thermodyn.*, 2007, **39**, 561.

168. S. Horstmann, H. Gardeler, R. Bölts and J. Gmehling, *J. Chem. Eng. Data*, 1999, **44**, 539.

169. B. S. Bjola, M. A. Siddiqi and P. Svejda, *J. Chem. Eng. Data*, 2001, **46**, 1167.

170. M. L. McGlashan, D. Stubley and H. Watts, *J. Chem. Soc. (A)*, 1969, 673.

171. P. J. Howell and D. Stubley, *J. Chem. Soc. (A)*, 1969, 2489.

172. R. Tanaka, S. Murakami and R. Fujishiro, *J. Chem. Thermodyn.*, 1974, **6**, 209.

173. T. M. Letcher and P. K. Naicker, *J. Chem. Thermodyn.*, 2001, **33**, 1027.

174. (a) F. Kohler, *Monatsh. Chem.*, 1960, **91**, 1113; (b) F. Kohler, *Monatsh. Chem.*, 1969, **100**, 1151.

175. J. A. Pople, *Disc. Faraday Soc.*, 1953, **15**, 35.

176. J.-P. E. Grolier, A. Inglese, A. H. Roux and E. Wilhelm, in *Chemical Engineering Thermodynamics*, S.A. Newman, ed., Ann Arbor Science Publishers, Ann Arbor, 1982, pp. 483–486.

177. D. Patterson, *J. Solution Chem.*, 1994, **23**, 105.

178. F. Kohler and J. Gaube, *Polish J. Chem.*, 1980, **54**, 1987.

179. E. Wilhelm, *Thermochim. Acta*, 1990, **162**, 43.

180. J. C. Cobos, *Fluid Phase Equilib.*, 1997, **133**, 105.

181. B. D. Djordjević, M. Lj. Kijevčanin, A. Ž. Tasić and S. P. Šerbanović, *J. Serb. Chem. Soc.*, 1999, **64**, 801.

182. M. Lj. Kijevčanin, A. B. Djordjević, I. R. Grgurić, B. D. Djordjević and S. P. Šerbanović, *J. Serb. Chem. Soc.*, 2003, **68**, 35.

183. J. Troncoso, C. A. Cerdeiriña, E. Carballo and L. Romani, *Fluid Phase Equilib.*, 2005, **235**, 201.

184. A. B. Bhatia and D. E. Thornton, *Phys. Rev. B*, 1970, **5**, 3004.

185. W. H. Young, *Rep. Prog. Phys.*, 1992, **55**, 1769.

186. R. G. Rubio, M. Cáceres, R. M. Masegosa, L. Andreolli-Ball, M. Costas and D. Patterson, *Ber. Bunsenges. Phys. Chem.*, 1989, **93**, 48.

187. L. Andreolli-Ball, M. Costas, D. Patterson, R. G. Rubio, R. M. Masegosa and M. Cáceres, *Ber. Bunsenges. Phys. Chem.*, 1989, **93**, 882.

188. S. Pittois, B. Van Roie, C. Glorieux and J. Thoen, *J. Chem. Phys.*, 2005, **122**, 024504.

189. N. J. Utt, S. Y. Lehman and D. T. Jacobs, *J. Chem. Phys.*, 2007, **127**, 104505.

190. M. Campostrini, A. Pelissetto, P. Rossi and E. Vicari, *Phys. Rev. E*, 1999, **60**, 3526.

191. J. Zinn-Justin, *Phys. Rep.*, 2001, **344**, 159.

192. J.-P. E. Grolier and E. Wilhelm, *Fluid Phase Equilib.*, 1981, **6**, 283.

193. K. N. Marsh, *J. Chem. Thermodyn.*, 1977, **9**, 719.

194. G. C. Benson, P. J. D'Arcy and O. Kiyohara, *J. Solution Chem.*, 1980, **9**, 931.

195. J. B. Ott, C. E. Stouffer, G. V. Cornett, B. F. Woodfield, C. Guanquan and J. J. Christensen, *J. Chem. Thermodyn.*, 1987, **19**, 337.

196. J. A. Larkin and R. C. Pemberton, *Natl. Phys. Lab. Rep. Chem.*, 1976, 43.

197. E. Bose, *Z. Phys. Chem. A*, 1907, **58**, 585.

198. K. Tanaka, I. Fujita and M. Uematsu, *J. Chem. Thermodyn.*, 2007, **39**, 961.

199. K. Tamura, S. Tabata and S. Murakami, *J. Chem. Thermodyn.*, 1998, **30**, 1319.

200. K. Tamura, T. Sonoda and S. Murakami, *J. Solution Chem.*, 1999, **28**, 777.

201. H. Doi, K. Tamura and S. Murakami, *J. Chem. Thermodyn.*, 2000, **32**, 729.

202. K. Tamura and Y. Yamasawa, *J. Therm. Anal. Calorim.*, 2003, **73**, 143.

203. U. Würz, M. Grubić and D. Woermann, *Ber. Bunsenges. Phys. Chem.*, 1992, **96**, 1460.

204. V. Privman, P. C. Hohenberg and A. Aharony, in *Phase Transitions and Critical Phenomena, Vol.14*, C. Domb and J.L. Lebowitz, eds., Academic Press, London, 1990.

205. J.-Y. Huot, E. Battistel, R. Lumry, G. Villeneuve, J.-F. Lavallee, A. Anusiem and C. Jolicoeur, *J. Solution. Chem.*, 1988, **17**, 601.

206. Z. Nan, B. Liu and Z. Tan, *J. Chem. Thermodyn.*, 2002, **34**, 915.

207. C. Yang, P. Ma, F. Jing and D. Tang, *J. Chem. Eng. Data*, 2003, **48**, 836.

208. C. Casanova, E. Wilhelm, J.-P. E. Grolier and H. V. Kehiaian, *J. Chem. Thermodyn.*, 1981, **13**, 241.

209. J. Konicek and I. Wadsö, *Acta Chem. Scand.*, 1971, **25**, 1541.

210. M. B. Neumann, *Z. Phys. Chem. A*, 1932, **158**, 258.

211. S. S. Dhondge, C. Pandhurnekar and L. Ramesh, *J. Chem. Thermodyn.*, 2008, **40**, 1.

212. H. S. Frank and W. M. Evans, *J. Chem. Phys.*, 1945, **13**, 507.

213. E. C. W. Clarke and D. N. Glew, *Trans. Faraday Soc.*, 1966, **62**, 539.

214. P. D. Bolton, *J. Chem. Educ.*, 1970, **47**, 638.

215. E. Wilhelm, R. Battino and R. J. Wilcock, *Chem. Rev.*, 1977, **77**, 219.

216. A. Inglese and R. H. Wood, *J. Chem. Thermodyn.*, 1996, **28**, 1059.

217. L. Hnědkovský and R. H. Wood, *J. Chem. Thermodyn.*, 1997, **29**, 731.

218. M. Slavik, J. Šedlbauer, K. Ballerat-Busserolles and V. Majer, *J. Solution Chem.*, 2007, **36**, 107.

219. M. L. Origlia-Luster and E. M. Woolley, *J. Chem. Thermodyn.*, 2003, **35**, 1101.

220. Š. Hovorka, A. H. Roux, G. Roux-Desgranges and V. Dohnal, *J. Solution Chem.*, 1999, **28**, 1289.

221. Š. Hovorka, V. Dohnal, A. H. Roux and G. Roux-Desgranges, *Fluid Phase Equilib.*, 2002, **201**, 135.

222. D. Fenclová, S. Perez-Casas, M. Costas and V. Dohnal, *J. Chem. Eng. Data*, 2004, **49**, 1833.
223. H. Piekarski and A. Pietrzak, *J. Mol. Liq.*, 2005, **121**, 46.
224. S. Cabani, P. Gianni, V. Mollica and L. Lepori, *J. Solution Chem.*, 1981, **10**, 563.
225. A. V. Plyasunov and E. L. Shock, *Geochim. Cocmochim. Acta*, 2000, **64**, 439.
226. A. V. Plyasunov and E. L. Shock, *J. Chem. Eng. Data*, 2001, **46**, 1016.
227. A. V. Plyasunov, N. V. Plyasunova and E. L. Shock, *J. Chem. Eng. Data*, 2006, **51**, 276.
228. P. Walden, *Bull. Acad. Imper. Sci. (St. Petersburg)*, 1914, 405.
229. T. Welton, *Chem. Rev.*, 1999, **99**, 2071.
230. M. J. Earle and K. R. Seddon, *Pure Appl. Chem.*, 2000, **72**, 1391.
231. K. N. Marsh, J. A. Boxall and R. Lichtenthaler, *Fluid Phase Equilib.*, 2004, **219**, 93.
232. A. Heintz, *J. Chem. Thermodyn.*, 2005, **37**, 525.
233. J. D. Holbrey, W. M. Reichert, R. G. Reddy and R. D. Rogers, in *Ionic Liquids as Green Solvents*, K.R. Seddon and R.D. Rogers, eds., ACS Symposium Series 856, American Chemical Society, Washington, D.C., 2003, pp. 121–133.
234. C. P. Fredlake, J. M. Crosthwaite, D. G. Hert, S. N. V. K. Aki and J. F. Brennecke, *J. Chem. Eng. Data*, 2004, **49**, 954.
235. K.-S. Kim, B.-K. Shin, H. Lee and F. Ziegler, *Fluid Phase Equilib.*, 2004, **218**, 215.
236. M. E. Van Valkenburg, R. L. Vaughn, M. Williams and J. S. Wilkes, *Thermochim. Acta*, 2005, **425**, 181.
237. D. Waliszewski, I. Stepniak, H. Piekarski and A. Lewandowski, *Thermochim. Acta*, 2005, **433**, 149.
238. J. M. Crosthwaite, M. J. Muldoon, J. K. Dixon, J. L. Anderson and J. F. Brennecke, *J. Chem. Thermodyn.*, 2005, **37**, 559.
239. A. Diedrichs and J. Gmehling, *Fluid Phase Equilib.*, 2006, **244**, 68.
240. Z.-H. Zhang, Z.-C. Tan, L.-X. Sun, Y. Jia-zhen, X.-C. Lv and Q. Shi, *Thermochim. Acta*, 2006, **447**, 141.
241. O. Yamamuro, Y. Minamimoto, Y. Inamura, S. Hayashi and H. Hamaguchi, *Chem. Phys. Lett.*, 2006, **423**, 371.
242. I. Bandrés, B. Giner, H. Artigas, C. Lafuente and F. M. Royo, *J. Chem. Eng. Data*, 2009, **54**, 236.
243. R. Gomes de Azevedo, J. M. S. S. Esperança, V. Najdanovic-Visak, Z. P. Visak, H. J. R. Guedes, M. Nunes da Ponte and L. P. N. Rebelo, *J. Chem. Eng. Data*, 2005, **50**, 997.
244. R. Gomes de Azevedo, J. M. S. S. Esperança, J. Szydlowski, Z. P. Visak, P. F. Pires, H. J. R. Guedes and L. P. N. Rebelo, *J. Chem. Thermodyn.*, 2005, **37**, 888.
245. L. A. Davis and R. B. Gordon, *J. Chem. Phys.*, 1967, **46**, 2650.
246. T. F. Sun, S. N. Biswas, N. J. Trappeniers and C. A. Ten Seldam, *J. Chem. Eng. Data*, 1988, **33**, 395.

247. D. Waliszewski, *J. Chem. Thermodyn.*, 2008, **40**, 203.
248. G. García-Miaja, J. Troncoso and L. Romani, *Fluid Phase Equilib.*, 2008, **274**, 59.
249. M. Costas, M. Carceres Alonso and A. Heintz, *Ber. Bunsenges. Phys. Chem.*, 1987, **91**, 184.
250. H. Funke, M. Wetzel and A. Heintz, *Pure Appl. Chem.*, 1989, **61**, 1429.
251. L. P. N. Rebelo, V. Najdanovic-Visak, Z. P. Visak, M. Nunes da Ponte, J. Szydlowski, C. A. Cerdeiriña, J. Troncoso, L. Romani, J. M. S. S. Esperanca, H. J. R. Guedes and H. C. de Sousa, *Green Chem.*, 2004, **6**, 369.
252. A. A. Strechan, Y. U. Paulechka, A. G. Kabo, A. V. Blokhin and G. J. Kabo, *J. Chem. Eng. Data*, 2007, **52**, 1791.
253. G. García-Miaja. J. Troncoso and L. Romani, *J. Chem. Eng. Data*, 2007, **52**, 2261.
254. L. E. Ficke, H. Rodríguez and J. F. Brennecke, *J. Chem. Eng. Data*, 2008, **53**, 2112.
255. G. García-Miaja. J. Troncoso and L. Romani, *J. Chem. Thermodyn.*, 2009, **41**, 161.
256. G. García-Miaja. J. Troncoso and L. Romani, *J. Chem. Thermodyn.*, 2009, **41**, 334.
257. P. Nockemann, K. Binnemans, B. Thijs, T. N. Parac-Vogt, K. Merz, A.-V. Mudring, P. C. Menon, R. N. Rajesh, G. Cordoyiannis, J. Thoen, J. Leys and C. Glorieux, *J. Phys. Chem. B.*, 2009, **113**, 1429.
258. J. Thoen, in *Physical Properties of Liquid Crystals*, D. Demus, J. Goodby, G. Gray, H.-W. Spiess and V. Vill, eds., Wiley-VCH, Weinheim, 1997, pp. 208–232.
259. A. Heintz, J. K. Lehmann and C. Wertz, *J. Chem. Eng. Data*, 2003, **48**, 472.
260. M. Wagner, O. Stanga and W. Schröer, *Phys. Chem. Chem. Phys.*, 2003, **5**, 3943.
261. J. M. Crosthwaite, S. N. V. K. Aki, E. J. Maginn and J. F. Brennecke, *J. Phys. Chem. B*, 2004, **108**, 5113.
262. U. Domańska and A. Marciniak, *J. Chem. Thermodyn.*, 2005, **37**, 577.
263. M. E. Fisher, *J. Stat. Phys.*, 1994, **75**, 1.
264. G. Stell, *J. Stat. Phys.*, 1995, **78**, 197.
265. H. Weingärtner and W. Schröer, *Adv. Chem. Phys.*, 2001, **116**, 1.
266. W. Schröer and H. Weingärtner, *Pure Appl. Chem.*, 2004, **76**, 19.

CHAPTER 5

Heat Capacity of Non-electrolyte Solutions

AMR HENNI

Department of Industrial Systems Engineering, University of Regina, Saskatchewan, S4S 0A2, Canada

5.1 Introduction

Accurate measurements of heat capacities are needed in many areas of physics, chemistry, and chemical engineering for establishing energy balances, obtaining entropy and enthalpy values, or for studying phase transitions. The study of heat capacity is important from both an applied (calculate thermal regimes of reactors and choice heat carriers for examples) and from basic viewpoints (in view of the role of the heat capacity in solution thermodynamics). The improved sensitivity and accuracy of the calorimeters have provided a wealth of heat capacity (C_p) data. The estimation of second-derivative property (C_p) is one of the most demanding tests in checking the performance limits of a thermodynamic model or an equation.

The knowledge of the molar isobaric heat capacity, C_p, of liquid mixtures as a function of temperature and/or composition is a source of important information on the molecular structure. Weak interactions responsible for the aggregation of molecules or the formation of intermolecular complexes in the liquid phase are often examined through calorimetric techniques. However, in some cases the dependence of heat capacities on composition appears much too complicated to be analyzed directly without the concomitant study of volumes or enthalpies. A careful analysis of the experimental data is always necessary in order to show structural changes. This analysis must be supported by

Heat Capacities: Liquids, Solutions and Vapours
Edited by Emmerich Wilhelm and Trevor M. Letcher
© The Royal Society of Chemistry 2010
Published by the Royal Society of Chemistry, www.rsc.org

phenomenological models; thus the relevant thermodynamic quantities characterizing the structural changes can be determined from the knowledge of the laws governing their concentration dependence.[1] Infinite dilution partial molar properties and standard hydration quantities (Gibbs energies, enthalpies, heat capacities) provide valuable information on the water–organic solute interactions, and are very important for theoretical studies in solution chemistry, e.g., those concerning water structure, conformation of molecules, and solution effects in chemical reaction equilibria and kinetics.

This chapter reports on the measurements of heat capacity of non-electrolyte solutions as published in the literature. A number of studies dealing with linear, branched, cyclic, and aromatic solvents including carbon, halogen, oxygen, nitrogen, and sulfur groups are reported as published. The thermodynamic fundamentals of heat capacity and some experimental techniques used are only briefly discussed as they are described in a comprehensive manner in other chapters of this book. The wide variety of non-electrolyte mixtures of current interest makes the experimental study of their thermodynamic properties an enormous task, so developing theoretical models for their prediction is of paramount significance. In this chapter, as a result of space limitations, only selected systems (mainly binary) and a couple of theoretical models are very briefly presented.

5.2   Fundamentals

5.2.1   Isobaric and Isochoric Heat Capacity

The response of the internal energy to an isochoric change in temperature (T), and that of the enthalpy to an isobaric change in T define the isochoric and isobaric heat capacities:

$$C_v = \left(\frac{\partial U}{\partial T}\right)_{V,N} \tag{1}$$

$$C_p = \left(\frac{\partial H}{\partial T}\right)_{p,N} \tag{2}$$

Heat capacities are sensitive to changes in T, and generally increase with increasing T. Except near the gas–liquid critical point, they are weak functions of p and V. Applying the above definitions of C_V and C_p to the fundamental equations $dU = TdS - PdV$ and $dH = TdS + Vdp$, respectively, we obtain the following expressions for the response of entropy to changes in temperature:

$$\left(\frac{\partial S}{\partial T}\right)_{V,N} = \frac{C_V}{T} \tag{3}$$

$$\left(\frac{\partial S}{\partial T}\right)_{P,N} = \frac{C_p}{T} \tag{4}$$

Since C_p and C_V are positive, S must always increase with both isometric and isobaric increases in T.

C_V and C_p are second temperature derivatives of Helmholtz free energy (A) and Gibbs free energy (G), which are established using the following fundamental relations:

$$dA = -SdT - PdV \text{ and } dG = -SdT + Vdp \tag{5}$$

$$C_V = -T\left(\frac{\partial^2 A}{\partial T^2}\right)_{V,N} \tag{6}$$

$$C_p = -T\left(\frac{\partial^2 G}{\partial T^2}\right)_{P,N} \tag{7}$$

$$\left(\frac{\partial C_p(p,T)}{\partial p}\right)_T = -T\left(\frac{\partial^2 V(p,T)}{\partial T^2}\right)_P \tag{8}$$

$$C_p - C_V = \frac{VT\alpha^2}{\beta} \tag{9}$$

The thermal expansivity and isothermal compressibility[†] are defined as:

$$\alpha = \frac{1}{V}\left(\frac{\partial V}{\partial T}\right)_P \tag{10}$$

$$\beta = -\frac{1}{V}\left(\frac{\partial V}{\partial p}\right)_T \tag{11}$$

5.2.2 Partial Molar Heat Capacity

The partial molar heat capacity ($C_{p,2}$) of solute 2 in a binary mixture is defined by the following equation:

$$C_{p,2} = \left(\frac{\partial C_p}{\partial n_2}\right)_{T,p,n_1} \tag{12}$$

Using the definition of C_p, the partial molar heat capacity of component 2 can be related to partial molar enthalpy of component 2 in the following way:

$$C_{p,2} = \left(\frac{\partial C_p}{\partial n_2}\right)_{T,p,n_1} = \left[\frac{\partial}{\partial n_2}\left(\frac{\partial H}{\partial T}\right)_P\right]_{T,p,n_1} = \left[\frac{\partial}{\partial T}\left(\frac{\partial H}{\partial n_2}\right)_{p,T,n_1}\right]_P = \left[\frac{\partial H_2}{\partial T}\right]_p \tag{13}$$

[†] In this chapter the isothermal compressibility is represented by the symbol β_T and not by κ_T as was recently recommended by IUPAC.

where H_2 is the partial molar enthalpy of the solute (component 2). The partial molar heat capacity is therefore the third derivative of Gibbs free energy:

$$C_{p,2} = \left(\frac{\partial C_p}{\partial n_2}\right)_{T,P,n_1} = -T\left(\frac{\partial^3 G}{\partial n_2 \partial T^2}\right)_{P,n_1} \tag{14}$$

5.2.3 Apparent Molar Heat Capacity

The apparent molar heat capacity $(C_{\varphi,2})$ of solute 2 in a binary mixture is:

$$C_{\varphi,2} = M_2 c_p + \left(\frac{c_p - c_{p,1}}{m_2}\right) \tag{15}$$

The partial molar heat capacities $(C_{p,2})$ of solute 2 in a binary mixture is:

$$C_{p,2} = C_{\varphi,2} + m_2\left(\frac{\partial C_{\varphi,2}}{\partial m_2}\right)_{T,p} \tag{16}$$

where M_2 is the solute molecular weight; m_2 is the molality of solution; c_p and $c_{p,1}$ are the specific heat capacities of the solution and pure solvent.

In terms of mole fraction the above equation becomes (for a binary mixture):

$$C_{p,2} = C_{\varphi,2} + x_2 x_1 \left(\frac{\partial C_{\varphi,2}}{\partial x_1}\right)_{T,p} \tag{17}$$

where x_1 is the mole fraction of the solvent.

Excess molar functions are often used to describe the thermodynamic properties of two-component liquid mixtures. The sign and shape of these functions provide information about the intermolecular interactions and the structure of solutions.[4] The excess molar heat capacity is defined as:

$$C_p^E = C_p(T,p,x) - C_p^{id}(T,p,x) \tag{18}$$

$$\text{with } C_p^{id}(T,p,x) = \sum_{i=1}^{n} x_i C_{p,\text{pure } i}(T,p) \tag{19}$$

5.3 Experimental Arrangements

In most experimental techniques in calorimetry, a sample under investigation is heated or cooled and the amount of exchanged heat and the corresponding temperature change are measured. When the process is carried out under constant pressure, the isobaric heat capacity C_p is obtained:

$$C_p = T\left(\frac{\partial S}{\partial T}\right)_p = \left(\frac{\partial H}{\partial T}\right)_p \tag{20}$$

Three calorimetric techniques are used to measure heat capacities namely: heat compensating, heat accumulating and heat flux exchange technique.

5.3.1 Heat Compensating Techniques

The principle of heat compensating techniques is the determination of energy or power required for compensating the heat flow rate through a sample. There are two kind of calorimeters based on this principle: isoperibol[2] and Dewar vessel[3,4] calorimeters.

5.3.2 Heat Accumulating Techniques

The principle of heat accumulating techniques is based on the measurement of the temperature change produced by the heat applied on the sample. Adiabatic,[5–9] bomb[10] and drop[11,12] calorimeters are based on this principle.

5.3.3 Heat Flux Exchange Techniques

The principle of the heat flux exchange technique is based on the measurement of the difference of temperature between the specimen and its surroundings produced by the heat. Differential scanning,[13–28] Calvet-Tian type,[29] and flow[30–41] calorimeters are based on this principle.

5.4 Heat Capacities of Aqueous Non-electrolytes

The strong intermolecular forces between water molecules produce an extensive three-dimensional network of hydrogen bonds and gives water distinctive properties. The small size of the water molecule, its 1:1 ratio of donor to acceptor sites, and its high dipole moment make H-bonding in water a highly cooperative phenomenon. The presence of non-ionic compounds in aqueous solutions in most cases results in significant changes to the long range and local water structure. Large thermodynamic effects occur in the presence of hydrophobic solutes.[30]

Heat capacities of aqueous mixtures are influenced by the geometry of the dissolved molecules, the nature of the polar groups in them, the hydrophobic character of the interaction of the molecules in the solution, and the proximity of the critical point of separation into layers.[42]

5.4.1 Heat Capacities of Aqueous Hydrocarbons

Hydrocarbons molecules in water stick together to minimize the area exposed to the water. This effect is called the hydrophobic effect, which is an entropy-driven process that seeks to minimize the free energy of a system by minimizing the interfacial surface between the hydrophobic molecules and water. Water around the hydrophobic surfaces adopts a more ordered and tightly bound set of hydrogen bonds by reducing conformational freedom and inducing iceberg formation, as reported by Gill and Wadsö. These well-formed hydrogen bonds are the reason for the high heat capacity associated with the hydrophobic effect. The size and shape of non-polar molecules and the nature of their interactions with water are related to the number of hydrogen atoms in the molecule.[43]

Partial heat capacities at infinite dilution of hydrocarbons for homologous series are linearly related to the number of CH_2 groups (n) and characteristic substituent group (a) as: $C_{p,2}^{\infty} = a + bn$, where $n \leq 6$.[44,45] An equation of state for the heat capacity in terms of the Random Network (RN) model structural parameters with contributions from H-bond bending, stretching, vibration terms is available. The heat capacities of hydration of the first four linear alkane series, methane, ethane, propane, and butane, were calculated by a combination of Monte Carlo simulations and the Random Network model of water.[46] The results show that the contribution from the water–water interaction and the solute–water interaction accounted for 45 % and 20 % of the experimental data, respectively.[47]

5.4.2 Heat capacities of Aqueous Ethers

Figure 5.1 shows the variation of c_P with respect to mole fraction of ethers and water as published in a study dealing with aqueous oligo (ethylene glycol) monomethyl ethers, C_1E_nOH, and dimethyl ethers, $C_1E_nC_1$, of various degrees of oligomerization. At high dilution the R-O-ROH moiety can be incorporated into the water structure almost without thermodynamic excess effects. At higher concentrations, the structure of water is greatly disturbed and association of the ethers becomes more important.[30]

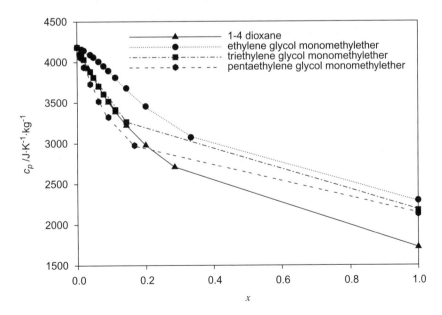

Figure 5.1 Specific heat capacity, c_p, of aqueous ethers at 298 K as a function of mole fraction, x of ethers or water.[30]

The thermodynamic behavior of ethers classifies them as structure breakers. The ether oxygen, owing to its inability to form cooperative hydrogen bonds with water, promotes in its neighborhood, regions of unbonded and very expansible water. When two ether oxygens are present, the structure breaking ability is greatly enhanced.[48] Cabani *et al.* concluded that in the neighborhood of the solute molecule, water possesses features that are not easily explainable in terms of limiting states (two-state model), having characteristics which can be evaluated from the properties of pure water.[49]

5.4.3 Heat Capacities of Aqueous Alkanolamines

Heat capacity data for aqueous alkanolamine solutions are needed for the design of heat exchangers, absorbers, and regenerators used in gas-treating processes. Figure 5.2 gives the trend of the molar heat capacities for different aqueous alkanolamines. The exothermic mixing process of amine and water indicates that the amine–water interactions are strong compared to the water–water and amine–amine interactions and are the main cause of the non-ideal behavior in these mixtures. The high values of enthalpies of mixing can lead to large increases in the temperature of the solution.[3] C_p value increases as the temperature and mole fraction of the amine increase. CH_2 and OH groups contribute to the increase in molar heat capacity with increasing temperature. The NH group exhibits its contribution to the molar heat capacity with a very

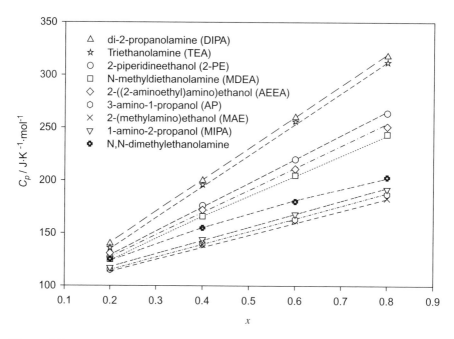

Figure 5.2   Molar heat capacity, C_p, of aqueous alkanolamine solutions at 323K, as a function of mole fraction, x of water.[25,34]

large temperature dependency, while the contribution of the N group is almost zero.[29]

The Treszczanowicz-Kehiaian (TK) model relates the apparent molar heat capacity of aqueous amine (diethylamine) solution (a proton-acceptor solvent), to the enthalpy change, equilibrium constants, ratio of molar volumes, and volume fraction of components. The formation of stronger hydrogen bond between water and amine favors the formation of complexes over a rapid increase of self-associated water species.[36]

5.4.4 Heat Capacities of Aqueous Dimethyl Sulfoxide, Acetone, and Acetamide

Figure 5.3 shows the concentration dependence of the apparent heat capacity of some non-electrolytes, with similar structures, in water. The apparent heat capacity value rapidly decreases to the intrinsic value (approximately equal to that of the pure substance) with acetone, dimethyl sulfoxide (DMSO), and acetamide. For Kiyohara *et al.*, this change is very small in the case of acetamide but significant in acetone. Even with acetone, there is little evidence of any structural hydrophobic–hydrophobic interactions. With strongly hydrophobic solutes, these hydrophobic interactions produce a positive contribution to the excess heat capacity, which opposes the decrease of the apparent heat capacity towards the intrinsic values; the net effect being a maximum or a positive hump in the apparent heat capacity. These solvents do not present such a behavior.[51] The effect on water of these solutes are comparable to those of dioxane, morpholine, and piperazine.[50] The data appears to show that what is

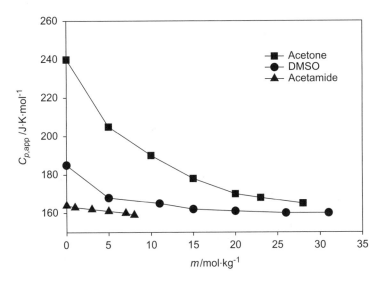

Figure 5.3 Apparent molar heat capacities, $C_{p,app}$, of aqueous dimethyl sulfoxide (DSMO), acetone and acetamide at 298 K as a function of molality, m.[51]

causing the high heat capacity at infinite dilution disappears at higher concentration. It is therefore concluded that acetamide, DMSO, and acetone are slightly hydrophobic in water, but the overall structural change is not large, especially in the case of acetamide. The increasing hydrophobic character from urea to acetone comes from the substitution of a hydrophilic NH_2 group by the hydrophobic CH_3 group. The smaller hydrophobic character of DMSO, compared to acetone, is a result of the higher dipole moment.[51]

5.4.5 Heat Capacities of Aqueous Amides

Usually miscible in water, amides cover a very wide range of dielectric constants. They are often used in studies of the interrelations between the solubility of drugs and the dielectric constants of the pure compounds and of their aqueous mixtures. Because of the lack of hydrogen bonding in the pure solvent, they are used to study the influence of solvent structure on the enthalpies of solution, and solvation of a third component as mentioned by de Visser *et al.* The interactions with the third compound can be examined in a systematic manner since it is possible to go from pure water to pure amide, *i.e.* from a very structured solvent to a solvent where specific structural effects are absent.[38] For de Visser *et al.*, the trends found in the apparent and partial molal heat capacities indicate that there is a specific interaction between the solute and water. By adding the amide, there is a possibility of hydrogen bonding between water molecules and the carbonyl group of the amide molecules. Hydrogen bonds between water and carbonyl oxygen are believed to be stronger than between water molecules.[38]

5.4.6 Aqueous Solutions of Hydroxy Compounds (mono-ols, glycols, polyols)

The properties of aqueous solutions of hydroxy compounds are of interest in many fields of enquiry. Among organic compounds, hydroxy derivatives stand out in virtue of their high solubility in water, and their thermodynamic properties in solutions often show abnormalities such as viscosity-composition maxima, or negative relative partial molar volumes.[52] In a comprehensive review of research dealing with these systems, Franks and Ives write that for monohydric alcohols in dilute aqueous solution, such peculiarities can be attributed to the bifunctional nature of the solute molecules. The hydrophobic hydrocarbon group appears to resist the pull into the solution exerted by the hydrophilic hydroxyl group, which, either as proton donor or acceptor, can hydrogen bond with the solvent molecules. A second hydroxyl group in the solute molecule (glycols) shifts the balance of competing influences in favor of "aqueous behavior" and the anomalies become less marked. Further hydroxylation, like in the case of sugars, eventually removes any anomaly making them among the most normal of solutes in water.[52]

More than any other property, heat capacity should be sensitive to the complexity of alcohol–water interactions, but unfortunately there is a scarcity of reliable data. Most of the alcohol systems fall in a class of aqueous solutions. The excess free energy G^E is positive, $(TS^E) > H^E$ (entropy-controlled mixing), the solution tends to unmix at high temperatures and the concentration dependence of the thermodynamic properties shows characteristic trends. The hydrophobic character of the co-solvent seems to be responsible for the general behavior of these solutions.[53] On the basis of intermolecular interactions; water is more similar to methanol than it is to any of the other alcohols. Thus, the properties of the water–methanol system are simpler and more regular than they are for the other systems. It was found that as the number of carbon atoms in the alcohol molecule increases, the solubility in water decreases. While methanol, ethanol, and the two propanols are all completely miscible with water, 1-butanol and the higher alcohols are only partially miscible. The rapid changes in the slopes of thermodynamic properties with respect to the mole fraction, for 1-propanol and for iso-propanol, in the range of (0.1–0.2) mole fraction, reflects the close approach to the formation of two phases in these systems.[93]

It is established that except for methanol, C_p for alcohol + water mixtures increases monotonically with alcohol mole fraction x. In dilute aqueous alcohol solutions, the variation of C_p with temperature is small and for the aqueous propyl alcohol (PrOH) system it passes through a shallow minimum near 313 K. It is also noteworthy that the dependence of C_p changes greatly with temperature at $x > 0.2$ but not $x < 0.2$ which remain essentially water-like in behavior; perhaps because of the resistance (up to a point) of the water structure to disruption. The MeOH + H$_2$O system, however, shows the peculiarity of C_p isotherms with maxima shifting to higher values with the rise of temperature. Franks and Ives[52] suggest that there is evidence that MeOH < EtOH < PrOH < ButOH is the order of increasing proton accepting ability.

Specific heat capacities, c_p, of (water + methanol) were measured from 323 K to 513 K at pressures from 2 MPa to 10 MPa. A steep rise in the isobaric heat capacity was observed when the experimental condition approach the critical temperature and pressure of pure methanol as the mixture gradually becomes richer in methanol.[53] Apparent molal heat capacities of aqueous tert-butanol are reported by de Visser *et al.*[54] The temperature dependence of the apparent molal heat capacity of tert-butanol is rather characteristic of hydrophobic solutes in water. At high concentration of solute, the apparent heat capacity data resemble those of surfactants, suggesting the existence of some microphasic separation. At low temperature and concentration, a significant hump occurs in the apparent heat capacity as a function of concentration (Figure 5.4), which according to de Vissier *et al.* is probably not due to the presence of OH group on the solute. Furthermore de Visser *et al.* noticed that many apparently hydrophobic aqueous non-electrolytes (ethers, esters, ketones, and amides) show a large initial decrease in the apparent heat capacity with concentration at 298 K in contrast to alcohols, carboxylic acids, and amines. They conclude that

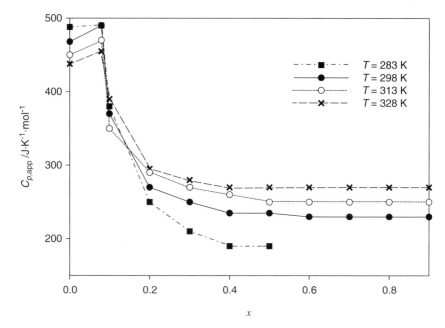

Figure 5.4   Apparent molar heat capacities, $c_{p,\mathrm{app}}$, of aqueous tert-butyl alcohol as a function of mole fraction, x of water, and temperature, T.[54]

alcohols may therefore not be typical of hydrophobic solutes as far as heat capacities are concerned.[54]

Origila-Luster *et al.*[55] studied the apparent molar heat capacity of dilute aqueous solution of ethanol, propan-1-ol, and propan-2-ol at temperatures from (278 to 393) K and at 0.35 kPa. In addition to the above mentioned solvents, 14 other solvents were studied (heptane, 2-butanone, cyclopentanone, cyclohexanone, 2,4-pentanedione, methyl acetate, methyl acetoacetate, ethyl acetoacetate, 2-methoxyethyl acetate, 2-ethoxyethyl acetate, methylmethoxy acetate, acetonitrile, propionitrile, and acrylonitrile). The results were extrapolated to determine the molal heat capacity at infinite dilution. The application of group contribution rules to the limiting partial molar heat capacities has been tested using the original method and parameters proposed by Cabani *et al.*[57] which were found to be somewhat satisfactory. Cabani *et al.*[57] have compiled limiting partial molar heat capacities for many alcohols in water at 298 K, and developed a group contribution method to predict many thermodynamic properties including the hydration heat capacities and limiting partial molar properties.

Inglese and Wood[58] reported the heat capacity of aqueous solutions of 1-propanol, butane-1,4-diol and hexane-1,6-diol between (300 to 525) K and 28 MPa. (Figure 5.5)

Pagé *et al.*[59] investigated and compared results between two systems: aqueous ethylene glycol (EG) and aqueous 2-methoxy ethanol (ME). As seen in

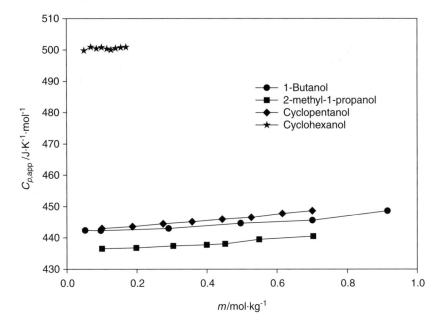

Figure 5.5 Apparent molar heat of capacity, $C_{p,\,app}$, of alcohol in water at 298 K as a function of molality, m of water.[56]

Figure 5.6 for water + ME, $C_p(ME)$ decreases sharply in the region and goes through a shallow minimum at low temperature. A very similar minimum was also observed with EG and thus cannot be assigned to effects associated with the greater hydrophobicity of ME relative to EG.

In contrast to C_p results, however, significant differences in C_v values of ME and EG are apparent only in a certain range. Beyond $x = 0.1$ the difference between $C_v(ME)$ and $C_v(EG)$ is less than the difference for the pure liquids. Here x refers to water. The behavior of C_p and particularly C_v thus provide support to the concept of enhanced enthalpy (or internal energy) fluctuations due to the solvation of the hydrophobic group (ME as opposed to EG); such fluctuations barely exist at $x > 0.1$.[59] Heat capacities of the aqueous system 2-butoxyethahol at 338 K and 348 K are reported by Wojtczakp *et al.*,[60] and the aggregation phenomena in this system was described using the two-point scaling theory.

Heat capacities of another glycol (1,4-butanediol) with water were measured over the entire mole fraction range over a temperature range (293 to 353) K.[61] With input from excess volume and viscosity deviation studies, the following effects were considered to be present: (a) expansion due to disruption of hydrogen bond in water and 1,4-butanediol, (b) contraction due to free volume difference of unlike molecules, and (c) contraction due to mutual association through hydrogen bond formation between water and 1,4-butanediol.

Homologous series of non-ionic surfactants being monoalkyl derivatives of polyoxyethylene glycol with a general formula {$C_nH_{2n+1}(OCH_2CH_2)mOH$} and

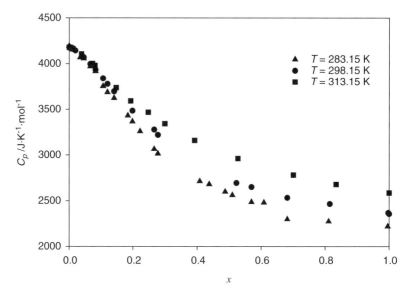

Figure 5.6   Molar heat capacity, C_p of aqueous 2-methoxy ethanol as a function of mole fraction, x, and temperature, T.[59]

abbreviated as C_nE_m, have frequently been studied in consideration of their numerous applications in industry and basic research. In aqueous solutions of these compounds, a typical miscibility gap as well as micro heterogeneous areas frequently appear and are due to the formation of molecular and micellar aggregates.[70] Piekarski *et al.* have studied aqueous systems of C_6E_2,[62] C_6E_5,[63] and C_6E_4.[18] The curves of C_p for $x_2 > 0.01$ have a similar character. At temperatures from 318 to 338 K, the curves show a monotonic drop in the values of the specific heat capacities with the increase in the amphiphile content within the whole range of mixture composition.

Piekarski and Tkaczyk explained that significant differences in the shape of C_p for temperature from 283 to 308 K, can be observed only in dilute solutions. The specific heat capacities initially decrease slightly, while within the composition range: $0.002 \leq x_2 \leq 0.0035$ one can observe their increase up to maximum value (Figure 5.7).

A decrease in temperature causes the maximum to grow and appear in the solution with a higher content of the amphiphile. Piekarski and Tkaczyk believe that in dilute solutions and at temperatures lower than 318 K, the increase in the C_6E_5 content in solution considerably changes the structure of the solution. At higher temperatures, these changes are insignificant, and result from the gradual addition of amphiphile molecules to already existing aggregates.[63]

For Hajji, some polyols show micellar behavior in aqueous solutions at concentrations greater than the critical micellar concentration (cmc). 1,2-alkanediols ($C_nH_{2n+2}O_2$ with $n = 5, 6, 7$), 1, 2, 3-alkanetriols ($C_nH_{2n+2}O_3$ with $n = 7, 8, 9$), and the geminated alkanetriols ($C_nH_{2n+2}O_3$ with $n = 8$ and 9) are

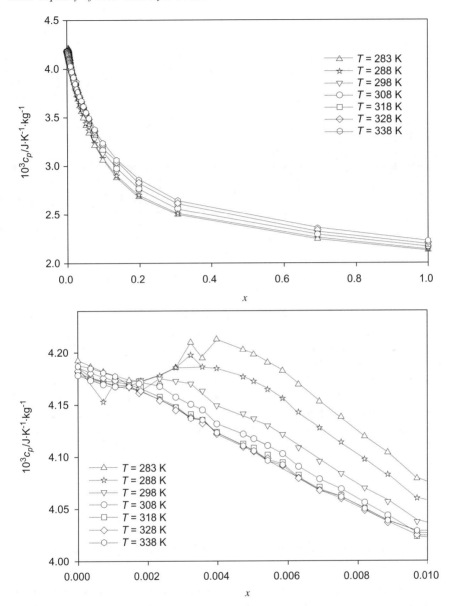

Figure 5.7 Specific heat capacity, c_p, of aqueous solutions of C_6E_5 as a function of mole fraction of water, x, and temperature, T.[63]

investigated by micro calorimetric techniques (Figure 5.8). Hajji presented the variations of the apparent molal volume with molality that clearly show the transition from the formation of true micelles, micelle-like aggregates, simple association, and self-association.[40] The critical micellar concentrations are determined with specific heat capacity measurements. The passage from the

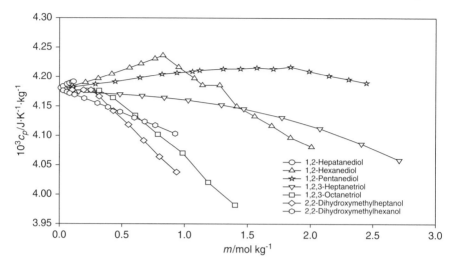

Figure 5.8   Specific heat capacity, c_p, of aqueous polyols as function of molality, m.[40]

dispersed environment to the organized environment gives a constant quantity of specific heat capacity (about $50\,J\,K^{-1}\,mol^{-1}$) only for the 1,2-hexanediol ($n = 6$), 1,2,3-octanetriol ($n = 8$) and 2,2-dihydroxymethyl-1-heptanol ($n = 9$) which form true micelles.[40]

Some aqueous alcohol solutions form azeotropes and the heat capacity of some azeotropic solutions of aqueous propan-1-ol,[64] aqueous propan-1-ol + benzene,[65] aqueous ethanol + benzene,[66] aqueous ethanol + toluene[67] have been reported. The heat capacities of ternary mixtures of aqueous ethanol + 1,2-ethanediol have also been reported. The mixtures exhibited complex phase transitions in the solid state.[68]

5.5   Heat Capacities of Non-electrolyte Mixtures

5.5.1   Heat Capacities of Mixtures of Ethers and Non-polar Solvents

Heat capacities of ether mixtures have weaker temperature dependence than those of alkanes.[69] Changes in C_p with temperature suggest that hydrophobic hydration is largest near the freezing temperature of water. Anomalies in the behavior of C_p values occur mainly in the water-rich region and increase with temperature. Positive deviation at higher temperature is due to anomalous energy fluctuations near the two-phase region.

There is no increase in anomaly near the lower critical solution temperature.[42] Tanaka and Saito showed that the values of the apparent molar heat capacity at infinite dilution are in linear relationship with n, but experience a jump due to a conformational change in the monomers between

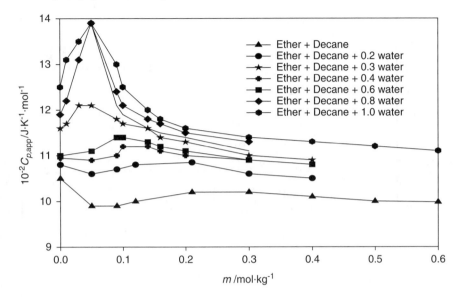

Figure 5.9 Apparent molar heat capacities, $C_{p,\text{app}}$ of polyoxyethylene glycol mon-decyl ether in decane and water as a function of molality of water, m.[31]

$n = 3$ and 4. It is interesting that the values are independent of the solvents. The conclusion drawn for the addition of water in polyoxyethylene glycol monodecyl ether (PEGMDE) in decane, heptanes (previous study) and cyclohexane depended on the values of n, m (molality) and the amount of water added (r).[31]

Tanaka and Saito explain that the emulsions become transparent, and swollen micelles are formed at higher molality where apparent heat capacity reaches a maximum. Very sharp peak suggests that the transition between the emulsion and the micellar phase is of a first-order type (Figure 5.9).[31] Tanaka and Saito found that the formation of aggregates is enhanced as the amount of water is increased, and that the maximum becomes sharper by increasing the amount of water. This shows that the aggregation is a pseudo-transition of second-order type. Further addition of water makes the peak sharper (first-order type). Transparent and swollen micelles are formed at higher molality where the apparent heat capacity reaches a maximum.

5.5.2 Heat Capacities of Hydrocarbon Mixtures

Heat capacities of hydrocarbon mixtures show more temperature dependence than those of ethers.[69] Information about the heat capacities of hydrocarbon mixtures are useful to examine the effects of charge and non-electrostatic interactions with water at high temperatures.[33] Heat capacities of p-xylene + n-decane,[23] benzene + acetone,[32] pentane + acetone,[37] benzene + n-alkanes ($n = 6$. 8, 10, 12, 14, 16),[69] toluene + n-alkanes,[69] p-xylene + n-alkanes ($n = 6$, 16)[69]

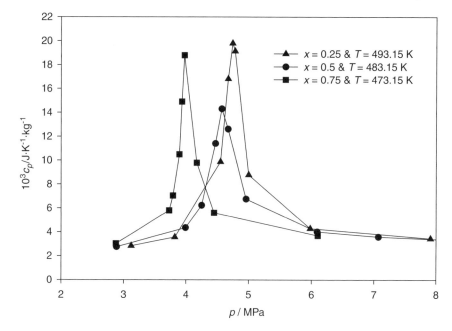

Figure 5.10   Specific heat capacities, c_p, of pentane and acetone mixtures as a function of mole fraction of water, x, pressure, p, and temperature, T.[33]

and benzene + heptane[69] systems are available in the literature. Figure 5.10 shows the specific heat capacity variation of heptane + acetone mixture with respect to pressure, temperature, and mole fraction. In sub-critical and super-critical regions, the isobaric heat capacities of binary systems of hydrocarbons shows large variations.[33] In other polar + alkane systems, molar specific heat capacity systems presents a small departure from ideality as long as the mixture is far away from its liquid-liquid critical point (x_c, T_c), which in many cases is found near room temperature. The molar heat capacity presents a strong enhancement, a fact which is reflected in the highly temperature-dependent W-shaped curves of the excess isobaric molar heat capacities. On the other hand, isobaric molar heat capacities for aromatic hydrocarbon + alkane mixtures often show small and negative deviations from ideality. Longer alkyl chain hydrocarbons have higher dispersive interactions between molecules, which imply that the mixing process breaks the more structured liquid, resulting in a higher contribution to excess properties. It is therefore suggested that when the temperature is increased, the structure of both liquids is easily broken by the mixing process.[17] Large positive excess enthalpies occur when the main mixing effect is the breaking of dipole-dipole interactions.[1]

   C_p values of hydrocarbon mixtures are calculated using residual heat capacities from equation of states (EOS) such as Reid *et al.*, Peng-Robinson, Patel-Teja, and Iwai *et al.* and others. Fitting a binary interaction parameter to the data could improve the C_p representation in the critical region by translating the pressure that corresponds to the C_p maximum. EOSs cannot predict the sharp increase in enthalpy with increasing temperature for mixtures.[33]

5.5.3 Heat Capacities of Alkanol and Alkane Mixtures

Alkanol + alkane mixtures are thermodynamically non-ideal due to changes in molecular interactions that accompany mixing. Different self-associations of alkanol through hydrogen bonding in the mixtures, as well as non-specific physical interactions between chemical species, and the packing effect actually present in the system lead to deviation in heat capacity from ideality.[70] Heat capacity of alcohol–hydrocarbon system are influenced by the system temperature,[71] pressure,[70] proximity to critical region,[71] types of bonds present in the component (degree of saturation),[72] and position of –OH groups (isomerism).[73]

Deviation from ideality for heat capacity is usually positive at temperatures close to room temperature under atmospheric pressure, and increase with increasing temperature. This trend is observed in the systems ethanol + dodecane, ethanol + tridecane,[71] ethanol + decane or undecane,[74] ethanol + heptane,[70] ethanol + benzene,[75] ethanol + toluene,[76] propan-1-ol + heptane, propan-1-ol + hexane (above room temperature),[77] propan-1-ol + propan-2-ol-hex-1-ene,[72] propan-1-ol + cyclohexane,[78] propan-1-ol + decane,[79] 1-propanol + dodecane or tridecane,[80] butan-1-ol + dodecane and finally butan-1-ol + dodecane.[73] Such behavior is explained with the partial breaking of he hydrogen-bonding structure of alcohol during mixing. The destruction of H-bond is higher at higher temperature.[74] More specifically, the fact that hydrogen bonds even exist in solution makes the structure of the mixtures to be weaker than that of pure 1-alkanol making the order decrease with temperature more rapidly for mixtures. Taking this into account, it usually results in a positive deviation in molar heat capacity from ideality.[84] The degree of deviation depends on the relative size of the alkane, and increases with the size of alkane (carbon number n) for some systems of ethanol + alkane and propan-1-ol + alkane.[74] This is due to the increased ability of destruction of 1-alkanol structures in the mixture with the size of the alkane.[74] In mixtures of the same components, this deviation is parabolic and composition dependent with a maximum located in the n-alkane rich region. However at lower temperatures, the hump decreases and even a minima (negative deviation) is seen to appear in propan-1-ol + hexane and + heptane systems.[78] Association models based on these ideas have been found to be good at predicting the values of excess properties. In ethanol + dodecane system, an inflection point is found at the medium-to-high mole fraction of ethanol and an M-shaped deviation curve is predicted in the immediate proximity of critical temperature of this system.[71]

The heat capacities at pressures up to 90 MPa and for temperatures ranging from (293 to 318) K for ethanol + heptane mixtures were calculated using the speeds of sound under elevated pressures, densities and heat capacities at atmospheric pressure.[70] The effect of pressure on deviation from ideality is rather small. Based on an earlier study, Dzida and Marczak mention that three main factors influence the variation of heat capacity with pressure: free volume for molecular vibration, intramolecular vibration, and rotational degrees of freedom. Pressure dependence of heat capacity is weaker for substances consisting of symmetrical molecules than for unsymmetrical ones.[70] The partial molar heat capacities of alkanols in alkanes are larger than molar heat capacities of pure 1-alkanols. Thus, the effect of dissolution of alkan-1-ol in alkane on its heat capacity is opposite to that caused by the increase of pressure. Consequently, as explained by Dzida and Marczak, the number of hydrogen bonds per mole of alcohol is lower in solution than in pure liquid.[70] The excess molar heat capacity of ethanol + heptane system decreases with increasing pressure as it shifts the equilibrium between $-C_2H_5OH$ entities from monomers to associates. The pressure therefore prevents H-bonds in the ethanol + heptane mixture from breaking. To demonstrate how moving away from ideal-associated mixtures to a solvent having weak interactions with alcohol will change the shape of excess molar heat capacity, an alkane has been replaced by an appropriate alkene. In that way, OH-π and much weaker π-π interactions have also been introduced into the mixture. Molar heat capacities are higher in the propan-1-ol + hex-1-ene system by approximately 3 J K^{-1} mol^{-1} than propan-1-ol + hexane.[70] However, the shape of the excess molar heat capacity curves are completely different. The curves in propan-1-ol + hex-1-ene[72] have only one extremum: a maximum (shifted towards lower mole fraction of alcohol). That kind of situation is hardly possible if complexes A-B can be formed in the mixture.

Molar heat capacity of propan-1-ol + cyclohexane was also measured in an adiabatic calorimeter over the whole composition range from 180 to 300 K. The magnitude of excess molar heat capacity is lower than in systems involving propan-1-ol + hexane and propan-1-ol + hex-1-ene at all temperatures. There is no visible shift of the maximum with composition whereas such shifts were found for all other mixtures.[77]

Heat capacity value varies with the position of the hydroxyl group in the alkyl chain of the alkanol.[81] This explanation was related to a decrease of the association capability when the hydroxyl group is located in a non-primary position. The steric hindrance over the hydroxyl group was found to be responsible for this fact rather than the debilitation of the H-bond energy.[81] Isobaric heat capacities per unit volume at (283 to 318) K were measured at atmospheric pressure and over the whole composition range in a comparative study of the thermodynamic behavior of dodecane + butan-1-ol and + butan-2-ol.[73] A method was suggested for calculating the heat capacity of 1-alkanol + alkane systems from their viscosity data taking into account the molecular association using the free-volume theory.[82] The calculated heat capacities are in good agreement with experimental data of six 1-alkanol + alkane systems (butan-1-ol + decane, pentan-1-ol + decane, ethanol + heptane, propan-1-ol + heptane, butan-1-ol + heptane, propan-1-ol + hexane) with a maximum deviation of 0.31 %.

5.5.4   Heat Capacities of Amide and Hydrocarbon Mixtures

Cope *et al.*[7] call amide and hydrocarbon mixtures as "adducts" because the guest hydrocarbon molecules fill the cavities in the host lattice of amides without leaving much space in order to maximize the attractive interactions between the guest and host molecules. Cope *et al.* explain that short straight-chain alkanes are not able to fulfill this requirement and do not form stable adducts, although the longer alkanes having 16 or more carbon atoms can, by virtue of their ability to coil and hence fill the cavities. Short alkanes having small branches have a larger cross-section than the straight-chain alkanes and alkenes, and are suitable adduct formers. Each amide with straight-chain alkanes and alkenes show thermodynamic anomalies in a particular tempera-ture range, being attributed to the onset of orientational disorder of the guest molecules as temperature is raised.[7] Figure 5.11 shows the heat capacity var-iation and the anomalous region for urea with hydrocarbons undecane, hex-adec-1-ene and dec-1-ene.[83] The same trend is found in thiourea-hydrocarbon adducts[7] and urea and alkene adducts.[8] The temperature at which the max-imum heat capacity occurs increases smoothly with the length of the hydro-carbon chain for the adducts. Comparing n-paraffin and 1-olefin adducts, the entropy contributions attributable to the re-orientation process, lending sup-port to the idea that the initial and final states involved are similar.[83] A sta-tistical mechanical Ising-type theory has been proposed by Parsonage and Pemberton in which these gradual transitions are interpreted in terms of the onset of reorientation of the guest molecules about their long axes.[84]

Interactions between the terminal groups of molecules play an important rule in the anomalous behavior of heat capacities.[8] The temperature and shape of such anomalies are modified by the presence and character of the guest molecules.[83]

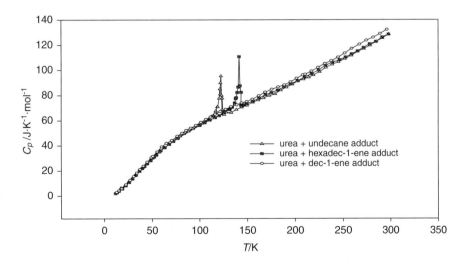

Figure 5.11   Molar heat capacities, C_p, of urea and hydrocarbon mixtures as a function of temperature, T.[83]

5.6   Modeling of Heat Capacities of Mixtures

5.6.1   Modeling of Heat Capacities Using EOSs

Modeling of the second derivatives of the thermodynamic potential (C_p) using EOS is a difficult task, since most of the equations have been developed to correlate and predict fluid phase equilibria.[85,86] In general, EOSs are not able to accurately reproduce phase behavior and second derivatives with a single set of adjustable parameters. The more complex the interactions, as in associations, the more difficult does the modeling task become. The Volume Translated Peng-Robinson (VTPR) group contribution EOS that has been shown, with some modifications, to be able to predict phase equilibria and heat capacities simultaneously, but cannot be employed to isolate the effect of association on fluid properties.[87] In this context, of special relevance, Lafitte *et al.*[88] and Llovell and Vega[89] were able to successfully describe second-order thermodynamic properties of many systems using the Statistical Associating Fluid Theory (SAFT model) in two different ways. C_p is very sensitive to association phenomena, with hydrogen bonding being the most representative case.[90,91] The SAFT (EOS) model has been modified several times and now appears in many versions with excess free energy mixing rules (PC-SAFT, SAFT-VR, SAFT-VR LJC, SAFT-VR Mie) (Figure 5.12).[92]

5.6.2   Solution Models

For mixtures, second-order excess properties can be modeled using many solution models such as the Wilhoit *et al.* model[93] the original Cabani model (Group Contributions) with updates,[57,97,98] and contributions from many other authors.[90,100] The most widely used solution models are: the Treszczanowicz-Kehiaian (TK),[101,102] with its many extensions such as ERAS (Extended Real Associated Solution Model)[94] and SERAS (Symmetrical Extended Real Associated Solution Model);[95] other solution models such as the UNIQUAC;[96] and the NRTL[96] models.

5.6.2.1   Group Contribution Models for Apparent Heat Capacities

Cabani *et al.*[57,97,98] proposed a method correlating the apparent heat capacities with molecular structure. The model is based mainly on the following considerations: the presence of a formation of a cavity inside the solvent; an interaction between the hydrocarbon part of the molecule and water and the hydrophobic centers and water; and an additional term used to account for interactions among centers when more than one hydrophilic group is present in the molecule. Three methods were used to find the group-contribution for apparent heat capacities of aqueous non-ionic organic compounds.

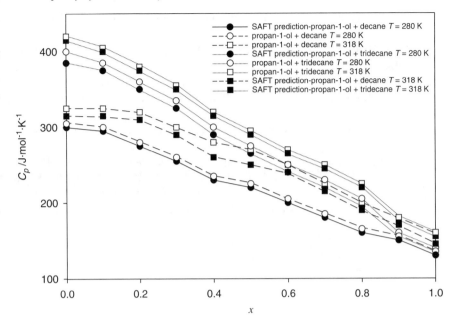

Figure 5.12 Prediction of the molar heat capacity, C_p, of propan-1-ol and hydrocarbons using SAFT EOS at various temperatures, T, and mole fraction of propan-1-ol, x.[92]

The molecular structures were described through a scheme of group contributions with five rules detailed in the original reference.[57] More contributing groups were added, by other researchers, to the original list.[90,99,100]

5.6.2.2 Treszczanowicz-Kehiaian (TK) model

The TK model[101] was originally developed to study self and cross associations in mixtures. The most common being the formation of hydrogen bonds. As explained by Costas and Patterson,[102] the model makes a distinction between nominal components and real components.

The model was extended to take into accounts complex formation.[103] Heintz *et al.* and others combined the effect of association with non-associative intermolecular interactions using the Flory EOS theory.[104–106] The model has been extended to ternary mixtures and has expressions for the physical and chemical contributions to the excess enthalpy and the excess volume.

References

1. G. Roux-Desgranges and A. H. Roux, *J. Mol. Liq.*, 1999, **81**, 3.
2. Virtual Institute for Thermal Metrology (Evitherm-http://evitherm. athena.as).

3. T. G. Zijlema, G. J. Witkamp and G. M. van Rosmalen, *J. Chem. Eng. Data*, 1999, **44**, 1335.
4. R. H. Weiland, J. C. Dingman and D. B. Cronin, *J. Chem. Eng. Data*, 1997, **42**, 1004.
5. N. Hadded and M. Bouanz, *J. Mol. Liq.*, 2007, **130**, 11.
6. M. Castagnolo, A. Inglese, G. Petrella and A. Sacco, *Thermochimica Acta*, 1981, **44**, 67.
7. A. F. G. Cope, D. J. Gannon and N. G. Parsonage, *J. Chem. Therm.*, 1972, **4**, 829.
8. D. J. Gannon and N. G. Parsonage, *J. Chem. Therm.*, 1972, **4**, 745.
9. Z. C. Tan, G. Y. Sun, Y. J. Song, L. Wang, J. R. Han, Y. S. Liu, M. Wang and D. Z. Nie, *Thermochimica Acta*, 2000, **352–353**, 247.
10. E. W. Hough, D. M. Mason and B. H. Sage, *J. Am. Chem. Soc.*, 1950, **72**, 5774.
11. J. Suurkuusk and I. Wadsö, *J. Chem. Therm.*, 1974, **6**, 667.
12. K. Yamaguchi and K. Itagaki, *J. Therm. Anal. Cal.*, 2002, **69**, 1059.
13. J. M. Pardo, C. A. Tovar, J. Troncoso, E. Carballo and L. Romaní, *Thermochimica Acta*, 2005, **433**, 128.
14. V. N. Kabadi, *Fuel*, 1996, **75**, 363.
15. C. A. Tovar, C. A. Cerdeiriña, D. González, E. Carballo and L. Romaní, *Fluid Phase Equilib.*, 2000, **169**, 209.
16. M. C. Righetti, F. Comelli and R. Francesconi, *Thermochimica Acta*, 1997, **294**, 179.
17. J. M. Pardo, C. A. Tovar, D. Gonzalez, E. Carballo and L. Romani, *J. Chem. Eng. Data*, 2001, **46**, 212.
18. H. Piekarski and M. J. Tkaczyk, *J. Therm. Anal. Cal.*, 2006, **83**, 541.
19. Y. J. Chen, T. W. Shih and M. H. Li, *J. Chem. Eng. Data*, 2001, **46**, 51.
20. L. F. Chiu and M. H. Li, *J. Chem. Eng. Data*, 1999, **44**, 1396.
21. Y. J. Chen and M. Li, *J. Chem. Eng. Data*, 2001, **46**, 102.
22. D. M. Swenson, S. P. Ziemer, M. B. Blodgett, J. S. Jones and E. M. Woolley, *J. Chem. Therm.*, 2006, **12**, 1523.
23. J. L. Valencia, J. Troncoso, J. Peleteiro, E. Carballo and L. Romaní, *Fluid Phase Equilib.*, 2005, **232**, 207.
24. A. Danilenko, V. Romanova, E. Kuleshova, Z. Parnes and E. Braudo, *Russ. Chem. Bull.*, 1998, **47**, 2134.
25. B. J. Kowalski, *J. Therm. Anal. Cal.*, 1988, **34**, 1321.
26. M. Origlia-Luster, T. G. Call and E. M. Woolley, *J. Chem. Therm.*, 2000, **32**, 847.
27. J. D. Sargent, T. L. Niederhauser and E. M. Woolley, *J. Chem. Therm.*, 2004, **36**, 603.
28. J. C. Cobos, I. Garcia and C. Casanova, *Fluid Phase Equilib.*, 1991, **69**, 223.
29. M. Mundhwa and A. Henni, *J. Chem. Eng. Data*, 2007, **52**, 491.
30. S. Schrödle, G. Hefter and R. Buchner, *J. Chem. Therm.*, 2005, **37**, 513.
31. R. Tanaka and A. Saito, *J. Coll. Int. Sc.*, 1990, **134**, 82.

32. Y. Ding, Q. Yu, R. Lin and H. Zong, *J. Chem. Therm.*, 1997, **29**, 1473.
33. K. Mulia and V. F. Yesavage, *Fluid Phase Equilib.*, 1999, **158–160**, 1001.
34. P. R. Tremaine, D. Shvedov and C. Xiao, *J. Phys. Chem. B.*, 1997, **101**, 409.
35. I. R. Tasker, S. K. Suri and R. H. Wood, *J. Chem. Eng. Data*, 1984, **29**, 193.
36. M. Costas and D. Patterson, *J. Chem. Soc., Faraday Trans. 1: Physical Chem. in Condensed Phases*, 1985, **81**, 2381.
37. P. K. Banipal, T. S. Banipal, J. C. Ahluwalia and B. S. Lark, *J. Chem. Therm.*, 2000, **32**, 1409.
38. C. de Visser, G. Perron, J. E. Desnoyers, W. J. M. Heuvelsland and G. Somsen, *J. Chem. Eng. Data*, 1977, **22**, 74.
39. P. K. Banipal, T. A. Banipal, J. C. Ahluwalia and B. S. Lark, *J. Chem. Therm.*, 2002, **34**, 1825.
40. M. S. Hajji, *J. Coll. Int. Sci.*, 2003, **257**, 364.
41. J. B. Ott and J. P. Sipowska, *J. Chem. Eng. Data*, 1996, **41**, 987.
42. G. Roux, G. Perron and J. E. Desnoyers, *Can. J. Chem.*, 1978, **56**, 2808.
43. S. J. Gill and I. Wadsö, *Proc. Natl. Acad. Sci. USA*, 1976, **73**, 2955.
44. S. J. Gill, N. F. Nichols and I. Wadsö, *J. Chem. Therm.*, 1976, **8**, 445.
45. S. J Gill, S. F. Dec, G. Olofsson and I. Wadso, *J. Phys. Chem.*, 1985, **89**, 3758.
46. K. A. Sharp and B. Madan, *J. Phys. Chem. B.*, 1997, **101**, 4343.
47. B. Madan and K. Sharp, *J. Phys. Chem. B.*, 1997, **101**, 11237.
48. S. Cabani, G. Conti, G. Martinelli and E. Matteoli, *J. Chem. Soc., Faraday Trans. 1*, 1973, **69**, 2112.
49. S. Cabani, G. Conti and E. Matteoli, *J. Sol. Chem.*, 1979, **8**, 11.
50. O. Kiyohara, G. Perron and J. E. Desnoyers, *Can. J. Chem.*, 1975, **53**, 2591.
51. O. Kiyohara, G. Perron and J. E. Desnoyers, *Can. J. Chem.*, 1975, **53**, 3263.
52. F. Franks and D. J. G. Ives, *Q. Rev. Chem. Soc.*, 1966, **20**, 1.
53. C. Dettmann, G. Ernst and H. Wirbser, *J. Chem. Therm.*, 2006, **38**, 56.
54. C. de Visser, G. Perron and J. E. Desnoyers, *Can. J. Chem.*, 1977, **55**, 856.
55. M. L. Origlia-Luster and E. M. Woolley, *J. Chem. Therm.*, 2003, **35**, 1101.
56. Å. Hovorka, A. H. Roux, G. Roux-Desgranges and V. Dohnal, *J. Sol. Chem.*, 1999, **12**, 1289.
57. S. Cabani, P. Gianni, V. Mollica and L. Lepori, *J. Sol. Chem.*, 1981, **10**, 563.
58. A. Inglese and R. H. Wood, *J. Chem. Therm.*, 1996, **28**, 1059.
59. M. Pagé, J. Huot and C. J. Jolicoeur, *J. Chem. Therm.*, 1993, **25**, 139.
60. L. Wojtczakp, H. Piekarski, M. Tkaczyk, I. Zasada and T. Rychtelska, *J. Mol. Liq.*, 2002, **95**, 229.
61. C. Yang, P. Ma and Q. Zhou, *J. Chem. Eng. Data*, 2004, **49**, 582.
62. H. Piekarski, M. Tkaczyk and M. J. Wasiak, *J. Therm. Anal. Cal.*, 2005, **82**, 711.

63. H. Piekarski and M. Tkaczyk, *Thermochimica Acta*, 2005, **428**, 113.

64. Z. Nan and Z. Tan, *J. Chem. Eng. Data*, 2005, **50**, 6.

65. Z. Nan and Z. Tan, *Thermochimica Acta*, 2004, **413**, 267.

66. Z. Nan and Z. Tan, *Thermochimica Acta*, 2004, **419**, 275.

67. Z. Nan and Z. Tan, *Fluid Phase Equilib.*, 2004, **226**, 65.

68. Y. Song, Z. Tan, S. Meng and J. Zhang, *Thermochimica Acta*, 2000, **352–353**, 255.

69. R. Burgdorf, A. Zocholl, W. Arlt and H. Knapp, *Fluid Phase Equilib.*, 1999, **164**, 225.

70. M. Dzida and W. Marczak, *J. Chem. Therm.*, 2005, **8**, 826.

71. J. Peleteiro, J. Troncoso, D. Gonzãlez-Salgado, J. L. Valencia, C. A. Cerdeiria and L. Romani, *Int. J. Thermophys.*, 2004, **25**, 787.

72. B. Kalinowska and W. Wóycicki, *J. Chem. Therm.*, 1985, **17**, 829.

73. J. Troncoso, J. L. Valencia, M. Souto-Caride, D. Gonzalez-Salgado and J. Peleteiro, *J. Chem. Eng. Data*, 2004, **49**, 1789.

74. J. Peleteiro, J. Troncoso, D. González-Salgado, J. L. Valencia, M. Souto-Caride and L. Romani, *J. Chem. Therm.*, 2005, **37**, 935.

75. Z. Nan, Q. Jiao, Z. Tan and L. Sun, *Thermochimica Acta*, 2003, **406**, 151.

76. M. J. Pedersen, W. B. Kay and H. C. Hershey, *J. Chem. Therm.*, 1975, **12**, 1107.

77. B. Kalinowska, J. Jedlinska, J. Stecki and W. Wóycicki, *J. Chem. Therm.*, 1981, **13**, 357.

78. B. Kalinowska and W. Wóycicki, *J. Chem. Therm.*, 1988, **20**, 1131.

79. J. Peleteiro, D. González-Salgado, C. A. Cerdeiriña and L. Romaní, *J. Chem. Therm.*, 2002, **34**, 485.

80. J. Peleteiro, D. González-Salgado, C. A. Cerdeiriña, J. L. Valencia and L. Romaní, *Fluid Phase Equilib.*, 2001, **191**, 83.

81. C. A. Cerdeirina, D. Gonzalez-Salgado, L. Romani, M. D. C. Delgado, L. A. Torres and M. Costas, *J. Chem. Phys.*, 2004, **120**, 6648.

82. E. Totchasov, M. Nikiforov and G. Alper, *Russ. J. Appl. Chem.*, 2006, **79**, 34.

83. A. F. G. Cope and N. G. Parsonage, *J. Chem. Therm.*, 1969, **1**, 99.

84. N. G. Parsonage and R. C. Pemberton, *Trans. Faraday. Soc.*, 1967, **63**, 311.

85. J. Greogorowicz, J. P. O'Connell and C. J. Peters, *Fluid Phase Equilib.*, 1996, **116**, 94.

86. E. A. Muller and K. E. Gubbins, *Ind. Eng. Chem. Res.*, 2001, **40**, 2193.

87. A. Diedrichs, J. Rarey and J. Gmehling, *Fluid Phase Equilib.*, 2006, **248**, 56.

88. L. Thomas, B. David, M. P. Manuel and D. J. Jean-Luc, *Chem. Phys.*, 2006, **124**, 024509-1-16.

89. F. Llovell and L. F. Vega, *J. Phys. Chem. B.*, 2006, **110**, 11427.

90. L. Andreoli-Ball, S. J. Sun, L. M Trejo, M. Costas and D. Patterson, *Pure Appl. Chem.*, 1990, **62**, 2097.

91. L. Andreoli-Ball, M. Costas, P. Paquet, D. Patterson and M. E. St.Victor, *Pure Appl. Chem.*, 1989, **61**, 1075.
92. F. Llovell, C. J. Peters and L. F. Vega, *Fluid Phase Equilib.*, 2006, **248**, 115.
93. R. C. Wilhoit, J. Chao, M. Carlo, *National Bureau of Standards Report, Eighteenth Report on A Survey of Thermodynamic Properties of the Compounds of the Elements CHNOPS*, 1970, 10291, 4 August, 1–90.
94. M. Dzida and M. ; P. Góralski, *J. Chem. Therm.*, 2006, **38**, 962.
95. A. Piñeiro, *Fluid Phase Equilib.*, 2004, **216**, 245.
96. Y. Demirel and H. O. Paksoy, *Thermochimica Acta*, 1997, **303**, 129.
97. S. Cabani, G. Conti, A. Martinelli and E. Matteoli, *J. Chem. Soc., Faraday Trans. 1*, 1973, **69**, 2112.
98. S. Cabani, G. Conti and E. Matteoli, *J. Sol. Chem.*, 1976, **5**, 125.
99. N. Nichols and I. Wadsö, *J. Chem. Therm.*, 1975, **7**, 329.
100. J. P. Guthrie, *Can. J. Chem.*, 1977, **55**, 3700.
101. H. Kehiaian and A. Treszczanowicz, *J. Bull. Acad. Chem. Fr.*, 1969, **5**, 1561.
102. M. Costas and D. Patterson, *J. Chem. Soc., Faraday Trans. 1: Physical Chemistry in Condensed Phases*, 1985, **81**, 635.
103. D. D. Deshpande, D. Patterson, L. Andreoli-Ball, M. Costas and L. M. Trejo, *J. Chem. Soc. Faraday Trans.*, 1991, **87**, 1133.
104. A. Heintz, *Ber. Bunsenges. Phys. Chem.*, 1985, **89**, 172.
105. H. Funke, M. Wetzel and A. Heintz, *Pure Appl. Chem.*, 1989, **61**, 1429.
106. A. Treszczanowicz and G. C. Benson, *Fluid Phase Equilib.*, 1985, **23**, 117.

CHAPTER 6

Heat Capacities and Related Properties of Vapours and Vapour Mixtures

CHRISTOPHER J. WORMALD

School of Chemistry, University of Bristol, Bristol, BS8 1TS, UK

6.1 Pioneering Measurements of the Heat Capacity of Gases

The measurement of the heat capacity of gases, using a flow calorimetric technique, was pioneered by Swann[1] in 1909 who adapted the flow apparatus which Callender and Barnes[2] used in 1902 to measure the heat capacity of liquids. This work was done about half a century later than the measurements of the cooling effect produced by expanding gases through a throttle conducted by Joule and Thomson[3] in 1852. In 1900 Buckingham[4] suggested constructing a flow calorimeter with a heater fixed in the throttle so that isothermal measurements of the enthalpy–pressure coefficient of gases could be made. This suggestion was adopted in 1937 by Collins and Keyes[5] who made outstanding measurements of the heat capacity and enthalpy–pressure coefficient of steam. In 1941 Pitzer[6] reported a recirculatory flow apparatus consisting of a boiler, which acted as a calorimeter for the measurement of the enthalpy of vaporisation of the liquid, connected in series with a vacuum jacketed flow calorimeter for the measurement of the isobaric heat capacity C_p of the vapour. The apparatus was capable of operation at pressures below atmospheric, and values of the ideal gas heat capacity C_p^0 were obtained by extrapolation to zero

Heat Capacities: Liquids, Solutions and Vapours
Edited by Emmerich Wilhelm and Trevor M. Letcher
© The Royal Society of Chemistry 2010
Published by the Royal Society of Chemistry, www.rsc.org

pressure. The C_p^0 values were compared with statistical mechanical calculations based on Boltzmann's Distribution Law and his assumption that the kinetic energy of a molecule is shared equally amongst all the degrees of freedom, principally translation, rotation and vibrational modes. The energy of a molecule was calculated by summing over all states and differentiation with respect to temperature. Einstein's theory of heat capacities was of particular value to the calculation of the vibrational heat capacity, and the entropy.

Before the advent of infrared spectroscopy about 1930 there was little experimental information about vibrational frequencies of diatomic and simple molecules, and vibrational frequencies were estimated by using them as adjustable parameters which were chosen to fit heat capacity measurements. The development of microwave spectroscopy in 1946 gave information about the moment of inertia and rotational states of polar molecules, and lead to the accurate calculation of rotational contributions to the heat capacity.

The industrial need for reliable heat capacities of vapours stimulated the Petroleum and Natural Gas division of the U.S. Bureau of Mines to construct an improved flow calorimeter[7,8] modelled on the Pitzer design. Waddington *et al.*[9] constructed a calorimeter capable of an accuracy of 0.1 % and used it to measure the heat capacity of many hydrocarbons. The flow calorimetric apparatus is shown in Figure 6.1.

The vacuum jacketed recycling vaporiser A is fitted with an electrical heater, and is mounted in an oil bath. The vacuum jacketed heat capacity calorimeter C which is fitted with a heater H is mounted in a second oil bath. Two platinum resistance thermometers were used to measure the temperature rise. The vapour is condensed at D and returned to the boiler. The flow rate is measured by condensing some of the vapour into a trap E, and weighing the amount collected over a measured time interval. A similar calorimeter constructed at the National Physical Laboratory by Hales *et al.*[10] was used to make extensive measurements on a series of oxygen-containing compounds. Flow calorimetric measurements on compressed gases can be made at supercritical temperatures and at pressures up to about 100 MPa.[11] Heat capacity measurements at even higher pressures are possible by confining the fluid in a cell and performing pressure controlled scanning calorimetry as described by Randzio.[12]

6.2   The Heat Capacity of Gases and Related Quantities

Heat capacities of gases are usually measured in a flow calorimeter which is often followed by a Joule–Thomson calorimeter. At high pressures a constant volume or variable volume cell is sometimes used used.[12] The constant volume heat capacity C_V is related to the internal molar energy U:

$$C_V = (\partial U/\partial T)_V \qquad (1)$$

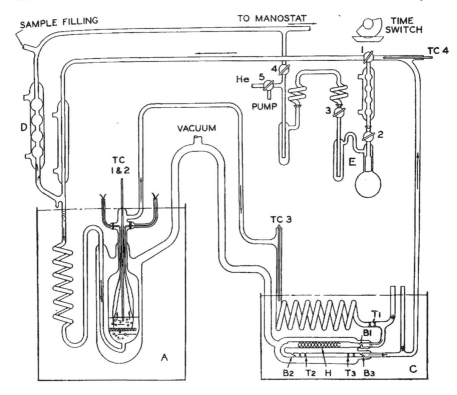

Figure 6.1   The heat capacity and enthalpy of vaporisation calorimeter of Waddington et al.[9]

while the constant pressure heat capacity C_p is related to the molar enthalpy H

$$C_p = (\partial H / \partial T)_p \qquad (2)$$

The internal energy is the sum of the translational, rotational, vibrational, electronic and nuclear spin energies of the separate gaseous molecules, plus the intermolecular energy due to pairwise, three-body, and higher order molecular interactions. The constant volume heat capacity for a real gas can be written as the sum of terms deriving from the intramolecular energies, and a term which derives from the intermolecular potential energy:

$$C_V = C_V^0 + \int_\infty^V T(\partial^2 p / \partial T^2)_V dV \qquad (3)$$

The ideal gas heat capacity C_V^0 can be obtained either by extrapolating measured heat capacities to zero pressure, or by calculating it from spectroscopic information using statistical mechanics. The integral term can be calculated from an equation of state. The molar heat capacities C_V and C_p are related through the equation:

$$C_p - C_V = TV\alpha^2/\kappa_T \tag{4}$$

where V is the molar volume, α is the coefficient of expansion, and κ_T is the isothermal compressibility. Measurements of C_p are of particular importance for the calculation of the enthalpy of gases. The change of enthalpy dH can be written as:

$$dH(p, T) = (\partial H/\partial T)_p dT + (\partial H/\partial p)_T dp \tag{5}$$

or

$$dH(p, T) = C_p dT + \phi dp \tag{6}$$

where $(\partial H/\partial T)_p = C_p$ is the isobaric heat capacity and $(\partial H/\partial p)_T = \phi$ is the iso-thermal enthalpy-pressure coefficient, or isothermal Joule–Thomson coefficient.

A similar quantity μ, the isenthalpic Joule–Thomson coefficient $(\partial T/\partial p)_H$ is often used. It is related to C_p and ϕ through the equation:

$$(\partial H/\partial T)_p (\partial p/\partial H)_T (\partial T/\partial p)_H = -1 \tag{7}$$

or

$$\phi = -\mu C_p \tag{8}$$

The isothermal Joule–Thomson coefficient ϕ, and the pressure derivative $(\partial C_p/\partial p)_T$ of the heat capacity can be related to the molar volume V, and hence the equation of state coefficients, by making use of the equations:

$$(\partial H/\partial p)_T = \phi = V - T(\partial V/\partial T)_p \tag{9}$$

and

$$(\partial^2 H/\partial p \partial T) = (\partial C_p/\partial p)_T = -T^2(\partial^2 V/\partial T^2)_p \tag{10}$$

For gases at low pressures it is advantageous to use the virial equation of state written as a polynomial in powers of the density n/V. For 1 mole of gas:

$$pV/RT = 1 + B/V + C/V^2 + D/V^3 + \dots \tag{11}$$

Written as a polynomial in powers of the pressure p the equation is:

$$pV/RT = 1 + B'p + C'p^2 + D'p^3 + \dots \tag{12}$$

where

$$B' = B, C' = (C - B^2)/(RT), \quad D' = (D - 3BC + 2B^2)/(RT)^2 \tag{13}$$

Using the pressure series form of the equation, the molar volume V is:

$$V = (RT/p) + B + C'p + \dots. \tag{14}$$

From Equation (10) we obtain:

$$\phi = (B - T\mathrm{d}B/\mathrm{d}T) + p(C' - T\mathrm{d}C'/\mathrm{d}T) \qquad (15)$$

From Equation (11) we obtain:

$$(\partial C_p/\partial p)_T = -T^2(\mathrm{d}^2B/\mathrm{d}T^2) - pT(\mathrm{d}^2C'/\mathrm{d}T^2) - .... \qquad (16)$$

In the limit of zero pressure we have:

$$\mathrm{Lim}_{p=0}(\partial H/\partial p)_T = \phi = B - T\mathrm{d}B/\mathrm{d}T \qquad (17)$$

and

$$\mathrm{Lim}_{(p=0)}(\partial C_p/\partial p)_T = \chi = -T^2(\mathrm{d}^2B/\mathrm{d}T^2) \qquad (18)$$

Measurements of B, ϕ and χ all give information about the intermolecular potential, and in this context measurements of the pressure dependence of the heat capacity are of particular importance. If spectroscopic measurements of the vibrational and rotational modes of a gas are available the intramolecular motion can be analysed and C_p^0 can be calculated from statistical mechanics. Comparison with experimental values of C_p^0 is of great interest as failure to agree indicates that there may be motion or a structural feature within the molecule which has not been correctly accounted for, and the assumed structure may have to be revised.

6.3 The Temperature Dependence of the Ideal Gas Heat Capacity

6.3.1 The Translational Contribution to the Heat Capacity

Details of the statistical mechanical calculations of the thermodynamic properties of gases are set out in text books by Alberty et al.[13] Noggle,[14] McQuarry[15] and Berry et al.[16] In this section key equations are summarised, and comparison of calculated and experimental C_p^0 values is made. The gas is assumed to be an assembly of independent particles each of which has translational, rotational, vibrational and electronic energy; nuclear energy can be ignored. If the various modes of motion are independent, and assuming that Boltzmann statistics apply, the degeneracy of an energy level is equal to the product of the degeneracies for the various modes. For a system of N_i particles of energy ε_i and degeneracy g_i the partition function q of a molecule in its ground state is:

$$q = q_t q_r q_v q_e \qquad (19)$$

Each of the partition functions has the same functional form, which, for the translational partition function q_t is:

$$q_t = \sum_i g_{it} \exp(-\varepsilon_{it}/kT) \tag{20}$$

where ε_{it} is the translational energy. We shall discuss contributions to the energy U due to translation U_t, rotation U_r, vibration U_V, and electronic excitation U_e in turn.

The molar internal energies of translation and rotation are:

$$U_t = NkT^2(\partial \ln q_t/\partial T)_V \quad \text{and} \quad U_r = NkT^2(\mathrm{d} \ln q_r/\mathrm{d}T) \tag{21}$$

Vibrational and electronic energies U_V and U_e are given by similar equations.

The translational partition function is given by:

$$q_t = (2\pi mkT/h^2)^{3/2}V \tag{22}$$

The molar translational energy and constant volume heat capacity are:

$$U_t = NkT^2(\partial \ln q_t/\partial T)_V = 3/2(NkT) \tag{23}$$

$$C_V = (\mathrm{d}U_t/\mathrm{d}T) = 3/2(R) \tag{24}$$

The molar translational enthalpy and constant pressure heat capacity are:

$$H_t = NkT[T(\partial \ln q_t/\partial T) + 1] = 5/2(NkT) \tag{25}$$

$$C_p = (\mathrm{d}H_t/\mathrm{d}T) = 5/2(R) \tag{26}$$

Experimental heat capacities of argon, krypton and xenon at different temperatures all lie on the same line which is independent of temperature and is in exact agreement with Equation (26).

For hydrogen, helium, and to a lesser extent for neon at low temperatures, experimental heat capacities are a few percent less that the calculated values. This is because at low temperatures the de Broglie wavelength $(h^2/2\pi mkT)^{1/2}$ becomes so large that the wavefunctions of the particles overlap, and Boltzmann's Distribution Law is not applicable.

6.3.2   Rotational Contributions to the Heat Capacity

The rotational energy of a rigid diatomic molecule of moment of inertia I is:

$$\varepsilon_r = J(J + 1)(h^2/8\pi^2 I) \tag{27}$$

Here J is the rotational quantum number, and rotational levels are $2J+1$ degenerate.

The rotational partition function is:

$$q_r = \sum_{j=0}^{\infty} (2J+1) \exp[-J(J+1)(h^2/8\pi^2 kT)] \qquad (28)$$

As rotational levels are closely spaced the sum can be replaced by an integral which gives the simple equation:

$$q_r = T/\sigma\,\theta_r \qquad (29)$$

The symmetry number σ is 1 for a heteronuclear diatomic molecule, and 2 for a homonuclear molecule. θ_r is the characteristic rotational temperature:

$$\theta_r = h^2/8\pi^2 I k \qquad (30)$$

For a non-linear polyatomic molecule there are three moments of inertia I_a, I_b and I_c, and three characteristic rotational temperatures each of which has the form of Equation (30). The rotational partition function for a non-linear polyatomic molecule is:

$$q_r = (\pi^{1/2}/\sigma)\,(T^3/\theta_a\theta_b\theta_c) \qquad (31)$$

The rotational energy U_r is calculated from Equation (21) and the rotational heat capacity is obtained by differentiation with respect to temperature. Except at very low temperatures the rotational energy levels are fully populated and for diatomic molecules, which have two degrees of rotational freedom, the rotational energy is $2/2\,RT$, and the rotational contribution to the heat capacity is $2/2\,R$. Non-linear molecules have three degrees of rotational freedom and the rotational contribution to the heat capacity is $3/2\,R$. Translational, rotational and vibrational contributions to C_V are shown schematically in Figure 6.2.

6.3.3   Vibrational Contributions to the Heat Capacity

To a good approximation, the vibration of diatomic molecules can be described in terms of a harmonic-oscillator rigid-rotator model.[16] The model works well when only the lowest vibrational levels are occupied. The energy of the lowest level, the ground state, is the zero-point energy $h\nu/2$. The vibrational energy of the system is measured from the ground state upwards, with vibrational quantum numbers $\nu = 0, 1, 2, 3 \ldots$ The partition function for a vibrational mode is:

$$q_V = \sum_{V=0}^{\infty} e^{-V(h\nu/kT)} = 1/(1 - e^{-h\nu/kT}) \qquad (32)$$

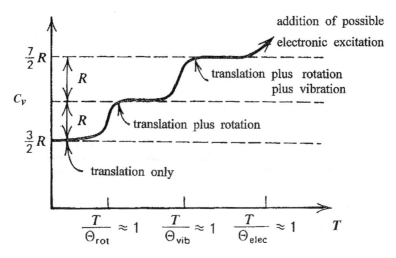

Figure 6.2 A schematic diagram showing how the translation rotation and vibrational and electronic modes contribute to the heat capacity of a gas.[16]

The group hv/k is the characteristic vibrational temperature θ_V. The vibrational energy U_V and heat capacity C_{vib} are:[13]

$$U_V = Nk \sum_i \theta_{Vi}/[(\exp(\theta_{Vi}/T) - 1)] \tag{33}$$

$$C_{vib} = Nk \sum_i [(\theta_{Vi}/T)^2 \exp(\theta_{Vi}/T)]/[\exp(\theta_{Vi}/T) - 1]^2 \tag{34}$$

For diatomic molecules at low temperatures C_{vib} approaches zero, and at high temperatures it levels off at R, 8.314 J mol^{-1} K^{-1}. Noggle[14] shows the heat capacities for nitrogen and chlorine to illustrate this behaviour. For linear polyatomic molecules of n atoms there are $(3n - 5)$ vibrational modes, and for non-linear molecules there are $(3n - 6)$ modes. The vibrational heat capacity of the molecule is the sum of the C_{vib} values calculated for each of the modes. Figure 6.2 shows how the vibrational contribution to the heat capacity increases most rapidly in the region where $T = \theta_v$.

Scott *et al.*[17] have measured the heat capacity of benzene vapour and made comparison with calculated values. Even at moderate temperatures (400 K) the heat capacity calculated on the harmonic oscillator model is lower than the experimental heat capacity by 1 %. Most of this difference can be attributed to the anharmonicity of the vibrations.

6.3.4 Electronic Contributions to the Heat Capacity

For most stable molecules electronic contributions to the heat capacity can be neglected. A notable exception is nitric oxide NO which has an electronic ground state with degeneracy $g_0 = 2$ and above it lies an excited state also with degeneracy $g_1 = 2$. The gap ε/hc between the levels is 121.1 cm^{-1}. The partition function is:

$$q_{elec} = g_o + g_1 \exp(-\varepsilon/kT) \tag{35}$$

The electronic contribution to the heat capacity has a maximum at about 80 K. Noggle[14] plots the total heat capacity as a function of temperature and shows that a maximum is followed by a minimum at 300 K, after which the heat capacity increases and approaches the equipartition value as the vibrational levels are populated. Agreement between the calculated and experimental heat capacities is excellent.

6.3.5   Contributions to the Heat Capacity Due to Nuclear Spin

At temperatures below 300 K, the heat capacity of the homonuclear molecules H_2, D_2 and T_2 show unusual behaviour,[14–16] in the case of deuterium exhibiting a maximum. The effect was explained when it was realised that two forms of hydrogen, ortho and para, are possible. Each hydrogen atom has nuclear spin $\frac{1}{2}$, and within a molecule coupling of the nuclear spins can be either parallel (ortho-hydrogen) or opposed (para-hydrogen). In ordinary hydrogen the two forms exist in a 3:1 ratio, and the heat capacity is that of the mixture. The explanation of this behaviour was a triumph for quantum mechanics.

6.3.6   Hindered Rotation Contributions to the Heat Capacity

In molecules such as ethane an important internal degree of freedom is the rotation of the methyl groups at the ends of the carbon–carbon bond. Pitzer *et al.*[18] suggested that this rotational motion is not free, but hindered. At low temperatures the hydrogen atoms of one methyl group lie between the hydrogen atoms of the other, and there is torsional vibration. As the temperature is raised the amplitude increases until the kinetic energy exceeds the potential energy of repulsion, and rotation of one methyl group with respect to the other takes place. The rotation is hindered by the potential energy barrier, but becomes freer at high temperatures. The potential energy of internal rotation in ethane V_r can be represented approximately by:

$$V_r = (V_o/2) \cdot (1 - \cos 3\phi) \tag{36}$$

where V_o is the height of the potential barrier and ϕ is the angle of skew. Assuming that the vibrational modes are known, subtraction of the calculated ideal gas heat capacity from that determined experimentally yields the torsional contribution to the heat capacity and hence the height of the potential barrier. Pitzer[19] and Wilson[20] examined various theoretical models of the barrier and concluded that for ethane at 298 K the contribution to the heat capacity from hindered rotation is 8 J mol^{-1} K^{-1} and V_o is 12 kJ mol^{-1}. This was confirmed by

Weiss *et al.*[21] who observed the IR torsional spectrum for CH_3CD_3. McQuarrie[15] shows how the hindered rotational contribution to the heat capacity changes with barrier height, and gives a table of barrier heights for different molecules. For chlorofluorocarbons the barrier heights are considerably higher than for ethane, the torsional motion is more complex and the contribution to the heat capacity is larger. When the amplitude of torsional oscillation is small it is adequate to assume a simple harmonic oscillator model, but when $kT > V_o$ the inadequacies of this model are more apparent. Another problem is that the molecule is not a structure consisting of independent oscillators, coupling between vibrational and torsional modes can be significant.

6.3.7 Pseudorotation in Cyclic Molecules with Puckered Rings

In 1941 Aston *et al.*[22] measured the entropy of cyclopentane down to 11.1 K and compared it with the entropy calculated from spectroscopic data; good agreement was not found. Pitzer *et al.*[7,8] then made gas phase heat capacity measurements and found that the calculated heat capacities differed by an amount greater than experimental error. A theoretical analysis of the possible vibrational modes lead the authors to conclude that the cyclopentane ring was not planar but puckered with either one or two atoms out of the plane of the ring. It was suggested that a ripple, which they described as pseudo rotation, circulated round the puckered ring. As cyclopentane has no permanent dipole moment spectroscopic evidence was hard to find. Harris *et al.*[23,24] produced a general theory of puckering in five-membered rings, and reported spectroscopic measurements on the polar molecule tetrahydrofuran which confirm the pseudo rotational motion.

6.4 The Pressure Dependence of the Heat Capacity of Gases

6.4.1 Non-ideal Behaviour of Gases

In the limit of zero pressure measurements of the isothermal Joule-Thomson coefficient ϕ (Equation (17)) and of the pressure dependence of the heat capacity χ (Equation (18)), both are functions of the second virial coefficient, and measurements of either quantity yields values of B. Figure 6.3 shows a selection of $(C_p - C_p^0)$ values derived from measurements of the heat capacity C_p of benzene,[17,25] acetone[26] and methanol[27,28] made over a range of pressure at several temperatures.

That the curves for methanol ($\mu = 1.87$ D) are much steeper than those for acetone ($\mu = 2.88$ D) can be interpreted as being due to the formation of hydrogen bonds between the methanol molecules. For the non-polar fluid benzene the line through the points is almost straight.

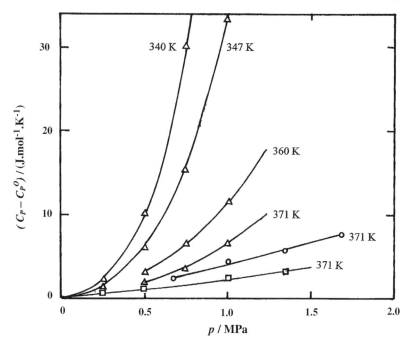

Figure 6.3 $(C_p - C_p^0)$ for ($\bigcirc$) acetone,[26] ($\square$) benzene,[17,25] and ($\triangle$) methanol[27,28] at selected temperatures.

Figure 6.4 (right) shows values of $-T^2\mathrm{d}^2B/\mathrm{d}T^2$ derived from $\mathrm{d}C_p/\mathrm{d}p$ measurements for benzene, acetone and methanol. The lower curve is for methanol, and the upper curve is for benzene. Values of B can be obtained by fitting the values of $-T^2\mathrm{d}^2B/\mathrm{d}T^2$ to an empirical equation based on a square well potential, or by fitting them to Kihara or Stockmayer potentials. To obtain best accuracy it is usual to include measurements of B in the data set and find parameters which simultaneously fit both sets of measurements.[25] Figure 6.4 (left) shows values of B for the three gases obtained in this way. It can be seen that at high temperatures B for methanol is small, but at low temperatures more hydrogen bonds form and it becomes more negative than B for acetone and benzene. Similar behaviour is exhibited for water and higher alcohols, and is attributed[27,28] to the association of the vapour into clusters of 2, 3 or 4 molecules *etc*.

6.4.2 Association of Like Molecules in the Gas Phase

It is evident from Figure 6.3 that the heat capacity isotherms for methanol have much greater curvature than those for benzene. Lambert[29] and Woolley[30] attributed this behaviour to hydrogen bonding between pairs, and to a lesser extent clusters of 3 or 4 methanol molecules. To a first approximation each cluster can be regarded as behaving ideally. The formation of dimer, trimer *etc*

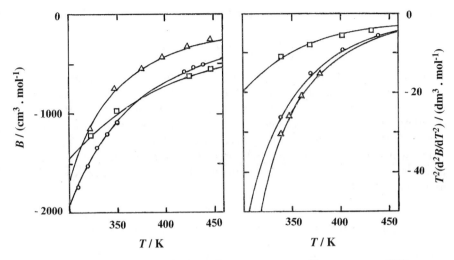

Figure 6.4 (right) Values of $T^2(d^2B/dT^2)$ for ($\bigcirc$) acetone,[26] ($\square$) benzene,[17,25] and ($\triangle$) methanol[27,28] obtained from measurements of dC_p/dp. (left) Values of B from pVT measurements. The equations which fit the measurements of B also fit the measurements of $T^2(d^2B/dT^2)$.

can be described in terms of equilibrium constants K_2, K_3, K_4 *etc*, and the total pressure is simply the sum of the partial pressures p_x of the clusters. When the summation for p_x is inserted into a series for the total pressure the compressibility factor takes the form:[30]

$$pV/RT = 1 - K_2 p + (3K_2^2 - 2K_3)p^2 \\ + (12K_2K_3 - 10K_2^3 - 3K_4)p^3 + \text{...} \tag{37}$$

Comparison with Equation (12) yields the second and third pressure series virial coefficients:

$$B' = -K_2(RT) \quad \text{and} \quad C' = (4K_2^2 - 2K_3)(RT)^2 \tag{38}$$

Each equilibrium constant is a function of temperature of the form:

$$\ln\{K(T_2)/K(T_1)\} = -(\Delta H/R)(1/T_2 - 1/T_1) \tag{39}$$

Where ΔH is the enthalpy of dimerisation, trimerisation *etc*. Following a suggestion made by Woolley[30], Ginell[31] subtracted a temperature independent excluded volume b so that the equation of state becomes:

$$p(V - b)/(RT) = \sum_{n=1}^{\infty} N_n \tag{40}$$

where N_n is the number of n-mers. The observed second virial coefficient is now:

$$B(\text{obs}) = B' = b - K_2(RT) \tag{41}$$

Wormald[32] pointed out that b should be a function of temperature and suggested that the second virial coefficient of a suitable homomorph should replace b. For example fluoromethane is an appropriate homomorph for methanol as it has almost the same dipole moment, polarisibility, shape and size. Analysis of B and $-T^2 d^2 B/d T^2$ for methanol by Weltner et al.[27] shows that an association model which assumes dimers and tetramers fits experiment better than a dimer-trimer model. This conclusion is supported by infra-red spectroscopy[33] and molecular orbital calculations.[34]

6.4.3   The Enthalpy of Mixing of Gases at Low Pressures

The heat capacity of a mixture can, in principle, be measured in the same kind of calorimeter that is used for pure component measurements. For binary mixtures, greater accuracy can be obtained by measuring the excess molar enthalpy H_m^E in a flow mixing calorimeter and deriving the excess heat capacity C_p^E from the temperature dependence:

$$C_p^E = (\partial H_m^E/\partial T)_p \tag{42}$$

The first flow mixing calorimeter for gases at pressures close to atmospheric reported by Wormald[35] was of differential design. Subsequent designs[36-38] incorporate many improvements. An advantage of the differential arrangement is that any Joule-Thomson effects or kinetic effects or heat leaks are automatically cancelled out. Furthermore exothermic mixing experiments can be performed by adjusting the energy supplied to the heater in the second calorimeter until the temperature rise matches that in the first calorimeter. Details of the differential apparatus are shown in Figure 6.5. The pure gases or

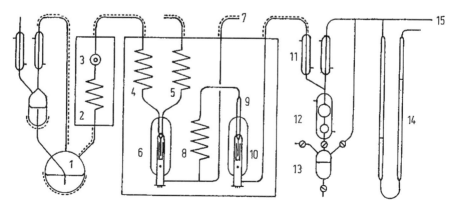

Figure 6.5   A differential flow mixing calorimeter for gases at low pressures.[38]

vapours pass through heat exchange coils 4 and 5 into mixing calorimeter 6. The gas mixture flows through heat exchange coil 8 into a second calorimeter 10 and then condensers 11 and calibrated bulbs 12 where the condensate flow rate is measured. Thermocouples in the two calorimeters are connected in series to make the instrument differential.

Measurements of H_m^E are a valuable source of information about second virial cross coefficients B_{12}, particularly for associated fluids for which pVT measurements are spoiled by adsorption errors. For a gas at low densities, and neglecting third virial coefficients, the excess molar volume is:

$$V_m^E = x_1 x_2 (2B_{12} - B_{11} - B_{22}) \tag{43}$$

Wormald[35,36] showed that H_m^E is given by:

$$
\begin{aligned}
H_m^E = {} & x_1 x_2 p (2\phi_{12} - \phi_{11} - \phi_{22}) \\
& - (p^2/RT)(B\phi - x_{11}B_{11}\phi_{11} - x_{22}B_{22}\phi_{22})
\end{aligned}
\tag{44}
$$

where ϕ is given by Equation (17). $B\phi$ is the product of B and ϕ for the mixture, where:

$$B = x_1^2 B_{11} + 2x_1 x_2 B_{12} + x_2^2 B_{22} \quad \text{and} \quad \phi = x_1^2 \phi_{11} + 2x_1 x_2 \phi_{12} + x_2^2 \phi_{22} \tag{45}$$

Cross terms B_{12} and ϕ_{12} are calculated from potential parameters for the pure components using suitable combining rules. For many non-associating gas mixtures, both polar and non-polar, values of H_m^E calculated from Equation (44) are in excellent agreement with experiment.[39,40]

6.4.4 The Enthalpy of Mixing of Associated Gases

If component 2 is self-associated B_{22} and ϕ_{22} are for the homomorph and:

$$B = (x_1^2 B_{11} + 2x_1 x_2 B_{12} + x_2^2 B_{22}) - x_2^2 K_{22} RT \tag{46}$$

$$\phi = (x_1^2 \phi_{11} + 2x_1 x_2 \phi_{12} + x_2^2 \phi_{22}) + x_2^2 K_{22} \Delta H_{22} \tag{47}$$

If the unlike molecules associate then B_{12} and ϕ_{12} are calculated from pair potential parameters for the two pure components, no homomorph is assumed.

$$B = (x_1^2 B_{11} + 2x_1 x_2 B_{12} + x_2^2 B_{22}) - x_1 x_2 K_{12} RT/2 \tag{48}$$

$$\phi = (x_1^2 \phi_{11} + 2x_1 x_2 \phi_{12} + x_2^2 \phi_{22}) + x_1 x_2 K_{12} \Delta H_{12}/2 \tag{49}$$

When either Equations (46) and (47) or (48) and (49), are inserted into Equation (44), experimental values of H_m^E give ΔH and K at the reference temperature $T = 298.15$ K.

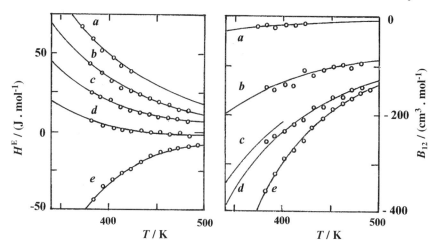

Figure 6.6 The enthalpy of mixing of (a) $(0.5H_2O + 0.5Ar)$,[47] (b) $(0.5H_2O + 0.5H_2S)$,[44] (c) $(0.5H_2O + 0.5SO_2)$,[45] (d) $(0.5H_2O + 0.5HCl)$,[43] (e) $(0.5H_2O + 0.5NH_3)$[46] (left) and values of B_{12} obtained from the measurements (right). For clarity values of B_{12} for H_2O-SO_2 are not shown.

An extensive set of measurements on binary mixtures containing benzene or cyclohexane mixed with 14 polar fluids[41] has been made using this technique, the object being to obtain ΔH_{12} for (benzene + polar fluid). Another set of measurements on 28 binary mixtures[42] containing steam up to high pressures has been successfully fitted using an association model. Finally an all glass plug-in mixing calorimeter[43] has been developed and used to make measurements on corrosive mixtures such as H_2O plus HCl,[43] H_2S,[44] SO_2,[45] and NH_3.[46] These measurements made at $p = 0.1013$ MPa are shown in Figure 6.6. Measurements on $(0.5H_2O + 0.5Ar)$[47] are also shown for comparison. Values of B_{12} obtained from the measurements are shown in the right-hand section of the figure. For $H_2O + HCl$, the enthalpy of hydrogen bond formation is 0.83 times that of the $H_2O + H_2O$ hydrogen bond, for $H_2O + H_2S$ the ratio is 0.74, for $H_2O + SO_2$ it is 0.92, and for $H_2O + NH_3$ it is 1.22.

6.5 The Enthalpy of Mixing of Gases at High Pressures

Beenakker *et al.*[48] were first to measure H_m^E for simple gas mixtures up to 15 MPa, with an accuracy of about 5 percent. Lee *et al.*,[49] Hejmadi *et al.*,[50] Klein *et al.*,[51] Wormald *et al.*[52] and Ba *et al.*[53] reported gas phase measurements made with mixing calorimeters of improved design. Naumowicz *et al.*[54] reported measurements on mixtures containing ammonia made using a high pressure differential apparatus. Wormald *et al.*[55] developed a high temperature (698 K) high pressure (25 MPa) flow mixing calorimeter and reported extensive measurements on 28 mixtures containing steam.[56] At pressures up to 10 MPa these

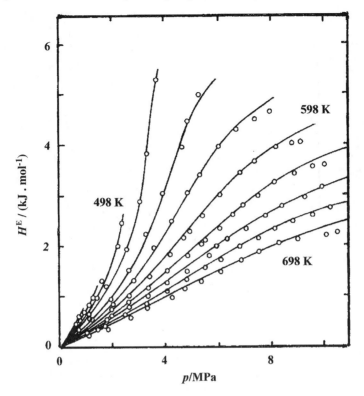

Figure 6.7   The enthalpy of mixing, H_m^E, for $(0.5H_2O + 0.5C_6H_{14})$ up to 10 MPa. The curves through the points were calculated from a cubic equation of state.[59]

could be fitted by a virial equation of state.[57] By modifying a cubic equation of state developed by Kubic[58] so that it gave a good fit to the residual enthalpy of steam at pressures up to 20 MPa, Wormald *et al.*[59] achieved a good fit to all the H_m^E measurements up to 698 K and 15 MPa. The fit to measurements on (water + hexane) are shown in Figure 6.7.

Measurements on (water + ethane)[60] and (water + carbon dioxide)[61] at pressures from (15 to 25.5) MPa could not be well fitted by the cubic equation, but a two fluid corresponding states model suggested by Rowlinson[62] and developed by Fenghour *et al.*[63] was found to be an excellent fit. The results are shown in Figure 6.8.

6.6   Heat Capacities of Gases at High Pressures

Ernst *et al.*[64] have described high pressure flow calorimetric apparatus for the measurement of the isobaric heat capacity of pure gases and gas mixtures. The calorimeter is capable of operation up to about 100 MPa and 450 K. The

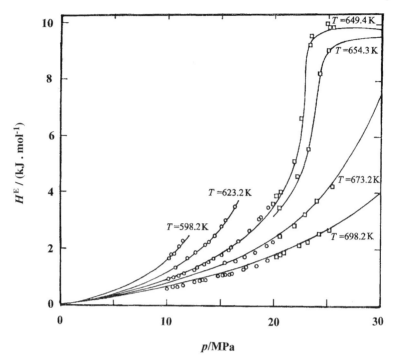

Figure 6.8 The enthalpy of mixing, H_m^E, of $(0.5H_2O + 0.5CO_2)$ plotted against pressure.[61] The curves through the points were calculated from a two fluid corresponding states model.[62,63]

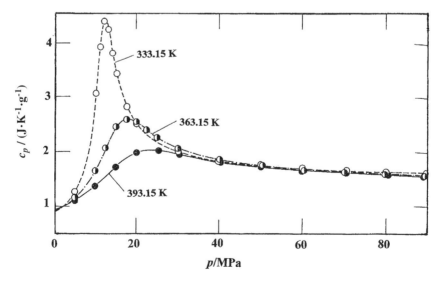

Figure 6.9 The specific heat capacity c_p of carbon dioxide up to 90 MPa.[11]

apparatus was tested by making measurements[11] on supercritical carbon dioxide at 333 K, 363 K and 393 K at pressures up to 90 MPa, as shown in Figure 6.9.

Another version of the apparatus[65] was constructed with a throttling calorimeter for Joule-Thomson coefficient measurements mounted in series with a calorimeter for isobaric heat capacity measurements. This calorimeter was used to make measurements on the mixture $(0.85CH_4 + 0.15C_2H_6)$ and on CF_3CHFCF_3 at pressures up to 15 MPa and 423 K.

Pressures of 100 MPa are about the limit of what can be achieved by flow calorimetry. To obtain data at higher pressures constant volume cells mounted in a pressure controlled scanning calorimeter can be used. Randzio[12] has constructed several calorimeters of this type, some of them designed for pressures as high as 400 MPa.[66] These calorimeters open the way for much future work.

References

1. W. F. G. Swann, *Proc.Roy.Soc.*, 1909, **82A**, 147.
2. H. L. Callender and H. T. Barnes, *Phil Trans.*, 1902, **199**, 55–149.
3. J. P. Joule and W. Thomson, *Phil.Mag.*, 1852, **4**, 481.
4. R. A. Buckingham, *An outline of the theory of thermodynamics, N.Y.*, 1900, **127**.
5. S.C. Collins and F. G. Keyes, *Proc. Am. Acad. Arts. Sci.*, 1937, **72**, 283.
6. K. S. Pitzer, *J. Am. Chem. Soc.*, 1941, **63**, 2413.
7. R. Spitzer and K. S. Pitzer, *J. Am. Chem. Soc.*, 1946, **68**, 2537.
8. J. E. Kilpatrick, K. S. Pitzer and R. Spitzer, *J. Am. Chem. Soc.*, 1947, **69**, 2483.
9. G. Waddington, S. S. Todd and H. M. Huffman, *J. Am. Chem. Soc.*, 1947, **69**, 22.
10. J. L. Hales, J. D. Cox and E. B. Lees, *Trans. Faraday Soc.*, 1963, **59**, 1544.
11. G. Ernst, G. Maurer and E. Wiederuh, *J. Chem. Thermodyn.*, 1989, **21**, 53.
12. S. L. Randzio, in *Solution Chemistry,* K. N. M. Marsh and P. A. G. O'Hare, eds., IUPAC Chemical Data Series No 39, Blackwell, Oxford, 1994.
13. R. A. Alberty and R. J. Sibley, *Physical Chemistry,* Wiley, NY, 1992.
14. J. H. Noggle, *Physical Chemistry,* Little, Brown & Co, Boston, 1985.
15. D. A. McQuarrie, *Statistical Mechanics,* Harper and Row, NY, 1976.
16. R. S. Berry, S. A. Rice and J. Ross, *Physical Chemistry,* OUP, 2000.
17. D. W. Scott, G. Waddington, J. C. Smith and H. M. Huffman, *J. Chem. Phys.*, 1947, **15**, 565.
18. J. D. Kemp and K. S. Pitzer, *J. Chem. Phys.*, 1936, **4**, 749.
19. K. S. Pitzer, *Disc.Faraday Soc.*, 1951, **10**, 66.
20. E. B. Wilson Jr., *Adv. Chem. Phys.*, 1959, **2**, 367.
21. S. Weiss and G. E. Leroi, *J. Chem. Phys.*, 1986, **48**, 962.

22. J. G. Aston, S. C. Schumann, H. L. Fink and P. M. Doty, *J. Am. Chem. Soc.*, 1941, **63**, 2029.
23. D. O. Harris, G. G. Engerholm, C. A. Tolman, A. C. Luntz, R. A. Keller, H. Kim and W. D. Gwinn, *J. Chem. Phys.*, 1969, **50**, 2438.
24. G. G. Engerholm, A. C. Luntz, W. D. Gwinn and D. O. Harris, *J. Chem. Phys.*, 1969, **50**, 2446.
25. C. J. Wormald, *J. Chem. Soc. Faraday Trans.*, 1975, **71**(1), 726.
26. R. E. Pennington and K. A. Kobe, *J. Am. Chem. Soc.*, 1957, **79**, 300.
27. W. Weltner and K. S. Pitzer, *J. Am. Chem. Soc.*, 1951, **73**, 2606.
28. J. F. Counsell and D. A. Lee, *J. Chem. Thermodyn.*, 1973, **5**, 583.
29. J. D. Lambert, G. A. H. Roberts, J. S. Rowlinson and V. J. Wilkinson, *Proc. Roy. Soc.*, 1949, **A196**, 113.
30. H. W. Woolley, *J. Chem. Phys.*, 1953, **21**, 236.
31. R. Ginell, *J. Chem. Phys.*, 1955, **23**, 2395.
32. M. Massucci, A. P. du'Gay, A. M. Diaz-Laviada and C.J. Wormald, *J. Chem. Soc. Faraday Trans.*, 1992, **88**, 427.
33. R. G. Inskeep, F. E. Dixon and H. M. Olson, *J. Mol. Spectrosc.*, 1960, **5**, 284.
34. L. S. Curtiss, *J. Chem. Phys.*, 1977, **67**, 114.
35. C. J. Wormald, *Proc. 1st Int. Conf. Calorimetry and Thermodynamics*, Warsaw, 1969, p. 601.
36. C. J. Wormald, *J. Chem. Thermodyn.*, 1997, **9**, 901.
37. C. J. Wormald, *J. Chem. Thermodyn.*, 1997, **29**, 701.
38. J. A. Doyle, J. C. Mayer and C. J. Wormald, *Z. Phys. Chem. N. F.*, 1981, **124**, 12.
39. C. J. Wormald, E. J. Lewis and D. J. Hutchings, *J. Chem. Thermodyn.*, 1979, **11**, 1.
40. J. Doyle, D. J. Hutchings, N. M. Lancaster and C. J. Wormald, *J. Chem. Thermodyn.*, 1997, **29**, 677.
41. C. J. Wormald, *J. Chem. Thermodyn.*, 2000, **32**, 1091.
42. C. J. Wormald, *Proc. 13th Int. Conf. Props Water and Steam*, NRC Press, Toronto, 1999, p. 355.
43. C. J. Wormald, *J. Chem. Thermodyn.*, 2003, **35**, 417.
44. C. J. Wormald, *J.Chem. Thermodyn.*, 2003, **35**, 1019.
45. C. J. Wormald, *J. Chem. Thermodyn.*, 2003, **35**, 91.
46. C. J. Wormald and B. Wurzberger, *J. Chem. Thermodyn.*, 2001, **33**, 1193.
47. P. Richards and C. J. Wormald, *Z. Phys. Chem. N. F.*, 1981, **128**, 35.
48. J. J. M. Beenakker, B. van Eijnsbergen, M. Knoester, K. W. Taconis and P. Zandbergen, *Proc. Symp. Thermophys. Prop.*, 1965, **3**, 114.
49. J. I. Lee and A. E. Mather, *J. Chem. Thermodyn.*, 1970, **2**, 881.
50. A. V. Hejmadi, D. L. Katz and J. E. Powers, *J. Chem. Thermodyn.*, 1971, **3**, 483.
51. R. R. Klein, C. O. Bennett and B. F. Dodge, *AIChE J.*, 1971, **17**, 958.
52. C. J. Wormald, K. L. Lewis and S. Mosedale, *J. Chem. Thermodyn.*, 1977, **9**, 27.

53. Le B. Ba, S. C. Kallaguine and R. S. Ramalaho, *J. Chem. Eng. Data*, 1982, **27**, 436.

54. E. Naumowicz and W. Woycicki, *J. Chem. Thermodyn.*, 1984, **16**, 1081.

55. C. J. Wormald and C. N. Colling, *J. Chem. Thermodyn.*, 1983, **15**, 725.

56. C. J. Wormald, *Proc.13th Int. Conf. Water and Steam*, NRC Press, Toronto, 1999, p. 355.

57. N. M. Lancaster and C. J. Wormald, *J. Chem. Soc. Faraday Trans.*, 1988, **84**, 3159.

58. W. L. Kubic, *Fluid Phase Equilibria*, 1982, **9**, 79.

59. C. J. Wormald and N. M. Lancaster, *J. Chem. Soc. Faraday Trans.*, 1989, **85**, 1315.

60. C. J. Wormald, M. J. Lloyd and A. Fenghour, *Int. J. Thermophysics*, 2000, **21**, 85.

61. C. J. Wormald, M. J. Lloyd and A. Fenghour, *J. Chem. Thermodyn.*, 1997, **29**, 1253.

62. J. S. Rowlinson and I. D. Watson, *Chem. Eng. Sci.*, 1969, **24**, 1565.

63. A. Fenghour, W. A. Wakeham and J. T. R. Watson, *High Temperatures High Pressures*, 1994, **26**, 241.

64. G. Ernst, B. Keil, H. Wirbser and M. Jaeschke, *J. Chem. Thermodyn.*, 2001, **33**, 601.

65. H. Wirbser, G. Brauning, J. Guntner and G. Ernst, *J. Chem. Thermodyn.*, 1992, **24**, 761.

66. S. L. Randzio, *Pure Appl. Chem.*, 1991, **63**, 1409.

CHAPTER 7

Heat Capacity of Electrolyte Solutions

ANDREW W. HAKIN AND MOHAMMAD M. H. BHUIYAN

Department of Chemistry & Biochemistry, University of Lethbridge, 4401 University Drive, Lethbridge, Alberta, T1K 3M4, Canada

7.1 Introduction

As a second derivative property of the Gibbs energy, the heat capacity of a system is a thermodynamic property that yields valuable insights into the structure of a system. With respect to electrolytes in aqueous solution this translates into information related to solute–solvent and solute–solute interactions. In identifying possible experimental approaches to obtaining heat capacity data for aqueous electrolyte solutions we begin by recognizing that the change in standard state heat capacity, ΔC_p^{o}, is a first derivative property of the change in standard state enthalpy, ΔH°, with respect to temperature, T:

$$\Delta C_p^{\text{o}} = \left(\frac{\partial \Delta H^{\circ}}{\partial T}\right)_p \tag{1}$$

In other words, heat capacities may be calculated from the results of experiments which yield enthalpy data for a given process (obtained under conditions of constant pressure) as a function of temperature. For example, heat capacity data for aqueous electrolyte systems may be obtained from traditional calorimetric experiments in which enthalpies of solution[1-4] (or even enthalpies of dilution[5,6]) are measured as a function of temperature. Enthalpy is

Heat Capacities: Liquids, Solutions and Vapours
Edited by Emmerich Wilhelm and Trevor M. Letcher
© The Royal Society of Chemistry 2010
Published by the Royal Society of Chemistry, www.rsc.org

in turn, a first derivative property of the change in standard state Gibbs energy, ΔG°, with respect to temperature:

$$\Delta H^\circ = -T^2 \left(\frac{\partial \Delta G^\circ / T}{\partial T} \right)_p \tag{2}$$

Alternatively, values for ΔH° may be obtained using the temperature dependencies of equilibrium constants, K:

$$\Delta H^\circ = RT^2 \left(\frac{\partial \ln K}{\partial T} \right)_p \tag{3}$$

where R is the gas constant. Values for ΔC_p° may therefore be related to the temperature dependencies of equilibrium constants using the equation:

$$\Delta C_p^\circ = R \left[2T \left(\frac{\partial \ln K}{\partial T} \right)_p + T^2 \left(\frac{\partial^2 \ln K}{\partial T^2} \right)_p \right] \tag{4}$$

At least in theory, utilization of Equation (4) provides an alternative route to the determination of standard changes in heat capacity. However, both the enthalpy and equilibrium constant routes to heat capacities are hindered by the realization that uncertainties in data are magnified when subjected to differentiation. Therefore only high precision enthalpimetric data and extremely precise equilibrium constant data obtained as a function of temperature are suitable for the calculation of heat capacities. In this chapter, we concentrate on experimental approaches in which precise values for heat capacities are obtained directly from experiment. Such techniques offer, through integration, the potential of yielding extremely precise equilibrium constant data which, in turn, can be used to interpret the temperature dependent speciation of a system.

7.2 Experimental Approaches

Most modern calorimetric methods used to probe the thermodynamics of aqueous electrolyte solutions yield apparent molar heat capacities, $C_{p\phi}$. Apparent molar heat capacities are defined in terms of the heat capacity of the solution, $C_p(\text{solution})$, and the heat capacity of the pure solvent, C_{p1}^*:

$$C_{p\phi} = \frac{C_p(\text{solution}) - n_1 C_{p1}^*}{n_2} \tag{5}$$

where n_1 and n_2 define the number of moles of pure solvent and solute in the solution, respectively. Apparent molar heat capacities are related to the partial molar heat capacities of the solute, $\overline{C_p}$, using the equation:

$$\overline{C_p} = n_2 \left(\frac{\partial C_{p\phi}}{\partial n_2} \right)_{n_1} + C_{p\phi} \tag{6}$$

Under the condition that $n_2 = 0$ (often referred to as the infinite dilution condition in which the molality of the solute of the solution, m, is equal to zero), the partial molar property is equal to the apparent molar property:

$$\overline{C_p^o} = C_{p\phi}^o \qquad (7)$$

where $\overline{C_p^o}$ is the standard partial molar heat capacity of the solute and $C_{p\phi}^o$ is the apparent molar heat capacity at infinite dilution.

Prior to the early 1970s, when flow methods were introduced to measure heat capacities, the majority of heat capacity data reported in the literature for aqueous electrolyte systems were obtained from precise enthalpy measurements made at a number of temperatures. The latter data were routinely measured with batch calorimeters and solution calorimeters. The comprehensive reviews of Parker[7] and Latysheva[8] provide excellent summaries of the scope of heat capacity determinations for aqueous electrolyte solutions during this period. More recent advances in calorimetry permit rapid and precise heat capacity measurements over extended ranges of temperature and pressure.[9-13] This chapter will focus on two calorimetric techniques which have been central in advancing our knowledge of the heat capacities of aqueous electrolyte systems. These are the techniques of flow calorimetry and differential scanning calorimetry (DSC).

7.2.1   Flow Calorimetry

First developed in the early 1970s, the flow technique developed by Picker *et al.*[9] became, and in many respects has remained, the technique of choice for obtaining precise apparent molar heat capacities of dilute aqueous electrolyte solutions. Although the majority of measurements conducted with Picker Dynamic Microcalorimeters have been limited to ambient pressure conditions and temperatures in the range $278 \leq T \, (/\text{K}) \leq 343$, the flow technique itself has been improved upon and utilized in the construction of a number of high temperature and pressure flow calorimeters.[10,11,13] The latter instruments, are by necessity, more complicated in their design and construction but have paved the way for significant advances in the theoretical modeling and our understanding of dilute aqueous electrolyte systems under elevated temperature and pressure conditions. Given the central role that flow methods have played in developing our current knowledge of the heat capacities of aqueous electrolyte solutions, a brief description of the fundamentals of the flow calorimetric technique will be provided.

In flow calorimetry liquids flow through the calorimeter at a constant rate and are thermostated to a selected base temperature of interest before being subjected to a controlled temperature increase that is designed to increase the temperature of the liquid by an amount ΔT. Under these conditions a volumetric heat capacity, σ, may be calculated from the applied power, W_o, required to raise the temperature of the liquid by the increment ΔT and the known volumetric flow rate of the liquid, f:

$$\sigma = \frac{W_0}{f \Delta T} \tag{8}$$

The massic heat capacity, c_p, of the liquid of interest may be calculated from the experimentally determined volumetric heat capacity, σ, using the measured, or known, density of the liquid, ρ:

$$c_p = \frac{\sigma}{\rho} \tag{9}$$

Although the latter equations could be used to calculate heat capacities, the design of the Picker Dynamic Microcalorimeter (Figure 7.1) removes the necessity to actively monitor flow rate. The Picker calorimeter contains two calorimetric cells which are connected in series but separated by a delay line. This arrangement ensures that the liquid flow rates in the two cells are identical.

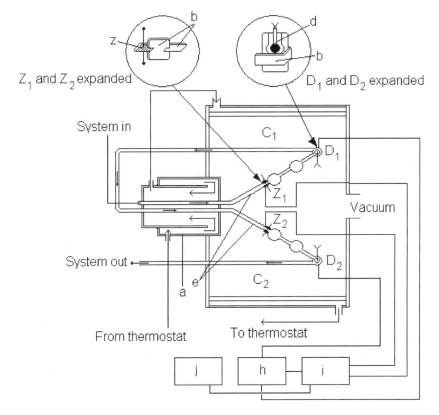

Figure 7.1 Schematic diagram of the Picker Dynamic Microcalorimeter: (a), pre-thermostated water jacket; (b), crimped stainless-steel tubing; (d), thermistor detectors; (e) stainless steel tubing; (h), digital voltmeter; (i), feedback circuit; (j), computer; (z), zener diode.[61]

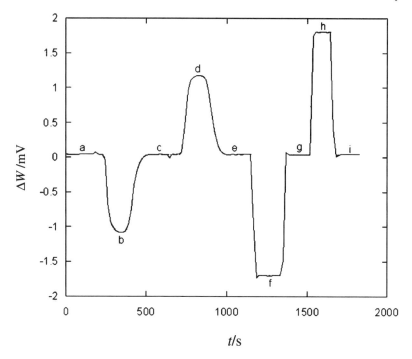

Figure 7.2   A plot of change in power, ΔW, versus time, t, for a typical calorimetric measurement performed using a Picker Dynamic Microcalorimeter.[61]

The instrument is calibrated by allowing solvent to flow through the system and adjusting and measuring the power (supplied via two zener diodes (z), Z_1 and Z_2) required to produce identical temperature increases in each cell. These temperature increases are typically $T = 2$ K or less and are automatically maintained by an electronic feedback system that constantly monitors the output resistances of the thermistors, D_1 and D_2.

A plot of a change in power versus time for a typical calorimetric measurement experiment is shown in Figure 7.2. An experimental measurement is initiated by allowing solvent to flow through system and monitoring the difference in power supplied to the two zener diodes. With pure solvent flowing through both cells the difference in power (recorded in terms of a voltage difference) is approximately zero and represents a baseline signal (baseline a in Figure 7.2). When a sample solution is introduced into the calorimeter the power supplied to zener diode Z_1 is automatically adjusted to once again ensure identical thermistor readings at the thermistors, D_1 and D_2. This difference in power $\Delta W_{s,w}$ (peak b in Figure 7.2) is related to the difference between the volumetric heat capacity of the sample solution, σ_s, and that of the pure solvent, σ_w:

$$\frac{\sigma_s - \sigma_w}{\sigma_w} = \frac{\Delta W_{s,w}}{W_o} \tag{10}$$

where W_o is the average power supplied to the zener diodes. The volumetric heat capacities are converted to massic heat capacities by introducing the densities of the solvent, ρ_w, and sample solutions, ρ_s, to yield:

$$\frac{c_{ps}\rho_s}{c_{pw}\rho_w} - 1 = \frac{\Delta W_{s,w}}{W_o} \tag{11}$$

where c_{pw} is the massic heat capacity of the pure solvent and c_{ps} is the massic heat capacity of the sample solution, respectively.

Once the sample solution has passed completely through the first cell of the calorimeter, the delay line, and the second cell of the calorimeter, voltage readings associated with $\Delta W_{s,w}$ once again return to close to baseline levels (baseline c in Figure 7.2). At this stage solvent is reintroduced into the system and a new signal, $\Delta W_{w,s}$, is recorded (peak d in Figure 7.2). This signal corresponds to the difference between the volumetric heat capacity of the solvent and the volumetric heat capacity of the sample solution:

$$\frac{\sigma_w - \sigma_s}{\sigma_s} = \frac{\Delta W_{w,s}}{W_o} \tag{12}$$

In terms of massic heat capacities, this equation may be written in the form:

$$\frac{c_{pw}\rho_w}{c_{ps}\rho_s} - 1 = \frac{\Delta W_{w,s}}{W_o} \tag{13}$$

Electrical calibrations of each calorimetric cell are required to quantify the changes in power associated with the recorded power differences $\Delta W_{s,w}$ and $\Delta W_{w,s}$. Electrical calibrations are performed by supplying known amounts of power to each zener diode in turn with only pure solvent flowing through the system (peaks f and h in Figure 7.2, respectively).

Commercial flow microcalorimeters of the Picker design require a correction for power loss, since a small amount of the electrical energy added to either cell via the zener diodes to increase the temperature of the sample solution or the solvent is lost to the environment. This heat loss may be accounted for in the determination of the massic heat capacities by the introduction of a heat loss correction factor. The method of Desnoyers *et al.*[14] for calculating heat loss correction factors that utilizes aqueous sodium chloride solutions has been widely used.

For aqueous solutions, apparent molar heat capacities, $C_{p\phi}$, may be calculated from experimentally determined massic heat capacities using the usual equation:

$$C_{p\phi} = M \cdot c_p + \frac{c_p - c_{pl}}{m} \tag{14}$$

where M and m are molar mass of the solute and the molality of the sample solution, respectively.

Precise values for apparent molar heat capacities should be accompanied by an analysis of uncertainty. Uncertainties in calculated values for $C_{p\phi}$, $\delta C_{p\phi}$, may be approximated using the standard propagation of error method:

$$\left(\delta C_{p\phi}\right)^2 = \left(\frac{\partial C_{p\phi}}{\partial c_p}\right)^2 \left(\delta c_p\right)^2 + \left(\frac{\partial C_{p\phi}}{\partial m}\right)^2 \left(\delta m\right)^2$$

$$+ \left(\frac{\partial C_{p\phi}}{\partial c_{p1}}\right)^2 \left(\delta c_{p1}\right)^2 + \left(\frac{\partial C_{p\phi}}{\partial M}\right)^2 \left(\delta M\right)^2 \tag{15}$$

However, if it is assumed that the uncertainties associated with the molar mass of the solute, δM, and massic heat capacity of the solvent, δc_{p1}, are negligible then the equation can be simplified to:

$$\left(\delta C_{p\phi}\right)^2 \approx \left(\frac{mM + 1}{m}\right)^2 \left(\delta c_p\right)^2 + \left(\frac{c_p - c_{p1}}{m^2}\right)^2 \left(\delta m\right)^2 \tag{16}$$

For a Picker Dynamic Microcalorimeter (Picker calorimeter) the uncertainty in the experimental massic heat capacity, δc_p, is typically $7 \times 10^{-2}\,\mathrm{J\,K^{-1}\,kg^{-1}}$ and therefore the uncertainty associated with the sample solution concentration becomes an important factor to minimize.

Further consideration of Equations (11) and (13) clearly highlights the need for precise density data in the determination of apparent molar heat capacities for dilute aqueous electrolyte solutions. With this in mind, many experimental setups of Picker calorimeters also include a high precision vibrating tube densimeter. The densimeter is usually placed in series with the calorimeter such that solution samples pass through the densimeter prior to entering the calorimeter, thus minimizing the total volume of sample solution required for measurements.

As indicated previously, the dynamic flow design of the Picker calorimeter was seminal to the field and paved the way for other flow calorimeters designed to operate under elevated temperature and pressure conditions. The starting point in this evolution were the designs of Smith-Magowan and Wood[10] and Rogers and Pitzer.[11] Instruments designed to function under elevated temperature and pressure conditions are necessarily more complex in their design and greater attention must be given to accounting for heat and power loss. Readers interested in learning further details of heat loss models are referred to the work of Carter and Wood,[15] White and Wood,[16] White and Downes,[17] and Downes, Hakin, and Hedwig.[18]

7.2.2 Differential Scanning Calorimetry

Alternatives to the use of flow calorimetric techniques for the determination of precise heat capacities for dilute aqueous electrolyte solutions have been limited. However, in 1997 Woolley[12] reported a technique that utilized a commercial, fixed-cell, power-compensation, differential-output, temperature-scanning calorimeter. In this technique the difference between the calorimetric outputs of the twin cells are found to be proportional to the difference in the volumetric heat capacities of the liquids in the cells. Three series of experiments are necessary to determine the heat capacities of aqueous sample solutions. In the first, the differential calorimetric signal is measured with both cells filled with pure water. This is referred to as a "baseline" experiment. In the second, the differential calorimetric signal is recorded over the desired temperature range and pressure when pure water is placed in the reference cell and a solution of known volumetric heat capacity is placed in the sample cell. This is referred to as a "calibration" experiment and is ideally performed at the same scan rate, r, as the baseline experiment. When combined with the results of the baseline experiment, the known densities and massic heat capacities of the calibration solution and pure water, the calibration experiment permits the calculation of calorimeter calibration factors, (f/f°)

$$(f/f^\circ) = (r/r^\circ)(c_p\rho - c_{p1}\rho_1)/(\Delta P_{cal} - \Delta P_{base}) \qquad (17)$$

In this equation, ΔP_{cal} and ΔP_{base} are the differences in power applied to the heaters in the calorimetric cells to maintain them at the same temperature during the "calibration" and "baseline" experiments, respectively.

The third series of experiments involves measuring the differential calorimetric signal with water in the reference cell and the sample solution of interest in the sample cell. When combined with the results of the baseline experiment, and the calculated calorimeter calibration constants the results of the third experiment may be used with Equation (17) to calculate the massic heat capacities of the sample solution of interest at temperatures within the scanned temperature range. It is noted, however, that this calculation is once again dependent upon knowing the precise densities of the sample solution, ρ, over the same range of temperature and the desired experimental pressure. When used to analyze a series of dilute NaCl(aq) solutions[12] the DSC technique was shown to be capable of producing volumetric heat capacities with relative precisions in the range 10^{-5} to 5×10^{-5} over the temperature range $283.15 \leq T\,(/K) \leq 393.15$ and $p = 0.35\,MPa$.

In addition to operating at temperatures above ambient, DSC instruments fitted with appropriate cooling accessories have been used to measure the heat capacities of aqueous electrolyte solutions in cold-stable and supercooled states.[19]

7.2.3 Other Experimental Approaches

There are, of course, numerous other experimental methods available for the determination of the heat capacities of aqueous electrolyte solutions under non-ambient temperature and pressure conditions. These include the batch calorimetry techniques of isoperibol calorimetry and drop calorimetry. Indeed, many of these techniques have been utilized to obtain heat capacities of electrolyte solutions under extremely corrosive high temperature and pressure conditions. The scope of this chapter does not permit further exploration of these techniques but the interested reader is directed to the work of Simonson and Mesmer[20] that provides a comprehensive overview of the area.

7.3   Some Theoretical Considerations

7.3.1   The Chemical Potential

The partial molar Gibbs energy is of such central importance to the field of chemical thermodynamics that it is routinely identified as the chemical potential, μ.[21] The chemical potential of an ion i, μ_i, in aqueous solution under conditions of constant temperature and pressure may be defined by the equation:

$$\mu_i = \mu_i^{\circ} + RT \ln a_i \tag{18}$$

where μ_i° is the chemical potential of a defined standard state and a_i is the activity of ion i. Alternatively if the activity of the ion i is expressed as the product of its molality, m_i, and its activity coefficient γ_i then:

$$\mu_i = \mu_i^{\circ} + RT \ln \left(m_i \gamma_i / m^{\circ} \right) \tag{19}$$

By extension, the chemical potential of a simple, fully dissociated, electrolyte $v_+ M^{z^+} v_- X^{z^-}$ (where v_+ and v_- are stoichiometric coefficients and z^+ and z^- are charge numbers) of molality m may be written as:

$$\mu_{MX} = \mu_{MX}^{\circ} + v_+ RT \ln \left(\frac{v_+ m \gamma_+}{m^{\circ}} \right) + v_- RT \ln \left(\frac{v_- m \gamma_-}{m^{\circ}} \right) \tag{20}$$

where γ_+ and γ_- are identified as the activity coefficients of the cations and anions contained within the salt MX and $m^{\circ} = 1 \, \mathrm{mol\,kg^{-1}}$. Alternatively, if a mean ionic activity coefficient, $\gamma_{\pm}$, is defined in the usual way and a quantity Q is defined by the equation:

$$Q = \left(v_+^{v_+} v_-^{v_-} \right)^{1/v} \tag{21}$$

where $v = v_+ + v_-$, then Equation (20) may be written in the form:

$$\mu_{MX} = \mu_{MX}^{\circ} + vRT \ln \left(\frac{Qm \gamma_{\pm}}{m^{\circ}} \right) \tag{22}$$

By definition, in the limit $m \to 0$ then $\gamma_{\pm} = 1.0$. When applied to Equation (22), this condition defines an ideal solution in which there are no ion–ion interactions. It is therefore clear that a better understanding of ion–ion interactions in real aqueous salt solutions may be obtained by appropriate theoretical developments of equations which define the temperature and pressure dependence of the mean ionic activity coefficient. In this regard, the theory described by Debye and Hückel,[22] from which $\gamma_\pm$ values can be calculated for dilute aqueous salt solutions, arguably represents one of the most significant breakthroughs in the field of aqueous electrolyte solution theory.

7.3.2 Debye–Hückel Theory

The most readily recognized form of the Debye–Hückel equation is:

$$\log \gamma_\pm = \frac{-A_\gamma |z_+ z_-| \sqrt{I}}{1 + aB_\gamma \sqrt{I}} \tag{23}$$

where a is an ion size parameter (in Angstroms), and A_γ and B_γ are solvent and temperature dependent constants.[21] The ionic strength, I, of the electrolyte solution is defined by the equation:

$$I = 0.5 \sum m_i z_i^2 = \omega m \tag{24}$$

where the symbol ω is used to identify a valency factor.

In very dilute aqueous electrolyte solutions, Equation (23) can be approximated by the Debye–Hückel Limiting Law in which the mean ionic activity coefficient is predicted to have a linear dependence on the square root of the ionic strength of the solution:

$$\log \gamma_\pm = -A_\gamma |z_+ z_-| \sqrt{I} \tag{25}$$

Accepting the general applicability of Debye–Hückel theory, the concentration dependence of any partial molar property, $\bar{Y}$, may be described by an equation of the form:

$$\bar{Y} = \overline{Y^o} + \overline{A_Y} \omega I^{1/2} \tag{26}$$

where $\overline{Y^o}$ is the standard partial molar property and $\overline{A_Y}$ is the associated Debye–Hückel limiting slope. It follows using simple thermodynamic relationships that the limiting slope for partial molar heat capacities, $\overline{A_C}$, is related to second differential of A_γ with respect to temperature:

$$\overline{A_C} = 2RT^2 \left(\frac{\partial^2 A_\gamma}{\partial T^2}\right)_p \tag{27}$$

However, as indicated previously, apparent molar properties and not partial molar properties lend themselves more readily to experimental determination. Partial molar properties are related to apparent molar properties, Y_ϕ, using the ionic strength by a general equation of the form:

$$\bar{Y} = Y_\phi + I\left(\frac{\partial Y_\phi}{\partial I}\right)_{T,p} \tag{28}$$

In an analogous form to Equation (26), it may be shown that for dilute aqueous electrolyte solutions:

$$Y_\phi = \overline{Y^o} + \omega A_Y I^{1/2} \tag{29}$$

where $A_Y = (2/3)\,\overline{A_Y}$. For studies of more concentrated solutions this equation may be extended in the following manner:

$$Y_\phi = \overline{Y^o} + \omega A_Y I^{1/2} + B_Y I + C_Y I^{3/2} \tag{30}$$

where B_Y and C_Y are adjustable parameters which may be obtained by fitting the equation to experimentally determined apparent molar data. It is noted that Equation (30) (or forms of Equation (30)) has provided a convenient method for the determination of standard partial molar heat capacities. Indeed, this approach is observed in many early (and many current) investigations of the apparent molar heat capacities of aqueous electrolyte solutions. In comparing $\overline{C_p^o}$ values for electrolytes obtained in this manner, the reader is urged to pay attention to the source of values for Debye–Hückel limiting slope data used in the various analyses. Values for limiting slopes obtained from different sources may result in small variations in reported infinite dilution values. The limiting slopes reported by Archer and Wang[23] are now widely utilized. However, other studies have used values reported by Ananthaswamy and Atkinson[24] and others.[25–27]

An alternative form of the extended Debye–Hückel equation is attributed to Güntelberg:[28]

$$\log \gamma_\pm = -A_\gamma |z_+ z_-| \frac{I^{1/2}}{1 + I^{1/2}} \tag{31}$$

Using this equation as the starting point Equation (29) takes the form:

$$Y_\phi = \overline{Y^o} + \frac{3}{2}\omega A_Y \left(\frac{1}{\varLambda} - \frac{\sigma}{3}\right) I^{1/2} + B_Y I + C_Y I^{3/2} \tag{32}$$

where

$$\varLambda = 1 + I^{1/2} \tag{33}$$

and

$$\sigma = 3 \frac{\left(\varLambda - \frac{1}{\varLambda} - 2\ln(\varLambda)\right)}{I^{3/2}} \tag{34}$$

These extended equations have been successfully utilized by numerous authors to model the ionic strength dependences of apparent molar heat capacities of aqueous electrolyte solutions.

Several studies[29,30] have also probed the temperature dependences of the adjustable parameters B_Y and C_Y. Various forms of the temperature dependence of these parameters have been investigated but in general it is found that a reasonable starting point is to assume a simple power series relationship of the general form:

$$B_Y = \sum_{i=1}^{j} a_i T^{i-1} \tag{35}$$

$$C_Y = \sum_{i=1}^{k} b_i T^{i-1} \tag{36}$$

where a_i ($i = 1$ to j) and b_i ($i = 1$ to k) are fitting parameters. Various forms of these equations should be investigated with the final forms being determined on the basis of the statistical significance of each parameter.

7.3.3 Chemical Relaxation

Any discussion of heat capacities of aqueous electrolyte solutions requires consideration of the effects of temperature on any equilibrium processes present in the system which result in changes in the relative concentrations of species present in the system. Woolley and Hepler[31] are credited with naming 'relaxation effects'; however, other authors have also contributed significantly to the discussion.[32,33] The following represents a brief introduction to the phenomenon of chemical relaxation and its impact on measured apparent molar heat capacities.

Consider a system in a state of chemical equilibrium that is maintained at constant temperature T and constant pressure p. The total enthalpy of the system, H, may be represented as the sum of the partial molar enthalpies of all components within the system. For a system with j components this may be written as:

$$H = \sum_{i=1}^{j} n_i \overline{H_i} \tag{37}$$

where n_i is used to identify the number of moles of each component i and $\overline{H_i}$ is the partial molar enthalpy of each component i in the system. Changes in the enthalpy of the system resulting from a change in system temperature at constant pressure leads to a definition of the heat capacity:

$$C_p = \left(\frac{\partial H}{\partial T}\right)_p = \sum_{i=1}^{j} n_i \left(\frac{\partial \overline{H_i}}{\partial T}\right)_p + \sum_{i=1}^{j} \overline{H_i} \left(\frac{\partial n_i}{\partial T}\right)_p \tag{38}$$

In circumstances where there is no change in composition of the system with change in temperature, the final term of the equation is zero. However, for those systems which undergo changes in composition, as a result of a change in temperature, the final term of Equation (38) clearly plays a role in determining the total heat capacity. The import of the latter term is highlighted when one considers (as detailed previously in this chapter) that heat capacities are often experimentally determined by imposing small temperature increases on the system of interest under constant pressure conditions.

In terms of apparent molar heat capacities:

$$C_{p\phi} = C_{p\phi}^{\text{exp}} - \frac{\left(\frac{\partial \overline{H}}{\partial T}\right)_p^{\text{relax}}}{m} \tag{39}$$

where $\left(\partial \overline{H}/\partial T\right)_p^{\text{relax}}$ identifies a relaxation contribution to the apparent molar heat capacity. Under conditions in which this relaxation term is found to be zero (as is the case observed for many simple, completely dissociated electrolytes in aqueous solution under close to ambient temperature and pressure conditions), the apparent molar heat capacity of the system, $C_{p\phi}$, is equal to an experimentally determined apparent molar heat capacity $C_{p\phi}^{\text{exp}}$. Following the method of Woolley and Hepler,[31] Equation (39) may be rewritten in the form:

$$C_{p\phi} = C_{p\phi}^{\text{exp}} - \sum_{i=1}^{j} \Delta \overline{H}_{\text{rxn}(i)} \left(\frac{\partial \alpha_i}{\partial T}\right)_{p,m(\text{rxn}(i))} \tag{40}$$

where $\Delta \overline{H}_{\text{rxn}(i)}$ is the change in partial molar enthalpy associated with each equilibrium reaction i and $(\partial \alpha_i/\partial T)_{p,m(\text{rxn}(i))}$ is used to identify the change in extent of each reaction i with change in temperature. Readers who wish to explore the effects of chemical relaxation on the apparent molar heat capacities of aqueous electrolyte systems which exhibit high degrees of speciation are directed to the study of the aqueous rare earth sulfate systems reported by Marriott *et al.*[34]

7.3.4 Pitzer Equations

Modeling the concentration dependences of heat capacities of more concentrated electrolyte solutions requires an increase in the range of the Debye–Hückel equations by the use of virial expansions. By analogy with Equation (19), the excess Gibbs energy of an ion i in an aqueous solution of the salt MX, G_i^{E}, may be written as:

$$G_i^{\text{E}} = RT \ln \gamma_i \tag{41}$$

The total excess Gibbs energy of the system, $G_{\text{total}}^{\text{E}}$, may be defined by applying a virial expansion to a simple version of the Debye–Hückel equation:

$$G_{\text{total}}^{\text{E}} = RTn_w \left(f(I)_{\text{D-H}} + \sum_i \sum_j \lambda_{ij} m_i m_j + \sum_i \sum_j \sum_k \mu_{ijk} m_i m_j m_k \right) \quad (42)$$

In this equation, the $f(I)_{\text{D-H}}$ term is a form of the Debye–Hückel equation, n_w identifies the number of moles of pure solvent (water), and the concentration terms, m_i, m_j, and m_k identify the molality of each species (i, j, or k) in solution. In addition, the second virial coefficient is identified as λ_{ij} and the third virial coefficient as μ_{ijk}. These coefficients describe pair interactions between the ions i and j and the triplet interactions between ions i, j, and k, respectively. The total excess heat capacity of the system may be obtained by differentiating the total excess Gibbs energy with respect to temperature under the condition of constant pressure:

$$C_{p,\text{total}}^{\text{E}} = -T \left(\frac{\partial^2 G_{\text{MX}}^{\text{EX}}}{\partial T^2} \right)_p \quad (43)$$

The resulting Pitzer ion-interaction equation that can be used to model apparent molar heat capacities for the salt $v_+ M^{z+} v_- X^{z-}$ in aqueous solution takes the form:

$$C_{p\phi} = \overline{C_p^{\circ}} + \frac{v|z_+ z_-| A_J \ln(1 + bI^{1/2})}{2b} - 2v_+ v_- RT^2 (mB_{\text{MX}}^J$$

$$+ (v_+ z_+) m^2 C_{\text{MX}}^J) \quad (44)$$

where the constant b is assigned a value of $1.2\,\text{kg}^{1/2}\,\text{mol}^{-1/2}$, the constant C_{MX}^J is found to be independent of ionic strength and the parameter B_{MX}^J is found to be ionic strength dependent. The ionic strength dependence of the B_{MX}^J parameter is modeled using the equation:

$$B_{\text{MX}}^J = \left[\beta^{(0)J} + 2\beta^{(1)J} f(I) \right] \quad (45)$$

where $\beta^{(0)J}$ and $\beta^{(1)J}$ are fitting parameters and the ionic strength function $f(I)$ is defined by the equation:

$$f(I) = \frac{[1 - (1 + \alpha I^{1/2}) \exp(-\alpha I^{1/2})]}{\alpha^2 I} \quad (46)$$

where $\alpha = 2.0\,\text{kg}^{-1/2}\,\text{mol}^{-1/2}$. As with Equation (30), the temperature dependence of the apparent molar heat capacity may be modeled by developing equations which attempt to model the temperature dependences of the constant C_{MX}^J and the $\beta^{(0)J}$ and $\beta^{(1)J}$ fitting parameters.

There is an extensive literature that describes the application of Pitzer ion-interaction equations to apparent molar volumes and heat capacities of aqueous electrolyte systems over a range of temperatures at ambient pressure and a detailed review is unwarranted in this chapter. However, the work of Criss and Millero[35,36] is notable in this regard as it not only provides a summary of apparent molar heat capacity data for a large number of aqueous electrolyte systems but also consistently models the data with Pitzer equations. It is also noted that Pitzer ion-interaction equations have been used with good success to model the thermodynamic parameters associated with aqueous mixed electrolyte solutions and also the thermodynamic properties of aqueous electrolyte systems under elevated temperature and pressure conditions. For detailed discussions of the Pitzer equations readers are referred to the work of Pytkowicz[37] and Pitzer.[38,39]

7.3.5 The Semi-empirical Model of Helgeson–Kirkham–Flowers (HKF)

Attempts to estimate and model the standard thermodynamic properties of aqueous electrolyte solutions at increasingly elevated temperatures and pressures have resulted in the creation of semi-empirical models. Of significant importance in this regard is the work of Helgeson and colleagues.[40–43] An extensive literature discusses the Helgeson–Kirkham–Flowers (HKF) model and therefore this chapter will focus only on how equations can be developed within the HKF framework which model the temperature and pressure dependences of standard partial molar heat capacities of aqueous electrolyte solutions.

A starting point for the application of the HKF model to standard partial molar heat capacities of electrolyte solutions is the separation of the property into a solvation, $\Delta \overline{C^o_{ps}}$, and a non-solvation, $\Delta \overline{C^o_{pn}}$, contribution:

$$\overline{C^o_p} = \Delta \overline{C^o_{ps}} + \Delta \overline{C^o_{pn}} \tag{47}$$

Development of the solvation contribution to the standard partial molar heat capacity begins by defining the change in Gibbs energy, $\Delta \overline{G}^o_{k,s}$, associated with the solvation of a salt k as:

$$\Delta \overline{G}^o_{k,s} = \omega_{e,k} \left(\frac{1}{\varepsilon} - 1 \right) \tag{48}$$

where ε is the dielectric constant of the solvent and $\omega_{e,k}$ is a Born parameter. Further, the Born parameter for a salt k is identified as the sum of the Born parameters of the ions, $\omega_{e,j}$, contained within the salt (e.g., $\omega_{e,Y(ClO_4)_3} = \omega_{e,Y^{3+}} + 3\omega_{e,ClO_4^-}$) and expressions for $\omega_{e,j}$ are based on Born equations which in turn contain contributions from the ionic radius of each ion, $r_{e,j}$. In the original HKF model the ionic radii of cations and anions were related to the crystallographic radii of the ions, $r_{x,j}$ using the relationships:

$$r_{e,j}(\text{cations}) = r_{x,j} + 0.94|z_j| \tag{49}$$

and

$$r_{e,j}(\text{anions}) = r_{x,j} \tag{50}$$

However, in the revised HKF model,[44] equations for $r_{e,j}$ were modified to become functions of temperature and pressure such that:

$$r_{e,j}(\text{cations}) = r_{x,j} + (0.94 + g)|z_j| \tag{51}$$

and

$$r_{e,j}(\text{anions}) = r_{x,j} + |z_j|g \tag{52}$$

In these equations, g is a term that is identified as having complex dependencies on temperature and pressure. Using expressions for $\omega_{e,j}$ utilized within the revised HKF model and the simple thermodynamic relations:

$$\Delta \overline{S}_s^{\,o} = -\left(\frac{\partial \Delta \overline{G}_s^{\,o}}{\partial T}\right)_p \tag{53}$$

$$\Delta \overline{C}_{ps}^{\,o} = T\left(\frac{\partial \Delta \overline{S}_s^{\,o}}{\partial T}\right)_p \tag{54}$$

the solvation contribution of the standard partial molar heat capacity may be defined as:

$$\Delta \overline{C}_{ps}^{\,o} = \omega_e TX + 2TY\left(\frac{\partial \omega_e}{\partial T}\right)_p - T\left(\frac{1}{\varepsilon} - 1\right)\left(\frac{\partial^2 \omega_e}{\partial T^2}\right)_p \tag{55}$$

In this equation, X and Y are identified as Born coefficients which are obtained from temperature derivatives of the dielectric constant of the solvent. This complex equation can be dramatically simplified by recognizing that at $T < 423\,\text{K}$ the temperature derivatives of ω_e are effectively zero. Under these conditions Equation (55) is simplified to:

$$\Delta \overline{C}_{ps}^{\,o} = \omega_e TX \tag{56}$$

The non-solvation contribution to the standard partial molar heat capacity, $\Delta \overline{C}_{pn}^{\,o}$, is described by the equation:

$$\Delta \overline{C}_{pn}^{\,o} = c_1 + c_2\left(\frac{1}{(T - \Theta)}\right)^2$$
$$- 2T\left(\frac{1}{(T - \Theta)}\right)^3\left(a_3(p - p_r) + a_4 \ln\left(\frac{\Psi + p}{\Psi + p_r}\right)\right) \tag{57}$$

where c_1, c_2, a_3, and a_4 are solute dependent coefficients, p_r identifies a reference pressure, and Θ and Ψ are solvent-dependent parameters. For water these parameters are assigned values of 228 K and 2600 bars, respectively.

Clearly, the full expression for the temperature and pressure dependence of the standard partial molar heat capacity of a salt in aqueous solution using the revised HKF model may be obtained by combining the Born equation derived solvation component (Equation (55)) with the empirical non-solvation contribution defined by Equation (57). However, under the conditions that $p = p_r = 1$ bar and $T < 423$ K, the HKF model yields the following simplified equation for the temperature dependence of the standard partial molar heat capacity of a salt in aqueous solution:

$$\overline{C_p^{\text{o}}} = c_1 + c_2 \left(\frac{1}{(T - \Theta)} \right)^2 + \omega_e TX \tag{58}$$

HKF equations have been utilized within the literature to model the temperature and pressure dependences of standard partial molar heat capacities (and standard partial molar volumes) of numerous aqueous electrolyte systems. For example, Hovey and Hepler utilized the semi-empirical HKF approach to model standard partial molar heat capacities for HNO_3(aq) and $HClO_4$(aq)[29] and H_2SO_4(aq)[45] whilst Xiao and Tremaine[46] applied the model to standard partial molar heat capacity data for $LaCl_3$(aq), $La(ClO_4)_3$(aq) and $Gd(ClO_4)_3$(aq). Readers interested in obtaining more detailed descriptions of the HKF model and its application to standard partial molar thermodynamic parameters at temperatures to $T = 1273$ K and pressures to $p = 5000$ bar are directed to the work of Shock and Helgeson.[47]

When used in combination, the Pitzer ion-interaction equations and the HKF model provide a powerful tool to model the apparent molar heat capacities of aqueous electrolytes over extended temperature and pressure ranges.

7.4 Studies of Apparent Molar Heat Capacities and Standard Partial Molar Heat Capacities of Aqueous Electrolyte Solutions

An exhaustive literature review of studies reporting apparent molar heat capacities or standard partial molar heat capacities of aqueous electrolyte solutions is not within the scope of this chapter. However, the interested reader is directed to the work of Criss and Millero,[35,36] Hepler and Hovey,[48] and the references contained therein, for an overview of studies completed using a Picker calorimeter at $T = 298.15$ K and $p = 0.1$ MPa. More generally, under conditions of ambient pressure and close to ambient temperature ($283.15 \leq T/$ K ≤ 328.15) the standard partial molar heat capacities of aqueous electrolyte systems are observed to increase with increasing temperature.

Studies of the apparent molar heat capacities of aqueous electrolyte solutions at elevated temperatures and pressures reported in the literature are limited compared with those conducted under ambient, or close to ambient, conditions. However, a listing of some of the more important studies is provided by Simonson and Mesmer[20] and is supplemented by the more recent work reported of Hnedkovsky et al.[13] In general, as temperature increases (at constant applied pressure) both the apparent and standard partial molar heat capacities of aqueous electrolyte solutions pass through a shallow maximum and then begin to rapidly decrease to very large negative values near the critical point of water until a very sharp minimum is reached. The observed very large and negative values are attributed to large interactions between ions and water in combination with large changes in the properties of water as its critical point is approached. Beyond this minimum, heat capacities increase very rapidly with increasing temperature to values close to zero. Readers looking to gain a greater understanding of the behavior of the apparent molar or standard partial molar heat capacities of aqueous electrolyte solutions close to the critical point are directed to some of the studies reported by Wood et al.[49-52]

This well-documented high-temperature behaviour of aqueous electrolyte solutions places severe operating criteria on calorimeters looking to provide precise heat capacity data. It is not surprising that relatively few research groups in the world have been able to successfully address the challenges that such work entails.

7.5 Calorimetry and Pressure Effects

The investigation of pressure effects using calorimetry has been strongly advocated by Randzio[53-55] who developed an instrument called a transitiometer. Transitiometers combine heat-flux, Calvet-type calorimeters with high-pressure control systems to create instruments capable of making calorimetric measurements at pressures up to $p = 400\,\mathrm{MPa}$. In general, the complexity and cost of the required instrumentation have been significant reasons why there have been relatively few studies of the effects of pressure on the heat capacities of aqueous electrolyte solutions reported in the literature.

However, recent advances in DSC instrumentation have made the technique of pressure perturbation calorimetry (PPC) more accessible. Interest in PPC stems from the fact that the technique provides a rapid and convenient approach to obtaining volumetric data from a measured calorimetric signal. From elementary thermodynamics one can show that the change in heat flow, ΔQ_{rev}, in a calorimetric cell as a function of pressure is directly proportional to the expansibility, α, of the system under investigation:

$$\Delta Q_{\mathrm{rev}} = -TV\alpha\Delta p \tag{59}$$

where Δp identifies a small pressure change and V is the volume of the system. The expansibility is an important thermodynamic parameter as it offers information that describes the volume change of the system as a function of temperature. Previously, volumetric data of this nature could only be obtained from measurements of density. Although initially applied to the investigation of aqueous systems of biochemical/biological interest, where it was used to probe protein unfolding and also to look at specific solvation effects of groups exposed at the protein/solvent interface,[56] PPC has recently been used to obtain precise expansibility data for aqueous electrolyte solutions.[57]

Although there have been some differences in opinion regarding the instrumental and methodological approaches used in PPC,[58-60] the technique clearly has great potential to further our understanding of aqueous electrolyte solutions.

References

1. C. M. Criss and J. W. Cobble, *J. Am. Chem. Soc.*, 1961, **83**, 3223.
2. E. C. Jakel, C. M. Criss and J. W. Cobble, *J. Am. Chem. Soc.*, 1964, **86**, 5404.
3. R. E. Verrall and L. Dickson, *J. Solution Chem.*, 1975, **5**, 203.
4. J. W. Cobble and R. C. Murray, *Trans. Faraday Soc.*, 1978, **64**, 144.
5. A. S. Levine and S. Lindenbaum, *J. Solution Chem.*, 1973, **2**, 445.
6. E. E. Berraducci, L. R. Morss and A. R. Miksztal, *J. Solution Chem.*, 1979, **8**, 717.
7. V. B. Parker, *Thermal properties of aqueous 1:1 electrolytes, NSRDS-NBS 2,* Government Printing Office, Washington, D.C., 1965.
8. V. A. Latysheva, *Russ. Chem. Rev.*, 1973, **42**, 803.
9. P. Picker, P.-A. Leduc, P. R. Philip and J. E. Desnoyers, *J. Chem. Thermodyn.*, 1971, **3**, 631.
10. D. Smith-Magowan and R. H. Wood, *J. Chem. Thermodyn.*, 1981, **13**, 1047.
11. P. S. Z. Rogers and K. S. Pitzer, *J. Phys. Chem.*, 1981, **85**, 2886.
12. E. M. Woolley, *J. Chem. Thermodyn.*, 1997, **29**, 1377.
13. L. Hnedkovsky, V. Hynek, V. Majer and R. H. Wood, *J. Chem. Thermodyn.*, 2002, **34**, 755.
14. J. E. Desnoyers, C. de Visser, G. Perron and P. Picker, *J. Solution Chem.*, 1976, **5**, 605.
15. R. W. Carter and R. H. Wood, *J. Chem. Thermodyn.*, 1991, **23**, 1037.
16. D. E. White and R. H. Wood, *J. Solution Chem.*, 1982, **11**, 223.
17. D. E. White and C. J. Downes, *J. Solution Chem.*, 1988, **17**, 733.
18. C. J. Downes, A. W. Hakin and G. R. Hedwig, *J. Chem. Thermodyn.*, 2001, **33**, 873.
19. D. G. Archer and R. W. Carter, *J. Phys. Chem. B*, 2000, **104**, 8563.
20. J. M. Simson and R. E. Mesmer, in *Solution Calorimetry: Experimental Thermodynamics,* **Vol. IV**, ed. K. N. Marsh and P. A. G. O'Hare, *IUPAC Chemical Data Series No. 39,* Blackwell Scientific Publications, Oxford, 1994, p. 243.
21. R. A. Robinson and R. H. Stokes, *Electrolyte Solutions,* Butterworths Publications Ltd., London, 1959.
22. P. Debye and E. Hückel, *Phys. Z.*, 1923, **24**, 185.

23. D. G. Archer and P. Wang, *J. Phys. Chem. Ref. Data*, 1990, **19**, 371.
24. J. Ananthaswamy and G. Atkinson, *J. Chem. Eng. Data*, 1984, **29**, 81.
25. D. J. Bradley and K. S. Pitzer, *J. Phys. Chem.*, 1979, **83**, 1599.
26. L. F. Silvester and K. S. Pitzer, *J. Phys. Chem.*, 1977, **81**, 1822.
27. J. J. Spitzer, I. V. Olofsson, P. P. Singh and L. G. Hepler, *Can. J. Chem.*, 1979, **57**, 2798.
28. E. Güntelberg, *Z. Phys. Chem.*, 1926, **123**, 199.
29. J. K. Hovey and L. G. Hepler, *Can. J. Chem.*, 1989, **67**, 1489.
30. J. K. Hovey and L. G. Hepler, *J. Phys. Chem.*, 1988, **92**, 1323.
31. E. M. Woolley and L. G. Hepler, *Can. J. Chem.*, 1977, **55**, 158.
32. M. J. Blandamer, J. Burgess and J. M. W. Scott, *J. Chem. Soc. Faraday Trans. I*, 1984, **80**, 2881.
33. K. S. Pitzer, R. N. Roy and L. F. Silvester, *J. Am. Chem. Soc.*, 1977, **99**, 4930.
34. R. A. Marriott, A. W. Hakin and J. A. Rard, *J. Chem. Thermodyn.*, 2001, **33**, 643.
35. C. M. Criss and F. J. Millero, *J. Phys. Chem.*, 1996, **100**, 1288.
36. C. M. Criss and F. J. Millero, *J. Solution Chem.*, 1999, **28**, 849.
37. R. M. Pytkowicz, Ed. *Activity Coefficients in Electrolyte Solutions*, Vol. I, CRC Press, Boca Ratan, 1979.
38. K. S. Pitzer, in *Activity Coefficients in Electrolyte Solutions*, Vol. I, ed. R. M. Pytkowicz, CRC Press, Boca Ratan, 1979, p. 157.
39. K. S. Pitzer, in *Thermodynamic Modeling of Geological Materials: Minerals Fluids and Melts*, Vol. 17, Reviews in Minerology, Mineral Society of America, 1987, p. 97.
40. H. C. Helgeson and D. H. Kirkham, *Am. J. Sci.*, 1974, **274**, 1089.
41. H. C. Helgeson and D. H. Kirkham, *Am. J. Sci.*, 1974, **274**, 1199.
42. H. C. Helgeson and D. H. Kirkham, *Am. J. Sci.*, 1976, **276**, 97.
43. H. C. Helgeson, D. H. Kirkham and G. C. Flowers, *Am. J. Sci.*, 1981, **281**, 1249.
44. J. C. Tanger IV and H. C. Helgeson, *Am. J. Sci.*, 1988, **288**, 19.
45. J. K. Hovey and L. G. Hepler, *J. Chem. Soc. Faraday Trans.*, 1990, **86**, 2831.
46. C. Xiao and P. R. Tremaine, *J. Chem. Thermodyn.*, 1996, **28**, 43.
47. E. L. Shock and H. C. Helgeson, *Geochim. Cosmochim. Acta*, 1988, **52**, 2009.
48. L. G. Hepler and J. K. Hovey, *Can. J. Chem.*, 1996, **74**, 639.
49. R. H. Wood, D. E. White, J. A. Gates, H. J. Albert, D. R. Biggerstaff and J. R. Quint, *Fluid Phase Equilib.*, 1985, **20**, 283.
50. R. H. Wood, *Thermochim. Acta*, 1989, **154**, 1.
51. J. A. Gates, R. H. Wood and J. R. Quint, *J. Phys. Chem.*, 1982, **4948**, 86.
52. A. V. Sharygin and R. H. Wood, *J. Chem. Thermodyn.*, 1997, **125**, 29.
53. S. Randzio, J.-P. Grolier and J. R. Quint, *Rev. Sci. Instrum.*, 1994, **65**, 960.
54. S. Randzio, in *Solution Calorimetry: Experimental Thermodynamics*, Vol. IV, ed. K. N. Marsh, P. A. G. O'Hare, IUPAC Chemical Data Series No. 39, Blackwell Scientific, Oxford, 1994, p. 303.

55. S. Randzio, *Thermochim. Acta*, 2000, **355**, 107.
56. L.-N. Lin, J. F. Brandts, J. M. Brandts and V. Plotnikov, *Anal. Biochem.*, 2002, **302**, 144.
57. K. Boehm, J. Rsgen and H.-J. Hinz, *Anal. Chem.*, 2006, **78**, 984.
58. S. Randzio, *Thermochim. Acta*, 2003, **398**, 75.
59. P. Kujawa and F. M. Winnik, *Macromolecules*, 2001, **34**, 4130.
60. J. F. Brandts and L.-N. Lin, *Thermochim. Acta*, 2004, **414**, 95.
61. K. M. Erickson, MSc Thesis, Department of Chemistry and Biochemistry, University of Lethbridge, Alberta, Canada, 2007.

CHAPTER 8

Scanning Transitiometry and its Use to Determine Heat Capacities of Liquids at High Pressures

STANISLAW L. RANDZIO

Polish Academy of Sciences, Institute of Physical Chemistry, Kasprzaka 44/52, 01-224 Warsaw, Poland

8.1 Introduction

The pressure variable has often been neglected in the thermodynamic investigation of liquids, although even at the beginning of the 20th century Bridgman had pointed out its role in testing theories of liquids.[1] In particular, the role of pressure as an independent variable is very important and thermodynamic functions are often determined through direct measurements of their derivatives against an independent variable. Pressure derivatives were usually neglected in such procedures, mainly because their magnitudes are much smaller than the corresponding temperature derivatives. However, since measurements of the pressure derivatives are performed under isothermal conditions, the thermal noise is at a constant level, which facilitates the experiment. This is extremely important in heat capacity measurements, where application of high pressures requires massive thick vessels which have a heat capacity much more important than the heat capacity of the sample under investigation. Thus, determination of heat capacity through measurements of other

Heat Capacities: Liquids, Solutions and Vapours
Edited by Emmerich Wilhelm and Trevor M. Letcher
© The Royal Society of Chemistry 2010
Published by the Royal Society of Chemistry, www.rsc.org

thermodynamic derivatives appears advantageous at high pressures. With this in mind, it is helpful to consider the following thermodynamic relation:[2,3]

$$C_P(T, P) = C_{P,P_R}(T) - T \int_{P_R}^{P} V(p, T) \left[\alpha^2 + (\partial\alpha/\partial T)_P\right] \mathrm{d}P \qquad (1)$$

where

$$\alpha(P, T) = (1/V)(\partial V/\partial T)_p \qquad (2)$$

is the coefficient of thermal expansion and

$$V(P, T) = V(P, T_R) \exp\left\{ \int_{T_R}^{T} \alpha(P, T)\mathrm{d}T \right\} \qquad (3)$$

is the volume (molar or specific) of the substance under investigation. From the above equations for the determination of the isobaric heat capacity over wide ranges of pressure and temperature, the coefficient of thermal expansion is the main experimental variable which should be measured as a function of both pressure and temperature. The other experimental variables needed are the volume isotherm $V(P, T_R)$ as a function of pressure at a low reference temperature T_R and the heat capacity $C_P(T, P_R)$ as a function of temperature at a low reference pressure P_R.

8.2 Scanning Transitiometry

Figure 8.1 presents schematically a relatively new technique called *scanning transitiometry*.[4-7]

Scanning transitiometry involves scanning with one of the three variables (P, V or T) with the second one kept strictly constant (see Figure 8.2). During the scanning the variations of the dependent variables and the associated heat flux are simultaneously recorded. From these two quantities and the scanned variable two thermodynamic derivatives, thermal and mechanical, are simultaneously determined for the system under study. Figure 8.3 presents four thermodynamic situations covered by scanning transitiometry, from which, using the respective state variables and the heat effect, one can determine four pairs of thermodynamic derivatives.[8] Each of the situations has a specific application which confirms their usefulness. For the purpose of this chapter, details will be limited to the first situation[9,10] with a short comment on the third situation.

The enthalpy differential of a pure substance is described by Equations (4) and (5):

$$\mathrm{d}H(P, T) = \left(\frac{\partial H}{\partial T}\right)_P \mathrm{d}T + \left(\frac{\partial H}{\partial P}\right)_T \mathrm{d}P \qquad (4)$$

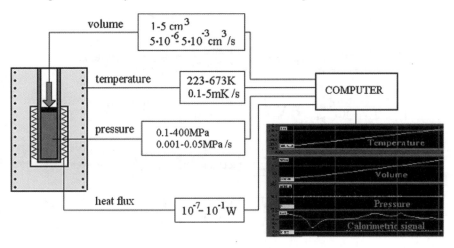

Figure 8.1 A schematic diagram of scanning transitiometry.

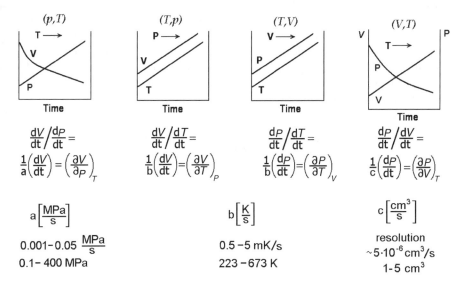

Figure 8.2 State variables in scanning transitiometry.

$$dH(P, T) = dQ + V dP \qquad (5)$$

When the temperature is kept constant and the pressure is varied as a linear function of time t such that: $dT = 0$, $P = P_o \pm bt$ and $dP = \pm b dt$, Equations (4) and (5) reduce to:

$$(\partial Q/\partial t)_T = q_T(P) = \pm \left\{ \left(\frac{\partial H}{\partial P} \right)_T - V \right\} b = \pm \left(\frac{\partial S}{\partial P} \right)_T bT = \pm \left(\frac{\partial V}{\partial T} \right)_P bT \qquad (6)$$

INPUTS   **OUTPUTS**

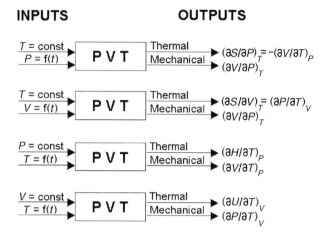

Figure 8.3   A thermodynamic scheme of scanning transitiometry.

where $q_T(P)$ is the heat flux measured at isothermal conditions and b is the rate of pressure variations. When the pressure variations are transferred to the calorimetric vessel through the substance under investigation (see Figure 8.4), its mass contained in the calorimetric detector changes during the course of the experiment and is equal to $V_E/V\,(P,T)$, where V_E is the internal volume of the calorimetric vessel directly exchanging heat with the calorimetric detector. Thus, Equation (6) for a given calorimeter can be transformed to:

$$q_T(P) = \pm\left(\frac{\partial V}{\partial T}\right)_P Tb\frac{V_E}{V} = \pm\alpha TbV_E \qquad (7)$$

When the pressure is varied as a series of n stepwise changes, the corresponding equation has the following form:

$$Q_n(P) = \pm\langle\alpha\rangle TV_E\Delta P_n \qquad (8)$$

where $\langle\alpha\rangle$ is the mean values of α for a given pressure step. Figure 8.4 presents an example of a calorimetric vessel adapted to measure α over wide pressure and temperature ranges.[3]

The vessel is closed with a cone plug and fixed in place with an internally threaded cover which also serves as a heat exchanger between the calorimetric vessel tubing and the calorimetric detector. During the pressure variations there is also a contribution to the thermal effect coming from the vessel itself. However, such a contribution is proportional to the thermal expansion of stainless steel from which the calorimetric vessel was made (α_{SS}).[10,11] Thus the full equation for the determination of α, for the linear pressure variations, has the following form:

$$\alpha = \alpha_{SS} - \frac{q_T(P)}{TbV_E} \qquad (9)$$

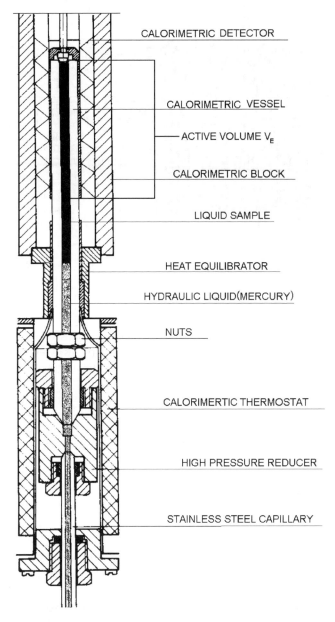

CALORIMETRIC DETECTOR

CALORIMETRIC VESSEL

ACTIVE VOLUME V$_E$

CALORIMETRIC BLOCK

LIQUID SAMPLE

HEAT EQUILIBRATOR

HYDRAULIC LIQUID(MERCURY)

NUTS

CALORIMERTIC THERMOSTAT

HIGH PRESSURE REDUCER

STAINLESS STEEL CAPILLARY

Figure 8.4 An experimental vessel for transitiometric measurements of thermal expansion coefficient for liquids.

or for the stepwise pressure variations:

$$\langle\alpha\rangle = \alpha_{SS} - \frac{Q_n(P)}{TV_E\Delta P_n} \tag{10}$$

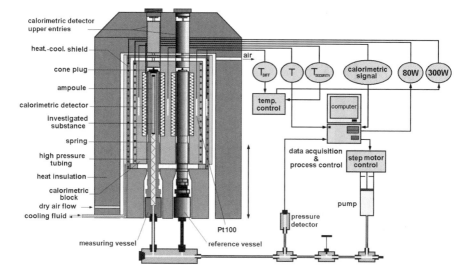

Figure 8.5   A schematic diagram of a scanning transitiometer.

The contribution from the stainless steel is rather small with respect to the thermal expansion of liquids and is only slightly dependent on both pressure and temperature. It is also worth adding that the pressure variations are realised with controlled displacements of a piston (or plunger) pump which are recorded simultaneously with the heat flux and the pressure variations. From both these displacements and pressure variations the compressibility is simultaneously measured (see the first situation in Figure 8.3).

Figure 8.5 presents a schematic diagram of a whole transitiometric installation. It consists of a calorimeter equipped with high-pressure vessels, a pVT system and LabView virtual instrument (VI) software. Two cylindrical calorimetric detectors ($\Phi = 17$ mm, $l = 80$ mm), made from 622 or 1244 thermocouples (chromel-alumel) each, are mounted differentially and connected to a nanovolt amplifier. The calorimetric detectors are placed in a metallic block, the temperature of which is directly controlled with a 22 bit digital feedback loop ($\approx 10^{-4}$ to 10^{-5}) K, which is part of the transitiometer software. The calorimeter block is surrounded by a heating-cooling shield. The temperature difference between the block and the shield is set at a constant value and is controlled by another controller. The temperature measurements, both absolute and differential, are made with calibrated Pt 100 sensors. The heaters are embedded in the outer surfaces of both the calorimeter block and the shield. The whole assembly is thermally insulated and enclosed in a stainless steel body fixed on a mobile stand, which allows the calorimeter to be moved up and down over the calorimetric vessels. When performing measurements near 273 K or below, dry air is pumped through the apparatus to prevent water condensation.

The calorimetric vessels are made from (0.8 to 0.47) cm internal diameter 316 stainless steel tubing and are fixed on a mounting table attached to the mobile stand. For very corrosive liquids the calorimetric vessels are made from has-telloy C22 alloy. Only the measuring vessel is connected to the pV line. The reference vessel acts only as a thermal reference, a stainless steel bar of appropriate dimensions is placed in it to balance the baseline of the differential calorimetric signal. The tubing of both measuring and reference vessels are connected to reducers, placed inside the calorimeter when it is in the lowered (measuring) position. The connections from the reducers to the manifold are made with thin stainless steel capillaries in order to reduce heat losses to the environment. The piston pump (total displaced volume is $9\,cm^3$) is driven by a stepping motor controlled by the transitiometer software (manual control is possible during preparatory operations). The pressure detector is a Viatran 245 transducer, of an appropriate pressure range with a relative uncertainty of 0.15% full scale deflection.

The pressure detector, the output of the calorimetric amplifier and the step-ping motor are connected to a NI PCI-MIO-16XE-50 multifunction board through a NI SCB-68 shielded connector block. The temperature measurements and digital control of the calorimetric block are performed through a serial port. The software, elaborated with the use of LabView language, performs as a Virtual Instrument (VI). It consists of ninety subVIs, each responsible for a particular function: pressure measurement, temperature measurement, counting the motor steps for recording the volume variations, measuring the calorimetric signal, *etc.* and each performs independently. However, all the subVIs form a hierarchical structure with a top window, where the experimenter can see simultaneously all four variables (pressure, P, volume variations, V, tempera-ture, T, and the heat flux, q, associated with the process under investigation) and the current status of the temperature and pressure control loops.

As was described previously,[11] the instrument can also easily be used to determine the reference isotherm $V(P,T_R)$, a parameter in Equation (3). The calorimetric vessel is filled with a known quantity of the liquid under investi-gation, the whole system is kept near the room temperature and compression/decompression cycles are realised by recording simultaneously the pressure and volume variations. Also the reference heat capacity isobar $C_p(T,P_R)$, a para-meter in Equation (1), can easily be determined in a transitiometer. To this end, light and small mass vessels adapted to low pressure work should be used. Such vessels can either be of classical construction described above, with the sample contained in a flexible ampoule as it is shown in Figure 8.5, or be of special construction where the separation of the sample from the hydraulic fluid is done with the help of a bellows or a membrane.[12] In both cases a gas should be used as the hydraulic fluid, which has a very low heat capacity with respect to a liquid. In this situation the third thermodynamic procedure in Figure 8.3 should be applied. In the case of an isobaric process ($dP = 0$; $T = T_o \pm at$ and $dT = \pm a dt$) Equations (1) and (2) reduce to:

$$(dQ/dt)_P = q_P(T) = \left(\frac{\partial H_m}{\partial T}\right)_P a = aC_P \qquad (11)$$

The above short description of the scanning transitiometric technique shows that it can easily be adapted to experimental determination of all the quantities needed to determine heat capacities of liquids under high pressures. Depending on the pressure and temperature ranges the relative uncertainty of the calorimetric measurements of α is near (1 to 3)%,[2–3,11] the relative uncertainty of the reference volume determination is near 0.6 %,[11] the relative uncertainty of the determination of the reference heat capacity can be as high as 0.3 %.[13] It is also worth adding that for many liquids there are precise literature data[14] for the saturated heat capacity over wide temperature ranges and thus the saturation line can be taken as a reference state for determination of pressure effects of the heat capacities of liquids. For such cases new experiments are not required.[3] It is also worth noting that the thermodynamic procedure used involves mainly integrating and adding the basic experimental data, which usually does not increase the total error. Thus, it can be concluded that the relative uncertainty of the determination of heat capacities of liquids under high pressures by the presented thermodynamic procedure and the scanning transitiometric technique is near ± 2 %.

8.3   Heat Capacities of Liquids under High Pressures

The behaviour of heat capacities of liquids under high pressures depends very much on their physicochemical nature. Thus, the analysis of the known data will be presented below separately for different kinds of liquids: simple; weakly associated; strongly associated; mixtures; *etc.*

8.3.1   Simple Liquids under High Pressures

In the absence of a quantitative theory for dense molecular liquids, the interpretation of experimental data for liquids over wide pressure and temperatures ranges can be done by comparison with the behaviour of simple liquids under similar conditions of pressure and temperature.[2] For a long time hexane has been proposed as such a model simple liquid.[15–17] On the basis of measurements performed in various laboratories, a collection of its thermophysical properties has been established, including numerical data for isobaric and isochoric heat capacities over the temperature range from 243.15 K to 503.15 K under pressures of up to 700 MPa.[3] Here, only an analysis of the data will be presented. Figure 8.6 presents the effect of pressure on the isobaric heat capacity of liquid hexane over middle and high temperature ranges. The most interesting observation is the existence of minima on the isotherms of the isobaric heat capacity. The minima are flattened and pushed to higher pressures with increasing temperature. Figure 8.7 shows that similar minima exist also on the isotherms of isochoric heat capacity, although they appear only at higher temperatures.

Before further analysis of this interesting observation we will try to verify the physical correctness of the analysed data. First of all the data can be compared with results of direct calorimetric heat capacity measurements. Figure 8.8

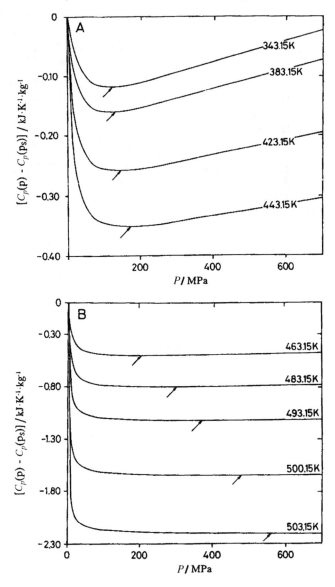

Figure 8.6 Pressure effect on the specific isobaric heat capacity of liquid hexane as a function of pressure over the middle-temperature range (A) and the high-temperature range (B). The arrows show the location of the minima in the curves.

presents, as solid lines, isotherms of isobaric specific heat capacity of hexane at 303.15, 403.15 and 503.15 K obtained from transitiometric data and, at discrete points from literature data[18] obtained by high-pressure flow calorimetry.

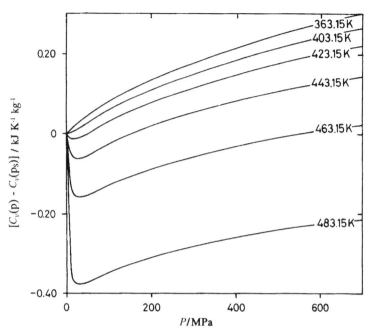

Figure 8.7   Pressure effect on the specific isochoric heat capacity of liquid hexane as a function of pressure at selected temperatures.

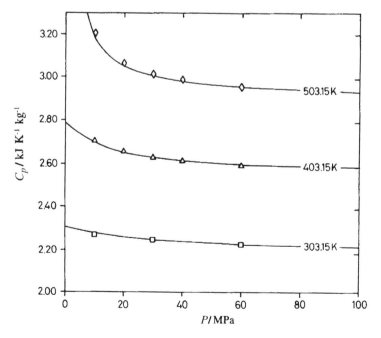

Figure 8.8   Comparison of specific isobaric heat capacity determinations by transitiometric procedure with flow calorimetric measurements from reference 18.

The agreement is within the error of the two techniques, *i.e.* 1 to 2 %; unfortunately, the flow calorimetry can only supply reliable data for measurements done under a pressure of a few dozen megapascals. Figure 8.9 presents isotherms of both C_P and C_V at 298.15 K obtained from transtiometric data (solid lines) and literature data[19] obtained in a high-pressure isoperibol heat capacity calorimeter. The literature data fall between the isobaric and the isochoric heat capacities. Analysis of the construction of the high-pressure heat capacity calorimeter and of the method of measurement[19] explains why the measured heat capacity is neither C_P nor C_V. The pressure in the calorimeter was established at the beginning of the measurement and not corrected for changes caused by heating during the heat capacity measurements. The measurement was thus not done at constant pressure. On the other hand the system was closed, but the measurement cannot be considered as a constant volume measurement because the calorimetric vessel was connected hydraulically to the pressure pump and the pressure changes caused by heating were partially compensated by compression of the hydraulic fluid. Thus, the results of these measurements must fall somewhere between constant-volume and constant-pressure conditions as indicated in Figure 8.9. However, it is worth noting that despite its unclear thermodynamic character the heat capacity isotherm measured in a high-pressure calorimeter is of great importance, because it also exhibits a minimum. Figure 8.10 presents isochoric heat capacities for liquid

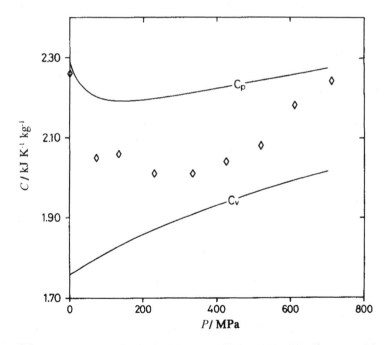

Figure 8.9 Comparison of specific heat capacities at 298.15 K from transitiometric data (lines) and determined with high-pressure calorimetric measurements (points) as described in reference 19.

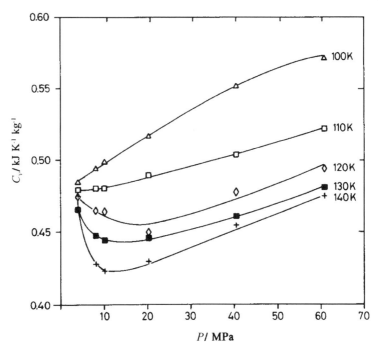

Figure 8.10   Isochoric heat capacity of liquid argon at selected pressures and temperatures as calculated from data in reference 20.

argon calculated from pVT equation of state and sound velocity data.[20] The results of these calculations show that over similar temperature ranges with respect to the critical temperatures, the shapes of C_V isotherms for liquid argon are similar to those obtained for hexane from transitiometric data. Similar minima on the heat capacity isotherms have also been observed for other simple liquids, such as butane and CO_2.[21] Thus, it can be concluded that the minima on the isotherms of heat capacities are an inherent property of dense simple liquids.

In an analysis of the minima in the isotherms of heat capacities of dense simple liquids one should consider both the physical and molecular causes of such behaviour. Figure 8.11 presents a set of isotherms of the coefficient of thermal expansion α for hexane. One can see that at low pressures α increases with temperature, while at high pressures the behaviour is opposite and α decreases with temperature. At pressure near 65 ± 2 MPa α becomes practically constant and does not depend on temperature (a crossing point of α isotherms). When inspecting Equation (1) one can easily conclude that the crossing point of α isotherms is the basic physical reason for the appearance of minima on the heat capacity isotherms, because at the pressure of the crossing point, $d\alpha/dT$ changes sign. The different relative magnitudes of $d\alpha/dT$ with respect to α^2 at various temperatures results in the minimum of the heat capacity isotherms, at higher temperatures to be flattened and pushed to higher pressures.

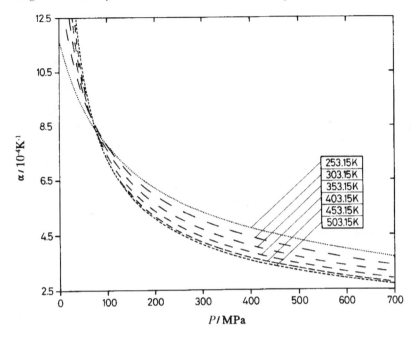

Figure 8.11   Thermal expansion coefficient (α) of hexane.

From the molecular point of view the change of sign of $d\alpha/dT$ can be interpreted as a result of a change in the shape of effective intermolecular potential in dense liquids caused by pressure.[22] Under the assumption that at high densities the main contribution to both heat capacity and thermal expansion comes from oscillations, the following interpretation can be given. A simplified model with an asymmetric intermolecular potential is presented in Figure 8.12. One can see that in such a model the distance between the equilibrium position r_e and the left (repulsive) branch of the potential is markedly smaller than the distance between the equilibrium position r_e and the right (attractive) branch of the potential, almost at each value of energy (situation *a*). So, the force acting on the molecule at distances greater than r_e is smaller than the force acting on the molecule at distances smaller than r_e. Thus, the average velocity of motion of molecules at distances greater than r_e is less than the average velocity at distances smaller than r_e. Accordingly, the molecule spends more time at distances $r > r_e$ than at distances $r < r_e$ and thus the mean distance apart is $r_e + x$ and not r_e. When the energy of vibration is further increased, by raising the temperature of the system, x increases by the same mechanism. On the macroscopic scale this is manifested by an increase in the dimension of the system. The amount by which x increases, when the energy is raised, depends on the shape of the effective intermolecular potential in which the molecule oscillates. There is an interesting possibility in which the shape of the potential is such that a linear relationship exists between the increase of the energy and the increase of x (situation *b*). In this case the thermal expansion coefficient

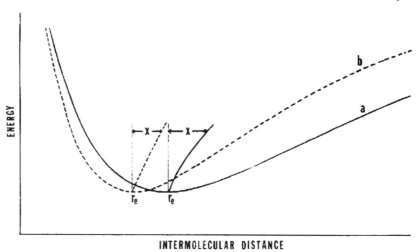

Figure 8.12 A model presentation of dependence of thermal expansion on temperature at normal pressure (a) and at high pressure where $d\alpha/dT = 0$ (b).

would be independent of temperature ($d\alpha/dT = 0$). In a real system this would appear as a crossing point of thermal expansion isotherms (see Figure 8.11).

The unique property of the crossing point of thermal expansion isotherms can be used as a criterion for verification of equations of state for simple liquids.[5,23] It was found that a reproduction of the crossing point of thermal expansion isotherms for both hexane and methane could be done only with equations of state with a modified repulsive contribution of the Carnahan-Starling type.[24] This conclusion and the previous attempt to relate the unique thermal properties of simple liquids under high pressures to the shape of the intermolecular potential have lead to a development of an equation of state with shifted Lennard-Jones pair potentials.[25] It was found that the best reproduction of the crossing point of thermal expansion isotherms for methane can be done with a soft sphere – van der Waals equation of state for a 8/4 pair potential (see Figure 8.13). Figure 8.14 presents results of calculations of isobaric heat capacity for methane performed with various equations of state.[25] The most important observation is that the minimum on the heat capacity isotherms can be reproduced only with equations of state which reproduce correctly the crossing point of thermal expansion isotherms.

8.3.2 Weakly Self-associated Liquids under High Pressures

On the basis of heat of dilution of quinoline in decane at 298.15 K it was concluded that quinoline can be considered as a weakly self-associated liquid.[27] Quinoline is also a component of coal-derived liquids and thus the determination of its properties over wide pressure and temperature ranges is important. With the use of the method and technique very similar to that described in this

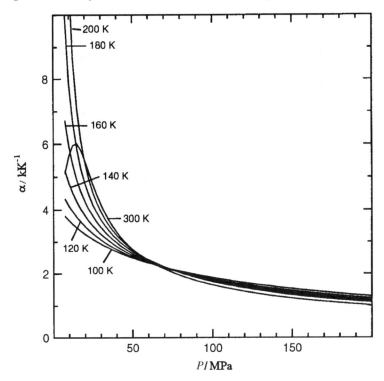

Figure 8.13 Pressure dependence of the thermal expansion coefficient of methane calculated from the soft sphere-van der Waals equation of state for a 8/4 pair potential.

chapter, such data have been determined over the temperature range from (303 to 500) K under pressures up to 400 MPa.[28] The behaviour of the thermal expansion coefficient (see Figure 8.15) and isobaric heat capacity (see Figure 8.16) of liquid quinoline as a function of pressure at various temperatures can be compared with the behaviour of hexane, a model simple non-associated liquid, discussed above. The thermal expansion coefficient of quinoline behaves like that of hexane. Both quinoline and hexane show a nearly unique crossing point of isotherms. At (60 ± 0.4) MPa for liquid quinoline, $\alpha = (6.50 \pm 0.02) \times 10^{-4}$ K^{-1} over the whole temperature range under study. However, the temperature dependence of α is greater at high pressures and lower at low pressures for quinoline than for hexane. Also, α for quinoline is significantly smaller than α for hexane. Thus, $d\alpha/dT$ is closer to the value of α in quinoline than in hexane and this numerical fact is reflected in differences in the shapes of isotherms of pressure effects on C_p for quinoline and hexane (compare Figure 8.17 with Figure 8.6). Because $d\alpha/dT$ changes sign at the crossing point, the isotherms of pressure effects on C_p for quinoline exhibit minima which move to higher pressures as the temperature decreases (opposite behaviour to that of hexane).

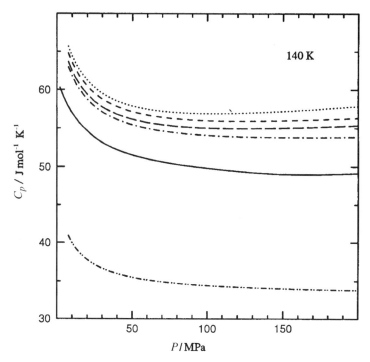

Figure 8.14 Isobaric heat capacity of methane as a function of pressure: ____-
experimental data;[26]---- soft sphere equation 12/6 potential; _ . _ . _ . _
soft sphere equation 8/4 potential; .... – Carnahan-Starling-van der
Waals equation with temperature independent size parameter; ------
Carnahan-Starling-van der Waals equation with temperature dependent
size parameter; -..-..- van der Waals equation.

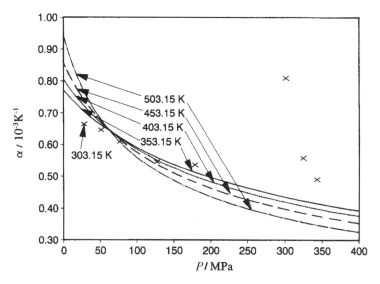

Figure 8.15 Thermal expansion coefficient (α) of quinoline.

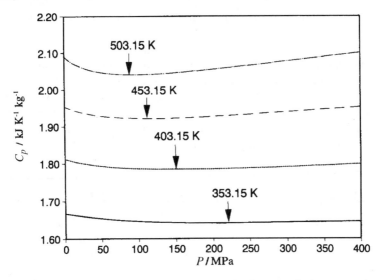

Figure 8.16 Isobaric specific heat capacity of quinoline. Vertical arrow points indicate minima.

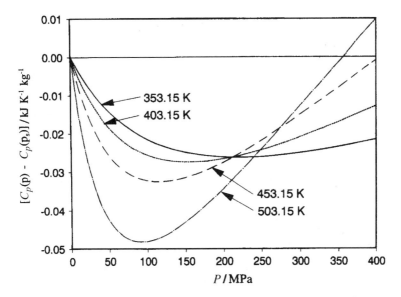

Figure 8.17 Isothermal pressure effect on isobaric heat capacity of liquid quinoline.

In exhibiting a minimum at all the temperatures covered by the experiments, the isobaric heat capacity of quinoline (see Figure 8.16) also behaves like that of hexane. However, the trend with temperature of the pressure at the minimum in the isotherms of the heat capacity is opposite for quinoline and hexane. For hexane the minima appear at higher pressures as the temperature is increased. For quinoline the minimum shifts to lower pressures as the temperature is increased.

The differences between quinoline and hexane can be explained by taking into consideration that quinoline is a weakly associated liquid. The heat capacities are thus the sum of the isobaric heat capacity of unassociated liquid quinoline and of a contribution from the enthalpy of association resulting from a shift in the self-association equilibrium. At high temperatures the associating bonds are broken, the contribution from the enthalpy of association becomes negligible and quinoline behaves like a simple liquid.

8.3.3 Strongly Self-associated Liquids under High Pressures

The self-association of *m*-cresol has been widely studied.[27,29] However, in those studies temperature was the intensive variable and thus the thermodynamic analysis was limited to temperature derivatives of the thermodynamic functions. Only through the use of the pressure controlled option of scanning transitiometry is one able to obtain a set of thermodynamic properties for *m*-cresol over wide temperature range from (300 to 503) K under pressures up to 400 MPa.[30]

Figures 8.18–8.21 present selected data with respect to the pressure influence on the heat capacities of *m*-cresol, which can be taken as a model behaviour for strongly self-associated liquids. *m*-Cresol is also an important component of coal-derived liquids and thus the knowledge of its properties over wide ranges of both pressure and temperature has a practical importance. As in the case of

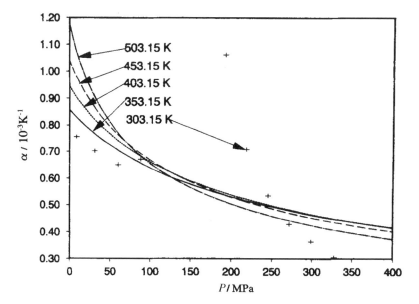

Figure 8.18 Thermal expansion coefficient (α) for *m*-cresol.

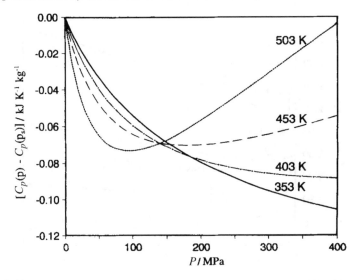

Figure 8.19 Isothermal pressure effect on isobaric heat capacity of liquid *m*-cresol.

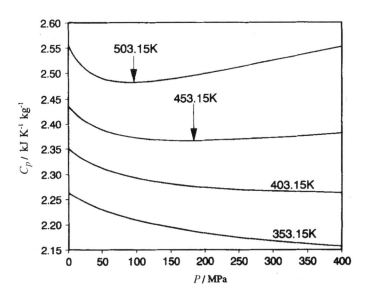

Figure 8.20 Isobaric specific heat capacity of *m*-cresol. Vertical arrow points indicate minima.

quinoline it is interesting to compare its properties with those of hexane, a model simple liquid.

Figure 8.18 shows that there is no unique crossing point of thermal expansion isotherms for liquid *m*-cresol. The isotherms do cross near 100 MPa, but the crossing points are temperature dependent. The isotherms cross at lower

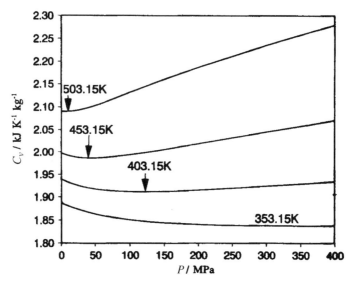

Figure 8.21   Isochoric specific heat capacity of *m*-cresol. Vertical arrow points indicate minima.

pressures as temperature increases. A unique crossing point may exist at some higher temperatures where *m*-cresol is unassociated, since the low temperature behaviour results from the self-association equilibrium.

Self-association also strongly affects the pressure effects on the heat capacities. Values of the isothermal pressure effects on the isobaric heat capacity of *m*-cresol at four temperatures are given in Figure 8.19. Opposite to the non-associated hexane and even to the weakly associated quinoline, the low temperature isotherms of thermal expansion for *m*-cresol do not show minima over the pressure range under investigation. Only the high temperature isotherms at 453 K and 503 K exhibit minima which are shifted to lower pressures with an increase in temperature; similarly to that what was previously observed for quinoline but opposite to that what was observed for simple non-associated liquids.

Figures 8.20 and 8.21 present respectively isobaric and isochoric heat capacities for *m*-cresol as a function of pressure at four selected temperatures. Comparison of the behaviour of isobaric and isochoric heat capacities of *m*-cresol with increasing pressure and temperature is also informative. At 353.15 K the isochoric heat capacity decreases or is constant with increasing pressure up to 400 MPa. With increasing temperature, the isotherms exhibit a minimum that shifts to lower pressures. Similar effects occur in isotherms of the isobaric heat capacity.

The behaviour of *m*-cresols can be explained by pressure and temperature dependent self-association equilibria. At $T \geq 453$ K the hydrogen bonds are mostly broken and *m*-cresol behaves like a liquid without strong intermolecular interactions. At lower temperatures, where hydrogen bonding is significant,

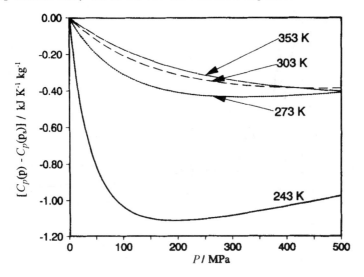

Figure 8.22 Isothermal pressure effect on isobaric heat capacity of liquid water.

increasing pressure shifts the equilibrium toward the state with lower volume. At low temperatures both the shift in hydrogen-bonded equilibria and the contribution from weak intermolecular interactions contribute to the pressure effect on the heat capacity.

Accurate data are also available for water as another example of strongly self-associated hydrogen-bonded liquid, obtained with the use of a similar procedure (see Figure 8.22).[31] There are strong similarities between the pressure effects on the heat capacity of *m*-cresol and water. In both liquids, the pressure effect decreases as the temperature increases and the isotherms of the pressure effects at different temperatures diverge, converge, cross at different points, and then diverge again as the pressure increases. As was discussed above, for simple liquids the isotherms of pressure effects on heat capacities at different temperatures always diverge throughout the pressure range studied.

Hexan-1-ol is another self-associated liquid for which the thermophysical properties have been determined over wide pressure and temperature ranges.[32] However, the mechanism of the self-association in hexan-1-ol must be different from the self-associated liquids discussed above. The thermal expansion isotherms practically do not exhibit a crossing point, and only at very high pressures are some irregular crossings observed. Also the isotherms of isobaric heat capacity and also those at the highest temperatures under investigation, do not exhibit minima. Figure 8.23 presents the temperature dependence of thermal expansion as a function of pressure at various temperatures. It is worth noting that the temperature derivative of thermal expansion exhibits a kind of a crossing point, but changes sign only at high pressures and temperatures. Consequently the isotherms of C_P do not exhibit minima, its pressure derivative is practically negative over the whole ranges of both pressure and temperature

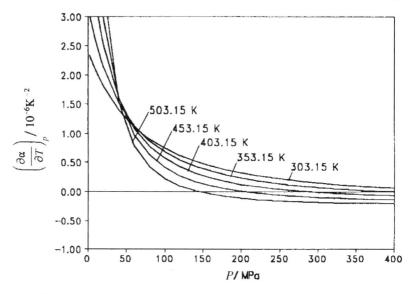

Figure 8.23 Selected isotherms of $d\alpha/dT$ for hexan-1-ol.

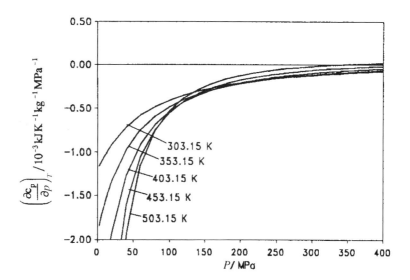

Figure 8.24 Selected isotherms of dC_P/dT for hexan-1-ol.

under investigation (see Figure 8.24). Although the pressure and temperature ranges of the derived data are rather large, they are not sufficient to yield final conclusions on the pressure–temperature behaviour of hexan-1-ol and its hydrogen bond structure. A further increase of pressure should be helpful to investigate the intersection of thermal expansion isotherms and to establish more in detail information on the character of its self-association.

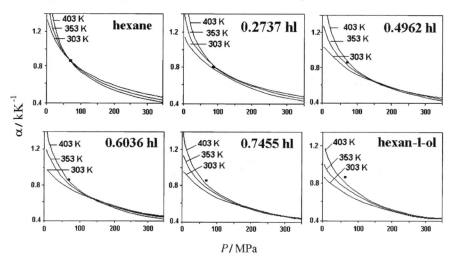

P / MPa

Figure 8.25 Comparison of isotherms of α for hexane and hexan-1-ol and for their binary mixtures. ■ is the crossing point of α isotherms for hexane, h1 refers to hexan-1-ol.

8.3.4 Liquid Binary Mixtures under High Pressures

The binary mixture (hexane + hexan-1-ol) has been extensively studied over wide ranges of pressures, temperatures and compositions.[33,34] Beside measurements on the pure components measurements have been performed for mixtures of several compositions. Figure 8.25 presents thermal expansion isotherms for selected compositions and compares them to those for the pure components. It is instructive to analyse the change of the behaviour of isotherms with an addition of successive amounts of hexan-1-ol, especially with respect to the crossing point (■) of thermal expansion isotherms for hexane. With the use of the transitiometric procedure described above it was possible to determine the influence of pressure on the heat capacity for both pure components and respective mixtures using Equation (12):

$$\left(\frac{\partial C_P(P,T)}{\partial P}\right)_T = -TV(P,T)\left[\alpha^2(P,T) + \left(\frac{\partial \alpha(P,T)}{\partial T}\right)_P\right] \quad (12)$$

From the collected data it was also possible to derive the pressure influence on the excess heat capacity with the use of the thermodynamic relation expressed in Equation (13):

$$\left(\frac{\partial C_P^E(P,T)}{\partial P}\right)_T = \left(\frac{\partial C_P^M(P,T)}{\partial P}\right)_T - x\left(\frac{\partial C_P^1(P,T)}{\partial P}\right)_T - (1-x)\left(\frac{\partial C_P^2(P,T)}{\partial P}\right)_T$$

$$(13)$$

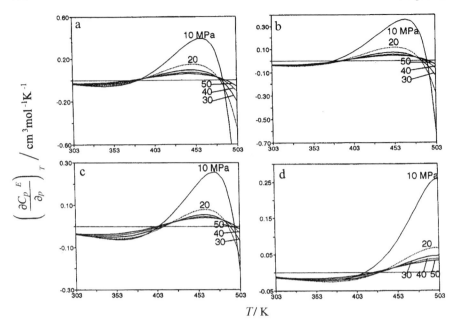

$$\left(\frac{\partial C_P^E}{\partial P}\right)_T / \text{cm}^3\text{mol}^{-1}\text{K}^{-1}$$

T/ K

Figure 8.26   Isobars of pressure derivatives of C_P^E for binary mixtures of hexane with hexan-1-ol over the low pressure region: a, $x = 0.1088$; b, $x = 0.2735$; c, $x = 0.4965$; d, $x = 0.7453$, where x is the mole fraction of hexan-1-ol.

where superscripts E, M, 1, 2 denote excess, mixture, hexane and n-hexanol respectively; x denotes the mole fraction of hexan-1-ol. Figures 8.26 and 8.27 present the pressure derivative of isobaric excess heat capacity for selected compositions of {hexane + hexan-1-ol} system over low and high pressure regions respectively.[34]

The behaviour of the pressure derivative of excess heat capacity for the binary mixtures of hexane with hexan-1-ol can be analysed in terms of the two effects: (i) breaking and reformation of H-bonds; and (ii) packing effect, as it is usually done for alkane-alkanol systems.[35] At least for moderate pressures the general behaviour is such that an application of pressure at a given temperature gives rise to a reformation of H-bonds, or in other words a pressure increase diminishes breaking of H-bonds. On the other hand, the temperature increase causes breaking of H-bonds. The thermal contribution of the packing effect of the alkane into the cavities of the H-bonded structure is negative and increases with decreasing P and increasing T.[35] When approaching the critical point of the alkane, other phenomena typical for this region must be also taken into consideration.[36] The compensation between those phenomena is clearly demonstrated in the isopleths of $(\partial C_P^E/\partial P)_T$ presented in Figures 8.26 and 8.27. For all concentrations there is a temperature region, where $(\partial C_P^E/\partial P)_T$ depends very little on both pressure and temperature, which means that that the actions of the two effects, H-bonding and packing, are compensating each other. Most probably, at lower temperatures the shifting of H-bond equilibria is the

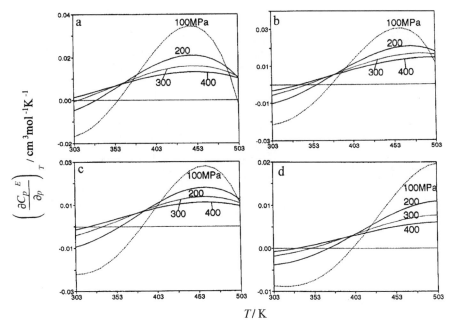

Figure 8.27 Isobars of pressure derivatives of C_P^E for binary mixtures of hexane with hexan-1-ol over the high pressure region: a, $x = 0.1088$; b, $x = 0.2735$; c, $x = 0.4965$; d, $x = 0.7453$, where x is the mole fraction of hexan-1-ol.

dominating effect. At higher temperatures the packing effect and near critical properties of hexane dominate, as shown by clear maxima, especially on lower-pressure isobars for mixtures at low concentrations of hexan-1-ol. This particular behaviour is shifted to higher temperatures with increase of concentration of hexan-1-ol. A similar behaviour was also observed for the pressure derivative of excess enthalpy.[34] It is also worth recalling that Brown *et al.*[37] have observed a similar maximum on the 5 MPa isobar of excess enthalpy for {0.2 propan-1-ol + propane} mixture when approaching the critical temperature of propane. The authors have explained this by a condensation of the near-critical liquid, but really gas-like, alkane into the liquid alkanol. It is also possible that the occurrence of such distinct maxima on isobars of $(\partial C_P^E / \partial P)_T$ is related to the presence of similar maxima on the C_P isobars of hexane above its critical temperature, as determined by flow calorimetric measurements.[38]

The binary system (hexane + hexan-1-ol) analysed above presents a specific type of binary mixtures in which only one component is self-associated and no specific interactions occur between the two components. Binary mixtures of *m*-cresol with quinoline are examples of another type of binary liquid mixtures, in which the both components are self-associated and extremely strong interactions exist between the components. Figure 8.28 presents thermal expansion isotherms for mixtures of selected compositions and compares them with pure

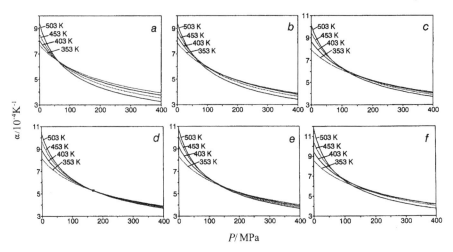

P/ MPa

Figure 8.28 Selected isotherm of α: quinoline (a), (0.1499 *m*-cresol + 0.8501 quinoline) (b); (0.5005 *m*-cresol + 0.4995 quinoline) (c); (0.6325 *m*-cre-
sol + 0.3675 quinoline) (d); (0.8501 *m*-cresol + 0.1499 quinoline) (e) and
m-cresol (f).

m-cresol and quinoline over wide ranges of pressure and temperature.[39] The
most striking observation is that mixtures of *m*-cresol with quinoline near a 2:1
mol ratio (Figure 8.28d) behave like liquids without association, *i.e.* the α
isotherms exhibit a unique crossing point at $P = 170$ MPa and $\alpha = (5.27 \pm 0.01)$
$\times 10^{-4}\,\mathrm{K}^{-1}$ over the temperature range from (353 to 503) K. Mixtures with
lower and higher concentrations of *m*-cresol behave similarly to *m*-cresol, *i.e.*
the α isotherms cross at lower pressures as the temperature increases. The
behaviour of the 2:1 mixture is explained by the very strong 2:1 association
between *m*-cresol and quinoline. The position of this equilibrium is not much
disturbed by changes in pressure and temperature over the ranges of pressure
and temperature investigated. It is also interesting to recall that near the 2:1
mol ratio, the excess enthalpy is at a maximum with a very large value of
$-7.7\,\mathrm{kJ\,mol}^{-1}$.[40] In the 2:1 liquid mixture, the liquid phase is composed of
strongly bound intermolecular complexes which behave like the molecules of a
liquid without association. Thus, the macroscopic properties of such a phase
are similar to those of nonpolar liquids. In mixtures with other than a 2:1 mol
ratio of *m*-cresol to quinoline the equilibria among the 2:1 and 1:1 complexes
between *m*-cresol and quinoline[27] and self-associated polymers of *m*-cresol, *i.e.*
dimers, trimers, *etc.*,[29] preclude a unique crossing point of the α isotherms.
Figure 8.29 presents three isotherms of the pressure effect on the isobaric heat
capacity of an equimolar mixture of *m*-cresol and quinoline. One can see that
none of the isotherms reaches the minimum over the investigated ranges of
pressure and temperature, although at 453 K the equilibrium starts to be shifted
towards non-associated forms.

   As was mentioned above, the soft-sphere equation of state[23,25] best repro-
duces the crossing point of α isotherms for unassociated liquids. For linearly

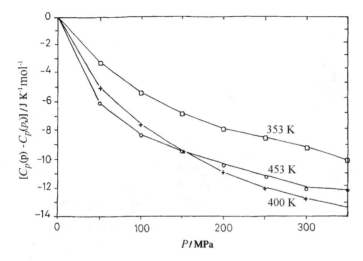

Figure 8.29 Isothermal pressure effect on isobaric heat capacity of liquid mixture (0.5005 *m*-cresol + 0.4995 quinoline).

associated liquids, such as hexan-1-ol and its binary mixtures with hexane, encouraging results have been obtained with an additional term accounting for the linear association[41] added to the soft-sphere equation of state. No similar attempts have yet been made to describe the specific type of association between *m*-cresol and quinoline with an equation of state, but the behaviour of $\alpha(P,T)$ for theses mixtures cannot be described by any existing equation of state.

8.3.5   Technological Liquids under High Pressures

Figure 8.30 presents a set of thermal expansion isotherms of a technological important liquid derived from coal in a hydrogenation plant.[42] Thermal expansion of coal-derived liquids as a function of pressure and temperature is interesting for two reasons. Firstly, from the point of view of basic knowledge it would be interesting to know, whether such complicated liquids have high-pressure-temperature behaviour similar to the behaviour of pure organic liquids. Secondly, the thermal expansion data over the wide temperature and temperature intervals can be used as a basis for derivation of other thermo-dynamic data of the liquids under investigation. The need of such data is evident, because the design of processes for the liquefaction of coal and upgrading of coal liquefaction products requires a knowledge of the thermodynamic properties of coal liquids over wide pressure and temperature intervals. The most interesting observation of results presented in Figure 8.30 is the existence of an unequivocal crossing point of isotherms at high temperatures. The low temperature isotherms exhibit unusual behaviour related most probably to a precipitation of high melting point components. A similar unusual behaviour is

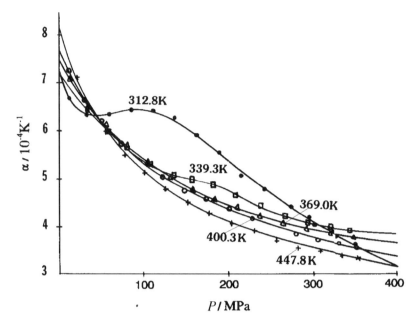

Figure 8.30 Thermal expansion coefficient (α) of a coal-derived liquid as a function of pressure.

also observed for the low temperature isotherms of isobaric heat capacity of the same coal-derived liquid (see Figure 8.31), which confirms the possible precipitation.[43] The three high temperature isotherms exhibit a behaviour observed before for simple liquids. The existence of the minimum in the isobaric heat capacity isotherms is thermodynamically consistent with the crossing point of thermal expansion isotherms.

8.4   Conclusions

Classical calorimetric measurements of the heat capacity of liquids under high pressures are extremely difficult to perform correctly because the heat capacity contribution of the massively thick experimental vessels, required to withstand the high pressure, is much more important than the contribution from the investigated sample itself. With the use of a scanning transitiometer it is possible to measure the thermal expansion coefficient, $\alpha(P,T)$, over wide ranges of pressure and temperature with a relative uncertainty of (1 to 3) %, reference volume isotherm, $V(P,T_R)$, at a low reference temperature T_R with a precision of 0.6 %a and the reference heat capacity isobar, $C_P(T,P_R)$, at a low reference pressure P_R with a relative uncertainty of 0.3 %. Using such data and a proper thermodynamic procedure it is possible to determine the heat capacity function for a liquid over wide pressure and temperature ranges with relative uncertainty of at least ±2 %, because the thermodynamic procedure requires mainly

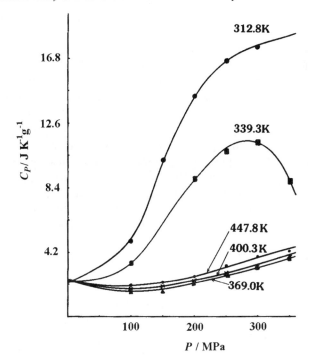

Figure 8.31 Isobaric specific heat capacity of a coal-derived liquid as a function of pressure at various temperatures.

integration and addition, which do not significantly enhance the total error. There are two additional advantages of such an approach. First, the determined quantities correspond exactly to their definitions and thus their thermodynamic significance is clear, what is not always the case in classical calorimetric techniques. Second, the thermal expansion coefficient is the main variable in thermodynamic formulation of the pressure influence on heat capacity. Thus, its direct experimental measurements are extremely helpful in explaining the nature of the pressure effects on heat capacities.

With the use of such an approach a great number of data on both thermal expansion coefficient and heat capacity have been collected for liquids of various natures. On this basis it is possible to make the following conclusions.

Simple liquids, *i.e.* liquids without strong intermolecular interactions, namely association, such as hexane, methane, exhibit a unique crossing point of thermal expansion isotherms, where $d\alpha/dT$ changes sign to negative, and minima on the heat capacity isotherms appear, which at higher temperatures are flattened and shifted to higher pressures.

Weakly self-associated liquids also exhibit the crossing point of thermal expansion isotherms, although its nature is a bit different from that of simple liquids ($d\alpha/dT$ is higher with respect to the α itself), and the minima in the heat capacity isotherms are shifted to lower pressures with an increase in temperature, because at higher temperatures the weakly associated complexes are broken.

Strongly self-associated liquids do not present the unique crossing point of α, at least under moderately high pressures up to 400 MPa, although the isotherms can cross and recross. The minima on the heat capacity isotherms appear only at high temperatures and are shifted to lower pressures with an increase in temperature due to the shift in the association equilibrium.

Non-interactive binary mixtures, where only one component is self-associated and there is no strong interactions between the components, exhibit a disappearance of the unique crossing point of α isotherms with an addition of the self-associated component. The isobars of pressure derivatives of heat capacity of such mixtures exhibit minima and maxima due to a compensation of two phenomena: (i) breaking and reformation of H-bonds; and (ii) packing effect. On approaching the critical point of one of the components maxima are observed which are typical for that region.

For interactive binary mixtures, where the two components are self-associated and there is a strong interaction between the components the most interesting observation is that at the composition where the strongest complex is formed the mixture behaves like a simple liquid exhibiting a unique crossing point of α isotherms, because the complexation is so strong that it cannot be disturbed by variations of either temperature or pressure.

Complex coal-derived liquids behave like simple liquids at high temperatures exhibiting the unique crossing point of α isotherms and the minima in the heat capacity isotherms, but at lower temperatures some precipitation of high melting point components is observed from both thermal expansion and heat capacity data.

Only soft-sphere equations of state can reproduce the crossing point of α isotherms and minima on the heat capacity isotherms for unassociated simple liquids. For linearly associated liquids and their binary mixtures with simple liquids, encouraging results have been obtained with an additional term accounting for the linear association added to the soft-sphere equation of state. No similar attempts have yet been made to describe with an equation of state the specific type of association between self-associated strongly interacting components.

References

1. P. W. Bridgman, *Proc. Am. Acad. Arts Aci.*, 1913, **4**, 3.
2. S. L. Randzio, Calorimetric determination of pressure effects, in *Experimental Themodynamics, Vol. IV; Solution Calorimetry*, K. N. Marsh and P. A. G. O'Hare, eds., IUPAC, Blackwell Science, 1994, pp. 303–324.
3. S. L. Randzio, J.-P. E. Grolier, J. R. Quint, D. J. Eatough, E. A. Lewis and L. D. Hansen, *Int. J. Thermophys.*, 1994, **15**, 415.
4. S. L. Randzio, *Pure & Appl. Chem.*, 1991, **63**, 1409.
5. S. L. Randzio, *Chem. Soc. Rev.*, 1995, **24**, 359.

6. S. L. Randzio, *Chem. Soc. Rev.*, 1996, **25**, 383.
7. S. L. Randzio, *Thermochim. Acta*, 2000, **355**, 107.
8. S. L. Randzio, *Thermochim. Acta*, 1997, **300**, 29.
9. S. L. Randzio, *Thermochim. Acta*, 1985, **89**, 215.
10. S. L. Randzio, J.-P. E. Grolier and J. Quint, *Rev. Sci. Instrum.*, 1994, **65**, 960.
11. S. L. Randzio, D. J. Eatough, E. A. Lewis and L. D. Hansen, *J. Chem. Thermodyn.*, 1988, **20**, 937.
12. S. L. Randzio and A. Kutner, *J. Phys. Chem. B*, 2008, **112**, 1435.
13. R. Paramo, M. Zouine, F. Sabron and S. Casanova, *Int. J. Thermophys.*, 2003, **24**, 185.
14. M. Zábranský, V. Růžièka Jr, V. Majer and E. S. Domalski, *J. Phys. Chem. Ref Data,* Monograph No. 6, American Chemical Society, Washington, D.C., 1996.
15. Ph. Pruzan, *J. Phys. Lett. (Paris)*, 1984, **45**, L273.
16. S. L. Randzio, *Thermochim. Acta*, 1987, **121**, 463.
17. Ph. Pruzan, *J. Chem. Thermodyn.*, 1991, **23**, 247.
18. A. A. Gerasimov and B. A. Grigoriev, *Izvest. Vysshikh Uchebn. Zavedenii Neft i Gaz*, 1978, **21**, 46.
19. I. Czarnota, *High Temp.-High Press.*, 1985, **17**, 543.
20. W. B. Street, *Physica*, 1974, **76**, 59.
21. C. Alba, L. Ter Minassian, A. Denis and A. Soulard, *J. Chem. Phys.*, 1985, **82**, 384.
22. S. L. Randzio, *Phys. Lett. A*, 1986, **117**, 473.
23. S. L. Randzio and U. K. Deiters, *Ber. Bunsenges. Phys. Chem.*, 1995, **99**, 1179.
24. N. F. Carnahan and K. E. Starling, *J. Chem. Phys.*, 1969, **51**, 635.
25. U. K. Deiters and S. L. Randzio, *Fluid Phase Equilib.*, 1995, **103**, 199.
26. U. Setzmann and W. Wagner, *J. Phys. Chem. Ref. Data*, 1991, **20**, 1061.
27. D. J. Eatough, S. L. Wolfley, L. J. Dungan, E. A. Lewis and L. D. Hansen, *Energy & Fuels*, 1987, **1**, 94.
28. S. L. Randzio, D. J. Eatough, E. A. Lewis and L. D. Hansen, *Int. J. Thermophys.*, 1996, **17**, 405.
29. E. M. Woolley, J. G. Travers, B. O. Erno and L. G. Hepler, *J. Phys. Chem.*, 1971, **76**, 359.
30. S. L. Randzio, E. A. Lewis, D. J. Eatough and L. D. Hansen, *Int. J. Thermophys.*, 1995, **16**, 883.
31. L. Ter Minassian, Ph. Pruzan and A. Soulard, *J. Chem. Phys.*, 1981, **75**, 3064.
32. S. L. Randzio, J.-P. E. Grolier and J. R. Quint, *Fluid Phase Equilib.*, 1995, **110**, 341.
33. S. L. Randzio, J.-P. E. Grolier and J. R. Quint, *Int. J. Thermophys.*, 1997, **18**, 733.
34. J.-P. E. Grolier and S. L. Randzio, *Fluid Phase Equilib.*, 1997, **133**, 35.
35. J. B. Ott, P. R. Brown and J. T. Sipowska, *J. Chem. Thermodyn.*, 1996, **28**, 397.

36. G. Morrison, J. M. H. Levelt-Sengers, R. F. Chang and J. J. Christensen, in *Supercritical Fluid Technology*, J. M. L. Penninger, M. Radosz, M. A. McHugh, V. J. Krukpnis, eds. Elsevier, Amsterdam, 1985, pp. 25–43.

37. P. R. Brown, J. D. Moore, A. C. Lewellen and J. B. Ott, *J. Chem. Thermodyn.*, 1996, **28**, 363.

38. A. A. Gerasimov, *Izvest. Vysshikh Uchebn. Zavedenii Neft i Gaz*, 1980, **23**, 61.

39. S. L. Randzio, L. D. Hansen, E. A. Lewis and D. J. Eatough, *Int. J. Thermophys.*, 1997, **18**, 1183.

40. H. Tschamler and H. Krischai, *Monatsch. Chem.*, 1951, **82**, 259.

41. U. K. Deiters, *Fluid Phase Equilib.*, 1993, **89**, 229.

42. S. L. Randzio and L. Ter Minassian, *Thermochim. Acta*, 1987, **113**, 67.

43. S. L. Randzio, *Thermochim. Acta*, 1987, **115**, 83.

CHAPTER 9

Speed of Sound Measurements and Heat Capacities of Gases

ANTHONY R. H. GOODWIN[a] AND J. P. MARTIN TRUSLER[b]

[a] Schlumberger Technology Corporation, Sugar Land TX, USA; [b] Department of Chemical Engineering, Imperial College London, London, UK

9.1 Introduction

Sound speed measurements offer a convenient and accurate route for determining the heat capacity of a gas. The speed of sound can usually be determined with very small random uncertainty and the systematic errors to which such measurements are exposed differ markedly from those encountered in conventional calorimetry.

For gases at pressures below a few MPa, we can identify three principal sources of systematic error in an experimental measurement of the speed of sound arising from:

(1) viscothermal boundary layers
(2) molecular thermal relaxation, and
(3) pre-condensation.

Each of these may be rendered small by the use of physically based models, appropriate experimental techniques, and/or a suitable pressure range for the measurements. Additional sources of systematic error are associated with characterization of the state of the gas in terms of temperature, pressure, and

Heat Capacities: Liquids, Solutions and Vapours
Edited by Emmerich Wilhelm and Trevor M. Letcher
© The Royal Society of Chemistry 2010
Published by the Royal Society of Chemistry, www.rsc.org

chemical composition. Again, these may be rendered small through appropriate choices of measurement technique.

The speed of sound in a gas depends primarily on the thermodynamic properties and, as described in Section 9.2, can be used to obtain the heat capacity as well as other thermodynamic quantities. For measurements in gases at moderate pressures (e.g. $0.01 \leq p/\text{MPa} \leq 10$), variable-frequency fixed-cavity resonators are the modern instruments of choice. In Section 9.3, we review cavity-resonator techniques, categorized by geometry. For further details concerning resonator and other methods of determining sound speed, and thus heat capacity, the reader should consult Van Dael,[1] Goodwin and Trusler,[2] Trusler,[3] and a four-volume series on the elastic properties of matter,[4-7] especially the contributions of Moldover et al.[8] and Giacobbe[9] that both describe resonators used to determine the speed of sound in gases.

9.2 Heat Capacity

Measurements of the speed of sound are one of the best ways of obtaining information on the heat capacity of fluids. In the perfect-gas limit, there is a simple and direct connection with the heat capacity, independent of other fluid properties, while in compressed fluids sound speed data may be combined with other measurements to determine heat capacities.

9.2.1 Speed of Sound in Gases

The speed of sound u in an isotropic Newtonian fluid is given by[10]

$$u^2 = (\partial p/\partial \rho)_s \tag{1}$$

where p is the pressure, ρ the mass density, and S entropy. Strictly, Equation (1) is valid only in the limits of vanishing amplitude, which is extremely easy to approach in practice, and vanishing frequency that is usually, but not always, realized.[3,10,11] For the discussion, we assume that a practical measurement of the speed of sound in a fluid is either identical with the zero-frequency limit or that it may be corrected to that limit as discussed in Section 9.3.

A connection between the speed of sound and the isochoric molar heat capacity, $C_{V,\text{m}}$, of the gas may be established by resolving Equation (1) into isothermal and isochoric terms as follows:

$$u^2 = \frac{1}{M}\left[\left(\frac{\partial p}{\partial \rho_n}\right)_T + \frac{T}{\rho_n^2 C_{V,\text{m}}}\left(\frac{\partial p}{\partial T}\right)_{\rho_n}^2\right] \tag{2}$$

where M is the molar mass, ρ_n is the amount-of-substance density, and T is the temperature. The isochoric heat capacity is further related to the perfect-gas isochoric heat capacity by

$$C_{V,\mathrm{m}} = C_{V,\mathrm{m}}^0(T) - \int_0^{\rho_n} (T/\rho_n^2)(\partial^2 p/\partial T^2)d\rho_n \tag{3}$$

and the equation of state of the gas is

$$p = \rho_n RT(1 + B\rho_n + C\rho_n^2 + \cdots) \tag{4}$$

where 1, B, C, ... are virial coefficients that depend only upon temperature and composition. Combining Equations (2) to (4) yields

$$u^2 = A_0(1 + \beta_a \rho_n + \gamma_a \rho_n^2 + \cdots) \tag{5}$$

where 1, β_a, γ_a, $\cdots$ are so-called *acoustic* virial coefficients that also depend upon temperature and composition only. The parameter A_0 (*i.e.* the square of the speed of sound in the limit of zero density) is given by

$$A_0 = \frac{RT(C_{V,\mathrm{m}}^{\mathrm{pg}} + R)}{C_{V,\mathrm{m}}^{\mathrm{pg}} M} \tag{6}$$

This quantity is an experimentally realizable perfect-gas property that forms the basis for several applications of sound speed measurements including the determination of perfect-gas heat capacities.

The second acoustic virial coefficient β_a is given by

$$\beta_a = 2B + 2(\gamma^{\mathrm{pg}} - 1)TB' + \{(\gamma^{\mathrm{pg}} - 1)/\gamma^{\mathrm{pg}}\}T^2B'' \tag{7}$$

where primes denote differentiation with respect to temperature, and the third acoustic virial coefficient is given by

$$\gamma_a = \{(\gamma^{\mathrm{pg}} - 1)/\gamma^{\mathrm{pg}}\}\{B + (2\gamma^{\mathrm{pg}} - 1)TB' + (\gamma^{\mathrm{pg}} - 1)T^2B''\}^2$$
$$+ (1/\gamma^{\mathrm{pg}})\left\{(1 + 2\gamma^{\mathrm{pg}})C + ([\gamma^{\mathrm{pg}}]^2 - 1)TC' + \tfrac{1}{2}(\gamma^{\mathrm{pg}} - 1)^2T^2C''\right\} \tag{8}$$

In these equations $\gamma^{\mathrm{pg}} = C_{p,\mathrm{m}}^{\mathrm{pg}}/C_{V,\mathrm{m}}^{\mathrm{pg}}$ where $C_{p,\mathrm{m}}^{\mathrm{pg}} = C_{V,\mathrm{m}}^{\mathrm{pg}} + R$ is the isobaric perfect-gas molar heat capacity.

Since the experimental quantities are nearly always (u, T, p), an expansion of u^2 in powers of pressure p is often more useful:

$$u^2 = A_0 + A_1 p + A_2 p^2 + \cdots \tag{9}$$

This relation may be obtained by eliminating ρ_n on the right hand side of Equation (5), making use of the series inversion of Equation (4), from which one finds the following relations for A_1 and A_2:

$$A_1 = A_0(\beta_a/RT) \tag{10}$$

and

$$A_2 = A_0(\gamma_a - B\beta_a)/(RT)^2 \tag{11}$$

For gases at low and moderate densities, Equation (4), for $p(T, \rho_n)$, and Equation (5), for $u^2(T, \rho_n)$, both converge rapidly. Typically, only the leading three or four terms of these equations are required to reduce the truncation error to the order of usual experimental errors.[12–14] Equation (9) for $u^2(T, p)$, also converges rapidly[15–17] but there is empirical evidence that, compared with Equation (5), convergence is less rapid at sub-critical temperatures.[13]

9.2.2 Perfect-gas Heat Capacity

Equation (6) provides one of the best means of evaluating the perfect-gas heat capacities of pure compounds that are sufficiently volatile and stable at the temperatures of interest to be studied in the gas phase.[12,13,15–37] Equation (6) can also be applied in thermal metrology.[38–41] Rearrangement of Equation (6) for $C_{V,m}^{\mathrm{pg}}$ gives

$$C_{V,m}^{\mathrm{pg}} = R/\{(MA_0/RT) - 1\} \tag{12}$$

The uncertainty $\delta C_{V,m}^{\mathrm{pg}}$ of $C_{V,m}^{\mathrm{pg}}$ is usually dominated by the uncertainty δM in the molar mass in which case

$$(\delta C_{V,m}^{\mathrm{pg}}/C_{V,m}^{\mathrm{pg}}) \approx \{\gamma^{\mathrm{pg}}/(\gamma^{\mathrm{pg}} - 1)\}(\delta M/M) \tag{13}$$

Since $\gamma^{\mathrm{pg}}/(\gamma^{\mathrm{pg}}-1)$ is at least 3.5 for a diatomic gas, and often much greater for polyatomic gases, the relative uncertainty of M is substantially amplified in determining the corresponding relative uncertainty of $C_{V,m}^{\mathrm{pg}}$. Thus knowledge or control of the molar mass of the gas will be paramount in determining the heat capacity, especially in relation to the role of chemical purity. To maintain the utmost purity of the sample under study a gas handling system for continuous purging of the resonator during the acoustic measurements has been reported.[42]

Measurements of the speed of sound are of course obtained at finite pressures and the precise determination of A_0 requires the evaluation of an appropriate number of acoustic virial coefficients of the gas. These can be used, as described in Reference 2, to determine the virial coefficients of the gas and their temperature derivatives, permitting the heat capacity of the gas at finite densities to be calculated from Equations (3) and (4).

As an illustration, we show in Figure 9.1 measured values of u^2 for pentane along the isotherm at $T = 300$ K at pressures in the range $5 \leq p/\text{kPa} \leq 42,$[21] together with the relative deviations of u^2 from a fit with Equation (6) truncated after the term in p^2. Clearly, this three-term fit was able to represent the experimental data to within a few parts in 10^6 and, despite small systematic deviations, provide a precise extrapolation to the limit of zero pressure with a relative statistical uncertainty below 10^{-5}. This led to a value of $C_{V,m}^{\mathrm{pg}}$ with a relative statistical uncertainty of order 10^{-4}, so that the overall uncertainty of $C_{V,m}^{\mathrm{pg}}$ was dominated by issues of chemical purity.

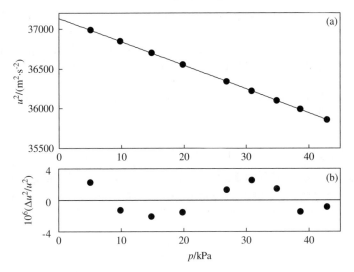

Figure 9.1 Speeds of sound squared in pentane at $T = 300$ K and extrapolation to zero pressure: •, experimental data; ——, fit with Equation (6) truncated after the term in p^2. (a) $u^2(p)$ along the isotherm; (b) relative deviations $\Delta u^2/u^2$ of u^2 from the fitted equation.

9.2.3 Binary Mixtures

In the specific case of a binary mixture $\{(1-x)A + xB\}$, where x is the mole fraction of B, A_0 is given by

$$A_0 = \left(\frac{RT(C_{V,A}^{\mathrm{pg}} + R)}{C_{V,A}^{\mathrm{pg}} M_A} \right) \left(\frac{1 + a_1 x}{1 + a_2 x + a_3 x^2} \right) \tag{14}$$

where

$$
\begin{aligned}
a_1 &= (C_{V,B}^{\mathrm{pg}} - C_{V,A}^{\mathrm{pg}})/(C_{V,A}^{\mathrm{pg}} + R), \\
a_2 &= (M_B/M_A) + (C_{V,B}^{\mathrm{pg}}/C_{V,A}^{\mathrm{pg}}) - 2, \\
a_3 &= [1 - (C_{V,B}^{\mathrm{pg}}/C_{V,A}^{\mathrm{pg}})][1 - (M_B/M_A)].
\end{aligned}
\tag{15}
$$

Equations (14) and (12) may be applied to determine the perfect-gas heat capacity of one of the components in a binary mixture of known composition when that of the other component is known, along with A_0 and both molar masses. This approach has been employed by Colgate and co-workers[43–48] for substances of low-volatility diluted in argon gas. In that way, a determination of $C_{V,\mathrm{m}}^{\mathrm{pg}}$ was possible at temperatures where the vapor pressure was insufficient to permit measurements on the pure substance.

Alternatively, a quadratic equation for x may be obtained from Equation (14) and solved to determine the composition of the mixture. Usually $A_0(x)$ is a single-valued function with just one root in the domain [0,1]. The precision with which A_0 may be determined in practice usually leads to a sensitive measure of the composition.[49,50] For $\{(1 - x)O_2 + xAr\}$, $A_0(x)$ is *not* a single-valued function. To use Equation (14) in this way requires the perfect-gas heat capacity and the molar mass of each component be known with appropriate uncertainty.

9.2.4   Compressed Gas Region

Numerical integration of the differential equations that link the speed of sound with other thermodynamic properties yield a rigorous route to the thermodynamic surface without restrictive assumptions about the equation of state. Initial conditions are required and usually comprise Z and $(\partial Z/\partial T)_p$ along one isotherm, in addition to the perfect-gas heat capacities obtained from the zero-density speed of sound. Here, $Z = p/\rho_n RT$ is the dimensionless compression factor of the gas, in terms of which the speed of sound is given by

$$u^{-2} = \left(\frac{M}{RTZ^2}\right)\left[\left\{Z - p\left(\frac{\partial Z}{\partial p}\right)_T\right\} - \left(\frac{R}{C_{p,\mathrm{m}}}\right)\left\{Z + T\left(\frac{\partial Z}{\partial T}\right)_p\right\}^2\right] \qquad (16)$$

Equation (16), combined with the relation

$$\left(\frac{\partial C_{p,\mathrm{m}}}{\partial p}\right)_T = -\left(\frac{R}{p}\right)\left\{2T\left(\frac{\partial Z}{\partial T}\right)_p + T^2\left(\frac{\partial^2 Z}{\partial T^2}\right)_p\right\} \qquad (17)$$

forms a non-linear second-order system of partial differential equations which can be solved numerically for $Z(T, p)$ and $C_{p,\mathrm{m}}(T, p)$ subject to the specified initial conditions.[51] The required data are the speed of sound as a function of temperature and pressure over the whole domain in which we wish to develop the solution, plus the initial conditions from which to start the numerical integration. The speed of sound is measured on closely spaced isotherms from the same upper bound p_{max} to a sufficiently low pressure for A_0 to be obtained by extrapolation. The solution will then be developed within a rectangular domain in (T, p) co-ordinates, extending down to zero pressure where $Z = 1$ and $u^2 = A_0$, without recourse to fitting functions, by means of numerical interpolation.[26,51–53] These methods have been applied successfully to methane,[26] ethane[54] and argon.[52] These methods would be useful in the investigation of new substances where it was impractical or uneconomic to make (p, ρ_n, T) measurements over the whole surface. The propagation of errors has been studied in some detail,[51–53] and a recursive equation solution method has been devised by Lago and Albo[55] that has the merit of permitting the uncertainties of the derived properties to be estimated in a rigorous way. Bijedic and Neimarlija have also considered alternative solution methods.[56]

9.3   Measurement of the Speed of Sound

In principle, the measurements of a single resonance frequency of a known mode of oscillation within a cavity of known dimension, or of a single time-of-flight over a known distance, is sufficient to determine the speed of sound. In practice, one usually prefers techniques providing redundancy in the form of measurements over different and resolved modes of oscillation, or over different frequencies and path lengths. This redundancy provides a means of identifying and reducing sources of error in the measurements. In general, standing-wave measurements in a metal-walled cavity or interferometer are preferred for gases because the greatly differing acoustic impedances of the medium and the wall lead to efficient reflection of sound and a simple connection between the measured resonance conditions and the speed of sound in the gas. Accordingly, we shall restrict the discussion to this class of measurement and review in this section generic factors that are common to all acoustic resonators and, in the following section, consider specific cavity geometries that have been deployed in practice.

To determine the perfect-gas heat capacity of a gas from sound speed measurements requires measurements at pressures below about 1 MPa. In this pressure regime, we can identify three significant sources of error in speed of sound measurements; these arise from precondensation, viscothermal boundary layers and molecular thermal relaxation. Fortunately, these can be rendered either negligible or small enough to model by selecting the appropriate experimental technique. We discuss each of these issues below.

Measurement of the sound speed at different frequencies provides a means of validating the working equations used to account for the viscothermal and non-rigid-boundary effects, and assist in identifying and reducing sources of error including the presence of pre-condensation and molecular thermal relaxation, both of which are frequency-dependent phenomena.

9.3.1   Dimensional Metrology

For variable-frequency fixed-geometry resonators the frequency measured for a particular mode yields the ratio of the sound speed to a characteristic dimension $D(T, p)$ of the cavity. To determine $C_{p,m}^{pg}(T)$ requires knowledge of that characteristic dimension in the limit of zero pressure, $D(T, p = 0)$, while the determination of acoustic virial coefficients requires only knowledge of the ratio $D(T, p)/D(T, p = 0)$. Values of the characteristic dimension can be obtained as a function of temperature from dialatometry,[38] dimensional microwave measurements,[39,40,57-60] or by calibration with a gas of known molar mass and perfect-gas heat capacity.[8] Simultaneous microwave measurements determine the ratio of speed of sound to speed of light and meet the requirement for a cavity that is dimensionally stable solely for the duration of the measurements on each isotherm; it also provides a measure of the complex electric permittivity of the gas. Moldover and co-workers[61-63] have suggested the use of quasi-spherical cavities to simplify the determination of the characteristic dimensions by microwave resonance measurements while preserving most of the acoustical advantages of the spherical geometry. The ratio $D(T, p)/D(T, p = 0)$ may usually be estimated from an elastic model making use of elastic constants from the literature.

In variable path length measurements, the speed of sound is determined from the displacement of one reflector required to tune the cavity from one resonance to another at fixed driving frequency. Thus only changes in length are required and these may be measured with an optical interferometer or a linear encoder; historically, mechanical micrometers were used to both generate and measure the displacement. The relative mechanical complexity of variable path interferometers have caused the technique to be largely abandoned in favor of variable frequency fixed cavity resonators.

9.3.2 Transducers and Gas Ports

All cavity resonators used for sound speed measurements have some method of introducing the gas, and one or, usually, two electroacoustic transducers to interconvert mechanical and electrical work.

In the case of fixed cavity resonators, the transducers should satisfy the following criteria:

(1) present a high acoustic impedance to the gas;
(2) be a small fraction of the surface area;
(3) have low power dissipation so as to not perturb thermal equilibrium within the cavity;
(4) operate over the desired temperature, pressure and frequency ranges; and
(5) be chemically inert.

The same criteria might also be applied to a variable path interferometer. However, it is usual in this case for the driving transducer to be much larger, typically covering one entire end face of the cavity. Then, depending in part on the type of drive transducer used, there are at least three alternative detection systems that might be deployed as follows:

(1) a second transducer conforming to the above criteria and used to monitor the sound pressure at another fixed location in the cavity;
(2) a measurement of the input electrical impedance or admittance of the driving transducer; or
(3) a measurement of the mechanical impedance or admittance of the driving transducer by means of an accelerometer.

The theoretical and practical aspects of transducers used for the measurement of sound speed are described in the literature.[2,3,64-68]

Tubes or other openings in the cavity wall are required to admit gas and these should be designed to minimize perturbations to the resonance while, at the same time, providing adequate flow, especially during evacuation prior to introduction of the sample. In many ways, the best solution is a valve built into the cavity wall which, when closed, plugs the port flush with the interior surface. Such an arrangement largely eliminates the perturbation for an opening in the wall, but at the expense of mechanical complexity. As we will see, a good alternative to a valve machined in the wall is to use a tube of small cross-section with a length and exterior termination chosen to simultaneously achieve high acoustic impedance at the resonance frequencies and a low impedance to flow. It is probably fair to say that the accuracy of a perfect-gas heat capacity derived from sound speed measurements is limited mainly by knowledge of the composition of the gas, and repeated flushing of the apparatus to help expel contaminants is preferred for all systems.

9.3.3 Measurement of Acoustic Resonances

The methods used to determine the resonance frequencies in a fixed cavity (steady-state) resonator have been documented elsewhere.[69,70] Typically, the amplitude and phase of the signal at the microphone is measured by means of a phase-sensitive detector at a number of discreet frequencies in the vicinity of the resonance, and a numerical analysis is employed to determine the resonance frequency, half width, and other parameters. In this way, resonance frequencies may be determined with a very small uncertainty so that random errors in the sound speed can be rendered small. A similar approach may be employed with a variable path interferometer in which case the data are collected over a small range of path lengths in the vicinity of a resonance, and the analysis leads to the path length at resonance. Transient measurements with cavity resonators are not well suited to precise determination of sound speed, but the resonance decay time (reverberation) may be obtained as detailed in references 23, and 71.

9.3.4 Viscothermal Boundary Layer and Shell Motion

For a non-thermally relaxing gas within a cavity of fixed geometry, operated at variable frequency, there are two key interactions that occur at the interface between the gas and the shell and which determine the range of densities over which precise sound speed measurements can be obtained. The first of these is the viscothermal boundary layers that exist at the cavity walls as a consequence of the non-zero viscosity and thermal conductivity of the gas and the boundary conditions that constrain the temperature and velocity fields at that surface. The separate viscous and thermal boundary layers each give rise to a fractional shift in the resonance frequencies of a cavity resonator, relative to an idealized case in which heat and momentum transfer at the wall are ignored, that varies like $(f\rho)^{-1/2}$ and hence becomes important at low frequencies f and low densities

ρ. The second key interaction between the gas and the wall is a mechanical one in which the wall responds elastically to the acoustic pressure acting there. Although the resulting wall motion is truly microscopic in an absolute sense, this effect also gives rise to a shift in the resonance frequencies, this time relative to a cavity with rigid walls. Typically, this effect varies in proportion to ρu^2 at frequencies far from a resonance of the cavity wall and hence is of growing importance as the gas pressure is increased. The effect also increases as the frequency of the gas resonance tends toward that of the shell to the extent that useful measurements cannot be made at any density.[2]

Each of these effects, discussed further below for specific cavity geometries, conforms more-or-less-well to a physical model from which corrections may be calculated. Accordingly, the associated errors are restricted to marginal failures of the model to account entirely for the phenomena and to the effects of errors in the parameters required to compute the corrections.

To minimize corrections for viscous and thermal boundary layers a resonator should have the smallest surface-to-volume ratio with smooth inner surfaces and be operated with a gas at pressure above 0.1 MPa and preferably greater than about 1 MPa because at lower pressures the boundary layers dominates the corrections that must be applied to the experimental resonance frequencies. To minimize the correction for wall motion, the wall should also be stiff and, at least for a sphere, the correction dominates at pressures above about 2 MPa.

Since the perfect-gas heat capacity is obtained from an extrapolation of u^2 to the limit of zero gas pressure, it is the viscothermal boundary layer corrections that dominate, and the result is essentially unaffected by the coupling of wall and shell motion. Fortunately,[2,16] with routine knowledge of the thermal conductivity and viscosity of the gas the sound of speed can be determined with a relative uncertainty of less than 10^{-5} at densities in the range (1 to 200) kg m^{-3} at pressures less than about 0.7 of the saturation pressure.[2,3] This uncertainty can be achieved even at pressures of 10 kPa with inefficient transducers when a spherical resonator, with its favorable surface-to-volume ratio, is deployed.

9.3.5 Molecular Thermal Relaxation

The other potential source of systematic error in sound speed measurements arises from molecular thermal relaxation in polyatomic gases,[10,72] which can give rise to dispersion and attenuation of sound. Provided sound speed measurements are performed at frequencies that are low compared with the inverse of the thermal relaxation time, the major influence of this phenomenon is on sound absorption. The thermal relaxation time is often dominated by the time required for the transfer of energy from translational to vibrational modes of molecular motion and, if the number of molecular collisions required to deactivate the vibration of a molecule by this mechanism is large and the time required long compared to the acoustic cycle, then dispersion can be significant.[72–74] Cottrell and McCoubrey[73] and Lambert[74] describe the basic mechanisms of inter- and intra-molecular energy transfer and review the experimental methods, including sound dispersion and attenuation measurements, used to study vibrational (and rotational) relaxation times.

The sound absorption coefficient α is usually represented as the sum of two terms: the first arising from the classical viscous and thermal dissipation mechanisms; and the other, $\alpha_{rel.}$, associated with molecular thermal relaxation. The classical contribution to the half width, which is present in all gases, is given by

$$\alpha_{cl} = (2\pi f^2/u^3)\{4D_v/3 + (\gamma - 1)D_t\} \tag{18}$$

where D_t and D_v are the thermal and viscous diffusivities respectively. For a gas with molar mass M, thermal conductivity κ, and shear viscosity η at mass density ρ, the thermal diffusivity is given by

$$D_t = \kappa M/(\rho C_{p,m}) \tag{19}$$

and the viscous diffusivity by

$$D_v = \eta/\rho \tag{20}$$

In determining the contribution to α from molecular thermal relaxation, it is assumed[72-74] that the vibrational-to-vibrational energy transfer is rapid compared with translational-to-vibrational exchange, so that the entire vibrational contribution to the heat capacity relaxes with a single time constant τ. The resulting contribution to the sound attenuation is then given by:[12]

$$\alpha_{rel} = (2\pi^2 f^2/u)(\gamma - 1)\Delta\tau \tag{21}$$

where $\Delta = C_{vib}/C_p$ and C_{vib} is the vibrational contribution to the heat capacity. Analysis of the resonance line-widths, after allowance for the other, usually much smaller, loss mechanisms allows the determination of the vibrational relaxation time, τ.[12] This value of τ can then be used to estimate the speed of sound in the limit of zero-frequency, u_0, from the measured speed of sound, u, at frequency f by means of the relation[12]

$$(u - u_0)/u = 2(\gamma - 1)\Delta(\pi f\tau)^2\{1 - \Delta(1 + 3\gamma)/4\} \tag{22}$$

Provided that $f \ll \tau^{-1}$, this correction is small. Expressions other than Equations (21) and (22) would be required if the gas relaxed with another intramolecular mechanism. In polyatomic gases, α_{rel} is comparable to or, especially for small rigid molecules such as CO_2 and CH_4, much larger than α_{cl}. Since the relaxation mechanism involves molecular collisions, the product $\tau\rho$ is approximately constant, so that τ, the frequency at which dispersion becomes significant, decreases with increasing density. Clearly the experimenter who intends to measure the speed of sound of a thermally relaxing gas must select a method that operates at sufficiently low frequency for the dispersion correction to be small.

9.3.6 Precondensation

The speed of sound is formally independent of the amount of substance and so is in principle immune from the effects of adsorption. Nevertheless, precondensation or adsorption on the walls of a cavity resonator affect sound reflection because the oscillating temperature and pressure fields associated with the sound wave drive oscillatory sorption and desorption at the surface, thereby causing a non-zero normal fluid velocity at the gas-wall interface with consequential dissipation and shifts in the resonance frequencies of the cavity. For an absorbed liquid film of a pure substance on a solid surface Mehl and Moldover[75] determined experimentally that the effect increased rapidly as the pressure increased towards the vapor pressure of the gas under study. Precondensation effects are a potential source of systematic error in many experiments designed to measure speed and attenuation of sound. It is primarily significant for the determination of acoustic virial coefficients but may also contribute to the uncertainty of the perfect heat-capacity obtained from $u(T, p \rightarrow 0)$. Moldover and Mehl[75] also derived a theoretical model but this cannot be applied quantitatively to practical resonators because of the influence of surface roughness. Accordingly, the experimenter must resort to judicial selection of the starting pressure for an isotherm. Fortunately, this does not usually introduce an error in the determination of A_0. The effect on both speed and attenuation of sound decreases with increasing frequency and thus measurement made over a range of operating frequencies can assist identification.[75]

The moderately high frequency of 500 kHz has been used with a double-transducer cylindrical interferometer[76,77] for sound speed measurements in compressed methane with good results even in the saturated vapor where precondensation would have invalidated a measurement in a cavity resonator operating at audio frequencies. An unguided resonator operating at MHz frequencies, when designed to carefully minimize diffraction errors, can provide the sound speed in fluids with a relative uncertainty of about 10^{-5} essentially unhindered by viscous, thermal, and precondensation effects.[78,79]

Another source of strong sound attenuation occurs at conditions close to the vapor-liquid critical point.[8,80,81] To measure the speed of sound under near-critical conditions requires an apparatus operating at low frequency and, in a gravitational field, of small height; an annular acoustic resonator may satisfy both of these criteria.[81]

9.4   Fixed Cavity Resonators

The fixed cavity resonator is the modern instrument of choice for sound speed measurements in gases. In this section, we consider the characteristics of the specific cavity geometries that have been adopted in practice.

9.4.1 Spherical Geometry

The spherical cavity is particularly attractive for sound speed measurements in gases at $p < 1$ MPa for the determination of heat capacity owing to the favorable volume-to-surface ratio in the sphere and the total absence of viscous damping at the surface for the radially symmetric modes. These factors result in the radial modes being characterized by higher quality factors than those found in any other geometry of similar volume and operating frequency, permitting resonance frequencies to be measured with a relative uncertainty of 10^{-6}. The resonance frequencies of the radial modes are also insensitive to geometric imperfections[82–84] so that typical manufacturing tolerances are acceptable and only the volume of the sphere is required for accurate absolute measurements of the sound speed. The typical relative uncertainty of u, obtained with a spherical resonator, at pressures below about 1 MPa is then of order 10^{-5} and the final uncertainty of the perfect gas heat capacity is usually restricted to the lower limit determined by imperfect knowledge of the molar mass.

Rayleigh[85] recognized the merits of the spherical cavity and the method was implemented by Bancroft,[86] Harris,[87,88] and Rudnick and collaborators.[89,90] It was Moldover and co-workers[15,42,71,91–100] who pioneered the use of spherical resonators for precise measurements of the properties of gases, and spawned many subsequent applications in the determination of perfect-gas heat capacities and thermophysical properties of pure gases and mixtures.[12,13–32,43–49, 51,52,54,75,101–138,] Brooks and Hallock[50] were the first to utilize the technique with Equation (14) to determine the composition of binary gaseous mixture, and the sensitivity of this approach has been demonstrated recently.[49] We now turn to a description of the working equations for a spherical resonator for the determination of perfect-gas heat-capacity and gas imperfections.

The zeroth-order description of the acoustic resonance $f_{l,n}^{(0)}$ of a fluid within a rigid spherical cavity of radius a is given by[85]

$$f_{l,n}^{(0)} = \nu_{l,n}[u/(2\pi a)]; \quad l = 0, 1, 2, \cdots; \quad n = 0, 1, 2, \cdots, \tag{23}$$

where $\nu_{l,n}$ is the n-th turning point of the spherical Bessel function of order l. We will consider only the particularly useful radial modes for which $l = 0$;[139] the uses of non-radial modes have been described elsewhere.[71,140] Moldover and co-workers[15,71] used first-order perturbation theory to account for the energy losses in the bulk of the gas and at the wall of the resonator. A complex resonance frequency $F_{0,n}$ for the n-th radial mode is defined by

$$F_{0,n} = f_{0,n} - ig_{0,n} = \nu_{0,n}[u/(2\pi a)] + \sum_j (\Delta f - ig)_j \tag{24}$$

where $f_{0,n}$ and $g_{0,n}$ are the observed resonance frequency and resonance half-width, and $\Sigma_j(\Delta f - ig)_j$ is the sum of a number of first-order perturbation (or correction) terms that account for deviations from the zeroth-order model. These corrections have been derived and discussed in great detail elsewhere;[2,3,95,96] nevertheless, we summarize here the corrections arising from the thermal boundary layer, non-zero elastic compliance of the spherical shell, and gas inlet tubes. Usually, the sum of the correction terms is estimated from the available physical models and thermophysical properties and, since it is small, quite approximate values may suffice.

The thermal-boundary-layer correction is given by[141]

$$
\begin{aligned}
(\Delta f - \mathrm{i}g)_t = & -(1+i)\{(\gamma-1)/(2a)\}(D_t f/\pi)^{1/2} \\
& + \mathrm{i}\{(\gamma-1)/(2a^2)\}(D_t/\pi) + (\gamma-1)f(l_t/a)
\end{aligned}
\tag{25}
$$

where $\gamma = C_{p,\mathrm{m}}/C_{V,\mathrm{m}}$ is the ratio of the molar heat capacities at constant pressure to that at constant volume, D_t the thermal diffusivity given by Equation (19) and l_t the thermal accommodation length. At pressure below about 1 MPa, Equation (25) normally dominates the corrections that must be applied to the experimentally determined resonance frequencies. In Equation (25), the first term is the classical expression describing the loss at a plane surface, the second term is the small correction for curvature of the surface, and the final term takes account at a phenomenological level of a temperature-jump at the gas-wall interface. For a gas with molar mass M and thermal conductivity κ at mass density ρ, the accommodation length is given by

$$
l_t = (\kappa/p)(\pi M T/2R)^{1/2}\{(C_{V,\mathrm{m}}/R)+1/2\}^{-1}\{(2-h_t)/h_t\} \tag{26}
$$

where h_t is the thermal accommodation coefficient between the particular gas and the resonator wall material. This parameter is sensitive to both the surface finish and the presence of an adsorbed layer of molecules and it cannot be determined *a priori*. The value of h_t can be estimated from the pressure dependence of the frequency,[141] with values of h_t near unity are typical for molecular gases and machined surfaces.[142,143]

For the determination of the perfect gas heat capacity from A_0 of Equation (6) the spherical resonator is typically operated at pressures <0.1 MPa and is also surrounded by a low pressure gas. In this case, the shell perturbation is only significant when the measurements of $u(T, p)$, are extrapolated to $p = 0$ from higher pressures. In that case, the correction can be obtained with sufficient accuracy from the expression derived for a perfectly elastic spherical shell by Greenspan,[144]

$$
\Delta f_s/f = A\rho u^2 \Big/ \Big\{1 - (f_{0,n}/f_{s,\mathrm{br}})^2\Big\} \tag{27}
$$

In Equation (27), $f_{s,\mathrm{br}}$ is the lowest radial resonance of the spherical shell, the so-called breathing mode, given by

$$
f_{s,\mathrm{br}} = \big\{(t^3-1)/[2\pi^2(t-1)(1+2t^3)]\big\}^{1/2}(u_s/a) \tag{28}
$$

and the compliance A is given by

$$A = (1 + 2t^3) / [2(t^3 - 2)\rho_s u_s^2]$$ (29)

Here, $t = b/a$ is the ratio of the inner to outer radii of the shell and u_s is the longitudinal wave speed in the wall material. The ratio of the breathing frequency of the shell, relative to the resonance frequencies of the gas, is controlled to a large extent by the ratio (u_s/u) of the speeds of sound in the wall material and the gas. For stainless-steel or aluminum shells containing gases heavier than helium, the lower four radial gas modes typically occur below $f_{s,br}$ at pressures <1 MPa and the relative corrections are on the order of 10^{-5}. At pressures >1 MPa, coupling between the motion in the gas and the shell become increasingly important and the generalized theory for the motion of a thick spherical shell reported by Mehl[145] is required to adequately determine the correction.

An orifice, with high acoustic and low pumping impedance, is required for the gas to enter and leave the spherical cavity. As mentioned above, a solution which eliminates the variation of the surface acoustic impedance is to machine a valve into the resonator wall.[38] A valve is also useful for gas mixtures, prepared outside of the resonator, if the gas phase separates on expansion or compression into the resonator.[146] The simple, and in most cases adequate, alternative is to use a small cross-section tube with a length and exterior termination chosen to simultaneously achieve high acoustic impedance at the resonance frequencies of interest and a low impedance to flow. This approach does require a correction to the resonance frequencies and contributions to the resonance half width but, by careful selection of dimensions, the former can be reduced to order $10^{-6} f$. In detail, a cylindrical tube of radius b and length L opening into the resonator gives rise to a perturbation

$$(\Delta f - ig)_o = (ub^2/8\pi a^3) \cot\{(kL + \alpha_{KH}L + \delta_L) + i(\alpha_{KH}L + \gamma_L)\}$$ (30)

where $k = 2\pi f/u$ is the propagation constant and

$$\alpha_{KH} = \left\{ (\pi f)^{1/2}/ub \right\} \left\{ D_v^{1/2} + (\gamma - 1)D_t^{1/2} \right\}$$ (31)

is the Kirchhoff-Helmholtz tube attenuation constant. In Equation (31), D_t and D_v are given by Equations (19) and (20) respectively. The parameters γ_L and δ_L describe the change in amplitude and phase at the end of the tube remote from the resonator. For an open flanged tube

$$\gamma_L = (kb)^2/2$$ (32)

and

$$\delta_L = 8kb/3\pi$$ (33)

while for a tube termination in a sealed cavity of volume V with dimensions small compared with the wavelength

$$\gamma_L = 0 \qquad (34)$$

and

$$\delta_L = -\arctan(\pi b^2 / kV) \qquad (35)$$

However, if the terminal cavity is not sealed then leakage conductance can greatly increase the terminal admittance so that

$$\gamma_L \approx \delta_L \approx 0 \qquad (36)$$

According to Equation (30), the perturbation arising from an open tube is approximately proportional to $\cot(v_{0n}L/a)$ at the radial resonance frequencies of the gas-filled spherical resonator. Since the eigenvalues v_{0n} of the $(0, n)$ radial modes are approximately $(n - 1/2)\pi$, with $n = 2, 3, \cdots$, when $L \approx a$ there is only a small perturbation to the radial resonance frequencies.[103,113]

A practical spherical resonator is usually constructed from two hemispheres joined at the equator and has transducers which are inserted into the wall as removable units. Both of these mechanical arrangements leave annular slits between the two hemispheres or the transducer housing and the wall. Trusler[2,147] has determined the effect on the radial resonance frequency and half width in a spherical cavity for a slit bounded by semi-infinite parallel flat surfaces. The perturbations arising from slots can be reduced to less than $10^{-6} f$ with relatively easily achieved machine shop tolerances of about 10 μm.

The choice of material from which to fabricate a resonator will be influenced by chemical compatibility with the gases to be studied, the range of temperature and pressures to be experienced, and the magnitude of the shell correction which, from Equations (29) and (27), is inversely proportional to $\rho_s u_s^2$. Spherical resonators of varying radii have been constructed from brass,[22] aluminum,[12,15,19,44,101,103,113] and stainless steel.[26,60,109,118,148] An example of a resonator formed from the latter and that also acted as a pressure vessel is shown in Figure 9.2. In this case, as in most others, the transducer ports were drilled in one hemisphere at a polar angle of $\pi/4$, separated by an azimuthal angle of π. This arrangement greatly reduces what would otherwise be serious overlap with the (3,1) non-radial mode, the resonance frequency of which is only about 0.5 % below that of the (0,2) mode. The (3,1) mode has a node at polar angle π and so is not detected on symmetry grounds when source and detector are at that angle. In practice, geometric imperfections and finite transducer area result in a non-zero signal at the detector for the (3,1) mode; nevertheless significant overlap with the (0,2) mode is usually avoided. Gas inlet and outlet tubes shown in Figure 9.2 were of internal diameter 2.4 mm placed in the bosses located at the poles and were of length $L \approx a = 45.7$ mm, thereby

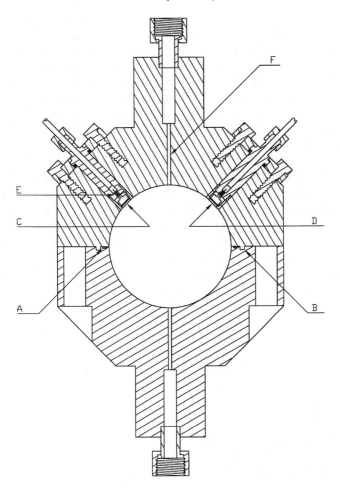

Figure 9.2   Cross-section through a spherical resonator fabricated as two hemispheres from a single cylindrical bar of type 316L stainless steel to form a cavity with a nominal internal radius of 45 mm. The acoustic transducers were fabricated to fit within ports and each housing was sealed with a silver o-ring with separate flanges and bolts. The gas under test was sealed by a glass-to-metal seals E so that it was exposed to the transducer back-plate and active diaphragms C and D, that were mounted flush with the resonators interior surface. The hemispheres were then machined at the equator to form an interlocking step B, which ensured accurate concentric alignment about the polar axis, and made metal-to-metal contact at the inner spherical surface. A groove suitable for a silver o-ring A was machined into the equatorial surface. To form the 2.4 mm diameter inlet tubes for fluid flow in each pole, a hole was drilled to a total of 45.7 mm F and the remaining 48 mm of the boss was opened out to a radius of 9 mm. A face-seal fitting was welded into the outer end of the larger diameter tube.

minimizing the correction given by Equation (30). Ripple *et al.*[42] report a tubing arrangement suitable for continuous purging, during acoustic measurements, for the purposes of maintaining extremely high gas purity. In this arrangement, the gas flowed through apertures placed to promote mixing but also designed to suppress acoustic noise generated by turbulence (which increases as the eighth power of the Reynolds number). For this purpose, a low-pass acoustic filter was formed from a duct of length $a/10$ and diameter 1.5 mm connecting the resonant cavity to a small volume of approximately 1 cm^3 in the wall of the spherical shell; this short duct also gives rise to a small perturbation to the acoustic modes of the cavity.

Fawcett[122] described an apparatus with four interconnected isothermal spherical resonators that could be used to provide both (u, T, p) and (p, V_m, T) data through either a Burnett or differential Burnett expansion on the same gas at temperatures between (293.15 and 303.15) K and pressures up to 10 MPa. This instrument would provide all of the information required to obtain both Z and $C_{p,m}$ in the compressed gas as described in Section 9.2.4. The apparatus was also claimed to provide the opportunity to determine the presence of precondensation effects from acoustic measurements on the same gas in spherical cavities of different surface area to volume ratios.[75] Measurements with helium were used to determine dimensional constants, including the radii, and measurements of the speed of sound in methane agreed with those reported in references 26 and 114 to within about $\pm 10^{-5}$ u. The apparatus was subsequently used for measurements of the speed of sound in natural gas mixtures at pressures up to 20 MPa and temperatures between (250 and 350) K.[122,123]

9.4.2   Hemispherical Geometry

In a spherical resonator, the transducers or (if used) waveguide terminations are typically located flush with the interior surface of the cavity where they couple, rather weakly, to the radial modes. Thus measurements with a spherical resonator may be constrained by the available signal-to-noise ratio at the low pressures required to determine the perfect-gas heat capacities according to Equation (12). For example, in the case of thermally relaxing gases, increased sound absorption leads to broader resonances and smaller amplitude; furthermore overlap with other modes may be a limitation.[12] The signal-to-noise ratio can be increased dramatically by locating the source transducer at the center of the resonator where the coupling to radial modes is maximal and non-radial modes are not excited efficiently. Unfortunately, the probe or tube needed to implement this source location adds an additional perturbation to the resonance frequencies that cannot be easily modeled. This problem can be addressed by resort to a hemispherical resonator as it is then easy to site the source in the center of the equatorial plane.

Angerstein[149] has developed and tested a hemispherical resonator. In reference 149, a transducer was placed in the center of a flat plate bolted to a hemispherical shell at the equator to form a pressure-tight vessel. The source transducer was coupled to the center of the flat plate by a waveguide of diameter 0.3 mm and length $L \approx a$, where a is the radius of the hemispherical cavity. This arrangement was chosen to reduce the perturbation arising from the non-zero cross-sectional area of the source. The detector was placed directly opposite the source in the spherical surface. The zeroth-order description for the motion of a fluid within a rigid hemisphere is identical to that of a sphere.[149] Thus, the ratio of the speed of sound to the resonators radius is determined from the measured frequency and the characteristic dimension of the cavity, the radius, is required to obtain the sound speed. However, the radial gas modes in a hemispherical cavity now include tangential fluid motion at the equator and the perturbations to the resonance frequencies also include the shear boundary layer and molecular slip as well as the thermal boundary layer.[149] The corrections for the opening in the equatorial boundary used as a gas inlet included the variation of the wave function over the area of the hole.[149] This expression is more complex than Equation (30) in which the wave function is constant on the spherical surface.[2] The description of the shell motion, which is a combination of a sphere and a plate, has not been developed but, by analogy with Equation (27), the magnitude of the correction should decrease with decreasing density and pressure. Further evaluation of the hemispherical resonator is required before this approach can be recommended for routine determination of perfect gas heat capacities for gases with strong sound absorption.

9.4.3 Cylindrical Geometry

Compared with spherical resonators, cylinders are easier to fabricate and offer an extra degree of freedom in the form of the ratio b/L, where b is the radius and L the length of the cylinder, which can be chosen to suit a particular application.[3] However, due to the less favorable surface-to-volume ratio and the presence of both viscous and thermal damping at the resonator walls, the resonances have a lower Q compared with a sphere of the same volume. In the cylinder, pure longitudinal, azimuthal and radial modes are available, as are compound modes with characteristics of all three. For longitudinal modes, all transducer positions on the end plates are equally efficient and a source located in the center of the end plate will couple effectively to all axisymmetric modes including all radial and mixed longitudinal modes but not azimuthal modes. To reduce or eliminate mode overlap, the lower-order resonances are preferred in measurements of the speed of sound while high-frequency longitudinal modes suffer smaller damping arising from the viscothermal boundary layers. However, excitation of a pure longitudinal mode, even when a piston-like source that covers the whole end face area is used, is not assured and the heat capacity obtained can contain an error that is difficult to estimate. Cylindrical resonators may operate as a fixed path length and variable frequency cavity[9,150–160] or as an interferometer with variable path length and fixed frequency.[76,77,161–171]

The zeroth-order working equation for a cylindrical cavity or interferometer of length L and radius b is

$$f_{l,m,n} = \left(\frac{u}{2L}\right)\left\{l^2 + \left(\frac{\chi_{m,n}L}{\pi b}\right)^2\right\}^{1/2} \tag{37}$$

Here, the positive integer l is the order of the longitudinal mode and $\chi_{m,n}$ is the n-th turning point of the cylindrical Bessel function of order m. The symmetry of each mode is determined solely by (l, m, n) so that $(l, 0, 1)$ refers to a longitudinal mode, $(0, m, 1)$ is an azimuthal mode, that circulates about the central axis, and $(0, 0, n)$ is a purely radial mode. When $m = 0$, $\chi_{m,n} = 0$, and the modes are non-degenerate; otherwise they are two-fold degenerate. The resonance frequency and half width of the modes in a cylinder are perturbed at the wall by viscothermal effects and holes. Viscous boundary loss occurs when the tangential component of the gas velocity is not zero; this is the case for a radial mode in a cylindrical cavity where the gas has tangential movement at the ends. For a longitudinal mode the gas moves along the side wall and for azimuthal modes the gas moves around the wall. Gillis[160] and Trusler[3] applied first order perturbation theory to determine the viscous and thermal contributions for a rigid cylinder so that Equation (37) becomes

$$f_{l,m,n} = \left(\frac{u}{2L}\right)\left\{l^2 + \left(\frac{\chi_{m,n}L}{\pi b}\right)^2\right\}^{1/2} + \Delta f_v + \Delta f_t \tag{38}$$

and the corresponding half widths are

$$g_{l,m,n} = g_v + g_t + g_b \tag{39}$$

In Equation (39), g_b is the bulk attenuation given by the sum of Equations (18) and (21). The frequency shifts Δf_v and Δf_t from the unperturbed resonance frequencies owing to viscous and thermal boundary layers, including molecular slip and temperature jump, are given by

$$-\Delta f_t = g_t(1 - 2l_t/\delta_t) \tag{40}$$

and

$$-\Delta f_v = g_v(1 - 2l_v/\delta_v) \tag{41}$$

respectively. In Equation (40) and Equation (41) l_t and l_v are the thermal and viscous accommodation lengths given by Equation 26 and

$$l_v = \left(\frac{\eta}{p}\right)\left(\frac{\pi RT}{2M}\right)^{1/2}\left(\frac{2 - h_v}{h_v}\right) \tag{42}$$

that describes molecular slip near the solid interface. In Equation (42), h_v is the momentum accommodation coefficient and, like h_t, is typically close to unity for molecular gases on machined surfaces. The δ_t and δ_v of Equations (40) and (41) are the thermal and viscous penetration lengths given by

$$\delta_t = \left\{ \kappa V_m / (\pi f C_{p,m}) \right\}^{1/2} = \left\{ D_t / (\pi f) \right\}^{1/2} \tag{43}$$

and

$$\delta_v = \left\{ \eta / (\pi f \rho) \right\}^{1/2} = \left\{ D_v / (\pi f) \right\}^{1/2} \tag{44}$$

When the path length of a cylindrical resonator is determined with argon and then used to calculate the perfect gas heat capacity at constant pressure of another gas the boundary layer correction is significant.[172] For example, for dimethylpropane uncertainties of about 1 % were introduced into the acoustically determined heat capacity when the calibration measurements were analyzed without recourse to the thermal boundary.[156,172]

Gillis[160] also reported an expression for the perturbation of a gas inlet tube located in the end face of the cylinder and found that in one practical situation it was less than $\pm 15 \times 10^{-6}$ and comparable with the uncertainty in determining u.[160] There is no exact expression to account for the frequency shift arising from shell motion. Trusler[3] has discussed the magnitude of the correction and the coupling but it is expected, based on the analysis for a sphere, the correction is proportional to fluid density and small provided the frequency does not coincide with the resonance of the shell.

The cylindrical resonator described by Gillis et al.[158] and Gillis[160] is shown in Figure 9.3. Gillis[160] used measurements with argon to determine the internal length of the cavity from purely longitudinal modes and the radius from purely radial modes and or pure azimuthal modes. The excitation of non-axisymmetric modes in a cylindrical resonator requires the source be offset from the center of an end plate or on one on the side walls. Gillis[160] had both a source and detector in the top plate. These were formed from diaphragms flush with the inner surface of the resonator and connected *via* waveguides to transducers located outside the thermostat. The diaphragms were used to separate the gas within the waveguide, which was argon, from that in the resonator. The operational frequency range of the cavity was restricted to between (1 and 8) kHz by the waveguides that permitted operation of the cavity over a wide temperature range with noxious substances with commercially available transducers. This resonator has been used to determine the thermophysical properties of numerous gases including those relevant to semiconductor processing.[33–37,173–183]

The sound speed can also be determined by tuning the cavity through successive longitudinal resonances, at fixed excitation frequency, by variation of the path length L. Unfortunately, this approach requires complex mechanical systems to move accurately a reflector and may also necessitate sliding pressure

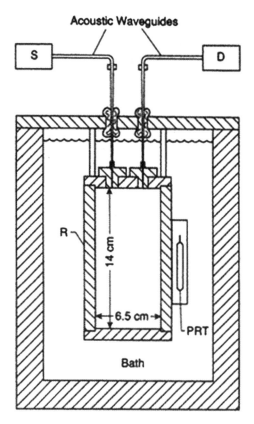

Figure 9.3   Schematic cross-section through the cylindrical resonator R located in a stirred fluid thermostat. The cylinder was 14 cm long and has an inner diameter of 6.5 cm. The temperature of the cavity was determined with a Platinum Resistance Thermometer (PRT). Waveguides connected the resonator to the source S and detector D of sound that were located outside the thermostat. Diaphragms flush with the upper inner surface of the cylindrical cavity separated the gas within the waveguide from that in the resonator.

seals.[152,153] Nevertheless, this approach has been applied with success in several forms. Henderson and Peselnick[162] used a single transducer interferometer of diameter 25.4 mm and a length varied by a micrometer over a range of about 75 mm operating at frequencies between (0.3 and 7) MHz to comprehensively study the dispersion and absorption of sound in CO_2 up to liquid densities. Gammon and Douslin[76,166] described a two-transducer cylindrical variable-path length interferometer operating at a frequency of 500 kHz with transducers separations varied by between (50 and 100) mm and determined with a micrometer to obtain sound absorption and speed with a precision of 10^{-5} u provided small corrections,[76,77,166,168] based on Del Grosso's calculations[184-186] were used to account for the so-called guided-mode dispersion arising from the excitation of compound modes.

Cylindrical resonators have been used to determine transport coefficients at temperatures up to 1000 K by Carey,[154] while Zuckerwar and Griffin[155] have described a cylindrical cavity of length about 17 m to measure the absorption of sound at low frequency-to-pressure ratios that cover the dispersion curve in nitrogen at frequencies from 10 Hz to 2500 Hz at pressures up to 10 MPa determined from the free decay of a standing wave. This apparatus was also used to study sound absorption in air.[187]

9.4.4   Annular Geometry

Low frequencies are required for gases where molecular thermal relaxation is slow and to study fluids close to the liquid-gas critical point. Moving the sound detector source to a node, as illustrated in Section 9.4.2, is one solution for highly attenuating fluids. Low frequencies can be achieved with a sphere of impractically large radius and long thin cylinders in which the ratio L/b is chosen so a few longitudinal modes occur at low frequencies albeit with low surface area-to-volume ratios and the ubiquitous unresolved modes. Alternatively, in a cylinder of radius $b > L$, the length, can be selected so the first few radial modes, azimuthal and mixed radial-azimuthal modes occur at lower frequency than the longitudinal modes. Unfortunately, mode overlap precludes resolution of these modes and introduces substantial error.

An attractive alternative is afforded by a cylindrical annulus or square-section toroidal resonator described by Buxton.[188] The first few azimuthal modes of the annulus occur at lower frequencies than the first radial mode albeit with greater half widths than in the corresponding cylinder owing to the less favorable surface-area-to-volume ratio. However, as shown in reference 188, the annular geometry is favorable: for example, for low-pressure measurements of the speed and adsorption of sound in thermally relaxing gases. In reference 188, the first five azimuthal modes $(0, m, 1)$ with $m = 1$ to 5, occurred at frequencies between (0.5 and 2.3) kHz when the cavity was filled with argon at a pressure of 100 kPa and a temperature of 300 K.[188] This frequency range allowed the study of relaxing gases at a pressure a factor of ten lower than for a practical sphere. Under the same conditions, the resonance quality factor $Q = 172 \{= f/(2g)\}$ for mode $(0, 1, 1)$ with argon. Buxton[188] applied first order perturbation theory to determine the viscothermal boundary layer corrections for a rigid annular resonator, which are essential for the determination of perfect-gas heat capacity, and an expression for the perturbation of the gas inlet tube. Shell motion has not been considered.

To determine the speed of sound and thus the perfect-gas heat capacity, from the measured resonance frequency, requires knowledge of three geometric parameters as a function of T and p. In the case of reference 188, these were determined from measurements with propene for which the resonance quality factor $Q \{= f/(2g)\}$ was significantly greater than in argon. Buxton[188] reported the speed of sound could be determined with a relative precision of about 2×10^{-5} using an annular resonator and perfect-gas heat capacity obtained with an uncertainty of about 0.1 %.

9.4.5   Cavity Resonators for Other Applications

A number of novel acoustic resonators have been explored for application in the determination of transport properties and relaxation times[2–9] and, although a digression from heat capacity, we briefly mention three of them because these physical properties are required by the acoustic model used to obtain sound speed from the measured frequency.

Moldover and co-workers[8] describe an acoustic determination of the Prandtl number ($P_r = \eta C_{p,m}/\kappa$) of gases from measurements of the ratio of the resonance frequency of odd ($l = 1, 3, 5, \ldots$) and even ($l = 2, 4, 6, \ldots$) longitudinal modes in a cylindrical cavity fitted with an insert comprising a honey-comb-shaped array of hexagonal ducts. This insert spanned the circular cross-section of the cavity approximately half way along. The measurement exploits the fact that, over mid cross-section of the unperturbed cavity, the odd-order longitudinal modes have maximal axial velocity and vanishing acoustic temperature oscillations, while the reverse is the case for the even-order modes. Thus primarily, the effect of the insert is it creates additional viscous damping for the odd modes and additional thermal damping for the even modes. A prototype instrument was described which, when calibrated for the effective area of the insert, led to the Prandtl number of another gas with a relative uncertainty of about 2 %. The thermal conductivity of the gas was then determined by combining the Prandtl number with acoustically or otherwise determined values of density, viscosity and heat capacity.

Measurements of half widths in an annular resonator may be used to advantage in the simultaneous determination of viscosity, thermal conductivity and vibration relaxation time for gases at low pressures as described in reference 188. In that work, SF_6 viscosity, thermal conductivity and relaxation times were determined that differed as $\rho \rightarrow 0$ from literature values by up to 5 % for both η and κ while the product $\tau\rho$, where τ is the relaxation time, varied between (1 and 15) % depending on the magnitude of $\tau\rho$.

The third example is the dumbbell-shaped double Helmholtz resonator described by Moldover and co-workers[8,189–192] (based on the work of Greenspan[193]) that operates at frequency, in which the gas oscillates through a duct connecting two chambers, on the order of 100 Hz with relatively low Q resonance. The resonator was primarily intended to determine the viscosity of a gas from the measured resonance half-widths and at pressure between (0.2 and 3.2) MPa was shown to provide viscosities that lie within about ±2 % of literature values. The low (about 100 Hz) frequency of the Helmholtz mode means the cavity can also be used to determine the speed of sound in gases that undergo vibrational relaxation. Measurements of the sound speed in methane were found to differ relatively by about ±2 × 10⁻⁴ from the literature value.[8] This uncertainty in sound speed can be used to provide perfect-gas heat capacity albeit with an uncertainty substantially greater than those determined with a cylindrical or spherical resonators.[8,192]

9.5 Summary

Measurements of the speed of sound in a gas, when extrapolated along an isotherm to the limit of zero pressure, provide a measure of the perfect-gas heat capacity, with an uncertainty comparable to those obtainable from flow-calorimetric measurements. Acoustic determination of perfect-gas heat capacity is best obtained with fixed cavity resonator which, except in the case of gases with high sound absorption, should preferably be of spherical geometry. The uncertainty of the acoustically derived perfect-gas heat capacity is typically controlled by imperfect knowledge of the molar mass, especially as a consequence of impurities. Other uncertainties, including systematic errors in temperature measurements, cancel to high order when the characteristic dimension of the cavity resonator is obtained by calibration with a gas of known molar mass and perfect-gas heat capacity. The application of a cavity resonator in measurements of the speed of sound is supported by well-developed working equations that model the physical system in a rigorous way. Sound speed measurements in compressed gases also provide useful information about heat capacities but some additional experimental data are required to extract this.

References

1. W. Van Dael, in *Experimental Thermodynamics, Volume II, Experimental Thermodynamics of Non-Reacting Fluids,* eds. B. Le Neindre and B. Vodar, Butterworths, London, 1975, Ch 11.
2. A. R. H. Goodwin and J. P. M. Trusler, in *Sound Speed in Experimental Thermodynamics Volume VI: Measurement of the Thermodynamic Properties of Single Phases,* eds. A. R. H. Goodwin, W. A. Wakeham and K. N. Marsh, Elsevier Science, Amsterdam, 2003, Ch. 6.
3. J. P. M. Trusler, *Physical Acoustics and Metrology of Fluids,* Adam-Hilger, Bristol, 1991.
4. *Handbook of Elastic Properties of Solids, Liquids, and Gases, Volume I, Dynamic Methods for Measuring the Elastic Properties of Solids,* M. Levy, H. E. Bass, and R. R. Stern, eds.-in-chief, A. G. Every, and W. Sachse, vol. eds., V. Keppens, supervising ed., Academic Press, New York, 2001.
5. *Handbook of Elastic Properties of Solids, Liquids, and Gases, Volume II, Elastic Properties of Solids: Theory, Elements and Compounds, Novel Materials, Technological Materials, Alloys, and Building Materials,* M. Levy, H. E. Bass, and R. R. Stern, eds.-in-chief, M. Levy, vol. ed., L. Furr, tech. Ed., V. Keppens, supervising ed., Academic Press, New York, 2001.
6. *Handbook of Elastic Properties of Solids, Liquids, and Gases, Volume III, Elastic Properties of Solids: Biological and Organic Materials, Earth and*

*Marine Scienc*es, M. Levy, H. E. Bass, and R. R. Stern, eds.-in-chief, M. Levy, vol. ed., L. Furr, tech. Ed., V. Keppens, supervising ed., Academic Press, New York, 2001.

7. *Handbook of Elastic Properties of Solids, Liquids, and Gases, Volume IV, Elastic Properties of Fluids: Liquids and Gases*, M. Levy, H. E. Bass, and R. R. Stern, eds.-in-chief, M. Levy, R. Raspet and D. Sinha, vol. ed., L. Furr, tech. Ed., V. Keppens, supervising ed., Academic Press, New York, 2001.

8. M. R. Moldover, K. A. Gillis, J. J. Hurly, J. B. Mehl and J. Wilhelm, in *Handbook of Elastic Properties of Solids, Liquids, and Gases, Volume IV, Elastic Properties of Fluids Liquids and Gases*, Ch. 12., M. Levy, H. E. Bass, and R. R. Stern, eds.-in-chief, M. Levy, R. Raspet and D. Sinha, vol. ed., L. Furr, tech. Ed., V. Keppens, supervising ed., Academic Press, New York, 2001.

9. F. W. Giacobbe, in *Handbook of Elastic Properties of Solids, Liquids, and Gases, Volume IV, Elastic Properties of Fluids: Liquids and Gases*, Ch. 13., M. Levy, H. E. Bass, and R. R. Stern, eds.-in-chief, M. Levy, R. Raspet and D. Sinha, vol. ed., L. Furr, tech. Ed., V. Keppens, supervising ed., Academic Press, New York, 2001.

10. K. F. Herzfeld and T. A. Litovitz, in *Pure and Applied Physics, Vol. 7, Absorption and Dispersion of Ultrasonic waves*, ed. H. S. W. Massey, Academic Press, London, 1959.

11. P. M. Morse and K. U. Ingard, *Theoretical Acoustics,* McGraw-Hill, New York, 1968, p. 233.

12. M. B. Ewing and J. P. M. Trusler, *J. Chem. Phys.*, 1989, **90**, 1106.

13. M. B. Ewing and J. P. M. Trusler, *Physica A*, 1992, **184**, 437.

14. J. P. M. Trusler and M. P. Zarari, *J. Chem. Thermodyn.*, 1995, **27**, 771.

15. M. R. Moldover and J. B. Mehl, *J. Chem. Phys.*, 1981, **74**, 4062.

16. M. B. Ewing and A. R. H. Goodwin, *J. Chem. Thermodyn.*, 1991, **23**, 1107.

17. M. B. Ewing and A. R. H. Goodwin, *J. Chem. Thermodyn.*, 1992, **24**, 301.

18. W. Lemming, *Fortschr.-Ber., VDI-Z*, 1986, **19**, 32.

19. M. B. Ewing, A. R. H. Goodwin, M. L. McGlashan and J. P. M. Trusler, *J. Chem. Thermodyn.*, 1987, **19**, 721.

20. M. B. Ewing, A. R. H. Goodwin, M. L. McGlashan and J. P. M. Trusler, *J. Chem. Thermodyn.*, 1988, **20**, 243.

21. M. B. Ewing, A. R. H. Goodwin and J. P. M. Trusler, *J. Chem. Thermodyn.*, 1989, **21**, 867.

22. A. R. H. Goodwin and M. R. Moldover, *J. Chem. Phys.*, 1990, **93**, 2471.

23. A. R. H. Goodwin and M. R. Moldover, *J. Chem. Phys.*, 1991, **95**, 5230.

24. A. R. H. Goodwin and M. R. Moldover, *J. Chem. Phys.*, 1991, **95**, 5236.

25. S. J. Boyes, M. B. Ewing and A. R. H. Goodwin, *J. Chem. Thermodyn.*, 1992, **24**, 1151.

26. J. P. M. Trusler and M. P. Zarari, *J. Chem. Thermodyn.*, 1992, **24**, 973.

27. T. Hozumi, T. Koga, H. Sato and K. Watanabe, *Int. J. Thermophys.*, 1993, **14**, 739.

28. T. Hozumi, T. Koga, H. Sato and K. Watanabe, *Int. J. Thermophys.*, 1994, **15**, 385.
29. T. Hozumi, H. Sato and K. Watanabe, *J. Chem. Eng. Data*, 1994, **39**, 493.
30. G. Esper, W. Lemming, W. Beckerman and F. Kohler, *Fluid Phase. Equilib.*, 1995, **105**, 173.
31. T. Hozumi, T. Koga, H. Sato and K. Watanabe, *J. Chem. Eng. Data*, 1996, **41**, 1187.
32. T. Hozumi, T. Koga, H. Sato and K. Watanabe, *Int. J. Thermophys.*, 1996, **17**, 587.
33. D. R. Defibaugh, K. A. Gillis, M. R. Moldover, G. Morrison and J. W. Schmidt, *Fluid Phase Equilibria*, 1992, **81**, 285.
34. D. R. Defibaugh, K. A. Gillis, M. R. Moldover, J. W. Schmidt and L. A. Weber, *Fluid Phase Equilibria*, 1996, **122**, 131.
35. D. R. Defibaugh, K. A. Gillis, M. R. Moldover, J. W. Schmidt and L. A. Weber, *Int. J. Refrig.*, 1996, **19**, 285.
36. J. J. Hurly, J. W. Schmidt and K. A. Gillis, *Int. J. Thermophys.*, 1997, **18**, 655.
37. J. J. Hurly, J. W. Schmidt and K. A. Gillis, *Int. J. Thermophys.*, 1997, **18**, 137.
38. M. R. Moldover, J. P. M. Trusler, T. J. Edwards, J. B. Mehl and R. S. Davis, *J. Res. Nat. Bur. Stand.*, 1988, **93**, 85.
39. M. R. Moldover, S. J. Boyes, C. W. Meyer and A. R. H. Goodwin, *J. Res. Nat. Inst. Stand. Tech.*, 1999, **104**, 11–46.
40. M. B. Ewing and J. P. M. Trusler, *J. Chem. Thermodyn.*, 2000, **32**, 1229.
41. A. R. Colclough, *Proc. Roy. Soc. Lond. A*, 1979, **365**, 349.
42. D. C. Ripple, D. R. Defibaugh, K. A. Gillis and M. R. Moldover, *Primary acoustic thermometer for use up to 800 K, in TEMPMEKO '99, Proceedings II of the 7th International Symposium on Temperature and Thermal Measurements in Industry and Science*, J. Dubbeldam and M de Groot, eds., NMi van Swinden Laboratorium, Delft, p. 418, 1999.
43. S. O. Colgate, A. Sivaraman and K. R. Reed, *Acoustic Determination of the Thermodynamic Reference State Heat Capacity of n-Heptane Vapor*, Gas Processors Association, Tulsa, OK, U.S.A. Report RR-109, 1987.
44. S. O. Colgate, A. Sivaraman and K. R. Reed, *Reference State Heat Capacities of Three C-8 Compounds*, Gas Processors Association, Tulsa, Ok, U.S.A., Report RR-123, 1989.
45. S. O. Colgate, A. Sivaraman and K. R. Reed, *Fluid Phase Equilib.*, 1990, **60**, 191.
46. S. O. Colgate, C. F. Sona, K. R. Reed and A. Sivaraman, *J. Chem. Eng. Data*, 1990, **35**, 1.
47. S. O. Colgate, A. Sivaraman and K. R. Reed, *J. Chem. Thermodyn.*, 1990, **22**, 245.
48. S. O. Colgate, A. Sivaraman and D. Tatro, in *Thermophysical Properties for Industrial Process Design, AIChE Symposium Series, Vol. 90, No. 298*, E.L. Gaden, series ed., T.B. Selover and C.-C. Chen, volume eds.,

G. Thomson and S. Watanasiri, volume co-eds., AIChE, New York, p. 34, 1994.

49. J. P. M. Trusler, W. A. Wakeham and M. P. Zarari, *Int. J. Thermophys.*, 1996, **17**, 35.

50. J. S. Brooks and R. B. Hallock, *Rev. Sci. Instrum.*, 1983, **54**, 1199.

51. A. F. Estrada-Alexanders, J. P. M. Trusler and M. P. Zarari, *Int. J. Thermophys.*, 1995, **16**, 663.

52. A. F. Estrada-Alexanders and J. P. M. Trusler, *Int. J. Thermophys.*, 1996, **17**, 1325.

53. T. Dayton, S. W. Beyerlein and A. R. H. Goodwin, *J. Chem. Thermodyn.*, 1999, **31**, 847.

54. A. F. Estrada-Alexanders and J. P. M. Trusler, *J.Chem. Thermodyn.*, 1997, **29**, 991.

55. S. Lago and P. A. G. Albo, *J. Chem. Thermodyn.*, 2008, **40**, 1558.

56. M. Bijedic and N. Neimarlija, *Int. J. Thermophys.*, 2007, **28**, 268.

57. J. C. Gallop and W. J. Radcliffe, *J. Phys. E: Sci. Instrum.*, 1981, **14**, 461.

58. J. Dominique, J. C. Gallop and W. J. Radciffe, *J. Phys. E: Sci. Instrum.*, 1983, **16**, 1200.

59. J. C. Gallop and W. J. Radcliffe, *J. Phys. E: Sci. Instrum.*, 1986, **19**, 413.

60. M. B. Ewing, J. B. Mehl, M. R. Moldover and J. P. M. Trusler, *Metrologia*, 1988, **25**, 211.

61. E. F. May, L. Pitre, J. B. Mehl, M. R. Moldover and J. W. Schmidt, *Rev. Sci. Instrum.*, 2004, **75**, 3307.

62. J. B. Mehl, M. R. Moldover and L. Pitre, *Metrologia*, 2004, **41**, 295.

63. L. Pitre, M. R. Moldover and W. L. Tew, *Metrologia*, 2006, **43**, 142.

64. W. P. Mason, *Piezoelectric Crystals and Their Applications to Ultrasonics,* Van Nostrand, New York, 1950.

65. D. A. Berlincourt, D. R. Curran and H. Jaffe, in *Physical Acoustics, Vol. I, Part A*, ed. W. P. Mason,Academic Press, New York, 1964, Ch. 3.

66. E. K. Sittig, in *Physical Acoustics Principles and Methods, Vol. IX*, eds. W. P. Mason and R. N. Thurston, Academic Press, New York, 1972, Ch. 5.

67. A. J. Zuckerwar, *J. Acoust. Soc. Am.*, 1978, **64**, 1278.

68. R. Zahn, *J. Acoust. Soc. Am.*, 1981, **69**, 1200.

69. J. B. Mehl, *J. Acoust. Soc. Am.*, 1978, **64**, 1523.

70. M. B. Ewing and J. P. M. Trusler, *J. Acoust. Soc. Am.*, 1989, **85**, 1780.

71. M. R. Moldover, J. B. Mehl and M. Greenspan, *J. Acoust. Soc. Am.*, 1986, **79**, 253.

72. J. P. M. Trusler, *Physical Acoustics and Metrology of Fluids,* Adam-Hilger, Bristol, 1991, Ch. 4.

73. T. L. Cottrell and J. C. McCoubrey, *Molecular Energy Transfer in Gases,* Butterworths, London, 1961.

74. J. D. Lambert, *Vibrational and Rotational Relaxation in Gases,* Clarendon Press, Oxford, 1977.

75. J. B. Mehl and M. R. Moldover, *J. Chem. Phys.*, 1982, **77**, 455.

76. B. E. Gammon and D. R. Douslin, *J. Chem. Phys.*, 1976, **64**, 203.

77. B. E. Gammon, *J. Chem. Phys.*, 1976, **64**, 2556.
78. V. A. Del Grosso and C. W. Mader, *J. Acoust. Soc. Am.*, 1972, **52**, 961.
79. V. A. Del Grosso and C. W. Mader, *J. Acoust. Soc. Am.*, 1972, **52**, 1442.
80. C. W. Garland, in *Physical Acoustics: Principles and Methods, Vol. VII*, eds. W.P. Mason and R. N. Thurston, Academic Press, New York, 1970, Ch. 2.
81. C. W. Garland and R. D. Williams, *Phys. Rev.*, 1974, **A10**, 1328.
82. J. D. Campbell, *Acustica*, 1955, **5**, 145.
83. J. B. Mehl, *J. Acoust. Soc. Am.*, 1982, **71**, 1109.
84. J. B. Mehl, *J. Acoust. Soc. Am.*, 1986, **79**, 278.
85. J. W. S. Rayleigh, *Theory of Sound*, Dover, New York. 1896, sec 331.
86. D. Bancroft, *Am. J. Phys.*, 1956, **24**, 355.
87. C. M. Harris, *J. Acoust. Soc. Am.*, 1963, **35**, 11.
88. C. M. Harris, *J. Acoust. Soc. Am.*, 1971, **49**, 890.
89. R. Keolian, S. Garrett, J. Maynard and I. Rudnick, *J. Acoust. Soc. Am. Suppl.*, 1973, **64**, S561.
90. R. Keolian, S. Garrett, J. Maynard and I. Rudnick, *Bull. Am. Phys. Soc.*, 1979, **24**, 623.
91. M. R. Moldover, M. Waxman and M. Greenspan, *High Temp. High Press.*, 1979, **11**, 75.
92. J. B. Mehl and M. R. Moldover, in *Proc. Eighth Symp. On Thermophysical Properties, Vol. 1 Thermophysical Properties of Fluids*, ed. J. V. Sengers, The American Society of Mechanical Engineers, New York, 1982, p. 134,.
93. M. R. Moldover and J. B. Mehl, in *Precision Measurements and Fundamental Constants II*, B. N. Taylor and W. D. Phillips, eds., Natl. Bur. Stand., Spec. Pub., 617, p. 281, 1984.
94. J. B. Mehl and M. R. Moldover, *Phys. Rev.*, 1986, **A 34**, 3341.
95. J. B. Mehl and M. R. Moldover, in *Topics in Current Physics Vol. 46*, ed. P. Hess, Springer-Verlag: Berlin, 1989, Ch. 4.
96. M. R. Moldover, J. P. M. Trusler, T. J. Edwards, J. B. Mehl and R. S. Davis, *Phys. Rev. Lett.*, 1988, **60**, 249.
97. M. R. Moldover, *IEEE Trans. Inst. Meas.*, 1989, **38**, 217.
98. J. B. Mehl and M. R. Moldover, *Phys. Rev.*, 1986, **A34**, 3341.
99. M. R. Moldover and J. P. M. Trusler, *Metrologia*, 1988, **25**, 165.
100. M. R. Moldover, S. J. Boyes, C. W. Meyer and A. R. H. Goodwin, *Primary acoustic thermometry from 217 K to 303 K*, in *TEMPMEKO '99, Proceedings of the 7th International Symposium on Temperature and Thermal measurements in Industry and Science*, J. Dubbeldam and M. de Groot, eds., NMi van Swinden Laboratorium, Delft, p. 412, 1999.
101. J. P. M. Trusler, Ph.D. Thesis, University of London, 1984.
102. S. O. Colgate, K. R. Williams, K. Reed and C. A. Hart, *J. Chem. Educ.*, 1987, **64**, 553.
103. A. R. H. Goodwin, Ph.D. Thesis, University of London, 1988.

104. S. O. Colgate, A. Sivaraman, C. Dejsupa and K. McGill, *Proceedings of the 1989 International Gas Research Conference,* Government Institutes, Rockville, 1990, p. 502.
105. M. B. Ewing, A. A. Owusu and J. P. M. Trusler, *Physica A*, 1989, **156**, 899.
106. K. C. McGill, Ph.D. Thesis, University of Florida, 1990.
107. S. O. Colgate, A. Sivaraman, C. Dejsupa and K. McGill, *Rev. Sci. Instrum.*, 1991, **62**, 198.
108. S. O. Colgate, A. Sivaraman, C. Dejsupa and K. McGill, *J. Chem. Thermodyn.*, 1991, **23**, 647.
109. S. J. Boyes, Ph.D Thesis, University of London, 1992.
110. S. O. Colgate, A. Sivaraman and C. Dejsupa, *Fluid Phase Equilib.*, 1992, **76**, 175.
111. S. O. Colgate, A. Sivaraman and C. Dejsupa, *Fluid Phase Equilib.*, 1992, **79**, 221.
112. S. O. Colgate, K. C. McGill, A. Sivaraman and D. Tatro, *Fluid Phase Equilib.*, 1992, **79**, 231.
113. M. B. Ewing and A. R. H. Goodwin, *J. Chem. Thermodyn.*, 1992, **24**, 531.
114. M. B. Ewing and A. R. H. Goodwin, *J. Chem. Thermodyn.*, 1992, **24**, 1257.
115. M. B. Ewing and A. R. H. Goodwin, *J. Chem. Thermodyn.*, 1993, **25**, 423.
116. M. B. Ewing and A. R. H. Goodwin, *J. Chem. Thermodyn.*, 1993, **25**, 1503.
117. J. P. M. Trusler, *J. Chem. Thermodyn.*, 1994, **26**, 751.
118. A. R. H. Goodwin, J. A. Hill and J. L. Savidge, in *Proceedings of the 1995 International Gas Research Conference, Vol. I*, ed. D. A. Dolenc, Government Institutes, Rockville, p. 681, 1995.
119. W. Beckerman and F. Kohler, *Int. J. Thermophysics*, 1995, **16**, 455.
120. J. P. M. Trusler and M. F. Costa Gomes, *The speed of sound in methane and two methane-rich gas mixtures,* Report prepared for GERG working group 1.3, Imperial College, London, 1995.
121. A. F. Estrada-Alexanders and J. P. M. Trusler, *J. Chem. Thermodyn.*, 1995, **27**, 1075.
122. D. Fawcett, Ph.D. Thesis, Murdoch University, Australia, 1995.
123. T. J. Edwards, D. J. Pack, D. Fawcet, R. D. Trengove and M. Resuggan, *Measurement and Prediction of Speed of Sound with Application to Gas Flow Metering in Australian Natural Gas,* Minerals and Energy Research Institute of Western Australia, Perth, Australia, Report No. 150, 1995.
124. A. F. Estrada-Alexanders, Ph.D. Thesis, University of London, 1996.
125. J. P. M. Trusler and M. P. Zarari, *J. Chem. Thermodyn.*, 1996, **28**, 329.
126. S. O. Colgate and A. Sivaraman, *Int. J. Thermophys.*, 1996, **17**, 15.
127. M. F. Costa Gomes and J. P. M. Trusler, *J. Chem. Thermodyn.*, 1998, **30**, 1121.
128. A. F. Estrada-Alexanders and J. P. M. Trusler, *J. Chem. Thermodyn.*, 1998, **30**, 1589.

129. A. F. Estrada-Alexanders and J. P. M. Trusler, *J. Chem. Thermodyn.*, 1999, **31**, 685.
130. J. F. Estela-Uribe and J. P. M. Trusler, *Int. J. Thermophys.*, 2000, **21**, 1033.
131. M. Grigiante, G. Scalabrin, G. Benedetto, R. M. Gavioso and R. Spagnolo, *Fluid Phase Equilib.*, 2000, **174**, 69.
132. K. Ogawa, T. Kojima and H. Sato, *J. Chem. Eng. Data.*, 2001, **46**, 1082.
133. G. Benedetto, R. M. Gavioso, R. Spagnolo, M. Grigiante and G. Scalabrin, *Int. J. Thermophys.*, 2001, **22**, 1073.
134. G. Scalabrin, P. Marchi, G. Benedetto, R. M. Gavioso and R. Spagnolo, *J. Chem. Thermodyn.*, 2002, **34**, 1601.
135. H. Sato, T. Kojima and K. Ogawa, *Int. J. Thermophys.*, 2002, **23**, 787.
136. M.-G. He and Z.-G. Liu, *Fluid Phase Equilib.*, 2002, **198**, 185.
137. M.-G. He, Z.-G. Liu and J. M. Yin, *Int. J. Thermophys.*, 2002, **23**, 1599.
138. T. Kojima, K. Okabe, K. Ogawa and H. Sato, *J. Chem. Eng. Data.*, 2004, **49**, 635.
139. *Royal Society Mathematical Tables*, Vol. 7, Bessel functions Part III Zeros and Associated Values, F. W. J. Oliver, ed., Cambridge University Press, Cambridge, 1960.
140. H. G. Ferris, *J. Acoust. Soc. Am.*, 1953, **25**, 47.
141. M. B. Ewing, M. L. McGlashan and J. P. M. Trusler, *Metrologia*, 1986, **22**, 93.
142. F. D. Shields and J. Faugh, *J. Acoust. Soc. Am.*, 1969, **46**, 158.
143. F. D. Shields, *J. Chem. Phys.*, 1975, **62**, 1248.
144. M. Greenspan, unpublished work reported in reference 15.
145. J. B. Mehl, *J. Acoust. Soc. Am.*, 1985, **78**, 782.
146. A. R. H. Goodwin, unpublished work, 1989.
147. J. P. M. Trusler, unpublished calculations reported in reference 38.
148. M. P. Zarari, Ph.D. Thesis, University of London, 1992.
149. J. L. Angerstein, Ph.D. Thesis, University of London, 1999.
150. T. H. Quigley, *Phys. Rev.*, 1945, **67**, 298.
151. D. H. Smith and R. G. Harlow, *Br. J. Appl. Phys.*, 1963, **14**, 102.
152. R. J. Quinn, A. R. Colclough and T. R. D. Chandler, *Phil. Trans. Roy. Soc.*, 1967, **A283**, 367.
153. A. R. Colclough, *Metrologia*, 1973, **9**, 75.
154. C. Cary, J. Bradshaw, E. Lin and E. H. Carnevale, *Experimental Determination of Gas Properties at High Temperatures and/or Pressures*, (Arnold Engineering Development Center, Arnold Air Force Station, TN 37389, USA) Report no. AEDC-TR-74-33, 1974.
155. A. J. Zuckerwar and W. A. Griffin, *J. Acoust. Soc. Am.*, 1980, **68**, 218.
156. M. B. Ewing, M. L. McGlashan and J. P. M. Trusler, *J. Chem. Thermodyn.*, 1985, **17**, 549.
157. B. A. Younglove and N. V. Fredrick, *Int. J. Thermophys.*, 1990, **11**, 897.
158. K. A. Gillis, M. R. Moldover and A. R. H. Goodwin, *Rev. Sci. Instrum.*, 1991, **62**, 2213.
159. F. W. Giacobbe, *J. Acoust. Soc. Am.*, 1993, **94**, 1200.

160. K. A. Gillis, *Int. J. Thermophys.*, 1994, **15**, 821.
161. W. G. Schneider and G. J. Thiessen, *Can. J. Res.*, 1950, **28A**, 509.
162. M. C. Henderson and L. Peselnick, *J. Acoust. Soc. Am.*, 1953, **29**, 1074.
163. M. Greenspan and M. C. Thompson Jr., *J. Acoust. Soc. Am.*, 1953, **25**, 92.
164. H. Plumb and G. Cataland, *Metrologia*, 1966, **2**, 127.
165. D. T. Grimsrud and J. H. Werntz Jr., *Phys. Rev.*, 1967, **157**, 181.
166. B. E. Gammon and D. R. Douslin, in *Proc. Fifth Symp. on Thermophysical Properties*, ed. C.F. Bonilla, American Society of Mechanical Engineers, New York, p. 107,1970..
167. A. R. Colclough, T. J. Quinn and T. R. D. Chandler, *Proc. R. Soc.*, 1979, **A368**, 125.
168. A. Sivaraman and B. E. Gammon, *Speed of sound measurements in natural gas fluids*, Gas Research Institute Report 86/0043, 1986.
169. M. S. Zhu, L. Z. Han, K. Z. Zhang and T. Y. Zhou, *Int. J. Thermophys.*, 1993, **14**, 1039.
170. Y.-Y. Duan, L. Shi, L.-Q. Sun, M.-S. Zhu and L.-Z. Han, *Int. J. Thermophys.*, 2000, **21**, 393.
171. C. Zhang, Y.-Y. Duan, L. Shi, M.-S. Zhu and L.-Z. Han, *Fluid Phase Equilib.*, 2001, **178**, 73.
172. M. B. Ewing, M. L. McGlashan and J. P. M. Trusler, *J. Chem. Thermodyn.*, 1986, **18**, 511.
173. D. R. Defibraugh, N. E. Carillo, J. J. Hurly, M. R. Moldover, J. W. Schmidt and L. A. Weber, *J. Chem. Eng. Data.*, 1997, **42**, 488.
174. K. A. Gillis, *Int. J. Thermophys.*, 1997, **18**, 73.
175. J. J. Hurly, *Int. J. Thermophys.*, 1999, **20**, 455.
176. J. J. Hurly, J. W. Schmidt, S. J. Boyes and M. R. Moldover, *Int. J. Thermophys.*, 1997, **18**, 579.
177. J. J. Hurly, D. R. Defibraugh and M. R. Moldover, *Int. J. Thermophys.*, 2000, **21**, 739.
178. J. J. Hurly, *Int. J. Thermophys.*, 2000, **21**, 185.
179. J. J. Hurly, *Int. J. Thermophys.*, 2000, **21**, 805.
180. J. J. Hurly, *Int. J. Thermophys.*, 2002, **23**, 455.
181. J. J. Hurly, *Int. J. Thermophys.*, 2002, **23**, 667.
182. J. J. Hurly, *Int. J. Thermophys.*, 2003, **24**, 1611.
183. N. R. Nannan, P. Colonna, C. M. Tracy, R. L .Rowley and J. J. Hurly, *Fluid Phase Equilib.*, 2007, **257**, 102.
184. V. A. Del Grosso, *Systematic Errors in Ultrasonic Propagation Parameter Measurements, Part 1, Effect of Free-Field Diffraction*, NRL Report 6026, 1964.
185. V. A. Del Grosso, *Systematic Errors in Ultrasonic Propagation Parameter Measurements, Part 2, Effects of Guided Cylindrical Modes*, NRL Report 6133, 1964.
186. V. A. Del Grosso,. *Systematic Errors in Ultrasonic Propagation Parameter Measurements, Part 3, Sound Speed by Iterative Reflection-Interferometry*, NRL Report 6409, 1966.
187. A. J. Zuckerwar and R. W. Meredith, *J. Acoust. Soc. Am.*, 1985, **78**, 946.

188. A. J. Buxton, Ph.D. Thesis, University of London, 1997.
189. K. A. Gillis, J. B. Mehl and M. R. Moldover, *Rev Sci. Instrum.*, 1995, **67**, 1850.
190. J. B. Mehl, *J. Acoust. Soc. Am.*, 1995, **97**, 3327.
191. J. B. Mehl, *J. Acoust. Soc. Am.*, 1999, **106**, 73.
192. J. Wilhelm, K. A. Gillis, J. B. Mehl and M. R. Moldover, *Int. J. Thermophys.*, 2000, **21**, 983.
193. M. Greenspan and F. N. Wimenitz, *An Acoustic Viscometer for Gases - I*, NBS Report 2658, 1953.

CHAPTER 10

Speed-of-Sound Measurements and Heat Capacities of Liquid Systems at High Pressure

TOSHIHARU TAKAGI[a] AND EMMERICH WILHELM[b]

[a] Department of Chemistry and Materials Technology, Kyoto Institute of Technology, Kyoto 606-8585, Japan; [b] Institute of Physical Chemistry, University of Wien, A-1090, Wien (Vienna), Austria

10.1 Introduction

For many years, ultrasound (about $20\,kHz$–$1\,GHz$) has been widely used in both scientific research and industrial applications such as cleaning, emulsification, promoting condensation, non-destructive metal testing, medical inspection and many others. On a more fundamental level, propagation and transmission of low-power acoustic waves in fluids at sufficiently low frequencies well below any dispersion region but at frequencies high enough to provide isentropic conditions, has served as an important source of information on thermodynamic properties.[1] Under these conditions, the speed of ultrasound u is a thermodynamic equilibrium property which is closely related to the isentropic compressibility κ_S:

$$\kappa_S \equiv \frac{1}{\rho}\left(\frac{\partial \rho}{\partial p}\right)_S = \frac{1}{\rho u^2} \tag{1}$$

where $\rho = M/V$ is the (mass-)density, M is the molar mass, V is the molar volume, p is the pressure, and S is the molar entropy. At high pressures, the

Heat Capacities: Liquids, Solutions and Vapours
Edited by Emmerich Wilhelm and Trevor M. Letcher
© The Royal Society of Chemistry 2010
Published by the Royal Society of Chemistry, www.rsc.org

direct experimental determination of thermodynamic properties such as the density, isobaric expansivity, isothermal compressibility and isobaric heat capacity is rather difficult. Over the past 30 years, as a result of improvements in electronic instrumentation and experimental techniques, the speed of sound can now be measured with sufficient precision over wide ranges of temperature *and* pressure, including the near critical region, to be useful for determining *indirectly* thermodynamic properties of liquids and liquid mixtures.[2,3] This is usually done by numerical integration and the technique has become important in many fields of science and technology. The effects of temperature T and pressure p on thermodynamic properties of fluids are frequently estimated with the help of equations of state which in turn are derived and improved through use of density data $\rho(T, p)$, vapour pressure data, caloric data, *etc*. As a result of advances in computer technology, impressive progress has been achieved in this field. This is evidenced by the publications of the IUPAC Thermodynamic Tables Project Centre, Imperial College, London, UK, which had undertaken the compilation of internationally agreed values of the equilibrium thermo-dynamic properties of liquids and gases of interest to both scientists and technologists. High-precision speed-of-sound measurements have proved to be of great value for the development of the Helmholtz equation of state for substances such as argon, oxygen, ethane, and methanol. High-precision sound speeds may be used advantageously to check the reliability of equations of state.[4]

Research in the general field of speed of sound in liquids at high pressure began with the work of Swanson[5] in 1934, who measured the speed of sound in some hydrocarbons at pressures up to 30 MPa at room temperature, using an acoustic interferometer. In the 1950s, with the development of a pulse method,[6–9] experimentation became easier and research in this field progressed, resulting in the publication of a large amount of high-precision speed-of-sound data at elevated pressures for pure water and many organic compounds. Much of this historical work has been reviewed by Oakley's group at Oakland University in 2003.[10,11] Furthermore, in 2005 Neruchev *et al.*[12] of Kursk State University presented an interesting review on the behaviour of the speed of sound along saturation lines for some organic compounds. Revival of interest in this field resulted in the publication of many measure-ments on a variety of substances, such as alkanes, ethers, alcohols and hydrofluorocarbons.[13–31]

Our contribution will cover the following topics: recent experimental tech-niques for speed-of-sound measurements at elevated pressures in liquids; the effect of temperature and pressure on the speed of sound in pure organic liquids and mixtures measured over wide ranges of temperature and pres-sure, including states near the liquid–vapour coexistence line; and the use of speed-of-sound measurements to derive important thermodynamic properties at elevated pressures, such as density, isobaric expansivity, isothermal com-pressibility, heat capacity at constant pressure, and heat capacity at constant volume.

10.2   Apparatus for the Speed-of-Sound Measurement in the Liquid Phase at High Pressure

Speed-of-sound measurements at elevated pressures require a significant amount of additional equipment which includes sophisticated pressure generating and measuring equipment, together with temperature control and temperature measuring equipment. This acts as a restraint in the field.

The pulse method has been widely used to measure the speed of sound in compressed liquids. The fundamental principle was developed by Pellam and Galt in 1946;[32] it involved measuring the time required for a short acoustic pulse to travel a known distance. In recent years, two significant improvements to this technique have been reported. One of them is the pulse-echo overlap method.[33–35] The excitation signal is delayed by a delay circuit and is made to superimpose (overlap) on the signal travelling in the sample; the sound speed is determined from the retardation time and the travelling path. The other technique is the phase comparison pulse-echo method, developed by Kortbeek *et al.*[36] It involves a single transducer held between two parallel plane reflectors. The sound speed is measured from the difference between the round trip transit times in two paths.[36,37] Recent improvements involving the sensitivity (resolution) and stability of the ancillary electrical (digital) instruments such as the pulse generator, the broadband amplifier, digital (storage) oscilloscope, *etc.*, have resulted in significant progress in this field.

In 1976 we constructed a pulse method apparatus using double X-cut quartz to fix the path length in order to measure high-pressure speeds of sound in organic liquids and their mixtures at pressures up to 180 MPa around room temperature.[38–48] Then in 1985, we devised the new sing-around experimental device, another improvement on a pulse method.[49] Many high-pressure acoustic interferometers have been constructed to suit the demands of experiments. Figure 10.1 shows the fixed-path acoustic interferometer which was

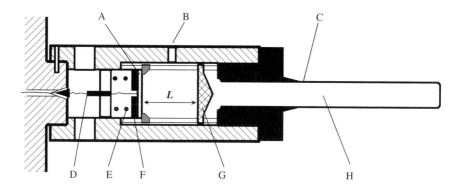

Figure 10.1   Acoustic interferometer employing a Teflon capsule, Takagi and Teranishi.[49] A, piezo-electric transducer; B, sample inlet; C, Teflon capsule (thickness 0.3 mm); D, electrode; E, spring; F, Teflon insulator; G, reflector; H, sample chamber; L, distance between transducer and reflector.

designed for the sing-around type instrument.[49] The transducer was a 20 mm diameter disc made of polarised piezoelectric ceramic (lead zircon titanate, PZT, 2 MHz, Tokin Co.) and the electrodes were vapour-deposited on either side. The transducer and reflector were fixed in parallel and housed in a cylinder made of SUS 316 stainless steel. In order to reflect the sound wave from the back end, a transducer was fixed using Teflon (an insulator), into which a spiral slot was cut, and the right-hand side of the reflector was pressed into it. This acoustic interferometer was placed in the stainless steel container mounted horizontally in a thermostat, and was suitable for measurements up to 300 MPa. The pressure was transmitted through a Teflon capsule (0.3 mm thickness) and through transmission oil in the pressure vessel. The interferometer had an internal volume of about 25 cm^3 and was designed for easy handling (washing, sample filling, *etc.*). However, since the volume change of the capsule is restricted, this apparatus is unsuitable for measuring over a wide range of temperatures and pressures, especially for fluids with large expansion coefficients.

Around 1990 it became evident that atmospheric ozone was being destroyed by chlorine containing refrigerants and it became imperative to measure thermodynamic properties of a new class of refrigerants: *hydrofluorocarbons*. In order to measure the speed of sound in these liquids over wide ranges of temperature (240 K to 350 K) and pressure (up to 35 MPa), a new high-pressure acoustic interferometer was constructed using SUS306 stainless steel,[55] with a piston cylinder for separating the oil and the sample (see Figure 10.2). The volume change, created by the moving piston in the sample chamber, was about 20 cm^3 (*i.e.* from 40 cm^3 to 60 cm^3). The transducer and the reflector were both fixed in their positions as in the previous example. The schematic diagram of the apparatus is shown in Figure 10.3. The acoustic interferometer was immersed in a liquid thermostat which contained a mixture of ethylene glycol and water (45/55 wt %) and was controlled to within ± 20 mK. The temperature was measured using a quartz thermometer (Tokyo Denpa Co., DMT-600B, resolution ± 0.001 K) and calibrated to within ± 5 mK against a standard platinum thermometer (ITS-90). The pressure was generated by a hand oil pump (Hikari High Pressure Co. KP-1R), using silicon oil with a kinematic viscosity of 1×10^{-6} m^2 s^{-1} and was transmitted to the sample through a sample-oil separator in a high-pressure vessel. The pressure in the sample was measured directly using two precision strain gauges (Nagano Keiki Co., KH-17): the one for a maximum pressure of 5 MPa and with an uncertainty of ± 0.003 MPa was calibrated against a quartz crystal pressure transducer (Paroscientific Co., 730-31K-10), and the other for a maximum of 35 MPa with an uncertainty of ± 0.035 MPa was calibrated against a precision manometer (Tsukasa Sokken Co., HP-22-G). In this way it was possible to measure the pressure directly. The gauges, valves and pipes (1 mm ID, 3 mm OD) were placed inside the thermostat in order to reduce the measurement error caused by a temperature gradient when measuring over a wide range of temperature. The pressure O-ring seal used in the piston cylinder was made of a special silicon rubber blend (Mitsubishi Cable Ind. Ltd.) which could cope with a large

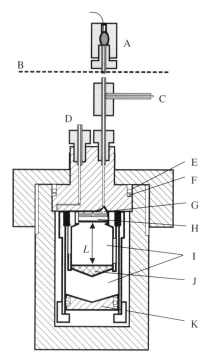

Figure 10.2   Acoustic interferometer employing a free piston, Takagi.[55] A, electrode;
B, water level; C, tube leading to sample injection, pressure transducer;
D, tube leading to the pressure generator line; E, O-ring; F, backup ring;
G, spring; H, piezoelectric transducer; I, sample chamber; J, reflector; K,
moveable piston; L, distance between transducer and reflector.

temperature range. The electrode of the high-pressure container was insulated
using a low-permittivity polymer (*e.g.* polyimide resin) processed into a cone
shape. The design of the high-pressure ultrasonic interferometer was commented on by Zak *et al.*[34]

The sing-around instrument is now commercially available (Ultrasonic Eng.
Co., UVM-2). The principle behind the instrument is a short acoustic pulse,
excited by the transducer, which travels over the fixed distance L in the sample.
When the wave, after reflection, returns to the transducer, the signal is detected
by the same transducer through a gate (window shift: $0.6\,\mu s–400\,\mu s$, wide:
$0.2\,\mu s–400\,\mu s$) which has been opened in anticipation by a window circuit. An
automatic gain controller and a delay line ($63\,\mu s \times N$, $N = 1 \sim 16$, stability: less
than $\pm 1\,ns\,min^{-1}$) were employed to avoid the distortion of signals caused by
absorption of the acoustic wave and the interference of multiple echoes. The
next acoustic short pulse was generated with a delay time τ ($256\,\mu s$ or $511\,\mu s$)
after the arrival of the reflected wave. In the present work, the repetition period
t including the delay time τ was measured using a universal counter (Advantest
Co., TR-5822, with a resolution of $0.1\,ns$) as the average of 1000 periods.

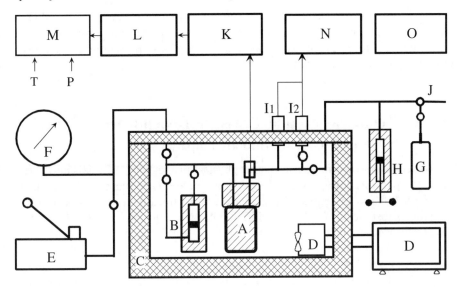

Figure 10.3 Schematic diagram for measuring the speed of sound in the liquid phase
at elevated pressures, Takagi.[55] A, high-pressure acoustic interferometer;
B, piston cylinder; C, thermostat; D, radiator; E, hand oil pump; F,
pressure gage; G, sample bomb; H, screw syringe; I_1 and I_2, pressure
transducers; J, vacuum line; K, sing-around unit; L, universal counter
and monitor; M, personal computer; N, power supply for pressure
transducers and indicator; O, temperature controller and quartz
thermometer.

The ultrasonic speed u was obtained by measuring the repetition period
between the first, t_1, and the second, t_2, echoes (including the delay time τ) of a
short acoustic pulse travelling the distance L between the transducer and reflector:

$$u = \frac{2L}{t_2 - t_1} \tag{2}$$

The value of L for this acoustic interferometer was determined from precision speed-of-sound measurements in pure benzene and tetrachloromethane at
298.15 K and 0.1 MPa. The literature data we used were as follows: for C_6H_6,
$u = 1299.9\,\mathrm{m\,s^{-1}}$ and for CCl_4, $u = 921.1\,\mathrm{m\,s^{-1}}$ respectively. The effect of
temperature on L was calculated from the expansion coefficient of stainless steel
(SUS304) which was taken as $13.6 \times 10^{-6}\,\mathrm{K^{-1}}$ (273 K to 373 K). The effect of
pressure on the dimensions of the fixed path was neglected in the pressure range
up to 50 MPa. The uncertainty of the results obtained with these instruments in
the temperature range 243.15 K to 333.15 K and for pressures up to the freezing
pressure or 35 MPa, was about $\pm 0.15\,\%$ to $\pm 0.25\,\%$ in the high density region.

The acoustic interferometer discussed above was also designed to measure
vapour pressures or bubble point pressures. The pressure was measured by
monitoring the difference between the liquid and the vapour phase absorption

of the acoustic wave used for sound-speed measurements. When the liquid level in the two-phase system was located close to the bottom position of the transducer, a large absorption was observed and the pressure was measured using a precision strain gauge.[57,61]

10.3 Temperature and Pressure Dependences of the Speed of Ultrasound in Liquids

The speed of ultrasound u is a thermophysical property of materials and varies with temperature and pressure. In the liquid phase, u decreases with increasing temperature and increases with increasing pressure. For about 50 pure liquid organic substances and binary liquid mixtures, and several chloro- and/or hydro-fluorocarbon refrigerants, the speed of sound has been measured either by the pulse technique at temperatures around room temperature and at pressures up to 180 MPa, or by the sing-around method at temperatures in the range 243 K to 333 K and at pressures from below the saturation line to about 30 MPa.[38–67]

The pressure dependence of u for a few representative liquids far below their respective critical temperature T_c is presented in Figure 10.4. Figure 10.4(a) shows the pressure dependence of the speed of sound in benzene and tetra-chloromethane.[65] The measurements were carried out from 283.15 K to 333.15 K, and for pressures up to the freezing pressure or 30 MPa. In all this

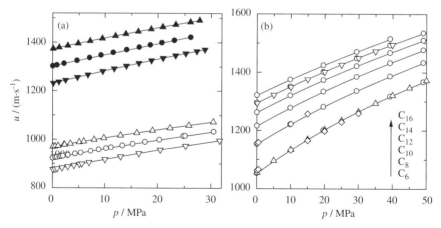

Figure 10.4   Pressure dependence of the speed of sound $u(T,p)$ in the liquid phase of (a) benzene (filled symbols) and tetrachloromethane (open symbols), and of (b) several alkanes. (a) ▲, ●, ▼: C_6H_6; △, ○, ▽: CCl_4; ▲, △: 283.15 K; ●, ○: 298.15 K; ▼,▽: 303.15 K; Takagi *et al.*[65] (b) alkanes at 303.15 K: C_6, hexane; C_8, octane; C_{10}, decane; C_{12}, dodecane; C_{14}, tetradecane; C_{16}, hexadecane. ○, Khasanshin and Shchemelev;[18] ◇, Takagi and Teranishi;[46] △, Daridon *et al.*;[13] ▽, Daridon and Lagourette.[15]

work it is important that the measurements are done on samples of high-grade purity. In this way the reliability of the experimental equipment can be tested. The results for these liquids are relatively easy to obtain. We note that in general the ultrasonic speed in a liquid consisting of pseudospherical molecules, *e.g.* CCl$_4$, is somewhat smaller than in a liquid composed of comparable nonspherical molecules.

Near the critical point the speed of sound is expected to approach zero weakly as $|T/T_c - 1|^{-\alpha/2}$, where $\alpha = 0.110$ is the exponent describing the critical divergence of the molar heat capacity C_V at constant volume (along the critical isochore).

The alkanes are perhaps the most frequently investigated substances in the field of ultrasonics in compressed liquids. Knowledge of the thermo-dynamical or thermophysical properties of alkanes is very important in process design in the petrochemical industry. Khasanshin and Shchemelev's group (Mogilev Institute of Technology, Belarus)[18-21] have measured the speed of sound in a number of liquid alkanes from 303 K to 433 K and at pressures up to 100 MPa by the pulse-echo overlap method with an uncertainty of about ±0.1 %. In Daridon and Lagourette's laboratory (Pau University, France), the speed of sound in liquid alkanes has also been measured, though the focus was on alkanes (and related compounds) of higher molar mass, and mixtures therefrom, over a wide temperature (273 K to 373 K) and pressure range (up to 150 MPa) related to the conditions encountered in petro-leum fields.[13-16] The speed of sound in alkanes increases with increasing carbon number, and the influence of pressure on u decreases as indicated graphically in Figure 10.4(b). Other major contributions in this field origi-nated in the van der Waals Laboratory at the University of Amsterdam, The Netherlands.[36,68-73]

In addition to investigating conventional thermodynamic properties of mixtures, such as the excess molar enthalpy, the excess molar volume, the excess molar heat capacity at constant pressure, *etc.*, Benson[74] and many other authors[75-82] also included the excess isentropic compressibility κ_S^E derived from ultrasonic speed measurements, hoping to thereby gain improved insight into intermolecular interactions. Expanding experiments to include high-pressure measurements of the speed of ultrasound in mixtures is another important step in this direction. Figure 10.5(a) shows the speed of sound as a function of composition (x denotes the mole fraction of cyclohexane) in the mixture $(1-x)C_6H_6 + xC_6H_{12}$ at 303.15 K and several pressures.[41] The mole fraction x_{min} where the ultrasound speed has a minimum value is in the cyclohexane-rich region, and with increasing pressure x_{min} shifts toward the benzene-rich region. In the case of nonpolar + nonpolar mixtures, this phenomenon can usually be qualitatively explained through use of an appropriate version of the isolated binary collision theory:[83] any increase of the mean free path L_{free} between molecular collisions contributes to a decrease of the sound speed. We note that in the movable-wall cell theory

$$L_{\text{free}} = (M/N_A\rho)^{1/3} - \sigma_{\text{eff}} \tag{3}$$

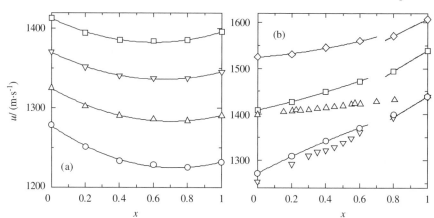

Figure 10.5 Composition dependence (*x* denotes mole fraction) of the speed of sound
u for (a) the binary liquid mixture benzene + cyclohexane, and (b) for
three mixtures involving nitrobenzene as one component, respectively, all at 303.15 K. (a) $(1-x)C_6H_6 + xC_6H_{12}$: ○, 0.1 MPa; △, 10 MPa;
▽, 20 MPa; □, 30 MPa; Takagi.[41] (b) $(1-x)C_6H_6 + xC_6H_5NO_2$:
○, 0.1 MPa; □, 30 MPa; ◇, 60 MPa; Takagi and Teranishi.[43]
$(1-x)C_6H_5CN + xC_6H_5NO_2$: △, 0.1 MPa; Takagi and Teranishi.[51]
$(1-x)C_6H_5Cl + xC_6H_5NO_2$: ▽, 0.1 MPa; Takagi and Teranishi.[44]

where N_A is the Avogadro constant, and σ_{eff} is an effective molecular size
parameter (effective hard core diameter).[84–86] With increasing pressure, these
concave curves become more shallow (and more symmetric).

Binary mixtures involving nitrobenzene have been found to show an interesting composition dependence of the speed of sound. Figure 10.5(b) illustrates
this for the mixture $(1-x)C_6H_6 + xC_6H_5NO_2$ up to about 60 MPa.[43] At
0.1 MPa, the speed of sound increases smoothly with composition along a
slightly convex curve in the benzene rich region, but near $x = 0.7$ it deviates
significantly from this smooth curve. With rise in pressure, in the benzene rich
region $u(x)$ gradually becomes concave (at about 15 MPa, $u(x)$ is linear). In
several other binary mixtures involving strongly polar molecules, such as
chlorobenzene + nitrobenzene,[44] and benzonitrile + nitrobenzene,[51] $u(x)$ exhibits a similar irregular composition dependence. Thermal expansion data point
into the same direction. While highly interesting, a full understanding of these
phenomena awaits future investigation.

The speed of sound for the binary mixture heptane + ethanol has been reported by Dzida, Zak and Ernst[27] as a function of composition, temperature and pressure. The pressure dependence of $u(x)$ at 298.15 K is reproduced in Figure 10.6(a).
At atmospheric pressure, the curve is strongly concave. Since the sound speeds in
pure heptane and in most mixtures with ethanol increase more rapidly with
increasing pressure than those in pure ethanol and mixtures very rich in ethanol,
$u(x)$ flattens out at elevated pressures. In Figure 10.6(b) the composition effect on
the speed of sound is given for benzene + benzonitrile.[66] Here, the curve is convex,

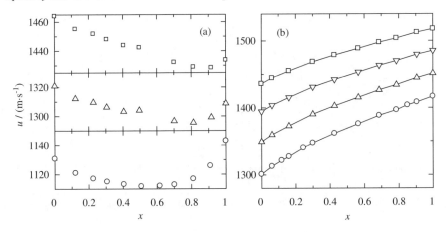

Figure 10.6 Composition dependence (x denotes mole fraction) of the speed of sound u in the binary liquid mixtures heptane + ethanol and benzene + benzonitrile at 298.15 K. (a) $(1-x)n\text{-}C_7H_{16} + xCH_3CH_2OH$: ○, 0.1 MPa; △, 30 MPa; □, 60 MPa; Dzida *et al.*[27] (b) $(1-x)C_6H_6 + xC_6H_5CN$: ○, 0.1 MPa; △, 10 MPa; ▽, 20 MPa; □, 30 MPa; Takagi and Wilhelm.[66]

especially at low pressures. The dipole moment of benzonitrile is one of the largest found for an organic compound, and the resulting strong intermolecular interactions may be responsible for this behaviour.

Following the international attention devoted to the ozone depletion potential of some refrigerants in the late 1980s (restricted use according to the Montreal Protocol of 1987), we started measuring the speed of sound in various liquid chlorine-free, halogenated refrigerants and their mixtures. The experimental results for the speed of sound in pentafluoroethane (HFC-125), CHF_2CF_3,[58] as a function of temperature and pressure, is summarised in Figures 10.7(a) and (b). The critical temperature (T_c = 339.17 K) and the critical pressure (p_c = 3.618 MPa) of this refrigerant are significantly lower than those found for most organic compounds. In the vicinity of the saturation line, $(\partial u/\partial p)_T$ increases with increasing temperature, the increase becoming more pronounced when approaching T_c. Close to the critical point, sound absorption also increases rapidly, and measurements become more and more difficult, as reflected by the blank area in Figure 10.7(a). The liquid-vapour equilibrium properties of refrigerants along the saturation curve are of technological importance. For the reasons indicated above, at higher temperatures the values of the speed of sound at saturation, $u_{sat} = u(T, p_{sat})$, where p_{sat} denotes the vapour pressure, have been estimated by extrapolation. As can be seen in Figure 10.7(b), these extrapolated results agree well with the results obtained via dynamic light scattering:[87] u_{sat} decreases smoothly with increasing temperature and reaches a minimum but non-zero value at the critical point (see also above).

Knowledge of the thermodynamic properties of hydrofluorocarbon mixtures is essential for the air-conditioning and heat-pump industry, but it is also of

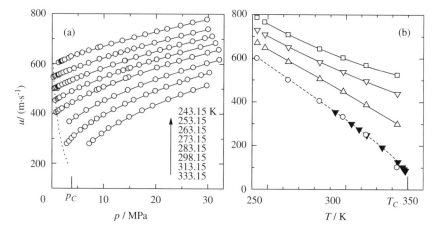

Figure 10.7 Pressure, p (a) and temperature, T (b) dependences of the speed of ultrasound, u in the liquid refrigerant pentafluoroethane (HFC-125), Takagi.[58] (a) ○, experimental results obtained at different temperatures as indicated in the figure (243.15 K to 333.15 K).[58] ------, locus of sound speeds $u_{sat} = u(T,p_{sat})$ in liquid HFC-125 at saturation pressure $p_{sat}(T)$; p_c denotes the critical pressure. (b) Temperature dependence of isobaric sound speeds:[58] △, 10 MPa; ▽, 20 MPa; □, 30 MPa; -----, ○, sound speeds $u_{sat} = u(T,p_{sat})$ in the saturated liquid; T_c denotes the critical temperature. ▼, sound speeds u_{sat} determined by Kraft and Leipertz[87] via dynamic light scattering.

interest to solution chemists. The speed of sound, as a function of composition and pressure, is given in Figure 10.8(a) for the liquid mixture pentafluoroethane + 1,1-difluoroethane at 298.15 K.[64] The bubble point pressure at different temperatures was measured with the same apparatus, and the results are shown in Figure 10.8(b).[67] They agree well with values calculated with REFPROP from NIST.[88] The bubble point pressure (solid curve) and the dew point pressure (dotted curve) in the figure have been calculated with the Peng-Robinson equation of state.[89]

10.4   Speed of Sound and Thermodynamic Properties

As already pointed out in the introduction, the speed of sound at sufficiently low frequencies well below any dispersion region but at frequencies high enough to provide isentropic conditions is closely related to several important thermodynamic properties. For instance, the isothermal compressibility $\kappa_T \equiv \rho^{-1}(\partial\rho/\partial p)_T$ is given by

$$\kappa_T = \kappa_S + \frac{TM\alpha_p^2}{C_p\rho} \qquad (4)$$

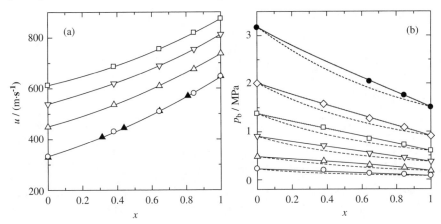

Figure 10.8 Composition dependences (x denotes the mole fraction) of the speed of sound, u, (a), and the bubble point pressure, p_b, (b), respectively, in the system pentafluoroethane (HFC-125) + 1,1-difluoroethane (HFC-152a), $(1-x)CHF_2CF_3 + xCHF_2CH_3$. (a) Experimental sound speeds at 298.15 K: $\bigcirc$, saturated liquid; $\triangle$, 10 MPa; $\triangledown$, 20 MPa; $\square$, 30 MPa; Takagi *et al.*[64] $\blacktriangle$, calculated with REFPROP.[88] (b) Experimental bubble point pressures $p_b(T,x)$: $\bigcirc$, 243.15 K; $\triangle$, 263.15 K; $\triangledown$, 283.15 K; $\square$, 298.15 K; $\Diamond$, 313.15 K; $\bullet$, 333.15 K; Takagi *et al.*[67] ——, Bubble point pressures and -------, dew point pressures were calculated with the Peng-Robinson[89] equation of state.

where $\alpha_p \equiv -\rho^{-1}(\partial\rho/\partial T)_p$ denotes the isobaric expansivity, and C_p is the molar heat capacity at constant pressure. Equation (4) provides the basis for the *indirect* determination of κ_T from the measured density, speed of sound (and hence κ_S), isobaric expansivity and molar heat capacity at constant pressure. Conversely, one may determine κ_S from p-ρ-T and C_p data, though usually with distinctly smaller accuracy. In Figure 10.9(a), the temperature dependence of the isentropic compressibility at atmospheric pressure of pure liquid benzene as obtained via Equation (1) from speed-of-sound measurements,[49,92,93] $\kappa_S(u)$, is compared with (I) the isentropic compressibility as calculated via Equation (4) from p-ρ-T data[90] and C_p data, $\kappa_S(c)$, and with (II) the isentropic compressibility measured *directly* using a piezometer, $\kappa_S(d)$. Because of inherent experimental difficulties, only relatively few direct measurements have been reported in the literature.[94–98] Generally, at 298.15 K the $\kappa_S(d)$ values are slightly higher (by about 4 %) than the $\kappa_S(u)$ results. The $\kappa_S(c)$ values at 0.1 MPa, calculated with Equation (4) using experimental values of C_p, and α_p and κ_T as obtained from p-ρ-T data, agree reasonably well with $\kappa_S(u)$ in the temperature range covered (see solid line). In Figure 10.9(b) the pressure dependence of the isentropic compressibility $\kappa_S(u)$ at 298.15 K and 323.15 K, respectively, is compared with that of $\kappa_S(c)$ for pressures up to 50 MPa: agreement is highly satisfactory throughout. This is a consequence of the fact that the p-ρ-T behaviour of benzene has been exceedingly well investigated (there are about

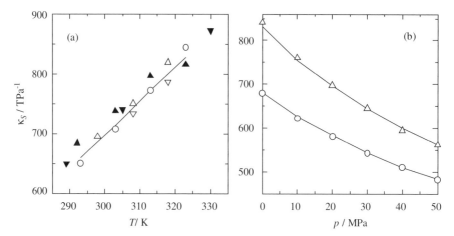

Figure 10.9 Temperature and pressure dependence of the isentropic compressibility κ_S of liquid benzene. (a) Temperature dependence at atmospheric pressure. ▲, ▼, values determined *directly* using a piezometer, $\kappa_S(d)$: ▼, Philip,[96] ▲, Staveley et al.;[97] ○, △, ▽, values calculated from the speed of ultrasound, $\kappa_S(u)$: ○, Takagi and Teranishi,[49] ▽, Desphande and Bhatgadde,[92] △, Masood et al.;[93] —, curve was calculated from p-ρ-T data,[90] $\kappa_S(c)$. (b) Pressure dependence of $\kappa_S(u)$: experimental values[49] are at 298.15 K, ○, and 323.15 K, △. Curves were calculated using p-ρ-T data reported by Gibson and Kincaid,[90] $\kappa_S(c)$.

50 papers on this topic in the literature).[91] Here, we used the high-precision values of Gibson and Kincaid.[90]

From Equation (4), at frequencies where dispersion is not important, the isothermal pressure dependence of the density is given by

$$\left(\frac{\partial \rho}{\partial p}\right)_T = \frac{1}{u^2} + \frac{TM\alpha_p^2}{C_p} \tag{5}$$

This equation may serve as the starting point for an alternative to the direct experimental route to high-pressure p-ρ-T data and related quantities. It may be integrated to give

$$\rho(T,p) = \rho(T,p_0) + \int_{p_0}^{p} u^{-2} \mathrm{d}p + TM \int_{p_0}^{p} \alpha_p^2 C_p^{-1} \mathrm{d}p \tag{6}$$

where $\rho(T,p)$ is the density at temperature T and pressure p, and $\rho(T,p_0)$ is the density at T and at a conveniently selected reference pressure p_0. At low temperatures (*i.e.* below the normal boiling point temperature), usually $p_0 = 0.1$ MPa, while above the normal boiling point temperature a compatible higher reference pressure has to be selected. The first integral can be evaluated directly after fitting the experimental speed-of-sound data $u(T,p)$ with suitably selected polynomials or Padé approximants. To evaluate the second integral, one may use

$$\left(\frac{\partial C_p}{\partial p}\right)_T = -\frac{TM}{\rho}\left[\alpha_p^2 + \left(\frac{\partial \alpha_p}{\partial T}\right)_p\right] \qquad (7)$$

and

$$\left(\frac{\partial \alpha_p}{\partial p}\right)_T = -\left(\frac{\partial \kappa_T}{\partial T}\right)_p \qquad (8)$$

in conjunction with a successive integration algorithm and reliable data on the temperature dependence of the density and the heat capacity at p_0, that is on $\rho(T,p_0)$ and $C_p(T,p_0)$. Thus, starting at temperature T from pressure p_0, the density $\rho(T,p)$, the isobaric expansivity $\alpha_p(T,p)$, the molar heat capacity $C_p(T,p)$ at constant pressure, and the isothermal compressibility $\kappa_T(T,p)$ may be determined.

Pioneers in this field include Davis and Gordon[2] (compressive behaviour of liquid mercury up to 1300 MPa), Kell and Whalley,[3] who published precise density data of pure liquid water in the range 273.15 K to 423.15 K for pressures up to 100 MPa, and Millero *et al.*,[99] who also presented an equation of state of pure water valid in the range 273.15 K to 373.15 K, for pressures from 0.1 MPa to 100 MPa. His work is based on Kell's[100] very precise low-pressure data (density and isothermal compressibility) for liquid water from 273.15 K to 423.15 K. The most recent comprehensive compilation of critically evaluated thermodynamic properties of water (IAPWS-95 formulation) has been presented by Wagner and Pruß.[4] More recently, Biswas *et al.* determined densities, isobaric expansivities, isothermal compressibilities, and heat capacities at constant pressure for heptane,[73] cyclohexane,[69] benzene,[69] toluene,[73] methanol,[71,72] and ethanol.[70] Daridon *et al.*[13–17] determined ρ, κ_S and κ_T for a number of alkanes and related compounds from high-pressure sound-speed measurements made over wide ranges of temperature (273.15 K to 373.15 K) and pressure (up to 150 MPa). Khasanshin *et al.*,[18–21] did much the same over a wider range of temperatures, from 293 K to 433 K, and for pressures up to 140 MPa and Dzida *et al.*[26–29] derived accurate values of ρ, C_p, κ_S and κ_T for several alcohols from speed-of-sound measurements between 293.15 K and 318.15 K and up to 100 MPa.

In the case of water, the contribution of the second integral on the right-hand side of Equation (6) to the density can be disregarded in the neighbourhood of 277 K, and it is only about 4 % at 373 K and 100 MPa.[3] On the other hand, the contribution of the second integral to the density of organic liquids (for example octane) may amount to 10 % at 100 MPa in the temperature range 283 K to 333 K.[40] With increasing temperature, the calculation of α_p and C_p at high pressures using Equations (7) and (8) becomes less reliable, especially close

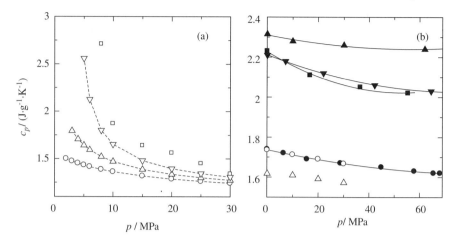

Figure 10.10 Pressure dependence of the specific heat capacity $c_p = C_p/M$ at constant pressure. (a) trifluoromethane:[23] ○, 250 K; △, 270 K; ▽, 290 K; □, 310 K. ------, REFPROP.[88] (b) ○, △, obtained via ultrasound measurements, at 298.15 K, and use of Equation (10); ○, benzene, △, tetrachloromethane; Takagi et al.[65] ●, ▲, ▼, ■, direct measurements; ●, benzene, at 298.15 K; Czarnota.[102] ▲, hexane, at 303.15 K; Randzio et al.[106] ▼, decane, at 298.8 K; Czarnota.[103] ■, octane, at 298.8 K; Czarnota.[104]

to the critical temperature. Guedes et al.[22–24] (Universidade Nova de Lisbon, Portugal) have measured sound speeds for some hydrofluorocarbons and their mixtures under the difficult conditions near the critical point, and have tried to estimate several thermophysical quantities. The effect of pressure on the isobaric heat capacity of trifluoromethane is shown in Figure 10.10(a) as obtained by Pires and Guedes.[23] The critical temperature and critical pressure for trifluoromethane are 299.29 K and 4.83 MPa, respectively. Since C_p diverges at the critical point as $|T/T_c - 1|^{-\gamma}$, where $\gamma = 1.239$ is the critical exponent, isotherms near T_c show a strong pressure dependence.

We have used another method for determining $\alpha_p(T,p)$ and $C_p(T,p)$ values[40] from high-pressure speed-of-sound measurements. This method is based on Equations (9) and (10) below which were suggested by Kato et al.[101] These equations were derived from a Taylor expansion, at constant temperature, of the volume around the reference pressure p_0 (say, atmospheric pressure) and yield $\alpha_p(T,p)$ and $C_p(T,p)$ at an elevated pressure $p = p_0 + \Delta p$:

$$\alpha_p(T,p) = \alpha_p(T,p_0) - \left[\frac{1}{\rho u^2}\left(\frac{\partial \gamma}{\partial T}\right) + \frac{\gamma \alpha_p}{\rho u^2} - \frac{2\gamma}{\rho u^3}\left(\frac{\partial u}{\partial T}\right)\right]_{p_0} \Delta p \qquad (9)$$

$$C_p(T,p) = C_p(T,p_0) - \frac{TM}{\rho(T,p_0)}\left[\alpha_p^2 + \left(\frac{\partial \alpha_p}{\partial T}\right)\right]_{p_0} \Delta p \qquad (10)$$

Here, $\gamma = \kappa_T/\kappa_S = C_p/C_V$, and C_V is the molar heat capacity at constant volume. When information is insufficient for a complete calculation in the spirit of Equation (6), these formulas should be useful for deriving α_p and C_p over narrow pressure ranges. In Figure 10.10(b) the heat capacity C_p of benzene and tetracloromethane, obtained by using Equation (10), is compared with the heat capacity measured directly with a high-pressure calorimeter.[102] The agreement is satisfactory.

One striking feature of the liquid state is the pressure dependence of C_p along isothermal paths. For instance, C_p of hexane,[105,106] indicated in Figure 10.10(b) together with results for decane[103] and octane,[104] decreases with increasing pressure, first rapidly and then *very* gently, to pass through a *shallow minimum* at *high* pressure: at 350 K the minimum is found at about 130 MPa. The minima move to higher pressures as T increases. Similar behaviour was observed for many other liquids.[72,73,107–112]

The precision of speed-of-sound measurements in liquids and liquid mixtures at high pressures is greater than of direct measurements of densities, isobaric expansivities and heat capacities at constant pressure. This makes the speed-of-sound method for the *indirect* determination of thermodynamics properties highly attractive. As a result it is expected that this field of experimental thermodynamics will prosper and be of great help in fundamental research as well as in applied chemistry and chemical engineering.

References

1. A. J. Matheson, *Molecular Acoustics,* John Wiley & Sons Ltd., London, 1971.
2. L. A. Davis and R. B. Gordon, *J. Chem. Phys.*, 1967, **46**, 2650.
3. G. S. Kell and E. Whalley, *J. Chem. Phys.*, 1975, **62**, 3496.
4. W. Wagner and A. Pruß, *J. Phys. Chem. Ref. Data*, 2002, **31**, 387.
5. J. C. Swanson, *J. Chem. Phys.*, 1934, **2**, 689.
6. E. H. Carnevale and T. A. Litovitz, *J. Acoust. Soc. Am.*, 1955, **27**, 547.
7. H. J. McSkimin, *J. Acoust. Soc. Am.*, 1957, **29**, 1185.
8. E. G. Richardson and R. I. Tait, *Phil. Mag.*, 1957, **2**, 441.
9. J. F. Mifsud and A. W. Nolle, *J. Acoust. Soc. Am.*, 1966, **28**, 469.
10. B. Oakley, G. Barber, T. Worden and D. Hanna, *J. Phys. Chem. Ref. Data*, 2003, **32**, 1501.
11. B. Oakley, D. Hanna, M. Shillor and G. Barber, *J. Phys. Chem. Ref. Data*, 2003, **32**, 1535.
12. Yu. A. Neruchev, M. F. Bolotnikov and V. V. Zotov, *High Temp.*, 2005, **43**, 266.
13. J. L. Daridon, B. Lagourette and J. -P. E. Grolier, *Int. J. Thermophys.*, 1998, **19**, 145.
14. J. L. Daridon, B. Lagourette and A. Lagrabette, *J. Chem. Thermodyn.*, 1998, **30**, 607.
15. J. L. Daridon and B. Lagourette, *High Temp. High Press.*, 2000, **32**, 83.

16. S. Dutour, B. Lagourette and J. L. Daridon, *J. Chem. Thermodyn.*, 2002, **34**, 475.
17. A. J. Queimada, J. A. P. Coutinho, I. M. Marrucho and J. L. Daridon, *Int. J. Thermophys.*, 2006, **27**, 1095.
18. T. S. Khasanshin and A. P. Shchemelev, *High Temp.*, 2001, **39**, 60.
19. T. S. Khasanshin, A. P. Shchamialiou and O. G. Poddubskij, *Int. J. Thermophys.*, 2003, **24**, 621.
20. T. S. Khasanshin, O. G. Poddubskii and A. P. Shchemelev, *J. Eng. Phys. Thermophys.*, 2004, **77**, 185.
21. T. S. Khasanshin, O. G. Poddubskij, A. P. Shchamialiou and V. S. Samuilov, *Int. J. Thermophys.*, 2006, **27**, 1746.
22. P. F. Pires and H. J. R. Guedes, *J. Chem. Thermodyn.*, 1999, **31**, 55.
23. P. F. Pires and H. J. R. Guedes, *J. Chem. Thermodyn.*, 1999, **31**, 479.
24. P. F. Pires, J. M. S. S. Esperanca and H. J. R. Guedes, *J. Chem. Eng. Data*, 2000, **45**, 496.
25. J. Szydlowski, R. Gomes de Azevedo, L. P. N. Rebelo, J. M. S. S. Esperanca and H. J. R. Guedes, *J. Chem. Thermodyn.*, 2005, **37**, 671.
26. M. Dzida and S. Ernst, *J. Chem. Eng. Data*, 2003, **48**, 1453.
27. M. Dzida, A. Zak and S. Ernst, *J. Chem. Thermodyn.*, 2005, **37**, 405.
28. M. Dzida and W. Marczak, *J. Chem. Thermodyn.*, 2005, **37**, 826.
29. E. Zorgbski and M. Dzida, *J. Chem. Eng. Data*, 2007, **52**, 1010.
30. M. F. Bolotnikov, V. N. Verveyko and M. V. Verveyko, *J. Chem. Eng. Data*, 2004, **49**, 631.
31. V. A. Gruzdev, R. A. Khairulin, S. G. Komarov and S. V. Stankus, *Int. J. Thermophys.*, 2008, **29**, 546.
32. J. R. Pellam and J. K. Galt, *J. Chem. Phys.*, 1946, **14**, 608.
33. E. P. Papadakis, *J. Acoust. Soc. Am.*, 1967, **42**, 1045.
34. A. Z. Zak, M. Dzida, M. Zorebski and S. Ernst, *Rev. Sci. Instrum.*, 2000, **71**, 1756.
35. J. L. Daridon, *Acustica*, 1994, **80**, 416.
36. P. J. Kortbeek, M. J. P. Muringer, N. J. Trappeniers and S. N. Biswas, *Rev. Sci. Instrum.*, 1985, **56**, 1269.
37. S. J. Ball and J. P. M. Trusler, *Int. J. Thermophys.*, 2001, **22**, 427.
38. T. Makita and T. Takagi, *Rev. Phys. Chem. Jpn.*, 1968, **38**, 41.
39. T. Takagi, *Rev. Phys. Chem. Jpn.*, 1978, **48**, 10.
40. T. Takagi, *Kagaku Kougaku Ronbunshu (Japanese)*, 1978, **4**, 1.
41. T. Takagi, *J. Chem. Thermodyn.*, 1980, **12**, 1183.
42. T. Takagi, *J. Chem. Thermodyn.*, 1981, **13**, 291.
43. T. Takagi and H. Teranishi, *J. Chem. Thermodyn.*, 1982, **14**, 1167.
44. T. Takagi and H. Teranishi, *J. Chem. Thermodyn.*, 1984, **16**, 591.
45. T. Takagi and H. Teranishi, *J. Chem. Thermodyn.*, 1984, **16**, 1031.
46. T. Takagi and H. Teranishi, *Fluid Phase Equilib.*, 1985, **20**, 315.
47. T. Takagi and H. Teranishi, *J. Chem. Thermodyn.*, 1985, **17**, 1057.
48. T. Takagi and H. Teranishi, *J. Chem. Eng. Data*, 1986, **31**, 105.
49. T. Takagi and H. Teranishi, *J. Chem. Thermodyn.*, 1987, **19**, 1299.

50. T. Takagi and H. Teranishi, *J. Chem. Eng. Data*, 1987, **32**, 133.
51. T. Takagi and H. Teranishi, *J. Chem. Thermodyn.*, 1988, **20**, 809.
52. T. Takagi, *Netsu Bussei (in Japanese)*, 1988, **2**, 101.
53. T. Takagi and H. Teranishi, *Int. J. Thermophys.*, 1989, **10**, 661.
54. T. Takagi, M. Kusunoki and M. Hongo, *J. Chem. Eng. Data*, 1992, **37**, 39.
55. T. Takagi, *High Temp.-High Press.*, 1993, **25**, 685.
56. T. Takagi and M. Hongo, *J. Chem. Eng. Data*, 1993, **38**, 60.
57. T. Takagi, *J. Chem. Eng. Data*, 1996, **41**, 1061.
58. T. Takagi, *J. Chem. Eng. Data*, 1996, **41**, 1325.
59. T. Takagi, *High Temp.-High Press.*, 1997, **29**, 135.
60. T. Takagi, *J. Chem. Eng. Data*, 1997, **42**, 1129.
61. T. Takagi, T. Sakura. T. Tsuji and M. Hongo, *Fluid Phase Equilib.*, 1999, **162**, 171.
62. T. Takagi and T. Sakura, *High Temp.-High Press.*, 2000, **32**, 89.
63. T. Takagi, K. Sawada, H. Urakawa, M. Ueda and I. Cibulka, *J. Chem. Eng. Data*, 2004, **49**, 1652.
64. T. Takagi, K. Sawada, H. Urakawa, T. Tsuji and I. Cibulka, *J. Chem. Eng. Data*, 2004, **49**, 1657.
65. T. Takagi, K. Sawada, H. Urakawa, M. Ueda and I. Cibulka, *J. Chem. Thermodyn.*, 2004, **36**, 659.
66. T. Takagi and E. Wilhelm, *unpublished results*.
67. Takagi, K. Sawada, J. H. Jun, H. Urakawa and T. Tsuji, 17th ECTP, Bratislava, 2005.
68. M. J. P. Muringer, N. J. Trappeniers and S. N. Biswas, *Phys. Chem. Liq.*, 1985, **14**, 273.
69. T. F. Sun, P. J. Kortebeek, N. J. Trappeniers and S. N. Biswas, *Phys. Chem. Liq.*, 1987, **16**, 163.
70. T. F. Sun, C. A. Ten Seldam, P. J. Kortebeek, N. J. Trappeniers and S. N. Biswas, *Phys. Chem. Liq.*, 1988, **18**, 107.
71. T. F. Sun, S. N. Biswas, N. J. Trappeniers and C. A. Ten Seldam, *J. Chem. Eng. Data*, 1988, **33**, 395.
72. T. F. Sun, J. A. Schouten and S. N. Biswas, *Ber. Bunsenges. Phys. Chem.*, 1990, **94**, 528.
73. T. F. Sun, S. A. R. C. Bominaar, C. A. Ten Seldam and S. N. Biswas, *Ber. Bunsenges. Phys. Chem.*, 1991, **95**, 696.
74. G. C. Benson and Y. P. Handa, *J. Chem. Thermodyn.*, 1981, **13**, 887.
75. E. Wilhelm, R. Schano, G. Becker, G. H. Findenegg and F. Kohler, *Trans. Faraday Soc.*, 1969, **65**, 1443.
76. E. Wilhelm, M. Zettler and H. Sackmann, *Ber. Bunsenges. Phys. Chem.*, 1974, **78**, 795.
77. E. Wilhelm, J.-P. E. Grolier and M. H. Karbalai Ghassemi, *Ber. Bunsenges. Phys. Chem.*, 1977, **81**, 925.
78. J.-P. E. Grolier, E. Wilhelm and M. H. Hamedi, *Ber. Bunsenges. Phys. Chem.*, 1978, **82**, 1282.

79. E. Wilhelm, J.-P. E. Grolier and M. H. Karbalai Ghassemi, *Thermochim. Acta*, 1979, **28**, 59.
80. K. Nishikawa, K. Tamura and S. Murakami, *J. Chem. Thermodyn.*, 1998, **30**, 229.
81. E. Aicart, M. Costas, E. Junquera and G. Tardajos, *J. Chem. Thermodyn.*, 1990, **22**, 1153.
82. S. Ernst and M. Dzida, *Fluid Phase Equil.*, 1998, **146**, 25.
83. W. M. Madigosky and T. A. Litovitz, *J. Chem. Phys.*, 1961, **34**, 489.
84. E. Wilhelm and R. Battino, *J. Chem. Phys.*, 1971, **55**, 4012.
85. E. Wilhelm, *J. Chem. Phys.*, 1973, **58**, 3588.
86. F. Kohler, J. Fischer and E. Wilhelm, *J. Molec. Structure*, 1982, **84**, 245.
87. K. Kraft and A. Leipertz, *Int. J. Thermophys.*, 1994, **15**, 387.
88. E. W. Lemmon, M. O. McLinden and M. Huber, NIST Standard Reference Database 23, REFPROP Ver. 7.0, 2002.
89. D. Y. Peng and D. B. A. Robinson, *Ind. Eng. Chem. Fundam.*, 1976, **15**, 59.
90. R. E. Gibson and J. K. Kincaid, *J. Am. Chem. Soc.*, 1938, **60**, 511.
91. I. Cibulka and T. Takagi, *J. Chem. Eng. Data*, 1999, **44**, 411.
92. D. D. Desphande and L. G. Bhatgadde, *J. Phys. Chem.*, 1968, **17**, 261.
93. A. K. M. Masood, A. M. North, R. A. Pethrick, M. Towland and F. L. Swinton, *J. Chem. Thermodyn.*, 1977, **9**, 133.
94. D. Tyrer, *J. Chem. Soc.*, 1913, **103**, 1675.
95. D. Tyrer, *J. Chem. Soc.*, 1914, **105**, 2534.
96. N. M. Philip, *Proc. Indian Acad. Sci.*, 1939, **A9**, 109.
97. L. A. K. Staveley, W. I. Tupman and K. R. Hart, *Trans. Faraday. Soc.*, 1955, **51**, 323.
98. J. Nývlt and E. Erdös, *Coll. Czech. Chem. Commun.*, 1961, **26**, 485.
99. C. T. Chen, R. A. Fine and F. J. Millero, *J. Chem. Phys.*, 1977, **66**, 2142.
100. G. S. Kell, *J. Eng. Eng. Data*, 1975, **20**, 97.
101. S. Kato, H. Nomura and Y. Miyahara, *Kagaku Kogaku (in Japanese)*, 1974, **38**, 369.
102. I. Czarnota, *J. Chem. Thermodyn.*, 1991, **23**, 25.
103. I. Czarnota, *J. Chem. Thermodyn.*, 1993, **25**, 639.
104. I. Czarnota, *J. Chem. Thermodyn.*, 1993, **25**, 355.
105. Ph. Pruzan, *J. Chem. Thermodyn.*, 1991, **23**, 247.
106. S. L. Randzio, J.-P. E. Grolier, J. R. Quint, D. J. Eatough, E. A. Lewis and L. D. Hansen, *Int. J. Thermophys.*, 1994, **15**, 415.
107. C. Alba, L. T. Minassian, A. Denis and A. Soulard, *J. Chem. Phys.*, 1985, **82**, 384.
108. V. M. Shulga, F. G. Eldarov, Yu. A. Atanov and A. A. Kuyumchev, *Int. J. Thermophys.*, 1986, **7**, 1147.

109. A. Asenbaum, E. Wilhelm and P. Soufi-Siavoch, *Acustica*, 1989, **68**, 131.

110. U. Setzmann and W. Wagner, *J. Phys. Chem. Ref. Data*, 1991, **20**, 1061.

111. S. L. Randzio, E. A. Lewis, D. J. Eatough and L. D. Hansen, *Int. J. Thermophys.*, 1995, **16**, 883.

112. S. L. Randzio, D. J. Eatough, E. A. Lewis and L. D. Hansen, *Int. J. Thermophys.*, 1996, **17**, 405.

CHAPTER 11

Heat Capacities and Brillouin Scattering in Liquids

EMMERICH WILHELM[a] AND AUGUSTINUS ASENBAUM[b]

[a] Institute of Physical Chemistry, University of Wien, Währinger Strasse 42, A-1090, Wien (Vienna), Austria; [b] Section for Experimental Physics, Department of Materials Research and Physics, University of Salzburg, A-5020, Salzburg, Austria

11.1 Introduction

Light scattering experiments in general are well suited for studying the dynamics of condensed matter, and in fact, since the advent of laser spectroscopy, a large part of the research activity in Brillouin scattering has been devoted to liquids.[1-8] In this contribution we will discuss the type of information which may be obtained from the experimentally determined Brillouin spectra, the focus being on heat capacities. Only *pure*, relatively simple (that is, *normal*) liquids without hydrogen bonds will be considered.

In Brillouin scattering experiments, laser light first passes through an appropriate polarising device which provides the vertical (V) polarisation of the incident beam. It then passes through the transparent liquid which scatters a very small fraction (about 10^{-6}) of the incident light. Light scattered at a scattering angle θ subsequently passes through a polarising analyser in vertical direction (V), and the frequency analysis is usually performed with a high-precision scanning Fabry–Perot interferometer. Finally, the light impinges on a suitable detector (photomultiplier).

Heat Capacities: Liquids, Solutions and Vapours
Edited by Emmerich Wilhelm and Trevor M. Letcher
© The Royal Society of Chemistry 2010
Published by the Royal Society of Chemistry, www.rsc.org

Brillouin scattering in pure liquids is the scattering of laser light by the always present thermally driven density fluctuations. Thermodynamically, these fluctuations may be described in terms of propagating pressure fluctuations at constant entropy, and non-propagating temperature or, equivalently, entropy fluctuations at constant pressure. Light scattered by the entropy fluctuations at constant pressure shows no frequency shift, and this component is referred to as the *Rayleigh peak*. As far back as 1922, Brillouin[9] suggested that light may also be scattered by thermal sound waves in liquids (that is, by propagating fluctuations at constant entropy). Indeed, two peaks are observed which are symmetrically located with respect to the unshifted Rayleigh peak, because scattering occurs from sound waves of frequency f travelling in opposite directions at the same hypersonic speed $v_H(f)$, and light scattered from them experiences a Doppler shift $\omega_B = 2\pi f$. The observed distribution of scattered light intensity as a function of frequency thus consists of *three* components: one at the frequency of incident light c/λ_0, where c is the speed of light and λ_0 is the vacuum wavelength of the incident laser light; one at $c/\lambda_0 + |\omega_B/2\pi|$; and one at $c/\lambda_0 - |\omega_B/2\pi|$. The whole spectrum is frequently called the *Rayleigh–Brillouin triplet*. As already indicated above, here it is assumed that the light is polarised with the electric vector being perpendicular to the plane defined by the incident laser beam and the observing direction, and that the scattered light is also observed in this polarisation (VV experiment). A different scattering spectrum is obtained when the incident light is vertically polarised but observations are made on the scattered light polarised in the plane (VH experiment): the depolarised component is in part caused by reorientation of anisotropic molecules. When Brillouin scattering is observed in liquid mixtures, scattering may also occur as a result of concentration fluctuations, whence information on diffusion coefficients may be obtained.

The frequency shift ω_B of the two Brillouin peaks is given by

$$\omega_B = \pm k \, v_H(f) \tag{1}$$

where

$$k = \frac{4\pi n}{\lambda_0} \sin(\theta/2) \tag{2}$$

denotes the absolute value of the wave vector transfer (that is transfer from incident to scattered light, due to the scattering geometry), $f = \omega_B/2\pi$ is the frequency of hypersound, and n is the refractive index of the liquid. These doublet peaks are also broadened due to various dissipative processes.[10-18] In addition, a detailed analysis shows a weak asymmetry of the two peaks,[2,3] which induces a slight pulling of their positions towards the Rayleigh peak. The discussion of Rayleigh–Brillouin scattering presented so far refers to simple (*monoatomic*) liquids. In liquids consisting of *polyatomic* molecules, the internal vibrational modes generally couple to the translational modes, thereby leading to relaxation mechanisms for the density fluctuations and thus to a broad,

TETRACHLOROMETHANE, *T* = 297.15 K, FSR = 10.00 GHz

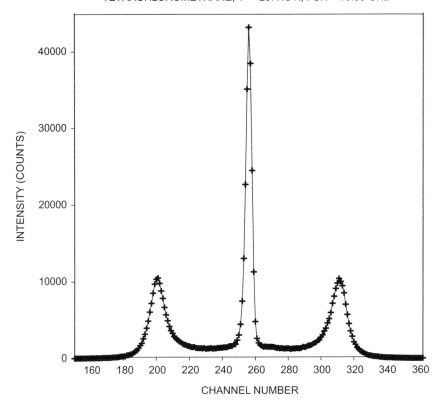

Figure 11.1   Rayleigh–Brillouin spectrum of liquid tetrachloromethane at 297.15 K and atmospheric pressure.[20] The free spectral range (FSR) was 10.00 GHz. Each channel number corresponds to a frequency interval of 97.66 MHz. The broad Mountain line is clearly visible.

unshifted central peak, the so-called *Mountain peak*.[10,13] It was first observed experimentally by Gornall *et al.*[19] and later corroborated by the more detailed study of Nichols *et al.*[15,17] Figure 11.1 shows a typical Rayleigh–Brillouin spectrum of a dense normal liquid, *i.e.* of tetrachloromethane,[20] where the broad Mountain peak, extending out to the two Brillouin peaks, is clearly visible. In passing we note that because of the different physical origins and because of the approximations used, the Rayleigh peak and the Brillouin peaks are usually investigated and discussed separately.

Rayleigh–Brillouin spectra contain highly valuable information on several important thermophysical properties.

- The separation ω_B of the Brillouin doublet from the Rayleigh peak is determined by the hypersound speed, *i.e.* the high-frequency speed of sound (GHz region)

$$v_H(f) = \omega_B/k \tag{3}$$

The hypersonic frequency region accessible *via* Brillouin scattering exceeds the usual ultrasonic region by at least two orders of magnitude. Thus, sound dispersion and sound absorption studies become feasible over very large frequency ranges.

- For simple liquids (liquid noble gases), the integrated intensities yield the ratio of the heat capacities $\kappa = C_P/C_V$ *via* the Landau-Placzek ratio:

$$\frac{I_R}{2I_B} = \frac{C_P - C_V}{C_V} = \kappa - 1 \tag{4}$$

Here, I_R is the integrated intensity of the central, unshifted Rayleigh peak, I_B is the integrated intensity of *one* Brillouin peak, C_P is the molar heat capacity at constant pressure P, and C_V is the molar heat capacity at constant molar volume V. For the integrated intensities, theory[2] yields $I_R \propto \rho^2 k_B T \beta_T (1-\kappa^{-1})$ and $2I_B \propto \rho^2 k_B T \beta_T \kappa^{-1}$, whence for the total integrated intensity $I = I_R + 2I_B$

$$I \propto \rho^2 k_B T \beta_T \tag{5}$$

is obtained. Here, $\rho = M/V$ is the (mass-) density, k_B is the Boltzmann constant, T is the temperature, M is the molar mass, and

$$\beta_T = \rho^{-1} \left(\frac{\partial \rho}{\partial P} \right)_T \tag{6}$$

is the isothermal compressibility.[†] For molecular (normal) liquids, the ratio of the intensity of the unshifted (central) components of the scattered light (Rayleigh *and* Mountain) to the intensity of the Brillouin components, is a rather complicated expression[2,3] of the following general form:

$$\frac{I_R + I_M}{2I_B} = (\kappa - 1)\frac{1 + A}{B} \tag{7}$$

that is to say, the ratio is greater than $(\kappa-1)$. However, for very short relaxation times it reduces to the classical Landau-Placzek ratio, Equation (4). Evidently, from Equation (7), if $C_V \approx C_P$ it will be rather difficult to observe the central Rayleigh peak. Liquid water, for temperatures around that of the density maximum, is such an interesting case.[21,22]

- The width of the Brillouin peaks (full width at half height, FWHH)) is closely related to the sound absorption coefficient which is governed by the shear viscosity η_s and the bulk or volume viscosity η_v. For non-metallic liquids, the heat conductivity contribution is usually *much* smaller. Thus,

---

[†] In this chapter the isothermal compressibility is represented by the symbol β_T and not by κ_T as was recently recommended by IUPAC. Similarly, the isentropic compressibility is represented by the symbol β_S and not by κ_S.

precise determination of the width yields η_v, provided the shear viscosity is obtained independently. In fact, Brillouin scattering and ultrasound absorption measurements[23,24] constitute the *only* experimental methods to obtain this important quantity. For molecular liquids, the introduction of a relaxing volume viscosity $\eta_v(f)$ is quite general and does not depend on any explicit specification of the internal relaxation processes (thermal relaxation) involved. These aspects are covered, for instance, in references 11, 13, 15–17, and 25–28. The width (FWHH) of the Rayleigh peak is closely related to the thermal diffusivity $D_T = lM/\rho C_P$. Thus, provided C_P is measured independently, the thermal conductivity[‡] l may be so determined.

While other approaches are possible,[12,18] our discussion of Brillouin scattering experiments on molecular (normal) liquids will proceed in terms of a frequency-dependent bulk (or volume) viscosity $\eta_v(f)$, which in turn may be related to a relaxing heat capacity.[15,26–33] A concise summary is contained in Section 11.3 of this chapter. Finally, we note that in a similar fashion shear relaxation may be discussed by introducing a frequency-dependent shear viscosity.[34] For simple liquids (like liquid argon), however, the associated time scale is beyond the range currently accessible to laser light Brillouin scattering.

11.2   Experiment

In a typical Brillouin scattering experiment an argon ion laser in single-frequency single-mode operation (for instance, $\lambda_0 = 514.67$ nm) is used as light source. Usually the laser beam with a power between 50 mW and 150 mW is focused into the liquid sample to increase the intensity of the light scattered from the scattering volume. To prevent the generation of a noticeable temperature gradient in the vicinity of the laser focus in the liquid sample, significantly higher powers should be avoided. For 90° scattering geometry, the liquid is contained in a cuvette with a quadratic cross section. Hexagonal cuvettes are used for 60° and 120°, while octagonal cuvettes are used for 45°, 90° and 135° scattering geometry.

In order to limit the imprecision of the experimentally determined hypersound speeds to about ±0.5 % for scattering at 90°, the experimental error of the scattering angle should not exceed ±15′. In our experiments the samples are thermostated to better than ±0.1 K, and the temperature is measured with a calibrated platinum resistance thermometer (PT 100).

The scattered light is analysed, for instance, with an electronically stabilised six-pass tandem Fabry-Perot interferometer (Sandercock type[35]) with a finesse of about 80. In older work also three-pass or five-pass Fabry-Perot interferometers were used with a finesse of about 40. Usually, a free spectral range (FSR) of about 15.00 GHz is selected. The scattered light is detected by a cooled photomultiplier and stored in the memory of a personal computer. Dark

---

[‡] In this chapter the thermal conductivity is represented by the symbol l and not by λ as was recently recommended by IUPAC.

current and background count rates are only a few counts per second. One scan of the six-pass tandem interferometer we currently use takes about 0.5 s, after which the parallelism of the Fabry-Perot mirrors is checked and then the mirrors are realigned. This technique guarantees the constancy of the spectral resolution over the measuring period. Furthermore, to achieve an acceptable signal-to-noise ratio, a large number of individual scans are added up with the unshifted laser frequency c/λ_0 as reference. This referencing/aligning procedure prevents broadening and distortion of the spectral components of the Rayleigh-Brillouin spectra due to the drift of the laser frequency as well as the drift of the interferometer pass frequency. This technique represents the standard for high-resolution, high-contrast Brillouin light scattering measurements. For details we refer to the original literature. Figure 11.2 shows a schematic representation of a state-of-the-art experimental Brillouin set-up.

A new optical superheterodyne light beating spectroscopy for Rayleigh-Brillouin scattering using frequency-tuneable lasers was recently developed by Tanaka and Sonehara.[36] It provides an extremely high frequency resolution (*ca.* 300 kHz), and has a high application potential in condensed matter physics (transparent liquids and solids).

At *high pressures*, Brillouin scattering is one of the most useful and frequently used spectroscopic methods for measuring sound speed and sound attenuation in liquids and solids, thereby yielding the pressure dependence of several important thermodynamic and mechanical quantities as well as of dynamical processes within the material. For our experiments on liquids at high pressures we used a hand-pump up to 700 bar (throughout this contribution the *pressure*

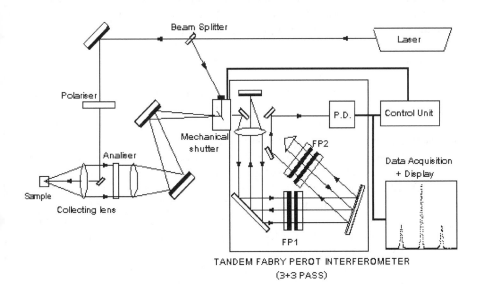

Figure 11.2 Experimental Brillouin scattering set-up with a six-pass tandem Fabry-Perot interferometer (FP1 and FP2), Sandercock type.[35] PD denotes the photon detector.

unit used is the bar $= 10^5$ Pa $= 10^5$ N m^{-2}), and a manually operated screw-press for the pressure range up to 2000 bar. A special fluid divider allowed pressure transmission from the pumping oil to the liquid under investigation without contaminating the sample. The pressure was measured with a Bourdon gauge, with a maximum inaccuracy of ± 10 bar, and with a temperature compensated manganin cell with a maximum inaccuracy of ± 12 bar. The design of a high-pressure sample chamber in combination with a non-scanning angle-dispersive Fabry-Perrot interferometer has been described by Koski *et al.*[37]

11.3 Theory

Generalised hydrodynamics[38–40] is at the heart of the analysis and discussion of experiments probing density-density correlation functions, such as light scattering,[2,3] neutron scattering,[6] and ultrasonics.[23,24] In light scattering experiments (VV), density fluctuations are probed, while orientational correlations are probed through the depolarised component of the scattered radiation (VH experiments). As pointed out above, only the former will be considered here.

Mountain[10–14] has derived the spectral distribution of light scattered by density fluctuations in dense fluids, considering monoatomic (simple) fluids as well as fluids with molecules possessing internal degrees of freedom weakly coupled to their translational degrees of freedom by a process characterised by only a single relaxation time. Nichols and Carome (NC)[15–17] went beyond Mountain's treatment of Brillouin scattering in relaxing liquids by (I) deriving more accurate mathematical expressions for the frequency spectrum of the scattered light intensity, and (II) extending the formalism to multiply relaxing liquids. In fact, the NC approach is used most frequently for analysing Brillouin spectra and has supplied the basis of our work in this field. Essentially, the spectral distribution can be calculated as the sum of *four* Lorentzians representing the two unshifted central lines (Rayleigh and Mountain) and the Brillouin doublet, and a non-Lorentzian correction which shifts the apparent Brillouin peaks *slightly* towards the centre and at the same time renders them *slightly* asymmetric. However, the total spectrum is still symmetric with respect to the central lines.

By least-squares fitting each experimental Brillouin spectrum with a theoretical NC-based spectrum convoluted with the instrumental function of the Fabry-Perot interferometer, the Brillouin shift ω_B, the relaxation time τ_v of the relaxing bulk (or volume) viscosity $\eta_v(f)$ and the non-relaxing bulk viscosity $\eta_{v,nr}$ are obtained simultaneously. The hypersonic speed $v_H = v_H(f)$ at sound frequency $f = \omega_B/2\pi$ is obtained from the Brillouin shift through use of Equations (2) and (3). In molecular liquids, the hypersonic speed is generally *not* equal to the *thermodynamic* low-frequency ultrasonic speed v_0 which is related to the isentropic compressibility β_S according to

$$v_0 = (\rho \beta_S)^{-1/2} \tag{8}$$

with β_S being defined by

$$\beta_S = \rho^{-1}\left(\frac{\partial\rho}{\partial P}\right)_S \tag{9}$$

Specifically, the *dispersion* of sound, for a single relaxation process, is given by

$$\frac{v_\infty^2}{v_0^2} - 1 = \left(\frac{v(f)^2}{v_0^2} - 1\right)\frac{1 + \omega^2\tau_v^2}{\omega^2\tau_v^2} \tag{10}$$

where v_∞ is the high-frequency limit of the sound speed, and $v(f)$ is the sound speed at frequency f, say $v_H(f)$. In normal liquids, where translation-vibration relaxation is the dominant feature, the dispersion is related to the molar heat capacities C_P and C_V, and the relaxing molar heat capacity C_i, *i.e.* the contribution to the heat capacity due to internal, vibrational degrees of freedom associated with the relaxation process:

$$\frac{v_\infty^2}{v_0^2} - 1 = \frac{(C_P - C_V)C_i}{(C_V - C_i)C_P} \tag{11}$$

The molar heat capacity at constant volume may be obtained from the calorimetrically determined C_P according to

$$\kappa = C_P/C_V = \beta_T/\beta_S \tag{12}$$

$$= 1 + TMv_0^2\alpha_P^2/C_P \tag{13}$$

where $\alpha_P = -\rho^{-1}(\partial\rho/\partial T)_P$ is the isobaric expansivity. Thus, in principle C_i may be obtained from the set of least-squares fit parameters ω_B [yielding $v_H(f)$ *via* Equation (3)] and τ_v.[28–31] However, in many cases the dispersion curve may not be known completely, that is $v_H(f)$ data may only be available *below* the relaxation frequency $1/(2\pi\tau_v)$, whence the corresponding uncertainty renders this approach less attractive from a practical point of view. It is for this reason that an alternative route may be adopted, in which experimental information on the *low-frequency sound absorption* is incorporated as follows.[27,28,31–33] The low-frequency sound attenuation α/f^2 in a liquid is given by[23]

$$\begin{aligned}\alpha/f^2 &= (\alpha/f^2)_s + (\alpha/f^2)_v + (\alpha/f^2)_{hc}\\ &= \frac{2\pi^2}{\rho v_0^3}\left[\frac{4}{3}\eta_s + \eta_v + lM\left(C_V^{-1} - C_P^{-1}\right)\right]\end{aligned} \tag{14}$$

where the subscripts s, v and hc indicate the contributions due to shear viscosity, bulk or volume viscosity, and heat conductivity to the total sound absorption (for most liquids the thermal conductivity term is much smaller than those due to shear or bulk viscosity). The bulk viscosity is thus accessible from experimental sound absorption data.

For single relaxation behaviour, the frequency dependence of the bulk viscosity is given by

$$\eta_v(f) = \eta_{v,nr} + \eta_{v,r}(f) \tag{15}$$

where $\eta_{v,nr}$ is the *nonrelaxing* contribution to $\eta_v(f)$, and

$$\eta_{v,r}(f) = \eta_{v,r}(0)/(1 + \omega^2 \tau_v^2) \tag{16}$$

Here,

$$\eta_{v,r}(0) = \lim_{f \to 0} \eta_{v,r}(f) = \eta_v(0) - \eta_{v,nr} \tag{17}$$

denotes the low-frequency value of $\eta_{v,r}(f)$. In turn, this quantity is related to the sound dispersion, and hence, *via* Equation (11), to the relaxing molar heat capacity:

$$\eta_{v,r}(0) = \rho v_0^2 \tau_v \left(\frac{v_\infty^2}{v_0^2} - 1 \right) \tag{18}$$

$$= \rho v_0^2 \tau_v \frac{(C_P - C_V)C_i}{(C_V - C_i)C_P} \tag{19}$$

Since $\eta_{v,nr}$ and τ_v are obtained by fitting experimental Brillouin spectra, and since the low-frequency value of the bulk viscosity, $\eta_v(0)$, is accessible, *via* Equation (14), from ultrasound absorption measurements, $\eta_{v,r}(0)$ and hence the sound dispersion v_∞^2/v_0^2 may be determined *via* Equation (18), and the relaxing heat capacity C_i *via* Equation (19). In turn, C_i so obtained may be compared with vibrational contributions calculated with the help of the Planck-Einstein equation,

$$C_{i,PE}/R = \sum_i \frac{g_i(hf_i/k_B T)^2 \exp(-hf_i/k_B T)}{[1 - \exp(-hf_i/k_B T)]^2} \tag{20}$$

where $R = N_A k_B$ is the gas constant, N_A is the Avogadro constant, h is the Planck constant, and f_i is the i-th fundamental vibrational frequency of the molecule with degeneracy g_i.

When identifying $(2\pi^2/\rho v_0^3)\eta_{v,r}(0)$ with the low-frequency vibrational contribution to sound absorption, that is when assuming

$$(\alpha/f^2)_v = (\alpha/f^2)_{v,nr} + (\alpha/f^2)_{vib} \tag{21}$$

the absorption due to a *single* relaxation process characterised by an *energy relaxation time* τ_e (and $2\pi f \tau_e \ll 1$) is given by

$$(\alpha/f^2)_{\text{vib}} = 2\pi^2 \tau_e v_0^{-1} \frac{(C_P - C_V)C_i}{C_P C_V} \tag{22}$$

with

$$\tau_e = \tau_v/(1 - C_i/C_V) \tag{23}$$

The formalism presented above may be extended to include *multiple* vibrational relaxation processes.[15–17,32,33,41,42] In this case, however, for the calculation of the appropriate $C_{i,\text{PE}}$ only a subset of the normal modes of the molecule has to be used.

We conclude this section by showing the explicit expressions for the full width at half height (FWHH) of the two Brillouin peaks, $\Delta\omega_B(\text{FWHH})$, and of the unshifted Rayleigh peak, $\Delta\omega_R(\text{FWHH})$:

11.3.1 Brillouin peaks

The width (FWHH) of each of the two peaks positioned at frequencies $(c/\lambda_0 + \omega_B/2\pi)$ and $(c/\lambda_0 - \omega_B/2\pi)$, respectively, is given by

$$\Delta\omega_B(\text{FWHH}) = 2\Gamma_B k^2 \tag{24}$$

$$\Gamma_B = \frac{1}{2\rho}\left[\frac{4}{3}\eta_s + \eta_v + \frac{lM}{C_P}(\kappa - 1)\right] \tag{25}$$

The quantity $D_v = \rho^{-1}\left(\frac{4}{3}\eta_s + \eta_v\right)$ is frequently called the longitudinal kinematic viscosity, and $D_T = lM/\rho C_P$ is known as the thermal diffusivity. Thus, Equation (25) may be written in a more compact form as

$$\Gamma_B = \frac{1}{2}[D_v + D_T(\kappa - 1)] \tag{26}$$

11.3.2 Rayleigh peak

The Rayleigh peak is positioned at frequency c/λ_0, and its width is given by

$$\Delta\omega_R(\text{FWHH}) = 2D_T k^2 \tag{27}$$

Experimental determination of the width of the central peak requires very high resolution, since $\Delta\omega_R(\text{FWHH})$ is usually of the order of 10 MHz.

11.4 Selected Results and Discussion

For normal fluids, quite a number of Brillouin scattering experiments have been reported in the literature, and have been discussed with special attention to the extraction of useful information about the hydrodynamic parameters, such as hypersound speed, transport coefficients, and relaxing heat capacity. For instance, Figure 11.3 shows the *dispersion* of the speed of sound in toluene at 293.15 K and 0.1 MPa.[30] We supplemented our own data with results reported by Oda *et al.*,[43] Takagi and Negishi,[44] Levy and D'Arrigo,[45] and Fleury and Chiao,[46] and then subjected the combined data set to least-squares smoothing according to

$$v_H(f)^2 = v_0^2 \left[1 + \frac{\omega^2 \tau_v^2}{1 + \omega^2 \tau_v^2} \frac{(C_P - C_V)C_i}{(C_V - C_i)C_P} \right] \tag{28}$$

As can be seen, the hypersound speed $v_H(f)$ at $f = 8.27$ GHz is *larger* by about 77 m s^{-1} than the thermodynamic sound speed at low frequencies, $v_0 = 1328$ m s^{-1}. Combining these results with our low-frequency data on sound speed and sound attenuation,[30] when using the MOUETA approach at 293.15 K we obtained $\tau_e = (39 \pm 6)$ ps and $C_i = 51$ J K^{-1} mol^{-1} (MOUXI yields essentially the same results). We note that the vibrational Planck-Einstein heat capacity $C_{i,PE} = 68.6$ J K^{-1} mol^{-1} is noticeably larger (corrections for anharmonicity contributions are negligibly small[47]). Qualitatively similar results have been reported by Levy and D'Arrigo.[45] Presumably, some of the vibrational modes of toluene relax at a much higher frequency, say, at 10 to 20 GHz. We note that in a recent paper, Rubio *et al.*[48] suggest that the relaxation process involves all but the lowest vibrational mode (*i.e.* that of 15 cm^{-1} assigned to the hindered rotation of the methyl group), though with a much smaller relaxation time.

Several groups have investigated whether the vibrational heat capacity of liquid benzene relaxes with a single relaxation frequency or not. From ultrasonics and Brillouin scattering studies, Nichols *et al.*[17] concluded that only a *double* relaxation of the vibrational heat capacity is consistent with experimental results. Specifically, at 298.15 K the heat capacity associated with the lowest vibrational mode (404 cm^{-1}, $g = 2$) relaxes with a single relaxation time $\tau_{v,2} = 36$ ps and a relaxation frequency of 4.4 GHz, while the *remaining* vibrational heat capacity relaxes with a second, longer relaxation time $\tau_{v,1} = 298$ ps and a relaxation frequency of 0.53 GHz. This conclusion agrees with those of other groups,[41,42,49–52] though the numerical values of the extracted relaxation times and relaxation frequencies differ somewhat (see also Eastman *et al.*[53]). We note that with increasing temperature the low-frequency relaxation broadens so that it may be more correct to describe the entire relaxation not any more as a double relaxation but as a broadened effectively single one.[29] A comparison of experimental results of C_i with values $C_{i,PE}$ calculated *via* Equation (20) is provided by Figure 11.4:[29] an effective single-step relaxation process was assumed, which results in only a relatively small error in v_∞.

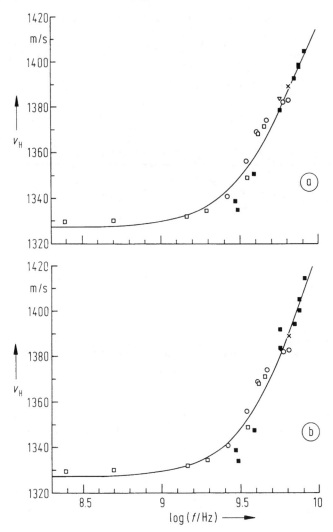

Figure 11.3 Dispersion of the speed of hypersound v_H in liquid toluene at 293.15 K and 0.1 MPa. The symbols (circles, squares, *etc.*) represent experimental results obtained either by us[30] (■) or taken from literature,[43-46] while the solid curves were obtained by least-squares fitting, according to Equation (28), of the combined data set. Our results were obtained by using two different, though in principle equivalent, theories suggested by Mountain: (a) MOUETA,[11] and (b) MOUXI.[12] Here, only MOUETA is presented in some detail.

The vibrational frequencies were taken from Shimanouchi.[54] Agreement seems to improve with increasing temperature.

This relaxation pattern of observing simultaneous relaxation of all but the energetically lowest vibrational mode(s) is by no means confined to benzene or

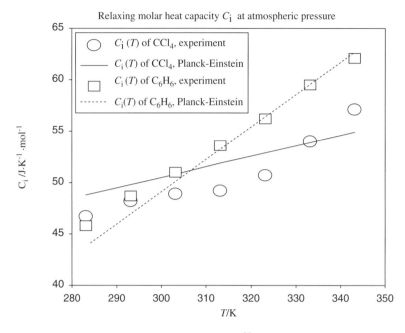

Figure 11.4 Comparison of experimental results[29] for the relaxing vibrational heat capacity C_i with values calculated *via* the Planck-Einstein Equation (20). Vibrational frequencies were taken from Shimanouchi.[54]

toluene, but has been also reported, for instance, for dichloromethane,[55,56] chloroform,[57] for several tetrachlorides[58-61] and for the three isomeric xylenes[62] (*ortho-, meta-* and *para*-dimethylbenzenes). Specifically, the results reported by Rubio *et al.*[62] on the xylenes suggest that the thermally excited low-energy vibrational modes do not contribute to the purely vibrational relaxation process. For *o*-xylene all modes are involved except the five of lowest energy, for *m*-xylene the four lowest do not participate, and for *p*-xylene the three lowest do not contribute. Relaxation times are similar to those found for toluene and are an order of magnitude smaller than those found for benzene.

In contradistinction to benzene, very few Brillouin experiments have been reported for heterocycles, such as pyridine. It has a molecular structure similar to that of benzene, and the interesting question whether the vibrational relaxation in pyridine is single or multiple was resolved by Takagi and Negishi[63] by measuring the speed of sound over a wide frequency range (at 293.15 K and 303.15 K): the speed of ultrasound at 3 MHz was determined with a pulse-echo-overlap apparatus; in the frequency range 70 MHz through 700 MHz both speed and absorption of ultrasound were measured with a high-resolution Bragg reflection method;[64-66] and the hypersound speed in the frequency range 3 through 7 GHz was determined by Brillouin scattering. Indeed, the same type of vibrational double relaxation as in benzene is found with relaxation frequencies close to those in benzene: the heat capacity associated with the lowest ($374\,cm^{-1}$) and the second lowest ($405\,cm^{-1}$) vibrational modes[67] relaxes in the hypersonic range at 3.9 GHz (at 293.15 K), while the heat capacity associated with *all* modes above, including the third lowest mode ($605\,cm^{-1}$), relaxes in the ultrasonic range at 762 MHz.

The Brillouin spectra of four liquid halogenated benzenes, that is of C_6H_5F, C_6H_5Cl, C_6H_5Br and C_6H_5I, have been investigated at 293.15 K, 298.15 K and 303.15 K by Inoue.[68] On the basis of relaxation caused by translational-vibrational energy exchange only, the speed-of-sound dispersion data could be analysed by assuming a single relaxation step: at 298.15 K, the relaxation frequency is 5.1 GHz for C_6H_5F, 6.3 GHz for C_6H_5Cl, 7.1 GHz for C_6H_5Br, and 3.3 GHz for C_6H_5I. Apparently, the dispersion in C_6H_5F is due to the vibrational relaxation of all vibrational modes, that in C_6H_5Cl and C_6H_5Br (where the energy of the lowest mode is *lower* than k_BT) is due to vibrational relaxation of all vibrational modes but the lowest, and that in C_6H_5I is due to vibrational relaxation of all but the three lowest vibrational modes.

Hypersonic sound attenuation in liquid benzonitrile, C_6H_5CN, has been measured between 298.75 K and 365.95 K by Goodman and Whittenburg.[69] Through careful analysis of their results, they concluded that (I) orientational relaxation of the shear viscosity is not important in the hypersonic frequency region covered, and (II) that any structural contribution to the bulk viscosity is very small. Thus they assigned the relaxation mechanism in benzonitrile to be due to the relaxation of the bulk viscosity due to relaxation of the heat capacity, which is consistent with what has been observed in other substituted benzenes. For Brillouin scattering experiments on acetonitrile between 278.15 K and 333.15 K, see Sassi *et al.*[70]

Sulfur dioxide, SO_2, is a triatomic, nonlinear (the bond angle $\varphi_{O=S=O}$ is 119.5°), dipolar molecule ($\mu = 5.44\times10^{-30}\,C\,m$). Its three fundamental eigenmodes are the bending mode of $518\,cm^{-1}$, the symmetric stretching mode of $1152\,cm^{-1}$, and the antisymmetric stretching mode of $1362\,cm^{-1}$. Most interestingly, liquid SO_2 shows a *clearly resolved double* vibrational relaxation process.[32,33,71,72] So far, such a well-separated double relaxational behaviour in the liquid state has *only* been found in liquid sulfur dioxide. At ambient temperatures and orthobaric pressure, the first relaxation step with a relaxation frequency of $f_1 = 22.5\,MHz$, is associated with the two stretching modes, the second relaxation step with a relaxation frequency of $f_2 = 1.6\,GHz$ is associated with the bending mode. The relaxing heat capacity extracted from our experimental results is in good agreement with the Planck-Einstein value. Figure 11.5 shows an experimental Brillouin spectrum of this interesting substance.[73]

For normal liquids without hydrogen bonds, where the molecules are closely packed, repulsive interactions not only determine, by and large, their structure[74,75] but play also a decisive role in the exchange of energy between external modes, *i.e.* translation and (external) rotation, and internal modes, *i.e.* vibration and/or internal rotation. Thus, besides temperature, the use of hydrostatic pressure as an experimental variable is indispensable for the study of static as well as dynamic properties of liquids. We note, however, that most of the

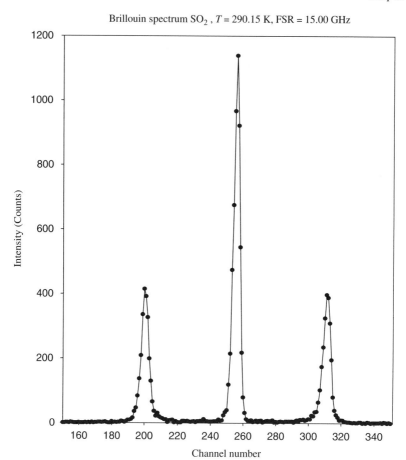

Brillouin spectrum SO_2, $T = 290.15$ K, FSR = 15.00 GHz

Figure 11.5 Experimental Rayleigh–Brillouin spectrum of liquid sulfur dioxide at $T = 290.15$ K and saturation pressure.[73] The free spectral range (FSR) was 15.00 GHz. Each channel number corresponds to a frequency interval of 62.58 MHz.

Brillouin scattering experiments on liquids have been performed at atmospheric pressure, with only a few at elevated pressures.[28,31,76–88] Unfortunately, not all the authors have analysed their results to gain insight into the associated dispersion behaviour/molecular relaxation processes.

In his seminal paper on the pressure dependence of the vibrational relaxation time in liquid tetrachloromethane, Sedlacek[76] used Brillouin scattering measurements at 298.15 K for pressures up to 1250 bar to determine $\tau_v(P)$ and $\tau_e(P)$ (for a complementary low-pressure study over a wide temperature range, see Asenbaum and Sedlacek[89]). By using the *isolated binary collision* (IBC) theory in conjunction with the moving-wall cell model[90,91] he was able to extract the *effective* collision diameter σ_{eff} of the CCl_4 molecule. In this theory, the relaxation time characterising vibrational deactivation is described by a factor due to the probability p for collisional transfer of energy per collision between the external and the internal (vibrational) modes, and a factor due to the collision frequency Z (thus $1/Z$ is the time between collisions):

$$\tau_e = \frac{1}{pZ} \tag{29}$$

This *ansatz* separates the relaxation dynamics contained in p (and assumed to be binary) and the translational dynamics contained in Z. Semi-empirically, one may use the approximation

$$Z = \frac{\langle u \rangle}{L_{free}} \tag{30}$$

where $\langle u \rangle = (8RT/M\pi)^{1/2}$ is the average molecular speed, and L_{free} is the mean free path in the liquid:

$$L_{free} = d - \sigma_{eff} = a(M/\rho N_A)^{1/3} - \sigma_{eff} \tag{31}$$

where $d = a(M/\rho N_A)^{1/3}$ is the mean centre-of-mass distance of the molecules, and a is a geometrical factor (for instance, $a = 2^{1/6}$ for the face-centred cubic lattice). We note, however, that the evaluation of L_{free} is by no means clear-cut, since in liquids the volume of the molecules themselves is a considerable part of the molar volume. That is to say, $y = N_A \sigma_{eff}^3 \pi/6V$ is of the order 0.5.[92,93] Thus, the value one obtains for the "free volume", and hence for L_{free}, depends upon the theory used. When using Equation (31), the energy relaxation time is given by

$$\tau_e = \frac{a(M/\rho N_A)^{1/3} - \sigma_{eff}}{\langle u \rangle p} \tag{32}$$

For the estimation of the collisional transfer probability in the liquid phase several approaches have been proposed, and the original articles should be consulted for details. The relaxation time $\tau_v(298.15\,\text{K},\ 1\,\text{bar}) = (68 \pm 5)\,\text{ps}$ is in excellent agreement with the result obtained by Stegemann *et al.*,[94] $\tau_e(298.15\,\text{K},\ 1\,\text{bar}) = (149 \pm 10)\,\text{ps}$, and the effective collision diameter $\sigma_{eff} = (5.31 \pm 0.04) \cdot 10^{-10}\,\text{m}$ is in good agreement with the diameter given by Wilhelm and Battino:[92] $\sigma = 5.37 \cdot 10^{-10}\,\text{m}$. Further improvements have to incorporate a temperature-dependent collision diameter as given by Wilhelm.[93] Because of the availability of more high-pressure data on auxiliary quantities as well as improved experimental equipment, Asenbaum and Hochheimer[79] determined Brillouin spectra of tetrachloromethane at pressures from 1 bar to 1500 bar and at three temperatures (298 K, 323 K and 348 K). Using the NC method,[15–17] they obtained the relaxation time τ_v of the bulk viscosity and the energy relaxation time τ_e [*via* Equation (23)]. At constant temperature, τ_e decreases with increasing density, *e.g.* at 298 K from about 145 ps at 1 bar to 105 ps at 1500 bar. At constant density, τ_e decreases with increasing temperature. From τ_e, using Equation (29) with a collision rate expression for hard spheres,[95] they obtained for the collision diameter $\sigma_{\text{eff}} = (5.24 \pm 0.05) \cdot 10^{-10}$ m for all temperatures.

For many years we have used various versions of the IBC model for discussing ultrasonic and hypersonic data, and so have many others, despite some critical remarks.[96–102] As Herzfeld[98] has successfully argued, some, if not the majority of these are based on semantic problems. Most of the energy transfer is due to collisions with energies several times larger than $k_B T$ (low-energy collisions are very ineffective). Therefore the (average) number of collisions per second should be replaced by the number of *high-speed* collisions per second, and since they are relatively rare, the time interval between them is long compared to that between every collision and to the period of internal vibration. As Zwanzig finally phrased it,[100] interference terms are actually unimportant in most cases, and the interactions may then be treated in a pairwise fashion. Thus the IBC approximation may be used with confidence for the description of vibrational energy relaxation in liquids.

Hypersound speeds in carbon disulfide, acetone and benzene were measured at 300.15 K in the pressure range from 1 bar to 1000 bar by Stith *et al.*[77] For benzene, when comparing their hypersound speeds at pressures up to 690 bar with ultrasonic speeds v_0 determined at high pressure,[103–105] they concluded that the dispersion first increases with increasing pressure and then decreases at high pressures, and that complete vibrational relaxation does not take place above 400 bar. This somewhat surprising result was challenged by Medina and O'Shea[78] who reported hypersonic speeds in benzene at five temperatures ranging from 298.45 K to 348.95 K and at pressures up to 3280 bar. In contrast to the results of Stith *et al.*, for all temperatures they observed a slight increase of sound dispersion with increasing pressure. Their study did not indicate any departure from complete vibrational relaxation at high pressures. The matter was finally settled through the careful study of Asenbaum and Hochheimer:[80] (I) The sound dispersion v_∞/v_0 due to the relaxation of the vibrational heat capacity was found to be nearly density independent, in agreement with the results of Medina and Shea;[78] (II) assuming that at high pressures the smaller relaxation time corresponds to the lowest vibrational mode, and the larger relaxation time represents all modes but the lowest (in complete analogy to the atmospheric pressure results), they used the more sophisticated NC approach[17] in the evaluation of their experimental Brillouin spectra. Within the limits of error, complete vibrational relaxation takes place over the entire pressure range.

As a complementary study to our low-pressure Brillouin study on toluene between 293.15 K and 313.15 K,[30] we have measured Brillouin spectra of liquid toluene at 303.15 K and at pressures from 1 bar to 1625 bar.[28,31] Using reliable literature data at low pressure for the density[106] and the heat capacity at constant pressure[107] in conjunction with high-pressure speed-of-sound data at low frequency (2 MHz),[108] high-pressure densities $\rho(T,P)$ were derived by a method previously described by Davis and Gordon[109] and, independently, by Vedam and Holton.[110] Specifically, Muringer *et al.*[108] used a successive integration algorithm based on

$$\rho(T, P) = \rho(T, P_{\text{ref}}) + \int_{P_{\text{ref}}}^{P} v_0^{-2} \mathrm{d}P + TM \int_{P_{\text{ref}}}^{P} \alpha_P^2 C_P^{-1} \mathrm{d}P \tag{33}$$

with the reference pressure $P_{\text{ref}} = 1$ bar. The remaining thermodynamic quantities α_P, β_T, C_V, $(\partial \alpha_P / \partial T)_P$, $(\partial C_P / \partial P)_T$ are easily obtained from standard relations, such as indicated by Equations (12) and (13), or by

$$\left(\frac{\partial C_P}{\partial P}\right)_T = -TV\left[\alpha_P^2 + \left(\frac{\partial \alpha_P}{\partial T}\right)_P\right] \tag{34}$$

thereby providing a consistent data set.

For the pressure dependence of the shear viscosity $\eta_s(T,P)$ at constant temperature we used a Tait-type representation of Kashiwagi and Makita's data:[111]

$$\ln[\eta_s(T,P)/\eta_s(T,P_{\text{ref}})] = E\ln[(D + P)/(D + P_{\text{ref}})] \tag{35}$$

where $P_{\text{ref}} = 1$ bar. Liquid toluene has been adopted by IUPAC as a standard reference material for thermal conductivity l.[112] It is for this reason, that for toluene a number of modern, comprehensive studies of thermal conductivity exist. Appropriate interpolation of the results of Kashiwagi et al.[113] measured in the range 273.15 K to 373.15 K and for pressures up to 2500 bar yielded the 303.15 isotherm.

Figure 11.6 shows the low-frequency speed of ultrasound v_0 and the speed of hypersound v_H of toluene as a function of pressure at 303.15 K. The experimental error limit of v_H is estimated to be $\pm 5\,\mathrm{m\,s^{-1}}$, that of v_0 is one order of magnitude less. We note that due to the pressure dependence of the refractive index, each hypersonic speed value refers to a slightly different wave vector, which may easily be calculated *via* Equation (2), which in turn yields the corresponding hypersound frequency $f = \omega_B/2\pi$ *via* Equation (1). Agreement with the recent results of Koski et al.[37] (they give a graphical presentation only!) is satisfactory.

Figure 11.7 shows the pressure dependence of the low-frequency bulk viscosity $\eta_v(0)$ and of the nonrelaxing bulk viscosity $\eta_{v,nr}$ of toluene at 303.15 K. The pressure dependence of the relaxation times τ_v and τ_e is presented in Figure 11.8. The experimental imprecision of τ_v is estimated to be $\pm 4\,\mathrm{ps}$, and τ_e should be reliable within $\pm 8\,\mathrm{ps}$.

The low-pressure limiting values of the relaxation times are in good agreement with those found in our low-pressure study[30] and with the results of Levy and D'Arrigo.[45] The weak pressure dependence of τ_v and τ_e is of particular note: it is in marked contradistinction to the results obtained from the high-pressure Brillouin studies on tetrachloromethane and benzene.[79,80]

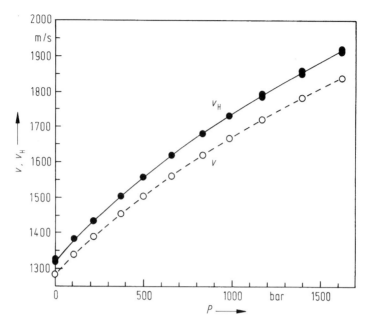

Figure 11.6 Pressure dependence of the experimental low-frequency speed of ultra-sound v_0 (○),[108] and the speed of hypersound v_H (●)[28] of toluene at 303.15 K. $v_H(303.15\,\text{K}, 1\,\text{bar}, 5.454\,\text{GHz}) = 1324.0\,\text{m s}^{-1}$, and $v_H(303.15\,\text{K}, 1625\,\text{bar}, 8.163\,\text{GHz}) = 1915.0\,\text{m s}^{-1}$.

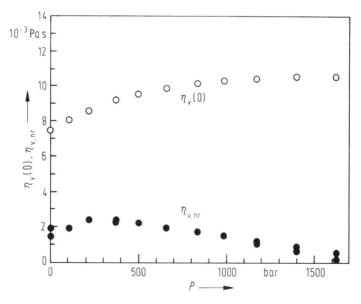

Figure 11.7 Pressure dependence of the low-frequency bulk viscosity $\eta_v(0)$, (○), and the nonrelaxing bulk viscosity $\eta_{v,\text{nr}}$, (●), of toluene at 303.15 K.[28]

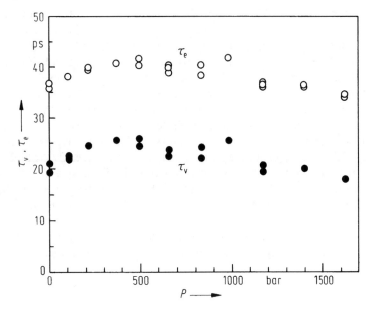

Figure 11.8 Pressure dependence of the relaxation times τ_v, ($\bullet$), and τ_e, ($\circ$), of toluene at 303.15 K.[28]

High-pressure Brillouin scattering studies on *simple* fluids, such as argon, are quite scarce. As pointed out by Fleury and Boon[114] this is rather surprising, since such studies may disclose some nonclassical behaviour (*i.e.* departure from the predictions of the Navier-Stokes equations) for very-high-frequency sound waves[115-117] due to the onset of frequency dependence of the dense fluid's transport coefficients. The dispersion phenomena should be most important at frequencies of the order of the reciprocal collision time, *i.e.* at 10^{12} Hz to 10^{13} Hz for dense simple fluids, which frequency range is not accessible for Brillouin experiments. However, as yet no model for estimating the frequency above which these new effects are expected to become more easily measurable is known. As a possible candidate, Fleury and Boon[114] suggested the *Lucas relaxation frequency*[23]

$$f_L = \frac{\rho v_H^2}{2\pi \left(\frac{4}{3}\eta_s + \eta_v\right)} \tag{36}$$

The speed of hypersound (*ca.* 3 GHz) in liquid argon was measured by Fleury and Boon[114] along the vapour-liquid equilibrium curve between 85 K and 100 K (that is at pressures smaller than 3.3 bar): v_H decreases linearly from 850 m s^{-1} at 85 K to 742 m s^{-1} at 100 K and is uniformly *lower*, by about 0.4 %, than the low-frequency (1 MHz) ultrasound speeds obtained under the same thermodynamic conditions.[118] This observation is in qualitative accord with theoretical predictions of a *negative dispersion* of the speed of sound at very high frequencies. On the other hand, Carraresi *et al.*[119] did *not* find any dispersion of the sound speed up to frequencies of 4 GHz and up to pressures of 2600 bar at 298 K. Because of the importance of the issue, Hochheimer *et al.*[87] undertook a high-pressure Brillouin scattering study of dense argon (and nitrogen) between 157.7 K and 296.8 K and pressures up to 5000 bar. Their results show negative speed dispersion, thus confirming the results of Fleury and Boon, but contradicting the findings of Carraresi *et al.* Apparently, the dispersion increases with increasing pressure. Recently, hypersound speeds in dense fluid argon have been reported by Jia *et al.*[88] between 293 K and 503 K, and for pressures up to the respective solidification pressure (at 293 K the solidification pressure of argon is about 13 kbar). Since only (very small) graphical representations of their results were provided, no quantitative conclusions can be drawn.

From Equations (1) and (2), the *angular* frequency shift of the two Brillouin peaks is related to the hypersonic sound speed in the medium by

$$\omega_B = \pm \frac{4\pi n v_H(f)}{\lambda_0} \sin\frac{\theta}{2} \tag{37}$$

Thus, determination of $v_H(f)$ as a function of temperature and pressure requires the knowledge of the temperature as well as the pressure dependence of the refractive index $n(T,P)$ at the respective laser frequency. In particular the pressure dependence is a formidable problem by itself, as indicated by the paucity of reliable data.[28] Here we present a brief list of selected articles presenting experimental results for the isothermal piezo-optic coefficient $(\partial n/\partial P)_T$ of liquids, that is references 120 through 135, the most comprehensive contribution undoubtedly coming from Vedam *et al.*[127–132] The pressure range extended up to 14 kbar , though for some liquids the maximum pressure was determined by the freezing of the liquid at the stated temperature (temperatures were either 293.15 K, or 298.15 K, or 303.15 K, and the wavelength was $\lambda_0 = 546.1$ nm). In order to get around this problem, special high-pressure cells have been designed[82–84,86,136,137] so that for a 90° scattering geometry, Equation (37) reduces to

$$\omega_B = \pm 2\sqrt{2}\pi\, v_H(f)/\lambda_0 \tag{38}$$

11.5 Concluding Remarks

The invention of the laser (1960) has greatly stimulated interest in optics in general, and in the scattering of laser light from liquids in particular. It was Raymond Mountain with his seminal article, *Spectral Distribution of Scattered Light in a Simple Fluid*,[10] who really opened up the field for the experimentalists.

His approach for the extraction of information from the spectrum of light scattered by density fluctuations in a dense fluid is essentially phenomenological, connecting the intensity and frequency distribution of the scattered light (Brillouin doublet, unshifted Rayleigh peak) to thermodynamic and transport properties of the fluid. For molecular liquids, he subsequently extended the hydrodynamic fluctuation theory to include internal, vibrational degrees of freedom, which then accounts for the observed appearance of the unshifted broad spectral feature known as the Mountain peak.

As we have seen, Rayleigh-Brillouin spectra contain a wealth of information on important thermophysical quantities, such as the speed of hypersound, the bulk or volume viscosity, the thermal conductivity, the relaxing heat capacity, and the relaxation time. Of course, the study of the influence of temperature is interesting, but so is also the influence of pressure: more work in this area would be highly desirable.

In this contribution, we have omitted an account of Brillouin scattering in self-associating liquids, such as the alcohols, in liquids near the liquid-vapour critical point, and in binary liquid mixtures. Space requirements for this book precluded their inclusion.

References

1. L. D. Landau and E. M. Lifshitz, (a) *Statistical Physics*, Addison-Wesley, Reading, Mass., 1958. (b) *Fluid Mechanics*, Addison-Wesley, Reading, Mass., 1959.
2. B. J. Berne and R. Pecora, *Dynamic Light Scattering*, Wiley, New York, 1976.
3. J. B. Boon and S. Yip, *Molecular Hydrodynamics*, McGraw-Hill, New York, 1980.
4. J. G. Dil, *Rep. Prog. Phys.*, 1982, **45**, 285.
5. B. Chu, *Laser Light Scattering: Basic Principles and Practice*, 2nd Edition, Academic Press, San Diego, 1991.
6. U. Balucani and M. Zoppi, *Dynamics of the Liquid State*, Clarendon Press, Oxford, 1994.
7. J. M. Ortiz de Zaráte and J. V. Sengers, *Hydrodynamic Fluctuations in Fluids and Fluid Mixtures*, Elsevier, Amsterdam, 2006.
8. U. Bafile, F. Barocchi and E. Guarini, *Condens. Matter Phys.*, 2008, **11**, 107.
9. L. Brillouin, *Ann. Phys. (Paris)*, 1922, **17**, 88.
10. R. D. Mountain, *Rev. Mod. Phys.*, 1966, **38**, 205.
11. R. D. Mountain, *J. Res. Natl. Bur. Stand.*, 1966, **70A**, 207.
12. R. D. Mountain, *J. Res. Natl. Bur. Stand.*, 1968, **72A**, 95.
13. R. D. Mountain, *CRC Crit. Rev. Solid State Sci.*, 1970, **1**, 5.
14. R. D. Mountain, *Adv. Molec. Relax. Processes*, 1976, **9**, 225.
15. W. H. Nichols and E. F. Carome, *J. Chem. Phys.*, 1968, **49**, 1000.
16. E. F. Carome, W. H. Nichols, C. R. Kunsitis-Swyt and S. P. Singal, *J. Chem. Phys.*, 1968, **49**, 1013.

17. W. H. Nichols, C. R. Kunsitis-Swyt and S. P. Singal, *J. Chem. Phys.*, 1969, **51**, 5659.
18. R. C. Desai and R. Kapral, *Phys. Rev. A*, 1972, **6**, 2377.
19. W. S. Gornall, G. I. Stegeman, B. P. Stoicheff, R. H. Stolen and V. Volterra, *Phys. Rev. Lett.*, 1966, **17**, 297.
20. A. Asenbaum, E. Wilhelm and C. Pruner, *private communication.*
21. J. Rouch, C. C. Lai and S. -H. Chen, *J. Chem. Phys.*, 1976, **65**, 4016.
22. J. Rouch, C. C. Lai and S. -H. Chen, *J. Chem. Phys.*, 1977, **66**, 5031.
23. K. F. Herzfeld and T. A. Litovitz, *Absorption and Dispersion of Ultrasonic Waves,* Academic Press, New York, 1959.
24. A. B. Bhatia, *Ultrasonic Absorption,* Oxford at the Clarendon Press, London, 1967.
25. R. Zwanzig, *J. Chem. Phys.*, 1965, **43**, 714.
26. A. Asenbaum, *Z. Naturforsch.*, 1983, **38a**, 336. In this article, the position of Mountain's theory MOUETA[11] vis-à-vis his other theory MOUXI[12] and the translational hydrodynamics theory of Desai and Kapral[18] is discussed.
27. A. Asenbaum, *J. Molec. Liq.*, 1983, **26**, 11.
28. A. Asenbaum, E. Wilhelm and P. Soufi-Siavoch, *Acustica*, 1989, **68**, 131.
29. A. Asenbaum and E. Wilhelm, *Adv. Molec. Relax. Interact. Processes*, 1982, **22**, 187.
30. A. Asenbaum, P. Soufi-Siavoch and E. Wilhelm, *Acustica*, 1989, **67**, 284.
31. A. Asenbaum, P. Soufi-Siavoch, R. Aschauer and E. Wilhelm, in *Laser Materials and Laser Spectroscopy,* W. Zhijiang and Z. Zhiming, eds. World Scientific, Singapore, 1989, pp. 277–279.
32. M. Musso, F. Aliotta, C. Vasi, R. Aschauer, A. Asenbaum and E. Wilhelm, *J. Molec. Liq.*, 2004, **110**, 33.
33. A. Asenbaum, R. Aschauer, C. Theisen, T. Fritsch and E. Wilhelm, *J. Molec. Liq.*, 2007, **134**, 55.
34. (a) R. Zwanzig and R. D. Mountain, *J. Chem. Phys.*, 1965, **43**, 4464; (b) R. D. Mountain and R. Zwanzig, *J. Chem. Phys.*, 1966, **44**, 2777.
35. (a) J. R. Sandercock, *J. Sci. Instrum.*, 1976, **9**, 566; (b) J. R. Sandercock, in *Light Scattering in Solids III,* M. Cardona and G. Güntherodt, eds., Vol. 51 of *Topics in Applied Physics*, Springer, Berlin, 1982.
36. (a) H. Tanaka and T. Sonehara, *Rev. Sci. Instrum.*, 2002, **73**, 1998; (b) T. Sonehara and H. Tanaka, *Rev. Sci. Instrum.*, 2002, **73**, 263.
37. K. J. Koski, J. Müller, H. D. Hochheimer and J. L. Yarger, *Rev. Sci. Instrum.*, 2002, **73**, 1235.
38. H. Mori, *Progr. Theoret. Phys.*, 1965, **33**, 423.
39. R. Zwanzig, *Ann. Rev. Phys. Chem.*, 1965, **16**, 67.
40. P. Schofield, in *Specialist Reports – Statistical Mechanics,* Vol. II, The Chemical Society, London, 1975.
41. H. C. Lucas, D. A. Jackson and H. T. Pentecost, *Opt. Commun.*, 1970, **2**, 239.

42. K. Takagi, P. -K. Choi and K. Negishi, *Acustica*, 1976, **34**, 336.
43. K. Oda, R. Hayakowa and Y. Wada, *Jpn. J. Appl. Phys.*, 1973, **12**, 1326.
44. K. Takagi and K. Negishi, *Ultrasonics*, 1978, **16**, 259.
45. C. Levy and G. D'Arrigo, *Mol. Phys.*, 1983, **50**, 917.
46. P. A. Fleury and R. Y. Chiao, *J. Acoust. Soc. Am.*, 1966, **39**, 751.
47. J. A. Draeger, *J. Chem. Thermodyn.*, 1985, **17**, 263.
48. J. E. F. Rubio, V. G. Baonza, M. Taravillo, J. Núñez and M. Cáceres, *J. Chem. Phys.*, 2001, **115**, 4681.
49. J. L. Hunter, E. F. Carome, H. D. Dardy and J. A. Bucaro, *J. Acoust. Soc. Am.*, 1966, **40**, 313.
50. E. F. Carome and S. P. Singal, *J. Acoust. Soc. Am.*, 1967, **41**, 1371.
51. J. L. Hunter, W. H. Nichols and J. Haus, *Phys. Lett.*, 1971, **37A**, 127.
52. M. J. Cardamone, W. Rhodes, D. Egan and N. Spotts, *J. Opt. Soc. Am. B*, 1985, **2**, 1612.
53. D. P. Eastman, A. Hollinger, J. Kenemuth and D. H. Rank, *J. Chem. Phys.*, 1969, **50**, 1567.
54. T. Shimanouchi, *Tables of Molecular Vibrational Frequencies. Part 1*, National Bureau of Standards, 1967.
55. K. Takagi, P. K. Choi and K. Negishi, *J. Acoust. Soc. Am.*, 1977, **62**, 354.
56. M. J. Cardamone, W. Rhodes, R. Atkinson and M. Yeastedt, *J. Opt. Soc. Am. B*, 1984, **1**, 779.
57. M. J. Cardamone, W. Rhodes and M. Yeastedt, *J. Chem. Phys.*, 1984, **80**, 966.
58. F. Barocchi and R. Vallauri, *J. Chem. Phys.*, 1969, **51**, 10.
59. D. Samios and T. Dorfmüller, *Mol. Phys.*, 1980, **41**, 637.
60. M. J. Cardamone and W. Rhodes, *Chem. Phys. Lett.*, 1982, **87**, 403.
61. M. J. Cardamone, W. Rhodes and M. Yeastedt, *Chem. Phys. Lett.*, 1984, **105**, 237.
62. J. E. F. Rubio, M. Taravillo, V. G. Baonza, J. Núñez and M. Cáceres, *J. Chem. Phys.*, 2006, **124**, 014503.
63. K. Takagi and K. Negishi, *J. Acoust. Soc. Am.*, 1979, **65**, 86.
64. E. I. Gordon and M. G. Cohen, *Phys. Rev.*, 1967, **153**, 201.
65. K. Takagi and K. Negishi, *Jpn. J. Appl. Phys.*, 1975, **14**, 29.
66. K. Takagi and K. Negishi, *Jpn. J. Appl. Phys.*, 1975, **14**, 149.
67. J. P. McCullough, D. R. Douslin, J. F. Messerly, I. A. Hossenlopp, J. C. Kincheloe and G. Waddington, *J. Am. Chem. Soc.*, 1975, **79**, 4289.
68. N. Inoue, *J. Phys. D: Appl. Phys.*, 1980, **13**, 1699.
69. M. A. Goodman and S. L. Whittenburg, *J. Phys. Chem.*, 1984, **88**, 5653.
70. P. Sassi, G. Paliani and R. S. Cataliotti, *J. Chem. Phys.*, 1998, **108**, 10197.
71. R. Bass and J. Lamb, *Proc. Roy. Soc. (London) Ser. A*, 1958, **243**, 94.
72. D. B. Fenner, D. E. Bowen and M. P. Eastman, *J. Chem. Phys.*, 1979, **71**, 4849.
73. A. Asenbaum and E. Wilhelm, *unpublished results*.
74. J. A. Barker and D. Henderson, *Rev. Mod. Phys.*, 1976, **48**, 587.

75. F. Kohler, E. Wilhelm and H. Posch, *Adv. Molec. Relax. Processes*, 1976, **8**, 195.
76. M. Sedlacek, *Z. Naturforsch.*, 1974, **29a**, 1622.
77. J. H. Stith, L. M. Peterson, D. H. Rank and T. A. Wiggins, *J. Acoust. Soc. Am.*, 1974, **55**, 785.
78. F. D. Medina and D. C. O'Shea, *J. Chem. Phys.*, 1977, **66**, 1940.
79. A. Asenbaum and H. D. Hochheimer, *J. Chem. Phys.*, 1981, **74**, 1.
80. A. Asenbaum and H. D. Hochheimer, *Z. Naturforsch.*, 1983, **38a**, 980.
81. M. Grimsditch, P. Loubeyre and A. Polian, *Phys. Rev. B*, 1986, **33**, 7192.
82. H. Shimizu, S. Sasaki and T. Ishidate, *J. Chem. Phys.*, 1987, **86**, 7189.
83. H. B. Bohidar, *J. Appl. Phys.*, 1988, **64**, 1810.
84. H. Bohidar, T. Jøssang and J. Feder, *J. Phys. D: Appl. Phys.*, 1988, **21**, S53.
85. S. N. Tkachev and J. D. Bass, *J. Chem. Phys.*, 1996, **104**, 1059.
86. H. Shimizu, N. Nakashima and S. Sasaki, *Phys. Rev. B*, 1996, **53**, 111.
87. H. D. Hochheimer, K. Weishaupt and M. Takesada, *J. Chem. Phys.*, 1996, **105**, 374.
88. R. Jia, F. Li, M. Li, Q. Cui, Z. He, L. Wang, Q. Zhou, T. Cui, G. Zou, Y. Bi, S. Hong and F. Jing, *J. Chem. Phys.*, 2008, **129**, 154503.
89. A. Asenbaum and M. Sedlacek, *Adv. Molec. Relax. Interact. Processes*, 1978, **13**, 225.
90. T. A. Litovitz, *J. Chem. Phys.*, 1957, **26**, 469.
91. W. M. Madigosky and T. A. Litovitz, *J. Chem. Phys.*, 1961, **34**, 489.
92. E. Wilhelm and R. Battino, *J. Chem. Phys.*, 1971, **55**, 4012.
93. E. Wilhelm, *J. Chem. Phys.*, 1973, **58**, 3558.
94. G. I. A. Stegeman, W. S. Gornall, V. Volterra and B. P. Stoicheff, *J. Acoust. Soc. Am.*, 1971, **49**, 979.
95. T. Einwohner and B. J. Alder, *J. Chem. Phys.*, 1968, **49**, 1458.
96. M. Fixman, *J. Chem. Phys.*, 1961, **34**, 369.
97. R. Zwanzig, *J. Chem. Phys.*, 1961, **34**, 1931.
98. K. F. Herzfeld, *J. Chem. Phys.*, 1962, **36**, 3305.
99. R. E. Nettleton, *J. Chem. Phys.*, 1962, **36**, 2226.
100. R. Zwanzig, *J. Chem. Phys.*, 1962, **36**, 2227.
101. H. K. Shin and J. Keizer, *Chem. Phys. Lett.*, 1974, **27**, 611.
102. D. W. Oxtoby, *Mol. Phys.*, 1977, **34**, 987.
103. J. F. Mifsud and A. W. Nolle, *J. Acoust. Soc. Am.*, 1956, **28**, 469.
104. E. G. Richardson and R. I. Tait, *Philos. Mag.*, 1957, **2**, 441.
105. T. Makita and T. Takagi, *Rev. Phys. Chem. (Japan)*, 1968, **38**, 41.
106. American Petroleum Institute, *Selected Values of Physical and Thermodynamic Properties of Hydrocarbons and Related Compounds*, Carnegie Press, Pittsburgh, Pa, 1953.
107. J. K. Holzhauer and W. T. Ziegler, *J. Phys. Chem.*, 1975, **79**, 590.
108. M. J. P. Muringer, N. J. Trappeniers and S. N. Biswas, *Phys. Chem. Liq.*, 1985, **14**, 273.
109. L. A. Davis and R. B. Gordon, *J. Chem. Phys.*, 1967, **46**, 2650.
110. R. Vedam and G. Holton, *J. Acoust. Soc. Am.*, 1968, **43**, 108.

111. H. Kashiwagi and T. Makita, *Int. J. Thermophys.*, 1982, **3**, 289.
112. C. A. Nietro de Castro, S. F. Y. Li, A. Nagashima, R. D. Trengove and W. A. Wakeham, *J. Phys. Chem. Ref. Data*, 1986, **15**, 1073.
113. H. Kashiwagi, T. Hashimoto, Y. Tanaka, H. Kubota and T. Makita, *Int. J. Thermophys.*, 1982, **3**, 201.
114. P. A. Fleury and J. P. Boon, *Phys. Rev.*, 1969, **186**, 244.
115. N. S. Gillis and P. D. Puff, *Phys. Rev. Lett.*, 1966, **16**, 606.
116. H. L. Frisch, *Physics*, 1966, **2**, 209.
117. B. J. Berne, J. B. Boon and S. A. Rice, *J. Chem. Phys.*, 1967, **47**, 2283.
118. W. Van Dael, A. Van Itterbeek, A. Cops and J. Thoen, *Physica*, 1966, **32**, 611.
119. L. Carraresi, M. Celli and F. Barocchi, *Phys. Chem. Liq.*, 1993, **25**, 91.
120. R. M. Waxler and C. E. Weir, *J. Res. Natl. Bur. Stand.*, 1963, **67A**, 163.
121. R. M. Waxler, C. E. Weir and H. W. Schamp Jr., *J. Res. Natl. Bur. Stand.*, 1964, **68A**, 489.
122. D. J. Coumou, E. L. Mackor and J. Hijmans, *Trans. Faraday. Soc.*, 1964, **60**, 1539.
123. E. Reisler and H. Eisenberg, *J. Chem. Phys.*, 1965, **43**, 3875.
124. G. Cohen and H. Eisenberg, *J. Chem. Phys.*, 1965, **43**, 3881.
125. H. Eisenberg, *J. Chem. Phys.*, 1965, **43**, 3887.
126. R. Josephs and A. P. Minton, *J. Phys. Chem.*, 1971, **74**, 716.
127. K. Vedam and P. Limsuwan, *Phys. Rev. Lett.*, 1975, **35**, 1014.
128. K. Vedam and P. Limsuwan, *Rev. Sci. Instrum.*, 1977, **48**, 245.
129. K. Vedam and P. Limsuwan, *J. Chem. Phys.*, 1978, **69**, 4762.
130. K. Vedam and P. Limsuwan, *J. Chem. Phys.*, 1978, **69**, 4772.
131. K. Vedam and P. Limsuwan, *J. Appl. Phys.*, 1979, **50**, 1328.
132. C. C. Chen and K. Vedam, *J. Chem. Phys.*, 1980, **73**, 4577.
133. A. J. Richard, K. T. McCrickard and P. B. Fleming, *J. Chem. Thermodyn.*, 1979, **11**, 93.
134. A. J. Richard and P. B. Fleming, *J. Chem. Thermodyn.*, 1981, **13**, 863.
135. T. Takagi and H. Teranishi, *J. Chem. Eng. Data*, 1982, **27**, 16.
136. C. H. Whitfield, E. M. Brody and W. A. Bassett, *Rev. Sci. Instrum.*, 1976, **47**, 942.
137. H. Bohidar, T. Berland, T. Jøssang and J. Feder, *Rev. Sci. Instrum.*, 1987, **58**, 1422.

CHAPTER 12

Photothermal Techniques for Heat Capacities

JAN THOEN AND CHRIST GLORIEUX

Laboratorium voor Akoestiek en Thermische Fysica, Departement
Natuurkunde en Sterrenkunde, Katholieke Universiteit Leuven,
Celestijnenlaan 200 D, B-3001, Leuven, Belgium

12.1 Introduction

The photoacoustic effect was accidentally discovered by A. G. Bell in 1880
during a series of experiments trying to transmit speech by modulating light
waves.[1] He found that a solid material subjected to intense modulated light
emitted sound of the same frequency as the modulation frequency. His discovery resulted in a series of new and exciting investigations by many important
scientists of that period.[2-7] It was soon discovered that the phenomenon was
not only limited to solids but also occurred in liquids and gases and resulted
from periodic heating (caused by absorption of the radiation) of the sample
under investigation. In spite of several theoretical attempts at that time the
phenomenon could only be explained for gaseous samples in a vessel. It was not
until almost a century later that the photoacoustic effect in condensed
matter was put on a correct theoretical basis in 1976 by A. Rosencwaig and
A. Gersho.[8] Since that time the fields of photoacoustics and the related photothermal field (direct detection of the temperature modulation in the sample)
have grown enormously. A multitude of ways of generating the effects has
emerged using all kinds of radiation, from laser light to particle beams. Like-
wise, the diversity in methods for the detection of the thermal or acoustical
waves has increased dramatically. One of the reasons for the popularity of this

Heat Capacities: Liquids, Solutions and Vapours
Edited by Emmerich Wilhelm and Trevor M. Letcher
© The Royal Society of Chemistry 2010
Published by the Royal Society of Chemistry, www.rsc.org

field is the wide applicability of these techniques. Also the development of lasers as convenient and powerful sources for localized energy deposition in samples has contributed to the success of the field. Applications include trace gas detection, absorption spectroscopy, generation of ultrasound (laser ultrasonics), measurements of static and dynamic thermal quantities, and thermal and acoustic waves for nondestructive evaluation of a variety of materials and structures. The earlier applications (and also those in the mid-1970s) were directed towards the absorption processes and called photoacoustic spectroscopy because the periodic heating effect (caused by the absorption of the radiation) was detected acoustically. In many subsequent applications the detection of the temperature oscillations in samples was non-acoustic (e.g. by photodeflection[9] or photopyroelectric[10]) and one speaks more appropriately of photothermal phenomena or techniques.

The photothermal techniques are ac techniques and the measured photothermal signal is characterized by its amplitude and phase. Both quantities contain information on the thermal parameters of the sample and by appropriate lock-in detection one is able to achieve an adequate signal to noise ratio by introducing only millikelvin temperature variations at the sample surface and correspondingly small dc temperature changes or gradients (by the absorption of the ac input of the optical radiation energy) in the sample. This makes the technique very attractive for high resolution thermal studies of phase transitions.

In this chapter we will focus our attention on the calorimetric applicability of photoacoustic and photothermal techniques to determine heat capacities of liquids. We will consider two cases, a standard photoacoustic microphone detection technique and different variants of photopyroelectric detection techniques. We will present examples for liquid phase transitions illustrating the high-resolution capabilities of these techniques.

12.2 Photoacoustic Technique

12.2.1 Gas Microphone Configuration and the Standard Model

In a photoacoustic experiment with gas microphone detection, the condensed matter sample (liquid or solid) is placed in an air tight volume with a gas (such as air) and also containing a microphone (see Figure 12.1). The photoacoustic signal (detected by the microphone) is caused by that fraction of the modulated light (entering the cell through a transparent window) that is converted to heat in the sample via non-radiative de-excitation processes. The periodic absorption of optical energy produces a fast decaying thermal wave in the sample but also generates an acoustical wave by the local contraction and expansion of the sample due to the periodic heating. In a photoacoustical experiment where the thermal wave dominates, by properly choosing the experimental configuration, the one-dimensional model of Rosencwaig and Gersho[8] (RG) can be used for the typical setup in Figure 12.1.

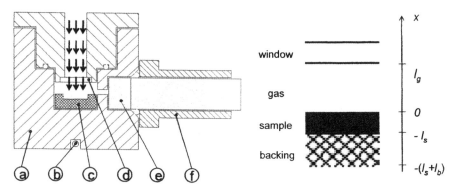

Figure 12.1 Left: schematic representation of a photoacoustic cell, (a): cell body, (b): thermometer, (c): liquid holder, (d): transparent window, (e): microphone, (f): microphone holder. Right: schematic cross-sectional view of a simple cylindrical photoacoustic cell with coordinates relevant for the standard one-dimensional model of Rosencwaig and Gersho.

In the RG model one assumes the incident light intensity to be uniform across the sample, resulting in the amplitude of the thermal source depending only on the depth in the sample and decaying exponentially with the distance from the surface. The right-hand part of Figure 12.1 gives a schematic cross-sectional view of a photoacoustic cell with relevant coordinates for the one-dimensional model are given. The microphone in contact with the gas in front of the sample has been omitted in the right-hand part. In the standard model it is also assumed that only the sample absorbs the modulated light, thus there is no heat production in the gas or the backing. For the properties of the gas, sample and backing we will further use the indices g, s and b, respectively. For harmonically modulated incident monochromatic light, at wavelength λ the following result holds for the power density produced at any point x in the sample due to light absorbed at that point.

$$H(x) = (1/2)\beta_s(\lambda)I_0(\lambda)(1 + e^{i\omega t})e^{\beta_s(\lambda)x} \qquad (1)$$

where I_0 is the light intensity, β_s the optical absorption coefficient of the sample and $\omega = 2\pi f$ the angular modulation frequency. The values of x in Figure 12.1 and Equation (1) range from $x = 0$ at the gas sample interface to $x = -l_s$ at the sample backing interface. The backing extends to $x = -(l_s + l_b)$ and the gas column to l_g. It is further assumed that l_g is much smaller than the wavelength of sound in the gas at the frequency $f = \omega/2\pi$ and that convection and radiation effects are negligible.

A solution for the temperature distribution in the cell can be obtained on the basis of thermal diffusion equations for the gas, the sample and the backing with the assumption of continuity of temperature and heat flux at the interfaces. Important for the production of the acoustical signal in the gas is the value of the (complex) temperature $\theta(0,t)$ at the gas-sample interface ($x = 0$), for which one arrives at the following expression[8]:

$$\theta(0, t) = F_0 + \theta_0 e^{i\omega t} \tag{2}$$

with

$$\theta_0 = E\left[\frac{(r-1)(b+1)e^{\sigma_s l_s} - (r+1)(b-1)e^{-\sigma_s l_s} + 2(b-r)e^{-\beta_s l_s}}{(g+1)(b+1)e^{\sigma_s l_s} - (g-1)(b-1)e^{-\sigma_s l_s}}\right] \tag{3}$$

In Equation (2) F_0 is the steady state increase of the temperature at $x = 0$. In Equation (3) $\sigma_s = (1+i)\alpha_s$, where α_s is the thermal diffusion coefficient equal to the inverse of the thermal diffusion length of the sample $\mu_s = (2\alpha_s/\omega)^{1/2}$ with the thermal diffusion coefficient $\alpha_s = \kappa/\rho C$ the ratio of the thermal conductivity κ and the product ρC of the density ρ and the specific heat capacity C. One further has $E = \eta_s \beta_s I_0/[2\kappa_s(\beta_s^2 - \sigma_s^2)]$, $r = \beta_s/\sigma_s$, $b = \kappa_b a_b/(\kappa_s a_s) = e_b/e_s$ and $g = \kappa_g a_g/(\kappa_s a_s) = e_g/e_s$, where $e = (\rho\kappa C)^{1/2}$ is the thermal effusivity and, η_s the efficiency of conversion of the optical energy into heat.

With a gas microphone detection photoacoustic setup the signal is caused by heat transfer at the sample-gas interface to a thin layer of gas, of the order of one thermal wavelength $2\pi\mu_g$. Indeed, at this distance from the interface the temperature variation in the gas is less than 0.5 % of that at the interface. This boundary layer of gas can be considered as a thermal piston creating the acoustical signal detected by the microphone in the cell wall. The pressure variation, δP, in the gas is obtained by assuming an adiabatic response to the piston,[8,11] resulting in:

$$\delta P = Q \exp(i\omega t - i\pi/4) \tag{4}$$

where

$$Q = \gamma_g P_0 \theta_0 / \left[(2)^{1/2} T_0 l_g a_g\right] = q e^{-i\psi} \tag{5}$$

with P_0 and T_0 the cell pressure and temperature and γ_g the ratio of the specific heats at constant pressure and constant volume of the gas. Because θ_0 is a complex quantity the acoustic signal can be written with an amplitude q and a phase ψ.

On the basis of Equations (3) to (5) it is in principle possible to arrive at the heat capacity and other thermal parameters of a sample. This, however, requires sufficient knowledge of the thermal parameters of the gas and the backing as well as of the optical parameters η_s and β_s of the sample. Moreover, the relation between the measured microphone signal and the thermal variations can be rather complicated because of complicated frequency responses. Most of these problems can be avoided by a calibration procedure using a reference material. On the other hand it is possible to substantially simplify the rather complicated Equation (3) by choosing proper measuring configurations. This possibility results from the fact that in photoacoustic as well as in photothermal experiments one disposes of three length scales which in many cases can be optimally chosen to arrive at the desired information. For a sample these are: its thickness l_s, the optical penetration depth $\beta_s^{-1}(\lambda)$ (at a given wavelength λ) and the thermal diffusion length $\mu_s = (2\kappa_s/\rho_s C_s \omega)^{1/2}$. For a sample with a given thickness l_s, μ_s can be modified by changing the modulation frequency ω and $\beta_s^{-1}(\lambda)$ by adjusting λ.

12.2.2 The Photoacoustic Technique for Liquid Crystal Heat Capacities

Apart from trace gas detection and gas spectroscopy, for thermal properties the photoacoustic technique with microphone detection is more suitable for solid samples. Nevertheless, high resolution measurements can be carried out on non-volatile liquids such as most liquid crystals.[12–18] For photoacoustic measurements in liquid crystals one usually works in the optically and thermally thick regime: $\exp(\sigma_s l_s)$ much larger than $\exp(-\sigma_s l_s)$ and $\exp(-\beta_s l_s)$. This results in substantial simplifications in Equation (3). It also means that $1/\beta_s$ and μ_s have to be much smaller than the sample thickness l_s. Then, any contribution of the backing disappears from the expressions. For the amplitude q and the phase ψ in Equation (5) one obtains[12]:

$$q = \frac{\gamma_g P_0 I_0 t (2t^2 + 2t + 1)^{-1/2}}{2\sqrt{2} T_0 l_g (1 + s) \kappa_g a_g^2} \tag{6}$$

$$\mathrm{tg}\psi = 1 + 1/t \tag{7}$$

where $t = \mu_s \beta_s/2$ and $s = a_s \kappa_s (\kappa_g a_g)^{-1}$. These two equations allow for the simultaneous determination of the heat capacity per unit volume ($\rho_s C_s$) and the thermal conductivity κ_s of the sample by solving for t from the phase of the signal in (7) and then s from the amplitude q in Equation (6). This results in:

$$\rho_s C_s = \beta_s \kappa_g a_g s t^{-1} \omega^{-1} \tag{8}$$

$$\kappa_s = 2\kappa_g a_g s t \beta_s^{-1} \tag{9}$$

However, in order to have enough phase sensitivity one should arrange the measurement conditions in such a way that $1/t = 2/(\mu_s \beta_s)$ is not too small compared to 1 in Equation (9). The thermal diffusion length μ_s can be controlled by changing the modulation frequency and β_s by choosing the appropriate wavelength of the modulated light. Since the final results in Equations (8) and (9) depend on the value of β_s at the chosen wavelength this quantity has to be measured in a separate experiment.[19] If one wants to convert the heat capacities per unit volume, $\rho_s C_s$, to specific heat capacities C_s per unit mass one also needs density data ρ_s. As already pointed out cell characteristics can be calibrated using a reference sample.[14,15]

Since the photoacoustic signal results from modulated heat flow in the sample (thermal wave) it is obvious that in addition to the heat capacity, the thermal conductivity plays a role. This means that for anisotropic materials, like liquid crystals, one has to know the direction in the mesophases (parallel or perpendicular) of the heat flow with respect to the director describing the orientation of the anisotropic molecules. In order to orient the liquid crystal molecules in the proper direction with respect to the heat flow direction (imposed by the direction of modulated optical beam), one can use magnetic fields parallel or perpendicular to the free surface of the liquid crystal samples. In Figure 12.2 results for the specific heat capacity are given for the liquid crystal octylcyanobiphenyl (8CB).[16] This compound has a smectic A (SmA) and nematic (N) phase and a normal isotropic (I) liquid phase (above 314 K). Two sets of data are displayed, one set without a magnetic field and a second one with the field parallel to the optical beam direction (perpendicular to the liquid surface). The data of both sets give the same result. In the upper part of the figure the anomalies in C_s associated with the I-N and N-SmA transitions are clearly visible and are in accordance with previously obtained results using adiabatic scanning calorimetry.[20] In the lower part of Figure 12.2 the results for the thermal conductivity are given, showing no difference between the two sets of data. This is somewhat surprising, but from further investigations we found out that for many types of liquid crystals a free surface is a strong ordering field that imposes strong homeotropic alignment in the sample.[14,21] Even imposing a strong field parallel to the liquid surface could not break the homeotropic alignment near the free surface, but resulted with depth in a gradual rotation of the director away from the normal to the free surface.[14,18]

Contrary to the heat capacity data, there is no indication of a critical anomaly for the thermal conductivity in our data for 8CB. Such an anomaly had previously been observed, photoacoustically, for SmA-N transitions of 8CB, 9CB and 8CB + 7CB , with 7CB and 9CB compounds of the same alkylcyanobiphenyl homologous series as 8CB.[15] For 9CB this result was subsequently confirmed in a photopyroelectric study.[17] Also for some other phase transitions (smectic A to smectic C and smectic A to hexatic B) thermal conductivity anomalies were obtained with an extension of the ac calorimetric technique.[22,23] However, more recent (photopyroelectric) investigations showed the absence of an anomalous behaviour in the thermal conductivity for all the above mentioned phase transitions.[24-27] This confirms that for reliable investigations of phase transitions in liquid crystals one not only needs high resolution temperature measurement and control, but also good control of the orientation of the liquid crystal director in photoacoustic and photopyroelectric experiments.

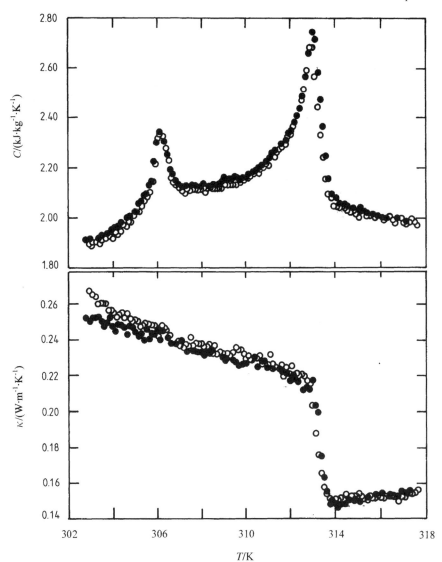

Figure 12.2 Photoacoustic results for the specific heat capacity (top) and thermal conductivity (bottom) of a octylcyanobiphenyl (8CB) liquid crystal sample. Solid dots are results without a magnetic field and open circles are results with the magnetic field perpendicular to the free surface of the sample.

12.3   Photopyroelectric Technique

12.3.1   Pyroelectric Detection

In general, the photopyroelectric technique is based on the use of a pyroelectric transducer to detect temperature variations caused by light-induced periodic heating of a sample-pyroelectric transducer assembly. If one considers a pyroelectric transducer (e.g. cut from a LiTaO$_3$ crystal) in a one-dimensional configuration (thickness much smaller than the lateral dimensions), with thickness l_p and surface area A, a change in temperature distribution $\theta(x,t)$ relative to an initial reference situation $\theta(x,t_0)$ will cause a change of polarization, that in turn induces a surface charge given by:

$$q(t) = \frac{pA}{l_p} \int_0^{l_p} dx[\theta(x,t) - \theta(x,t_0)] = \frac{pA}{l_p}\bar{\theta}(t) \tag{10}$$

where p is the pyroelectric coefficient of the transducer. This corresponds in a current detection configuration to $i = pA[d\bar{\theta}(t)/dt]$. Considering the equivalent electrical circuitry one arrives at a pyroelectric output voltage[10,28]

$$V(\omega) = H(\omega)pA\bar{\theta}(\omega)e^{i\omega t} \tag{11}$$

where $H(\omega)$ is an electrical transfer factor and $\bar{\theta}(\omega)$ the spatially averaged temperature variation in the detector. An expression for the averaged temperature over the detector $\bar{\theta}(\omega)$ in Equation (11) can be quite complicated and in general depends on all the layers involved in the one-dimensional stack considered. Initial theoretical results for a simple one-dimensional photopyroelectric model were obtained by Mandelis and Zver[10] in 1985 and generalized in 1989 (in a six layer model) by Chirtoc and Mihailescu[29] accounting for a larger variety of experimental configurations. In the early 1990s we developed a multilayer model for the direct calculation of the photothermal signal for a large (but finite) stack of layers very suitable for simulation purposes.[30,31] However, under properly chosen experimental conditions, it is possible to arrive at substantially simplified equations allowing, for a sample in contact with a pyroelectric detector, heat capacity and thermal conductivity determinations simultaneously, or each of these quantities separately. Before discussing specific setups and applications for heat capacity results, we will in the next section first consider more generally two basic configurations.

12.3.2 Two Basic Configurations

The most simple configuration in a photopyroelectric experiment aiming at thermal properties of a condensed matter sample is an isolated (suspended in air is a good approximation) one-dimensional stack of this sample in contact with a pyroelectric transducer subjected to a modulated light beam. However, one is then immediately confronted with the question of whether the sample or the transducer or both are irradiated and absorb light. There are two basic configurations to consider by assuming that either the sample is irradiated and absorbs all the light or the opposite situation where the transducer is irradiated and absorbs all the light. In Figure 12.3 the two cases are schematically depicted. In the left-hand part the so-called back detection photopyroelectric

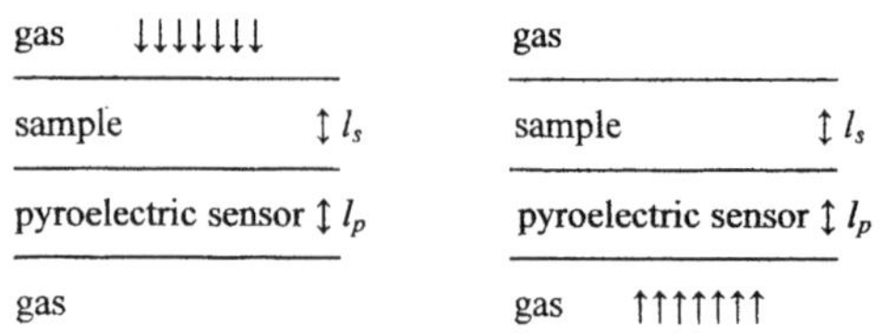

gas	↓↓↓↓↓↓↓	gas	
sample	↕ l_s	sample	↕ l_s
pyroelectric sensor ↕ l_p		pyroelectric sensor ↕ l_p	
gas		gas	↑↑↑↑↑↑↑

Figure 12.3 Left: Configuration for the back detection photopyroelectric technique (BPPE). Right: Configuration for the front detection photopyroelectric technique(FPPE). The arrows indicate the direction of the modulated optical beam.

(BPPE) configuration is given and in the right-hand part the front detection photopyroelectric (FPPE) configuration is given. In BPPE the thermal wave is generated at the top of the sample and detected at the back of the sample. In FPPE the thermal wave in the sample is generated (via the transducer) and detected at the front of the sample at the interface between sample and transducer. If one also takes the (infinitely thick) gas layers on both sides into account one can calculate $\bar{\theta}(\omega)$ in Equation (11) following the standard procedure[28–32] by solving the thermal diffusion equations for the different layers and imposing continuity of temperature and heat flux as is also done in photoacoustics.

For the case of the back detection photopyroelectric configuration (BPPE) of Figure 12.3 with an optically opaque sample and thermally thick gas layers, one obtains:

$$\bar{\theta}(\omega) = \frac{2I_0(e^{-\sigma_p l_p} - e^{\sigma_p l_p})}{\kappa_s \sigma_s \sigma_p l_p (h-1)}$$

$$\times \left[\begin{array}{l} (g+1)(h-1)(b_{ps}+1)e^{-(\sigma_s l_s + \sigma_p l_p)} + 2(b_{ps}+g)(h-1)e^{(\sigma_s l_s - \sigma_p l_p)} \\ + 2(g-1)(g+1)e^{(-\sigma_s l_s + \sigma_p l_p)} + (h-1)(g+1) \end{array} \right]^{-1}$$

(12)

where the indices s and p are for the sample and the pyroelectric transducer, respectively. As in Equation (3) one has I_0 the light intensity, l_s the sample thickness, l_p the thickness of the transducer, $g = e_g/e_s$, $h = e_g/e_p$, $b_{ps} = e_p/e_s$ and $\sigma_j = (1+\mathrm{i})a_j$.

For the case of the front photopyroelectric configuration (FPPE) of Figure 12.3 with a very opaque transducer ($l_p >> \beta_p^{-1}$) one obtains:

$$\bar{\theta}(\omega) = \frac{I_0}{2\kappa_p \sigma_p^2 l_p} \left[\frac{\begin{array}{l}(1 - e^{\sigma_p l_p})\left[(1 - b_{sp})(1-g)e^{-\sigma_s l_s} + (1 + b_{sp})(1+g)e^{\sigma_s l_s}\right] \\ +(e^{-\sigma_p l_p} - 1)\left[(1 + b_{sp})(1-g)e^{-\sigma_s l_s} + (1 - b_{sp})(1+g)e^{\sigma_s l_s}\right]\end{array}}{\begin{array}{l}\left[(1 + b_{sp})(1-g)e^{-\sigma_s l_s} + (1 - b_{sp})(1+g)e^{\sigma_s l_s}\right](h-1)e^{-\sigma_p l_p} \\ +\left[(1 - b_{sp})(1-g)e^{-\sigma_s l_s} + (1 + b_{sp})(1+g)e^{\sigma_s l_s}\right](1+h)e^{\sigma_p l_p}\end{array}} \right]$$

(13)

where $h = e_g/e_p$, $b_{sp} = e_s/e_p$, $g = e_g/e_s$ and also $\sigma_j = (1+\mathrm{i})a_j$.

In both cases the expression for $\bar{\theta}(\omega)$ is rather complicated. However, it is possible for many practical applications to arrive at substantially simplified expressions by proper choices for some parameters in Equations (12) and (13). For thermal measurement one will arrange for an optically very thick (very large β) sample or/and transducer. In that case one does not need information on the optical absorption coefficient. As will be seen further this sometimes requires the addition of another layer in the simple stacks of Figure 12.3. Moreover, for liquids (or gases) it is often also needed to fix l_s, the sample thickness. In the following sections different examples for direct or indirect heat capacity measurements will be described.

12.3.3 The BPPE Technique for Liquid Crystal Heat Capacities

In the back detection photopyroelectric configuration of Figure 12.3 the absorption of the light and thus the heating of the sample takes place at the top of the sample and the detection by a pyroelectric transducer with opaque metal electrodes takes place at the bottom (back) of the sample. As already indicated, the need to know the optical absorption coefficient β_s of the sample disappears for very large values of β_s. However, for most liquids or gases this is not the case for common optical wavelengths. To solve this problem one usually replaces the top gas layer in the BPPE configuration of Figure 12.3 by sufficiently thick transparent solid (e.g. quartz) with a very thin opaque coating on the side in contact with the fluid. In this way one can also fix the sample thickness conveniently.[17,19,21] In Equation (12) g then has to be replaced by $m = e_m/e_s$, with e_m the thermal effusivity of the medium above the sample (because the metal coating is thermally very thin its thermal contribution can be neglected). If one makes the sample sufficiently thin such that its thermal diffusion length becomes comparable with its thickness one has a quasi thermally thick sample regime. Equation (12) then reduces to[33]:

$$\bar{\theta}(\omega) = \frac{I_0 e_p e^{-\sigma_s l_s}}{i\omega \rho_p C_p l_p e_s (e_m/e_s + 1)(e_p/e_s + 1)} \tag{14}$$

Direct illumination of the pyroelectric transducer without the window (m) and the sample (s) gives:

$$\bar{\theta}(\omega) = \frac{I_0}{i\omega \rho_p C_p l_p} \tag{15}$$

Inserting Equations (14) and (15) into Equation (11) and making the ratio of the two resulting equations, results in the following expression for a calibrated voltage signal:

$$V_c(\omega) = \frac{e_p e^{-\sigma_s l_s}}{e_s (e_m/e_s + 1)(e_p/e_s + 1)} \tag{16}$$

This complex signal can be separated in an amplitude and a phase, resulting in:

$$|V_c(\omega)| = \frac{e_p e^{-l_s \sqrt{\omega/(2\alpha_s)}}}{e_s(e_m/e_s + 1)(e_p/e_s + 1)} \tag{17}$$

$$\varphi(\omega) = -l_s \sqrt{\omega/(2\alpha_s)} \tag{18}$$

From these last two equations the thermal diffusivity α_s and the thermal effusivity e_s of the sample can be obtained resulting in the following results for the specific heat capacity and thermal conductivity:

$$C_s = e_s/(\rho_s \sqrt{\alpha_s}) \tag{19}$$

$$\kappa_s = e_s/\sqrt{\alpha_s} \tag{20}$$

The BPPE technique has been extensively used to measure the temperature dependence of the heat capacity and the thermal conductivity near phase transitions of liquid crystals by our group and in particular by a group at the university of Roma 2 'Tor Vergata'.[17,19,21,24-27,34-43] As already pointed out, liquid crystalline phases are anisotropic and in the BPPE technique one imposes a temperature gradient in a direction normal to the layers in the stack. This requires a proper orientation of the liquid crystal director (parallel or perpendicular) with respect to the direction of the temperature gradient in order to arrive at meaningful results for heat capacity and thermal conductivity. Parallel or perpendicular orientations can be imposed by proper surface treatment of the solid surfaces in contact with the liquid crystal sample and/or by magnetic or electric fields. In Figure 12.4 specific heat capacity results near the nematic isotropic transition are given for the liquid crystal di-hexylazoxybenzene (6AB) obtained with a BPPE setup for parallel (planar) and perpendicular (homeotropic) orientations.[21,44] Proper alignment of a nearly 100 μm thick sample was achieved by surface treatment of the pyroelectric detector and the gold coated quartz cover plate. In addition further alignment was enforced by a magnetic field. As expected the specific heat capacity results are within experimental accuracy direction independent. The accompanying thermal conductivity data can be found elsewhere.[43,44]

12.3.4   The FPPE Technique for Liquid Crystal Heat Capacities

Ac calorimetry is a widely used technique that has already convincingly proven to be very useful in studying heat capacity behaviour of solid and liquid samples.[45-47] However, if one is mainly interested in heat capacity results and not so much in the thermal conductivity the measuring configuration in the FPPE technique (Figure 12.3) can be chosen in such a way that it operates as an ac calorimeter. It suffices to choose a sufficiently low modulation frequency.

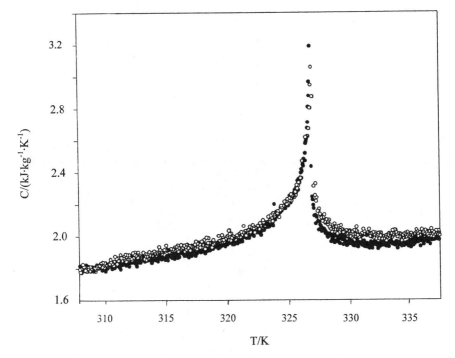

Figure 12.4 Specific heat capacity for the liquid crystal 6AB near the nematic-isotropic transition with planar (dots) and homeotropic (open circles) alignment in the nematic phase (below 327 K).

In that case both sample and pyroelectric detector are thermally very thin, $l_s \ll \mu_s$ and $l_p \ll \mu_p$. In this frequency range the output voltage of the detector depends only on the heat capacities of sample and transducer, because the thermal gradient over sample and detector is negligible and true calorimetric measurements can be performed. If in addition the effusivity e_g of the surrounding gas can be neglected compared to the effusivities of the sample (e_s) and the detector (e_p), the full expression of the one-dimensional model (Equation (13)) substantially reduces and the phase of the pyroelectric signal is constant and the amplitude is given by:[32]

$$|V(\omega)|_{norm} = H(\omega) \frac{\rho_p C_p l_p}{\rho_s C_s l_s + \rho_p C_p l_p} \tag{21}$$

The signal is normalized with the temperature dependence of the bare pyroelectric signal, to eliminate dependence on several sample independent factors in Equation (13). For measurements at very low measuring frequencies a small correction to Equation (21) can be applied for the contribution of the surrounding gas.[32]

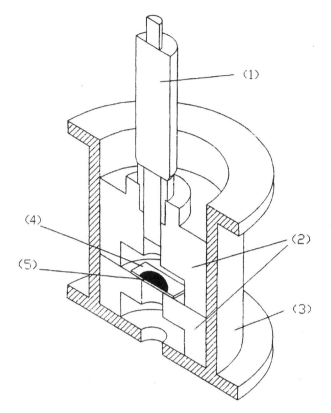

Figure 12.5   Schematic overview of a front detection photopyroelectric measuring
setup: (1) micrometer screw, (2) stainless steel positioners, (3) temperature controlled holder, (4) pyroelectric detector, (5) sample.

A schematic overview of the setup as it was used to arrive at absolute values
for the heat capacity of liquids is shown in Figure 12.5. The measuring cell
consists of two stainless steel parts that can be fixed to each other. The pyroelectric sensor on top of the lower part is placed so that the non-illuminated
parts are in good thermal contact with the cell. The temperature of the setup is
controlled by a copper shield with a heating coil and a thermistor close to the
pyroelectric detector. A second temperature controlled shield (not shown) is
mounted around the first one in order to shield external temperature variations.
The setup can operate from room temperature to 373 K with a stability of
1 mK. The sample is a drop of the liquid that is allowed to spread (or smeared
out) over the detector to a thickness l_s (at first unknown). Above the sample in
the gas phase a micrometer screw (as a thermal load) is placed with its end
surface parallel with the detector. By positioning the end of the micrometer
screw at different levels above the sample and carrying out frequency scans it is
possible to arrive at a value of l_s and derive absolute heat capacities per unit
volume.[32] In the top part of Table 12.1 (top 5 lines) results are given for the

Table 12.1 Results for some liquids from measurements with an air gap between sample and the thermal load (top five lines) and with the load in contact with the sample surface (bottom three lines) compared with literature data (at room temperature except for 8CB).

Substance	$l_s/\mu m$	$C_s/(J\,kg^{-1}\,K^{-1})$	$C_s/(J\,kg^{-1}\,K^{-1})$
Olive oil	63±3	2046±120	1970[a]
Silicone oil (AK250)	63±3	1480±110	1470[a]
8CB (at 318 K)	32±3	2400±200	2350[b]
Di-octylphtalate	86±4	1860±100	
Guaiacol	74±4	1976±110	
Ethyleneglycol		2340±300	2385[a]
Glycerol		2419±152	2430[a]
H_2O		4122±132	4183[a]

[a] Reference 48
[b] Reference 20

specific heat capacity of some liquids and compared with literature data. To convert the heat capacity per unit volume to per unit mass we used literature data for the density (relative uncertainty less than 1 %).

A drawback of the presence of the air gap is the need to prepare a sufficiently flat liquid layer on the detector (gold coated) surface. Depending on the surface tension of the liquid and on the wetting phenomena on the gold layer, the liquid might form a drop instead of a flat layer, rendering the method inapplicable unless a surface treatment is introduced. Another mode of operation involves the elimination of the air gap and bringing the micrometer screw head surface in contact with the sample (imposing a fixed sample thickness). This results of course in a different one-dimensional configuration. Although an ac calorimetric regime can never be attained with a thick solid load, from the frequency spectrum of the signal it is still possible to obtain values for C_s and κ_s. The lower part of Table 12.1 (bottom three lines) contains results for some liquids obtained in this way.[32,33]

The latter approach can be considered as the conventional FPPE technique with a thermally thick sample backing solid material, which has been used extensively in the past by others.[49–52] If one is mainly interested in the heat capacity and not in the thermal conductivity the method without a thermal backing has substantial advantage because pure ac calorimetry is possible at sufficiently low modulation frequencies. For liquid crystals this has the big advantage that one does not have to be concerned with the director homogeneity in the sample. We extensively used the setup of Figure 12.5 to investigate the heat capacity behaviour near phase transition lines in binary mixture of liquid crystals.[33,53] In particular we studied the nematic-smectic A transition lines in order to locate and characterize the heat capacity anomalies near tricritical points where the transition line crosses over from second order (for large nematic ranges) to first order (for narrow nematic ranges).[53] Very rich thermal behaviour was found for binary liquid crystal mixtures exhibiting

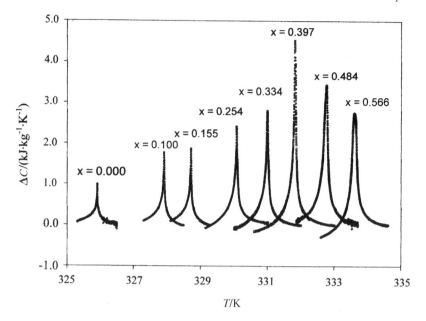

Figure 12.6   Overview of the excess specific heat capacity (background subtracted) near the nematic-smectic A transition for the binary system 7AB + 8AB for different mole fractions x of 8AB. A tricritical point occurs for $x \approx 0.397$.

so-called injected smectic A phases.[53] In Figure 12.6 heat capacity curves are given for a series of mixtures of 7AB (di-hepthylazoxybenzene) and 8AB (di-octylazoxybenzene) exhibiting a tricritical point for a mole fraction $x = 0.397$ of 8AB in the mixture. For lower mole fractions the transitions are second order, for larger concentrations they are first order.[33]

12.3.5   Critical Heat Capacities near Consolute Points of Binary Liquid Mixtures

Binary liquid mixtures near consolute points are characterized by large concentration fluctuations that affect many static and dynamic physical parameters. High-resolution studies of the heat capacity near upper or lower critical solution temperatures have been obtained by adiabatic (scanning) calorimetry or by ac calorimetry.[54–64] Recently, we also used a FPPE technique to investigate simultaneously the temperature dependence of the thermal effusivity, thermal conductivity and specific heat capacity.[65–68] Besides the advantage of the possibility of static and dynamic thermal parameters simultaneously, the FPPE technique also allows local measurements in a liquid sample, contrary to other calorimetric techniques that measure a sample as a whole. This is in particular useful to obtain separate values for coexisting phases. For this purpose we built a measuring cell schematically given in Figure 12.7.

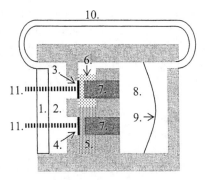

Figure 12.7   Schematic diagram of a measuring cell for binary liquid mixtures. (1) window, (2) air, (3) upper sensor, (4) lower sensor, (5) denser sample phase, (6) lighter sample phase, (7) backings, (8) air and sample vapor, (9) flexible teflon membrane, (10) connecting tube, (11) incident radiation.

The cell is equipped with two (300 μm thick) sensors at a distance of 100–300 μm of gold coated brass heat sinks (backings). The mixture also fills the gaps between the sensors and the backings. The purpose of the dual sensor configuration is to enable simultaneous measurements in the upper and lower coexisting phases upon phase separation. Of course in the homogeneous phase identical results should be obtained. For each of the sensor configurations one recognizes four layers: gas, sensor, sample and backing. This is identical to the situation in Section 12.3.4 with the micrometer screw in contact with the sample and corresponds to the configuration in the right part of Figure 12.3 with the top gas layer replaced by the backing. The current response of the sensor in such a configuration can in general be written as

$$I(t) = \frac{pI_0 A_p}{2l_p \rho_p C_p} \Gamma(\omega) e^{i\omega t} \tag{22}$$

Working out the one-dimensional model for the four layer case here yields (assuming $e_p \gg e_g$, which is well satisfied e. g. for air at atmospheric pressure) an expresion for $\Gamma(\omega)$ in terms of thermal and dimensional parameters of the different layers.[66] Although the full expression can be used for frequency scans, two particularly interesting cases can be considered: the thermal conductivity mode and the thermal effusivity mode. In the thermal conductivity mode one considers a thermally thin sample ($\mu_s \gg l_s$), by using a sufficiently low modulation frequency, and an excellent heat sink $e_b \gg e_s$. In this situation one arrives at:

$$\Gamma(\omega) = \frac{(b_{bp} + 1 + \sigma_s l_s b_{bs})(e^{\sigma_p l_p} - 1) + (b_{bp} - 1 - \sigma_s l_s b_{bs})(e^{-\sigma_p l_p} - 1)}{(b_{bp} + 1 + \sigma_s l_s b_{bs})e^{\sigma_p l_p} + (b_{bp} - 1 - \sigma_s l_s b_{bs})e^{-\sigma_p l_p}} \tag{23}$$

with $b_{bp} = e_b/e_s$ and $b_{bs} = e_b/e_s$.

In this expression the sample properties only appear through the combination $b_{bs}\sigma_s l_s$. This results in a pyroelectric signal that is only sensitive to the thermal resistance l_s/κ_s of the sample. In the effusivity mode one considers the opposite situation of a thermally very thick sample ($\mu_s \ll l_s$). The resulting expression for $\Gamma(\omega)$ is

$$\Gamma(\omega) = \frac{(b_{sp}+1)e^{\sigma_p l_p} + (b_{sp}-1)e^{-\sigma_p l_p} - 2b_{sp}}{(b_{sp}+1)e^{\sigma_p l_p} + (b_{sp}-1)e^{-\sigma_p l_p}} \tag{24}$$

In this expression the sample properties only appear through the ratio $b_{sp} = e_s/e_p$. Thus at sufficiently high modulation frequencies one obtains the thermal effusivity e_s of the sample. It should also be noted that the thermal properties of the sensor, backing, gas as well as l_s and l_p have to be known. These values can be obtained from the literature and/or calibration runs without sample or with reference samples.

Detailed FPPE measurements were carried for two systems with a lower critical point, 2,6-lutidine-water and n-butoxyethanol-water, and for two systems with an upper critical point, cyclohexane-nitroethane and aniline-cyclohexane.[65–68] In Figure 12.8 for aniline-cyclohexane thermal conductivity, κ, results are given in the left-hand part and thermal effusivity, e, results in the right-hand part. Combining these simultaneously measured κ and e data and independently measured[66] density data allows one to calculate the specific heat capacity, C, from $C = e^2/\rho\kappa$. In Figure 12.9 a comparison is made between C from the FPPE measurements and data from independent measurements by adiabatic scanning calorimetry (ASC) (see Chapter 13). In order to make the data in the homogeneous phase coincide, the FPPE data were shifted down by 0.5 K and up by about 4 % in value (well within absolute accuracy limits). It is

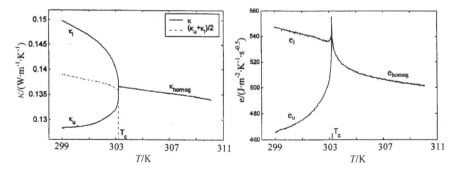

Figure 12.8 Left: Thermal conductivity of aniline-cyclohexane at the critical concentration in the homogeneous phase (above T_c) and in the two phase region (below T_c), κ_u and κ_l are for the upper and lower phase, respectively. Right: Corresponding simultaneously measured results for the thermal effusivity e.

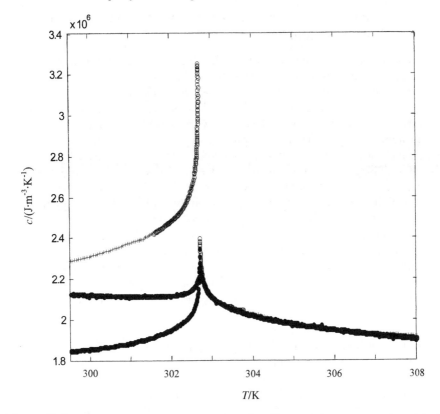

Figure 12.9 Heat capacity per unit volume of aniline-cyclohexane near the critical point. Solid dots are from FPPE measurements, pluses are ASC data over a wide temperature range and open circles are ASC results over a narrow range.

clear that in the homogeneous phase there is excellent agreement for the temperature dependence of both data sets. In the two-phase region there is apparently a large discrepancy. This is, however, entirely the result of the difference in measuring technique. In ASC the heat capacity is obtained for the sample as a whole and in the two-phase region reflects heat exchange for mixing (or demixing) of the compounds. For the local FPPE measurements one obtains two branches, one for each of the two coexisting phases and in this situation the mixing heat does not contribute to the κ and e results, because it is exchanged with the (temperature controlled) cell wall. The possibility to obtain heat capacities for coexisting phases separately is one of the nice aspects of the FPPE technique not available in standard calorimetry. Moreover, simultaneous results can be obtained for thermal conductivity and heat capacity. Information on the other systems studied and on the thermal conductivity aspect can be found elsewhere.[65–68]

12.3.6 Photopyroelectric Heat Capacity Spectroscopy of Glass Forming Liquids

Many physical parameters in supercooled liquids exhibit relaxation in a frequency range that is strongly temperature dependent. The frequency dependence of some thermal parameters in supercooled liquids has been known for more than two decades but is still a subject of theoretical and experimental investigations. Until we introduced photopyroelectric techniques about ten years ago,[69–73] apart from some photoacoustic investigations,[74–76] the most extensively used method was the so-called 3ω method introduced in 1985 by Birge and Nagel and subsequently used extensively by Nagel and co-workers.[77,78] In the 3ω and photoacoustic experiments, the frequency dependence of the effusivity (product of thermal conductivity κ and heat capacity per unit volume ρC) was measured, almost exclusively, up to a few kHz. Efforts to separate the frequency dependence of κ and C were only possible in the limited range between 0.01 and 10 Hz.[79,80] No evidence was found for frequency dependence of κ. Using PPE techniques we were able to extend the upper limit of the frequency range to 100 kHz for measurements of the thermal effusivity and could demonstrate frequency independence of κ up to 1.1 kHz. To achieve these results we used three different photopyroelectric configurations, schematically represented in Figure 12.10.

In the effusivity configuration the sensor and the sample have a joint interface and the sensor is in contact with air on the other side. The air and sample layers are thermally thick. Heating is accomplished by illuminating the optically opaque sample sensor interface. In this FPPE configuration Equation 24 can be used to arrive at values for e_s. In the thermal diffusivity configuration the heat is generated by illuminating the sample-window interface. This configuration is a BPPE one and corresponds to the configuration in the left-hand part of Figure 12.3 with the top gas layer replaced by a thermally thick window. The derivation of the relevant expressions is similar to the derivation of

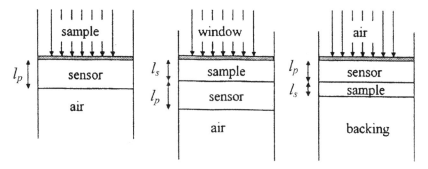

Figure 12.10   Left-hand site: effusivity configuration, middle: diffusivity configuration, right-hand side: thermal conductivity configuration. The dark layer is a thermally thin optically opaque coating.

Equation 12. In the FPPE thermal conductivity configuration heat is generated at the air-sensor interface. This configuration is conceptually identical to the one described in Section 12.3.5 for the measurements in binary liquid mixtures and Equation 23 applies.

We used these three configuration for a comprehensive photopyro-electric thermal spectroscopic study of the two glass forming liquids glycerol and propylene glycol over a broad frequency range between 0.01 Hz and 100 kHz and between room temperature and 173 K. No evidence for frequency dependence was found for the thermal conductivity and the observed frequency dependence of the thermal effusivity can be entirely ascribed to the heat capacity relaxation. As an example in Figure 12.11 plots for the real and imaginary part of e^2, which is equal to $\rho\kappa C$, are displayed as a function of frequency for a set of temperatures. The data could be well described by the very general empirical Havriliak-Negami expression for relaxation phenomena.[81] Further details on these investigations can be found elsewhere.[70–73]

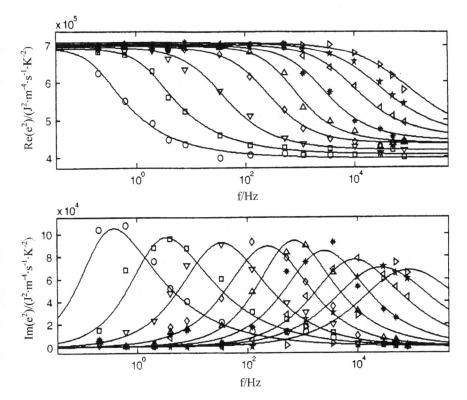

Figure 12.11 Frequency dependence of $e^2 = \rho\kappa C$ of glycerol at different temperatures from 198 K (left) to 238 K (right) in steps of 5 K. The lines are the best Havriliak-Negami fits.

References

1. A. G. Bell, *Am. J. Sci.*, 1880, **20**, 305.
2. J. Tyndall, *Proc. Roy. Soc. (London)*, 1881, **31**, 307.
3. W. C. Röntgen, *Phil. Mag.*, 1881, **11**, 308.
4. A. G. Bell, *Phil. Mag.*, 1881, **11**, 510.
5. W. H. Preece, *Proc. Roy. Soc. (London)*, 1881, **31**, 506.
6. M. E. Mercandier, *C. R. Acad. Sci. (Paris)*, 1881, **92**, 409.
7. Lord Rayleigh, *Nature*, 1881, **23**, 274.
8. A. Rosencwaig and A. Gersho, *J. Appl. Phys.*, 1976, **47**, 64.
9. W. B. Jackson, N. M. Amer, A. C. Boccara and D. Fournier, *Appl. Opt.*, 1981, **20**, 1333.
10. A. Mandelis and M. Zver, *J. Appl. Phys.*, 1985, **57**, 4421.
11. A. Rosencwaig, *Photoacoustics and Photoacoustic Spectroscopy*, John Wiley and Sons, New York, 1980.
12. M. Marinelli, U. Zammit, F. Scudieri and S. Martellucci, *Nuovo Cimento*, 1987, **9D**, 855.
13. C. Glorieux, E. Schoubs and J. Thoen, *Mater. Sci. Eng. A*, 1989, **122**, 87.
14. J. Thoen, C. Glorieux, E. Schoubs and W. Lauriks, *Mol. Cryst. Liq. Cryst.*, 1990, **191**, 29.
15. U. Zammit, M. Marinelli, R. Pizzoferrato, F. Scudieri and S. Martellucci, *Phys. Rev. A*, 1990, **41**, 1153.
16. J. Thoen, E. Schoubs and V. Fagard, in *Physical Acoustics: Fundamentals and Applications* eds. O. Leroy and M. E. Breazeale, Plenum Press, New York, 1992, p. 179.
17. M. Marinelli, U. Zammit, F. Mercuri and R. Pizzoferrato, *J. Appl. Phys.*, 1992, **72**, 1096.
18. C. Glorieux, Z. Bozoki, J. Fivez and J. Thoen, *J. Appl. Phys.*, 1995, **78**, 3096.
19. E. Schoubs, H. Mondelaers and J. Thoen, *J. Physique IV*, 1994, **C7**, C7–257.
20. J. Thoen, H. Marynissen and W. Van Dael, *Phys. Rev. A*, 1982, **26**, 2886.
21. E. Schoubs, Ph.D. thesis, 1994, Katholieke Universiteit Leuven, Belgium.
22. G. Nounesis, C. C. Huang and J. Goodby, *Phys. Rev. Lett.*, 1986, **56**, 1712.
23. E. K. Hobbie, H. Y. Liu, C. C. Huang, C. Bahr and G. Heppke, *Phys. Rev. Lett.*, 1991, **67**, 1771.
24. M. Marinelli, F. Mercuri, U. Zammit and F. Scudieri, *Phys. Rev. E*, 1996, **53**, 701.
25. M. Marinelli, F. Mercuri, S. Folietta, U. Zammit and F. Scudieri, *Phys. Rev. E*, 1996, **54**, 1604.
26. F. Mercuri, U. Zammit and M. Marinelli, *Phys. Rev. E*, 1998, **57**, 596.
27. F. Mercuri, M. Marinelli, U. Zammit, C. C. Huang and D. Finotello, *Phys. Rev. E*, 2003, **68**, 051705.
28. H. J. Coufal, R. K. Grygier, D. E. Horne and J. E. Fromm, *J. Vac. Technol.*, 1987, **A5**, 2875.
29. M. Chirtoc and G. Mihailescu, *Phys. Rev. B*, 1989, **40**, 9606.

30. C. Glorieux, J. Fivez and J. Thoen, *J. Appl. Phys.*, 1993, **73**, 684.
31. C. Glorieux, Ph. D. thesis, 1994, Katholieke Universiteit Leuven, Belgium.
32. J. Caerels, C. Glorieux and J. Thoen, *Rev. Sci. Instrum.*, 1998, **69**, 2452.
33. J. Caerels, Ph. D. thesis, 1999, Katholieke Universiteit Leuven. Belgium.
34. M. Marinelli, F. Mercuri, U. Zammit and F. Scudieri, *Int. J. Thermophys.*, 1998, **19**, 595.
35. M. Marinelli, F. Mercuri, U. Zammit and F. Scudieri, *Pys. Rev. E*, 1998, **58**, 5860.
36. F. Mercuri, A. K. Gosh and M. Marinelli, *Phys. Rev. E*, 1999, **60**, R6309.
37. M. Marinelli and F. Mercuri, *Phys. Rev. E*, 2000, **61**, 1616.
38. M. Marinelli, A. K. Gosh and F. Mercury, *Phys. Rev. E*, 2001, **63**, 061713.
39. F. Mercuri, U. Zammit, F. Scudieri and M. Marinelli, *Phys. Rev. E*, 2003, **68**, 041708.
40. F. Mercuri, U. Zammit and M. Marinelli, *Appl. Phys. Lett.*, 2004, **85**, 4642.
41. F. Mercuri, S. Paolini, U. Zammit and M. Marinelli, *Phys. Rev. Lett.*, 2005, **94**, 274801.
42. F. Mercuri, S. Paolini, U. Zammit, F. Scudieri and M. Marinelli, *Phys. Rev. E*, 2006, **74**, 041707.
43. J. Caerels, E. Schoubs and J. Thoen, *Liq. Cryst.*, 1997, **22**, 659.
44. J. Thoen and C. Glorieux, *Thermochim. Acta*, 1997, **304/305**, 137.
45. P. F. Sullivan and G. Seidel, *Phys. Rev.*, 1968, **173**, 679.
46. C. C. Huang, J. M. Viner and J. C. Novack, *Rev. Sci. Instrum.*, 1985, **56**, 1390.
47. C. W. Garland, *Thermochim. Acta*, 1985, **88**, 127.
48. U. Grigull and H. Sander, *Wärmeleitung,* Springer, Berlin, 1990.
49. D. Dadarlat. M. Chirtoc, C. Nematu, R. M. Candea and D. Bicanic, *Phys. Status Solidi A*, 1990, **121**, K231.
50. D. Dadarlat, H. Visser and D. Bicanic, *Meas. Sci. Technol.*, 1995, **6**, 1215.
51. M. Marinelli, F. Mercuri and U. Zammit, *Appl. Phys. Lett.*, 1994, **65**, 2663.
52. M. Marinelli, F. Mercuri, S. Paolini and U. Zammit, *Phys. Rev. Lett.*, 2005, **95**, 237801.
53. J. Caerels, C. Glorieux and J. Thoen, *Phys. Rev. E*, 2002, **65**, 031704.
54. J. Thoen, E. Bloemen and W. Van Dael, *J. Phys. Chem.*, 1978, **68**, 735.
55. E. Bloemen, J. Thoen and W. Van Dael, *J. Chem. Phys.*, 1980, **73**, 4628.
56. G. Sanchez, M. Meichle and C. W. Garland, *Phys. Rev. A*, 1983, **28**, 1647.
57. U. Würz, M. Grubić and D. Woermann, *Ber. Bunsenges. Phys. Chem.*, 1992, **96**, 1460.
58. L. V. Entov, V. A. Levchenko and V. P. Voronov, *Int. J. Thermophys.*, 1993, **14**, 221.
59. A. C. Flewelling, R. J. DeFonseka, N. Khaleeli, J. Partee and D. T. Jacobs, *J. Chem. Phys.*, 1996, **104**, 8048.
60. J. Thoen, J. Hamelin and T. K. Bose, *Phys. Rev. E*, 1996, **53**, 6264.
61. V. P. Voronov and V. M. Buleiko, *JETP*, 1998, **86**, 586.
62. P. F. Rebillot and D. T. Jacobs, *J. Chem. Phys.*, 1998, **109**, 4009.
63. E. R. Oby and D. T. Jacobs, *J. Chem. Phys.*, 2001, **114**, 4918.

64. A. W. Norwicki, M. Gosh, S. M. MecClellan and D. T. Jacobs, *J. Chem. Phys.*, 2001, **114**, 4625.

65. S. Pittois, G. Sinha, C. Glorieux and J. Thoen, in *Proceedings of the 27th International Thermal Conductivity Conference and 15th International Thermal Expansion Symposium*, 26–29 October 2003, Knoxville, TN (DEStech Publications, Lancaster, PA, 2004), pp. 163–173.

66. S. Pittois, B. Van Roie, C. Glorieux and J. Thoen, *J. Chem. Phys.*, 2004, **121**, 1886.

67. S. Pittois, B. Van Roie, C. Glorieux and J. Thoen, *J. Chem. Phys.*, 2005, **122**, 024504.

68. S. Pittois, Ph.D. thesis, 2004, Katolieke Universiteit Leuven, Belgium.

69. M. Chirtoc, E. H. Bentefour, C. Glorieux and J. Thoen, *Thermochim. Acta*, 2001, **377**, 105.

70. E. H. Bentefour, C. Glorieux, M. Chirtoc and J. Thoen, *Rev. Sci. Instrum.*, 2003, **74**, 811.

71. E. H. Bentefour, C. Glorieux, M. Chirtoc and J. Thoen, *J. Appl. Phys.*, 2003, **93**, 9610.

72. E. H. Bentefour, C. Glorieux, M. Chirtoc and J. Thoen, *J. Chem. Phys.*, 2004, **120**, 3726.

73. E. H. Bentefour, Ph.D. thesis, 2002, Katholieke Universiteit Leuven, Belgium.

74. B. Büchner and P. Korpiun, *Appl. Phys. B*, 1987, **43**, 29.

75. K. Madhusoodanan and J. Philip, *Phys. Rev. B*, 1988, **39**, 7922.

76. S. Kojima, in *Physical Acoustics Fundametals and Applications*, eds. O. Leroy and M. Breazeale, Plenum Press, New York, 1991, p. 399.

77. N. O. Birge and S. R. Nagel, *Phys. Rev. Lett.*, 1985, **54**, 2674.

78. N. O. Birge and S. R. Nagel, *Rev. Sci. Instrum.*, 1987, **57**, 1464.

79. N. Menon, *J. Chem. Phys.*, 1996, **105**, 5246.

80. N. O. Birge, P. K. Dixon and N. Menon, *Thermochim. Acta*, 1997, **304**, 51.

81. S. Havriliak and S. Negami, *Polymer*, 1967, **8**, 101.

CHAPTER 13

High Resolution Adiabatic Scanning Calorimetry and Heat Capacities

JAN THOEN

Laboratorium voor Akoestiek en Thermische Fysica, Departement Natuurkunde en Sterrenkunde, Katholieke Universiteit Leuven, Celestijnenlaan 200 D, B-3001, Leuven, Belgium

13.1 Introduction

Heat capacity measurements play an important role in monitoring the changes in energy content of condensed matter systems. As such, calorimetry is an indispensable technique for many scientific fields. Depending on the application envisioned, several different technical approaches with varying degrees of accuracy and precision have been developed. Over wide temperature ranges generally the classical Nernst heat pulse method is used.[1] The method offers quite good accuracy, but the application of this method is a tedious process and can only be used with increasing temperature. Although extensive auto-matization and the gradual availability of more and more powerful and easily programmable computers and sophisticated software have substantially reduced the burden, the classical heat-pulse-induced temperature-step method remains time consuming. During the last 50 years several new approaches, supported to a large extent by novel developments in electronic measurements instrumentation, have emerged, e.g. differential scanning calorimetry (DSC),[2,3] scanning transitiometry[4-6] (see also Chapter 8) and modulation techniques like ac calorimetry[7,8], the 3ω method[9] and more recently photoacoustic and

Heat Capacities: Liquids, Solutions and Vapours
Edited by Emmerich Wilhelm and Trevor M. Letcher
© The Royal Society of Chemistry 2010
Published by the Royal Society of Chemistry, www.rsc.org

photopyroelectric techniques (see Chapter 12 and references therein) and Peltier ac and Peltier tip calorimetry.[10,11]

A development beyond classical adiabatic heat-pulse-calorimetry took place at the end of the 1960s. In an effort to approach liquid–gas critical points much closer than in the pioneering work of Voronel and co-workers[12–14] (by the classical adiabatic method), Buckingham and co-workers[15–17] imposed a very slow dynamic constant heating (or cooling) rate on the thermal shield (in a classical type adiabatic calorimeter) surrounding the sample cell and forced the cell to follow with the same rate. By measuring the imposed rate the power applied (heating) to or extracted (cooling) from the cell, the heat capacity C (in the liquid-gas case at constant volume) is readily obtained from

$$C = T \frac{dS}{dT} = \frac{dQ}{dT} = \frac{dQ/dt}{dT/dt} = P/\dot{T} \tag{1}$$

with S the entropy, T the temperature, Q the supplied heat, t the time, P the supplied power and $\dot{T}$ the temperature scanning rate. If one considers the shield (forced to change its temperature at constant $\dot{T}$) as the reference 'sample', the setup is conceptually similar to the (power compensated) differential scanning calorimeter (DSC). There are, however, basic differences in design principles and area of applications. The DSC is useful when the energy of a transition is of greater interest than the detailed form of the specific heat curve. A commercial DSC (or modulated DSC) generally does not yield accurate absolute values of specific heat and by using high scanning rates (typically above $0.2\ \mathrm{K\ s^{-1}}$ to have a reasonable sensitivity) quite often operates out of thermodynamic equilibrium, in particular near fluctuations dominated phase transitions. Moreover, DSC cannot discriminate between second order (continuous) phase transitions and (weakly) first order ones.[18] Several of the limitations of DSCs have been eliminated in scanning transitiometry introduced in 1985 by Randzio by imposing very slow constant scanning rates in a high precision differential concept.[4] However, imposing constant rates remain a basic problem for high-resolution work at and near (weakly) first order transitions. Buckingham and co-workers called their apparatus a high precision-scanning-ratio-calorimeter (for use near phase transitions).[17] In order to cope with the critical slowing down near the liquid–gas critical point they imposed constant scanning rates as low as $10^{-6}\ \mathrm{K\ s^{-1}}$.

In the mid-1970s we built a similar four stage scanning calorimeter to measure with high resolution the heat capacity (at constant pressure) near the critical (consolute) point of binary and ternary liquid mixtures.[19–22] The construction of our calorimeter was such that in addition to different scanning modes it could also be used as a classical step calorimeter. Our first results were mainly obtained in the stepping way and with constant heating or cooling rates (constant $\dot{T}$).[19,20] The step data at that time were substantially better than the scanning results (at sub millikelvin per second rates) because of too large internal relaxation times (in the quite large) measuring cell. To solve this problem a new liquid cell with internal heating and a stirring mechanism (see below) was built. At that time we also realized that near the critical point it would be much easier to cope with the critical slowing down and the large increase of the heat capacity, by imposing a constant heating or cooling power to the cell instead of a constant heating or cooling rate as we and Buchingham *et al.* did before, *i.e.* keeping *P* constant and not $\dot{T}$ in Equation (1).[20,21] Later we will see that this change in operation mode is essential for the proper investigation of (weakly) first order phase transtions.[23,24]

Around that time, calorimeters similar to our adiabatic scanning calorimeter (ASC) were developed by other groups as well. In 1980 Würz and Grubić[25] described a three stage adiabatic calorimeter of the scanning ratio type and included a figure with heat capacity data near a liquid–liquid critical point obtained at constant scanning rates of 128.8 µK s^{-1} and 6.98 µK s^{-1}. Junod[26] described a setup with a continuous adiabatic (scanning) method for the graphical recording of the heat capacity of solids over the temperature range between 80 K to 320 K at moderate to fast scanning rates (typically around 10 mK s^{-1}). A microcomputer controlled ASC type apparatus for solid samples was described in a paper of 1981 by Lancaster and Baker.[27] After our introduction of adiabatic scanning calorimetry (ASC) for first order and second order phase transition studies in liquid crystals[20,23] it was also used for liquid crystal studies by Anisimov and co-workers.[28] Bessergenev *et al.* used different ASC modes of operation to study first and second order transitions in rear earth metals.[29] Lysek *et al.* described a scanning rate calorimeter (at rates of about 1 mK s^{-1}) for use in adsorption studies.[30] An ASC techniques similar to ours was used by Sirota to study phase transitions and super cooling of normal alkanes.[31,32] Schnelle and Gmelin introduced a high resolution ASC for small (solid) samples.[33] Apparently unaware of previous developments, Moon and Yeong proposed, in 1996, a so-called rate-scanning modified adiabatic calorimeter (MAC) (with scanning rate between 0.2 mK s^{-1} and 30 mK s^{-1}).[34,35] However, their setup is operationally the same as the previously well established standard ASCs used by several other groups. An ASC similar to ours for the study of liquid–liquid critical points was build by Jacobs and collaborators.[36]

In the rest of this chapter we will first present details about the operational modes that can and have been implemented in our adiabatic scanning calorimeters, present some details on construction as well as on controlling electronics and software. Finally, some typical results will be presented for liquid mixtures and liquid crystals.

13.2 Modes of Operation of an ASC

To measure the heat capacity of fluids in a scanning mode calorimeter, one needs at least a fluid cell and a surrounding (adiabatic) shield. However, for high resolution and slow scanning rates two or more shields are required. A practical way to illustrate the versatility of operational modes of a scanning calorimeter is to use the schematic representation in Figure 13.1 of a four stages

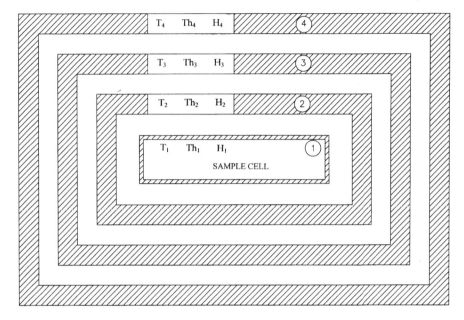

Figure 13.1   Schematic diagram of a four stages scanning calorimeter with centrally a liquid sample cell. T_i, Th_i and H_i ($i = 1$ to 4) are the temperatures, thermometers and heaters of the different stages.

calorimeter with centrally a sample cell surrounded by three different thermal shields. Each of the stages (1 to 4) has its own thermometer (Th_i), its own (electrical) heater (H_i) and is at its own temperature T_i. The stages are in very poor thermal contact and the space between the stages is vacuum pumped.

Considering Equation (1), one can easily recognize four practical modes of operation of an ASC. One can keep the nominator or the denominator constant and measure the other one, to arrive at the heat capacity C. Moreover, one can choose a positive or a negative value for the quantity kept constant. This results in heating at constant power or rate and in cooling at constant power or rate. Obviously, the different modes need quite different settings of the electronics and programming. In Table 13.1 the ideal working conditions of the four modes are summarized.

Let us first consider the two heating modes 1 and 2. For both heating modes ideally one wants zero heat transfer between stages 1, 2 and 3. The quantities P_1^l and P_2^l, representing the heat leaks from stage 1 and 2, respectively, are zero. For mode 1, one applies electrically a constant heating power P_1^e to stage 1, in mode 2 one applies a constant P_2^e to stage 2. In this case a constant rate $\dot{T}_2 = \dot{T}_1$ is imposed provided the heat capacity C_2 of stage 2 can be considered constant over a given temperature range. In Table 13.1 the temperature of the outside of the calorimeter (stage 4) is assumed to be kept at a constant temperature $T_4 = K_4$ and serves as a heat sink. This can e.g. be room temperature or the temperature of a water (or oil) bath or of a cryogenic liquid in a Dewar.

Table 13.1 Modes and working conditions of the scanning calorimeter of Figure 13.1

	mode 1 (heating) (constant power)	mode 2 (heating) (constant rate)	mode 3 (cooling) (constant power)	mode 4 (cooling) (constant rate)
$\Delta T_{12} = T_1 - T_2$	0	0	$K^a > 0$	$K > 0$
$\Delta T_{23} = T_2 - T_3$	0	0	$K > 0$	$K > 0$
$\Delta T_{34} = T_3 - T_4$	>0	>0	>0	>0
T_4	K_4^b	K_4	K_4	K_4
$\dot{T}_1$	P_1^e/C_1	$=\dot{T}_2$	P_1^e/C_1	$=\dot{T}_2$
$\dot{T}_2$	$=\dot{T}_1$	P_2^e/C_2	$=\dot{T}_1$	$(P_2^l-P_1^l)/C_2$
$\dot{T}_3$	$=\dot{T}_1$	$=\dot{T}_2$	$=\dot{T}_1$	$=\dot{T}_2$
P_1^e	$K > 0$	$\dot{T}_2 C_1$	0	$-P_1^l + \dot{T}_2 C_1$
P_1^l	0	0	$K<0$	$K<0$
P_2^e	$\dot{T}_1 C_2$	$K > 0$	$P_1^l + \dot{T}_1 C_2 - P_2^l$	0
P_2^l	0	0	$K<0$	$K<0$
P_3^e	$-P_3^l + \dot{T}_1 C_3$	$-P_3^l + \dot{T}_2 C_3$	$P_2^l + \dot{T}_1 C_3 - P_3^l$	$P_2^l + \dot{T}_2 C_3 - P_3^l$
P_3^l	<0	<0	<0	<0

The subscript number refers to the stage, the superscripts e and l refer to electrically supplied heating or leaking power, respectively.
$^a K$ represents a constant value for the corresponding quantity of the first column.
$^b K_4$ is the constant heat sink temperature.

If desired, one can also force stage 4 to follow the temperature of stage 3, but then one needs a heat sink at a temperature T_5. In order to obtain the heat capacity C_1 of stage 1 it is necessary to measure accurately the constant power P_1^e and to obtain $\dot{T}_1$ from a careful measurement of the temperature versus time evolution of stage 1. In mode 2 one has to measure accurately the changing power P_1^e. Provided the constant power P_2^e and the heat capacity $C_2(T)$ are known, one arrives at $C_1(T)$ from

$$C_1/C_2 = P_1^e/P_2^e \qquad (2)$$

In the cooling modes 3 and 4 the ASC is operated in such a way that in the stationary state, ΔT_{12} and ΔT_{23} are kept constant over the entire scanning range. This requires a temperature setting of the stage 4 below the final temperature of the run. It is also essential that the thermal resistance between stages is sufficiently large and constant. In mode 3 the power P_1^e is set equal to zero and the cooling rate $\dot{T}_1$ becomes inversely proportional to C_1 provided that the heat loss P_1^l is constant. In mode 4 the temperature difference is set to be similar, and P_2^e becomes zero. The sample cools down with a constant rate provided C_2 and the thermal resistances remain constant. The power P_1^e applied to stage 1 in order to keep ΔT_{12} constant has to be measured to arrive at C_1.

In order to satisfy the conditions of zero temperature difference between the stages 1, 2 and 3 in mode 1 and 2, sensitive servo systems are needed. This is particularly important for ΔT_{12}. This might, however, become very difficult for mode 2 near a second order phase transition where, because of the sharp increase of the heat capacity, the power needed to keep a constant rate increases very rapidly. Moreover, if the phase transition is first order it will not be possible to deliver instantaneously the finite amount of latent heat to the sample in stage 1, and T_1 will lag behind T_2. A similar basic problem arises also in scanning transitiometry, as well as for DSCs, that also operate similarly to mode 2. This problem is more intrusive for DSCs because of the rather large scanning rates. The opposite situation occurs in the first mode. At or near a phase transition the servo system will have to deliver less power to stage 2 because $\dot{T}_1$ will slow down. First order transitions also do not pose a problem because in principle the rate $\dot{T}_1$ stays zero at the transition for a time interval given by

$$\Delta t \equiv t_f - t_i = \Delta H_L / P_1^e \qquad (3)$$

with ΔH_L the latent heat of the transition. t_i and t_f are times during the scan at which the transition is reached and left, respectively. In fact the direct experimental result $T(t)$ immediately gives the enthalpy as a function of temperature by

$$H_1 = H_1(T_s) + P_1^e(t - t_s) \qquad (4)$$

with T_s the starting temperature of the scanning run at the time t_s.

The implementation of cooling runs with constant (negative) power (mode 3) and with constant (negative) rate (mode 4) is less obvious because one has to implement a constant leaking power P_1^l or P_2^l. This is done by keeping ΔT_{12} and ΔT_{23} constant and small to assure heat transfer (mainly radiation) does not change over the scanned temperature range. This has to be verified and usually involves a calibration to arrive at absolute values for the heat capacity or enthalpy. The cooling mode 3 is very complementary to heating mode 1 and also easily allows one to deal with first order transitions. In that respect the cooling mode 4 encounters the same problems as the heating mode 2.

Although the four modes described above are convenient ones, there are several other possibilities to scan. As long as one measures simultaneously at each instant the (variable) power and the (variable) rate, Equation (1) allows one to calculate the heat capacity. For cooling runs, however, it might not be very practical to have control of a varying leaking power.

The construction of an ASC is optimized for scanning; it can easily be operated as a normal heat pulse step calorimeter as well. This can be very practical for calibration purposes and for verification of absolute heat capacity values.

13.3   Design and Operational Implementation

Essential in the operation of a scanning calorimeter is the high-resolution control and measurement of the temperature. An essential feature of this type of calorimeter is the elaborate efforts one has to make to isolate the specimen holder from the laboratory by means of a precisely controlled thermal environment. Over the years we have constructed several apparatuses. Differences resulted from adaptation to study specific types of samples, extension of temperature ranges or to novel electronic controlling possibilities. The first setups we constructed in the 1970s mechanically very much resembled the schematic diagram given in Figure 13.1.[20–23] The thermal environment around the liquid cell consisted of three massive concentric copper cylindrical shields, thermally insulated from each other by vacuum space and thin low thermal conductivity supports. The polished inside and outside gold plated sample cell is suspended by thin nylon threads inside stage 2. The polished inside of stage 2 is also gold plated and its outside and the other stages are nickel plated. Stages 2 and 3 have been equipped with 10 Ω constantan heating wires having a small temperature coefficient of resistance. The heating wires are evenly distributed on all sides of the stages and wound in deep grooves and thermally anchored with a good thermal conductive and electrically insulating epoxy. To minimize thermal transfer between stages all electric feed through leads, on passing from one stage to another, are several thermal diffusion lengths long, and are neatly coiled in order not to touch the wall of either stage. These leads are also thermally anchored at each stage. The temperature of the exterior stage 4 is set by means of a temperature controlled water (or oil) flow through copper tubing soldered on the exterior of that stage.

All inner stages had several thermistors (with room temperature values between 10 and 100 kΩ) incorporated in the walls to measure and regulate the temperature of these stages. Stage 2 also contained a reference platinum resistance thermometer to allow *in situ* calibration of the thermistors. In the heating mode 1 it was possible to arrive at a negligibly small heat exchange P_1^l, compared to the P_1^e values imposed, by keeping $\Delta T_{12} < 1$ mK by means of the servo system displayed in Figure 13.2. This system consisted of an ac-bridge with two thermistors R_{1a} (on stage 1) and R_{2a} (on stage 2) and an inductive divider, followed by a phase sensitive detector system and a power amplifier delivering current to the heater H_2 on stage 2. Stage 3 was forced to follow the temperature of stage 2 very closely by means of a less elaborate dc-bridge and servo system incorporating another thermistor on stage 2 and one on stage 3. Cooling mode 3 is the constant cooling power (P_1^l) analogue of the constant heating power mode 1. In this case one imposes a constant (negative) heat leak (P_1^l) from stage 1 to stage 2. This can be achieved for slow rates (over a not too large temperature interval) by keeping the temperature of stage 2 a constant value ΔT_{12} below that of stage 1. In both these modes of operation the basic measured quantities are the temperature of the sample cell as a function of time $T_1(t)$ and the electric heating power P_1^e (mode 1) or the leaking power P_1^l (mode 3). P_1^e can easily be obtained from the measurements of the voltage over and the current through heater H_1. To arrive at P_1^l one needs heat leak calibration runs. It can, however be derived in a much easier way from a comparison with heat capacities from heating runs or from heat pulse measured ones in the same temperature range.

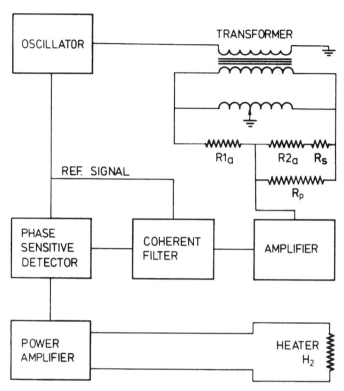

Figure 13.2   Diagram of an ac-bridge servo system between calorimeter stages 1 and 2. R_{1a} and R_{2a} are thermistors on stage 1 and 2, respectively. R_s and R_p are series and parallel resistors to bring the temperature coefficient of that bridge arm, with R_{2a}, closer to that of R_{1a}.

Although we occasionally used modes 2 and 4, modes 1 and 3 are the ones mainly used, because they not only allow us to obtain the heat capacity via Equation (1), but also directly the enthalpy $H(T)$ as a function of temperature via Equation (4). Knowing $H(T)$ is essential in order to be able to discriminate between first and second order phase transitions, which is quite often important for liquid crystals (see further).

In our initially built calorimeters the temperatures of the different stages were controlled by the elaborate Wheatstone bridge type setups as given in Figure 13.2 and described above, requiring careful calibration and regular readjustments. A typical mode 1 experimental run was always preceded by a long period of very careful temperature stabilization of stage 2 with a different thermistor (R_{2b}) in a separate bridge and heating system. The temperature of stage 3 was kept at a constant temperature difference (typically 50 mK) below that of stage 2. Stage 1 (inside stage 2) was allowed to drift freely until its temperature became equal to that of stage 2. The temperature stability could be kept better than 0.1 mK for periods of days. Once a stable condition was reached the bridge of the servo system S_{12} was set to zero and the power P_1^e was applied to H_1 on stage 1. The temperature versus time evolution of that stage was then measured via another thermistor R_{1b} with a six decade automatic ac potentiometer (ASL model 103), later on replaced by a 8 decade home made one. The rate of data collection (typically a reading every 5 to 20 s) and the time scale was set by means of an accurate digital clock. Data were collected via a teletype on paper tape for off-line analysis with a computer.

One of the drawbacks of measurements on liquids is the need for a (closed) sample holder. In order to have maximum sensitivity, the heat capacity of the holder should be as small as possible compared to that of the liquid. This can be realised easier with larger samples. With (thin walled) cells with a content of the order $100\ cm^3$ a ratio (holder to sample) of 0.2 is quite possible. However, for larger samples (sometimes with low thermal conductivity) one has to consider, even for slow scanning rates, a possible temperature difference of the order of the rate times the internal relaxation time. These relaxation times can be measured and corresponding temperature inhomogeneities estimated.[23] In order to substantially reduce possible T inhomogeneities stirring of the sample is very effective.

In Figure 13.3 a cross section of a cylindrical cell (with a possible liquid content of about $100\ cm^3$) with a built in stirring system and internal heating is given. This thin walled copper cell was gold coated inside and outside. The two electrical heaters are sealed 1 m meter long thin walled stainless steel tubes of 1 mm diameter containing a heating wire embedded in magnesium oxide insulating material. Stirring was achieved in the horizontally mounted cylindrical cell by including in the sample cell a thin walled, open ended stainless steel tube containing a gold plated copper ball that could roll back and forth in this tube by changing periodically (usually a few times per minute) the inclination of the plate supporting the calorimeter. This cell was mainly used for critical point studies in binary and ternary liquid mixtures.[19-23] In our early measurements on the liquid crystal octylcyanobiphenyl a similar cell as in Figure 13.3 was used, but scaled down by about a factor of four.[23,24] Later on for measurements in less common (delicate) liquid crystal compounds we scaled down much further by heating wires on the outside of the small cylindrical cells with thermistor embedded in the cell wall, but still containing a (small) stirring ball. These cells are typically made of tantalum or molybdenum for low heat capacity and good thermal conductivity. Presently, our smallest cell typically contains 200 to 300 mg.[37]

The appearance during the last ten to twenty years of fast PCs and measurement instrumentation with extensive interfacing capabilities as well as powerful software, has significantly simplified the design of ASCs as well their operation, and as a result, they can run, if desired, for weeks without human interference. The schematic diagram in Figure 13.4 provides an example of such a modern setup with a four stages ASC that can operate between room temperature and about 470 K. Stage 4 of this calorimeter is composed of a hot air oven and the outer thermal and vacuum shield of the actual calorimeter with three internal stages. The temperature of the oven is measured and controlled

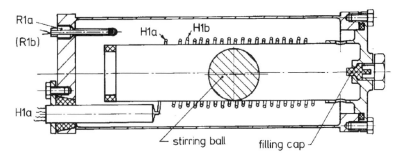

Figure 13.3   Diagram of a gold coated copper cylindrical sample holder (of about 100 cm^3) used for binary and ternary liquid mixture studies. H$_{1a}$ and H$_{2b}$ are two separate internal electrical heaters. R$_{1a}$ and R$_{1b}$ are temperature sensing thermistors for temperature measurement and control. Stirring can be achieved by rolling the stirring ball back and forth in the open ended thin walled central stainless steel tube (see text).

by means of thermistor Th$_4$ and computer controlled power delivery to the heater of the oven (with a temperature stability to within ± 1 K). In Figure 13.5 a diagram of the actual cylindrical calorimeter (horizontally placed in the oven) is given. The centrally placed sample cell can hold also a stirring ball (not shown). Stirring can here also be achieved by periodically changing the inclination of a plate supporting the oven.

13.4   Phase Transition Studies in Binary and Ternary Liquid Mixtures

In this section I will illustrate the application of ASCs for the detailed study of the critical behaviour of the heat capacity anomaly near consolute and plait points in binary and ternary liquid mixtures. Before giving some specific examples it is very useful to draw attention to some unique aspects of running an ASC in mode 1 (constant heating power) or mode 3 (constant cooling power) (see Table 13.1). Although one readily arrives at the heat capacity in Equation (1) from the (known) power and the temperature depending rate by numerical differentiation of the carefully measured temperature evolution, $T(t)$, of the sample with time, the fact that via Equation (4) $T(t)$ can immediately be transformed in an enthalpy versus temperature curve opens new possibilities for data analysis. This can best be demonstrated using Figure 13.6 with a generic enthalpy curve near a second order phase transition in a binary (or ternary) critical mixture at constant pressure.

   At temperature T with a corresponding enthalpy value $H(T)$, two quantities with dimensions (J kg^{-1} K^{-1}) of a specific heat capacity, C^s and C^c, are introduced. C^s correspond with the slope, dH/dT, of the enthalpy curve at T, at

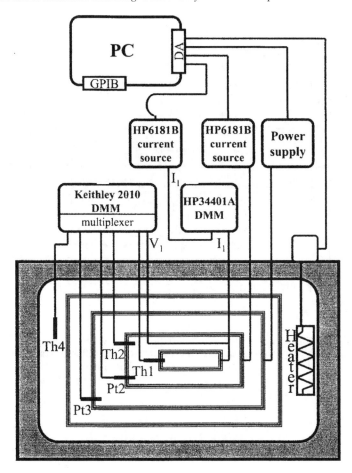

Figure 13.4 Schematic diagram of a four stages calorimeter with modern measurement and control instrumentation.

constant pressure and constant overall concentration for the systems considered here. The quantity C^c is defined as

$$C^c = \frac{H - H_c}{T - T_c} \qquad (5)$$

and thus corresponds to the slope of the chord connecting $H(T)$ at T with H_c at T_c. At second order phase transitions the limiting behaviour of the specific heat capacity at the critical point can be described by the following power law expression (see also Chapter 14 and references therein)

$$C^s = A|\tau|^{-\alpha} + B \qquad (6)$$

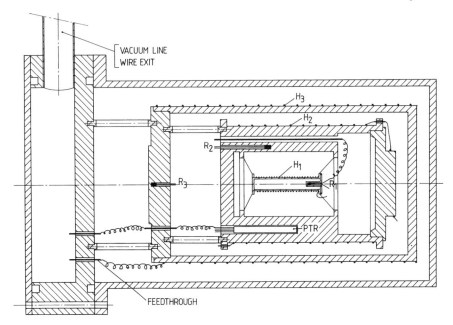

Figure 13.5 Schematic diagram of a adiabatic scanning calorimeter. Electric heaters and temperature sensing thermistors on the different stages are denoted by H and R. A platinum resistance thermometer for *in situ* calibration purposes is indicated by PTR. The whole calorimeter is placed in a hot air temperature controlled oven.

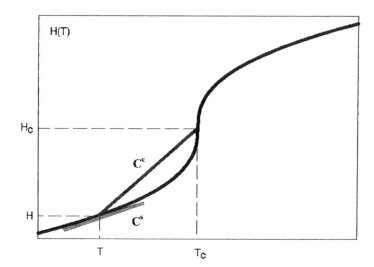

Figure 13.6 Generic enthalpy curve near a second order phase transition at $H_c(T_c)$. The slope of $H(T)$ at a temperature T corresponds to C^s and the slope of the chord between $H(T)$ and H_c corresponds to C^c.

with $\tau = (T - T_c)/T_c$, α the critical exponent, A the critical amplitude and B the background. It can easily be shown that C^c also has a power law behaviour of the form[21,24]

$$C^c = \frac{A}{1 - \alpha}|\tau|^{-\alpha} + B \tag{7}$$

Thus, both quantities have the same critical exponent and background term. There is, however, a different critical amplitude. Both Equations (6) or (7) can be used for fitting experimental data of the corresponding quantities to arrive at important values for the critical exponent α and amplitude A. It is usually better to use C^c instead of C^s, because in the latter case one has, in order to have high T resolution, to divide locally along the $H(T)$ curve small ΔH values by small ΔT values (corresponding to small ΔT steps in classical adiabatic calorimeter). In the former case for most of the data one divides quite large $(H - H_c)$ by large $(T - T_c)$, resulting in higher resolution. We have extensively used this approach to analyse the critical behaviour of several binary and ternary liquid systems.[21–23,38–40] In Figure 13.7 this is illustrated for the binary system triethylamine-heavy water.[21] In the top part the direct enthalpy data are displayed and in the lower part the derived values for the quantity C^c. These results are measured for the critical mole fraction and at constant pressure.

Although the quantity C^c is very useful for the analysis of critical behaviour, quite often it is more appropriate to use the thermodynamic specific heat capacity C^s directly obtained via Equation (1) (e.g. away from a critical point, and certainly for non-critical systems). ASC allows to obtain high resolution result for this quantity as well. Moreover, as already explained above, ASCs can also be operated as classical heat pulse adiabatic calorimeters. This is illustrated in Figure 13.8 where specific heat capacity results, obtained in the two ways, are given for some binary and ternary critical mixtures containing triethylamine.[22]

13.5 Phase Transition Studies in Liquid Crystals

Liquid crystals are composed of anisotropic organic molecules that do not melt in a single stage from the crystalline solid to an isotropic liquid.[41–43] They exhibit one or more mesophases between the solid and the isotropic liquid that have a symmetry intermediate between that of solids and isotropic liquids. The different liquid crystalline phases are characterized by the orientational order of the long molecular axes (for rod-like molecules) and by no or partial positional order of the centres of mass of the molecules. In the nematic phase there is only long range orientational order present, while the different types of smectic phases exhibit one-dimensional positional order in a typical layered structure. Phase transitions between the different phases in liquid crystals can be either first order or second order, where critical fluctuations may play an important role. During the last three decades an impressive amount of calorimetric results have been obtained, in many cases resulting in a better understanding of several

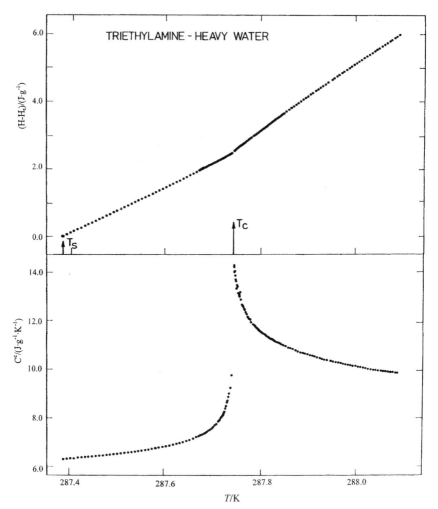

Figure 13.7   Temperature dependence of the enthalpy and of the quantity C^c defined in Equation (5) for the critical mole fraction of the binary liquid mixture triethylamine-heavy water exhibiting a lower critical solution temperature (homogeneous phase below T_c and two phase region above T_c).

phase transitions in liquid crystals. Several extensive reviews have been published[18,44–47] and an up to date account is given in Chapter 17. We have used ASC extensively for phase transitions studies in liquid crystals. The fact that with ASC one can distinguish between first and second order transition has played a significant role in our present understanding of several phase transitions. Here I only want to give some illustrative examples. For full details I refer to the cited reviews and to Chapter 17 and references cited therein. Many liquid crystalline compounds exhibit several phase transitions, quite often in limited temperature ranges. The compound heptyloxybenzylide-heptylaniline (7O.7)

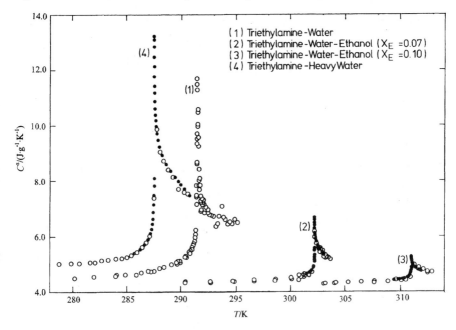

Figure 13.8 Specific heat capacity values at constant pressure and constant overall concentration for several binary and ternary mixtures containing triethylamine. The open symbols are obtained with the heat pulse method. The solid symbols are results obtained via Equation (1).

exhibits e.g. in the temperature range between 305 K and 358 K six major phase transitions. Using ASC we have studied this compound in detail.[48]

In Figure 13.9 an overview of the specific heat capacity results is displayed. As a function of decreasing temperature one encounters an isotropic (I) phase, a nematic (N) phase, a smectic A (A) and a smectic C (C) phase. These are all true liquid phases with different degrees of orientational and positional order. The B phase, the G phase and the crystal phase (not shown) below the G phase all posses long range three-dimensional positional order and are thus to be considered as (soft) crystal phases. With the exception of the smectic C to smectic A transition all the transitions are first order ones. The structure seen in the curve in the B phase is not noise but corresponds to four first order transitions with very small latent heats. High resolution X-ray scattering experiments have identified these as layer restacking transitions in this phase with very anisotropic degrees of stiffness.[49–51] Thus, strictly speaking, one should count ten phase transitions in the range between 305 K and 358 K. The ability of ASC to discriminate between types of phase transitions and accurately determine latent heats when present, is illustrated in Figure 13.10.

In a temperature range of about 0.5 K two phase transitions, AN and NI, are observed. The enthalpy in the left part of the figure clearly shows discrete (latent heat) jumps at these two first order transitions. The right part displays the corresponding specific heat capacity as derived via Equation (1). It should

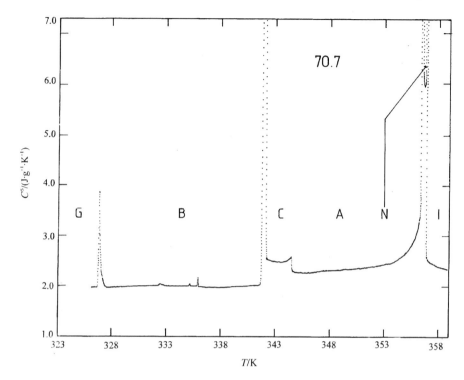

Figure 13.9   Specific heat capacity as a function of temperature for several phases and phase transitions of the liquid crystal compound heptyloxybenzylide-heptylaniline (7O.7). I is the normal isotropic liquid phase, N is the nematic phase, A is the smectic A phase, C is the smectic C phase, B and G are soft crystalline phases. The structure in the B phase corresponds to four layer restacking transitions (see text). All transitions are first order except the C to A transition.

also be noted that the enthalpy jumps are not perfectly vertical because of (slight) impurity broadening. The width of the corresponding two phase region is indicated on the T axis and by the vertical dashed lines.

As already pointed out above, near second order phase transitions both C^s and C^c can be used to arrive at the relevant information on the heat capacity anomaly by using either Equation (6) or (7) in a non-linear curve fitting procedure. However, by considering the difference $(C^c - C^s)$, above or below T_c, the (unimportant) background term drops out, resulting in:

$$C^c - C^s = \frac{\alpha A}{1 - \alpha} \tau^{-\alpha} \qquad (8)$$

Taking the logarithm of both sides of Equation (8) gives

$$\lg(C^c - C^s) = \lg(\frac{\alpha A}{1 - \alpha}) - \alpha \lg|\tau| \qquad (9)$$

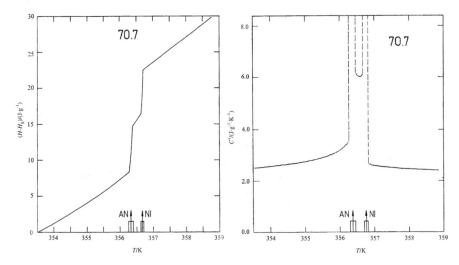

Figure 13.10 Enthalpy and specific heat capacity of the liquid crystal compound heptyloxybenzylidene- heptylaniline (7O.7) in the temperature range of the smectic A to nematic and the nematic to isotropic phase transitions.

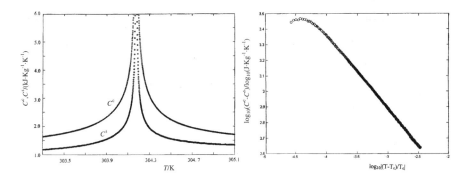

Figure 13.11 Left: temperature dependence of the quantities C^s and C^c over the smectic A to nematic phase transition of a mixture of the liquid crystal octylcyanobiphenyl (8CB) and cyclohexane(CH) with mole fraction $x_{CH} = 0.0451$. Right: double logarithmic plot of the difference $(C^c - C^s)$ versus the absolute value of the reduced temperature difference $(T - T_c)/T_c$ for the data the smectic A phase.

Thus one obtains a straight line with a (negative) slope immediately giving the critical exponent α. This is illustrated in Figure 13.11 for the AN transition in a mixture of the liquid crystal octylcyanobiphenyl (8CB) and cyclohexane (CH) (with mole fraction $x_{CH} = 0.0451$).[52,53] The left part of the figure gives the experimental values for C^s and C^c. In the right part their difference (below the transition) is displayed on a double logarithmic plot. Within experimental (enthalpy) resolution the transition is still second order but close to a tricritical point. Over a temperature range of about 20 mK the transition is slightly (impurity) broadened. The obtained critical exponent $\alpha = 0.49 \pm 0.03$ is close to the tricritical value of 0.50. Further details can be found in the cited references.

References

1. W. Nernst, *Ann. Phys.*, 1911, **36**, 395.
2. M. J. O'Neill, *Anal. Chem.*, 1964, **36**, 1238.
3. B. Wunderlich, *Thermal Analysis*, Academic Press, San Diego, 1990.
4. S. L. Randzio, *Thermochim. Acta*, 1985, **89**, 215.
5. S. L. Randzio, J.-P. E. Grolier and J. R. Quint, *Cal. Anal. Therm.*, 1990, **20–21**, 315.
6. S. L. Randzio, *Chem. Soc. Rev.*, 1996, **25**, 383.
7. P. Sullivan and G. Seidel, *Phys. Rev.*, 1968, **173**, 679.
8. C. W. Garland, *Thermochim. Acta*, 1985, **88**, 127.
9. N. Birge and S. Nagel, *Phys. Rev. Lett.*, 1985, **54**, 2674.
10. D. H. Jung, I. K. Moon and Y. H. Jeong, *Thermochim. Acta*, 2002, **391**, 7.
11. Y. J. Yun, D. H. Jung, I. K. Moon and Y. H. Jeong, *Rev. Sci. Istrum.*, 2006, **77**, 064901.
12. M. E. Bagatskii, A. V. Voronel and V. G. Gusak, *Sov. Phys. JETP*, 1963, **16**, 517.
13. A. V. Voronel, Yu. R. Chaskin, V. A. Popov and V. G. Simkin, *Sov. Phys. JETP*, 1964, **18**, 568.
14. A. V. Voronel, V. G. Goburnova, Yu. Chaskin and V. V. Shechekochikhina, *Sov. Phys. JETP*, 1966, **22**, 597.
15. C. Edwards, J. A. Lipa and M. J. Buckingham, *Phys. Rev. Lett.*, 1968, **20**, 496.
16. J. A. Lipa, C. Edwards and M. J. Buckingham, *Phys. Rev. Lett.*, 1970, **25**, 1086.
17. M. J. Buckingham, C. Edwards and J. A. Lipa, *Rev. Sci. Instrum.*, 1973, **44**, 1167.
18. J. Thoen, in *Physical Properties of Liquid Crystals* eds. D. Demus, J. Goodby, G. Gray, H.-W. Spiess and V. Vill, Wiley-VCH, Weinheim, 1997, pp. 208–232.
19. J. Thoen, E. Bloemen and W. Van Dael, *J. Chem. Phys.*, 1978, **68**, 735.
20. E. Bloemen, Ph. D. Thesis, 1979, Katholieke Universiteit Leuven, Belgium.
21. E. Bloemen, J. Thoen and W. Van Dael, *J. Chem. Phys.*, 1980, **73**, 4628.
22. E. Bloemen, J. Thoen and W. Van Dael, *J. Chem. Phys.*, 1981, **75**, 1488.
23. J. Thoen, E. Bloemen, H. Marijnissen and W. Van Dael, in *Proceedings of the 8th Symposium on Thermophysical properties*, Nat. Bur. Stand., 1981, Maryland, Am. Soc. Mech. Eng., New York, 1982, pp. 422–428.

24. J. Thoen, H. Marijnissen and W. Van Dael, *Phys. Rev. A*, 1982, **26**, 2886.
25. U. Würz and M. Grubić, *J. Phys. E: Sci. Instrum.*, 1980, **13**, 525.
26. A. Junod, *J. Phys. E: Sci. Instrum.*, 1979, **12**, 945.
27. P. C. Lancaster and D. P. Baker, *J. Phys. E: Sci. Instrum.*, 1981, **14**, 805.
28. M. A. Anisimov, V. P. Voronov, A. O. Kulkov and F. Kholmurodov, *J. Phys. (Paris)*, 1985, **46**, 2137.
29. V. G. Bessergenev, Yu. A. Kovalevskaya, I. E. Paukov and Yu. A. Shkredov, *Thermochim. Acta*, 1989, **139**, 245.
30. M. Lysek, P. Day, M. LaMadrid and D. Goodstein, *Rev. Sci. Instrum.*, 1992, **63**, 5750.
31. E. B. Sirota and D. M. Singer, *J. Chem. Phys.*, 1994, **101**, 10873.
32. E. B. Sirota, *J. Chem. Phys.*, 2000, **112**, 492.
33. W. Schnelle and E. Gmelin, *Thermochim. Acta*, 1995, **269/270**, 27.
34. I. K. Moon and Y. H. Jeong, *Rev. Sci. Instrum.*, 1996, **67**, 3553.
35. Y. H. Jeong, *Thermochim. Acta*, 1997, **304/305**, 67.
36. A. C. Flewelling, R. J. Fonseka, N. Khaleeli, J. Partee and D. T. Jacobs, *J. Chem. Phys.*, 1996, **104**, 8048.
37. G. Cordoyiannis, D. Apreutesei, G. Mehl, C. Glorieux and J. Thoen, *Phys. Rev. E*, 2008, **78**, 011708.
38. J. Thoen, J. Hamelin and T. K. Bose, *Phys. Rev. E*, 1996, **53**, 6264.
39. B. Van Roie, G. Pitsi and J. Thoen, *J. Chem. Phys.*, 2003, **119**, 8047.
40. S. Pittios, B. Van Roie, C. Glorieux and J. Thoen, *J. Chem. Phys.*, 2004, **122**, 024504.
41. P. G. de Gennes and J. Prost, *The Physics of Liquid Crystals,* Clarendon Press, Oxford, 1993.
42. G. Vertogen and W. H. de Jeu, *Thermotropic Liquid Crystals, Fundamentals,* Springer-Verlag, Berlin, 1988.
43. M. A. Anisimov, *Critical Phenomena in Liquids and Liquid Crystals,* Gordon and Breach, Philadelphia, 1991.
44. J. Thoen, in *NATO ASI Ser. B, Phase Transitions in Liquid Crystals* eds. S. Martellucci and A. N. Chester, Plenum, New York, 1992, pp. 155–174.
45. C. W. Garland, in *NATO ASI Ser. B, Phase Transitions in Liquid Crystals* eds. S. Martellucci and A. N. Chester, Plenum, New York, 1992, pp. 175–187.
46. J. Thoen, *Int. J. Mod. Phys. B*, 1995, **9**, 2157.
47. C. W. Garland, in *Liquid Crystals: Experimental Study of Physical Properties and Phase Transitions* ed. S. Kumar, Cambridge University Press, Cambridge, 2001, pp. 240–294.
48. J. Thoen and G. Seynhaeve, *Mol. Cryst. Liq. Cryst.*, 1985, **127**, 229.
49. J. Collett, L. B. Sorenson, P. S. Pershan, J. D. Litster, R. J. Birgeneau and J. Als-Nielsen, *Phys. Rev. Lett.*, 1982, **49**, 553.
50. J. Colett, L. B. Sorensen, P. S. Pershan and J. Als-Nielsen, *Phys. Rev. A*, 1985, **32**, 1035.

51. E. B. Sirota, P. S. Pershan, L. B. Sorensen and J. Collett, *Phys. Rev. A*, 1987, **36**, 2890.
52. K. Denolf, B. Van Roie, C. Glorieux and J. Thoen, *Phys. Rev. Lett.*, 2006, **97**, 107801.
53. K. Denolf, G. Cordoyiannis, C. Glorieux and J. Thoen, *Phys. Rev. E*, 2007, **76**, 051702.

Heat Capacities in the Critical Region

MIKHAIL ANISIMOV[a] AND JAN THOEN[b]

[a] Institute for Physical Science & Technology and Department of Chemical & Biomolecular Engineering, University of Maryland, College Park, MD 20742, USA; [b] Laboratorium voor Akoestiek en Thermische Fysica, Departement Natuurkunde en Sterrenkunde, Katholieke Universiteit Leuven, Celestijnenlaan 200 D, B-3001, Leuven, Belgium

14.1 Introduction

In early and mid 1960s Voronel and co-workers[1–4] made a major discovery in the thermodynamics of gas–liquid critical phenomena. For the first time, they performed accurate measurements of the isochoric heat capacity of several simple fluids by the pulse adiabatic method as close to the critical temperature as a few hundredths of a degree. However, most importantly, they made some necessary measures to eliminate, as much as it was possible at that time, the effects of fluid inhomogeneities caused by the anomalously large compressibility and, hence, an extreme sensitivity to all possible kinds of perturbations, such as gravity, temperature gradients, and impurities. Voronel and co-workers claimed that the isochoric heat capacity diverges at the critical point, although weaker than the isobaric heat capacity. The divergence of the isobaric heat capacity at the critical point of single-component fluids rigorously follows from the thermodynamic relation

$$C_p - C_V = \frac{T}{\rho^2} \left(\frac{\partial p}{\partial T} \right)_V^2 \left(\frac{\partial \rho}{\partial p} \right)_T, \tag{1}$$

Heat Capacities: Liquids, Solutions and Vapours
Edited by Emmerich Wilhelm and Trevor M. Letcher
© The Royal Society of Chemistry 2010
Published by the Royal Society of Chemistry, www.rsc.org

where C_p is the isobaric molar heat capacity, C_V is the isochoric molar heat capacity, T is the temperature, p is the pressure, V is the volume, and ρ is the molar density amount of substance per unit volume. Since the isothermal compressibility $\kappa_T = \rho^{-1}(\partial \rho / \partial p)_T$ is positive and infinite at the critical point while $(\partial p / \partial T)_V$ is finite, C_p diverges following the divergence of the isothermal compressibility. The statement that C_V also diverges at the critical point, though not forbidden by thermodynamics, was unexpected and at first considered as controversial. The classical Gibbs and van der Waals theory of critical phenomena, generalized by Landau,[5] predicts a discontinuity of the finite isochoric heat capacity upon crossing the two-phase boundary. Commonly used in engineering practice equations of state, from the van der Waals equation to most sophisticated ones, are based on the classical theory and all predict the finite isochoric heat capacity at the critical point.

Accurate experiments near critical points of fluids and fluid mixtures are challenging. The quality and reproducibility of experimental data collected in the critical region are often determined by the quality of the physical state of the system under investigation rather than by the resolution of the instrument. Reproducible distortions of "ideal" critical anomalies have often been sources of misinterpretation and confusion.[6] A measurement is always associated with a perturbation. In a typical thermal experiment on a fluid, the perturbation may be neglected if

$$\varepsilon \ll k_B T \qquad (2)$$

Where ε is the perturbation energy per molecule and k_B is Boltzmann's constant. In the critical region, where the susceptibility of fluids to external perturbations is anomalously large, the condition given by Inequality (2) is insufficient and must be replaced by a more severe constraint:

$$\frac{\varepsilon}{k_B T} \ll |\tau| \qquad (3)$$

where $\tau = (T - T_c)/T_c$ with T_c being the critical temperature. At the distance from the critical temperature (along the critical isochore) of about 0.3 K and $T_c = 300$ K, the requirement of obtaining undisturbed data is a thousand times more severe than under regular conditions!

Another important constraint is to ensure that the near-critical fluid is in thermodynamic equilibrium. The characteristic relaxation time t_R is very large in the critical region. The thermal relaxation time in a single-component near-critical fluid can be estimated as

$$t_R \simeq \frac{L^2 C_p}{\lambda} \qquad (4)$$

where L is a characteristic length scale (e.g. the linear size of the calorimeter cell) and λ is the thermal conductivity. Since, upon approaching the critical point, the isobaric heat capacity increases much faster than the thermal conductivity, the relaxation time increases and ultimately diverges at the critical point. To obtain thermodynamic equilibrium data in the critical region, one must be sure that the relaxation time is much smaller than the characteristic measurement time. Very slow thermal equilibration of fluids in the critical region is the major reason why the popular differential scanning calorimeter, despite its high resolution, is unable to provide quantitative information on the heat-capacity critical anomalies. As discussed in the previous chapter,[7] the most reliable and accurate calorimetric method to study near-critical anomalies in fluids is adiabatic slow-scanning calorimetry. With the scanning rate as low as 10^{-5}–10^{-6} K s^{-1}, it is possible to obtain equilibrium data for a sample with a relaxation time on the order of an hour.

Due to the anomalously large compressibility, the presence of gravity causes a significant inhomogeneity of the density and may dramatically affect the measurements of the heat capacity in the critical region. The effect of gravity on the apparent heat-capacity anomaly, which strongly depends on the height of the sample, is illustrated in Figure 14.1.

The gravity effect, as well as the equilibration time, can also be significantly reduced, but not completely eliminated, by proper stirring.[6] A comprehensive review of the effects of gravity on critical phenomena has been recently made by Barmatz *et al.*[9] More detailed discussions of specific requirements for obtaining undisturbed and reliable experimental information on near critical fluids can be found in the review of Voronel[6] and in the book of Anisimov.[10]

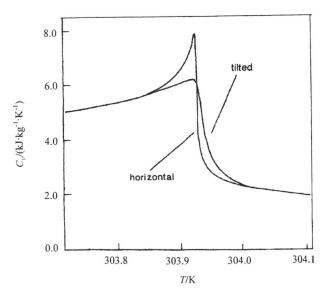

Figure 14.1   Isochoric heat capacity of carbon dioxide along the critical isochore measured in a disc-shaped sample for two positions, horizontal (1 mm height) and tilted (14 mm height), adapted from reference 8.

14.2   Isochoric Heat Capacity Near the Gas–Liquid Critical Point

With an increase of the resolution of adiabatic calorimetry and the overall quality of the heat-capacity measurements, by the early 1970s it became obvious that the divergence of the isochoric heat capacity at the gas–liquid critical point is a universal phenomenon for all single-component fluids. Figure 14.2 shows the isochoric heat capacity of argon on the critical isochore as a function of temperature obtained by an adiabatic pulse calorimeter.

In Figure 14.3 the same data are shown in semi-logarithmic scale. In the asymptotic critical region the experimental data follow a simple power law:

$$\frac{C_V}{R} = \frac{A^{\pm}}{\alpha}\,|\tau|^{-\alpha} + B \tag{5}$$

where α is the critical exponent, $A^{\pm}$ is the asymptotic critical amplitude with $\pm$ representing the condition above and below the critical temperature, B is the analytical background, and R is the gas constant.

The factor $1/\alpha$ in the definition of the amplitude of the power law is introduced to incorporate the possibility of the logarithmic divergence since

$$\lim \frac{|\tau|^{-\alpha}}{\alpha} \to -\ln|\tau| \text{ as } \alpha \to 0$$

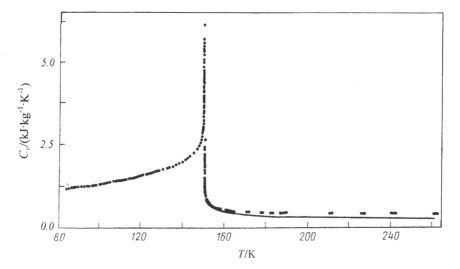

Figure 14.2   Isochoric heat capacity of argon along the critical isochore.[10,11] Solid curve is the asymptotic power law given by Equation (5), shown in the one-phase region (above the critical temperature).

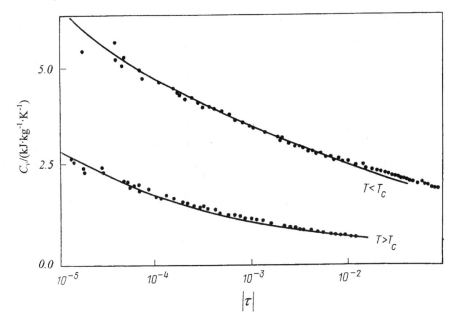

Figure 14.3 Isochoric heat capacity of argon along the critical isochore in a semi-logarithmic scale.[10,11] Solid curves represent the asymptotic power law (5) with the critical exponent $\alpha = 0.11$.

The solid curves in Figure 14.3 represent a fit to Equation (5) in the range from $\tau \simeq 10^{-2}$ to $\tau \simeq 2.10^{-5}$ with a fixed value of $\alpha = 0.11$. Some subtle deviations from the simple power law, noticeable in Figure 14.3 above the critical temperature, suggest that the asymptotic range may be even narrower.

Since late 1960s and early 1970s a number of fluids in the critical region have been tested by various experimental groups.[11–22] These experiments basically confirmed the validity of the power law given by Equation (5) and suggested a universal value of the critical exponent $\alpha \simeq 0.10$–0.12. While the critical amplitudes $A^{\pm}$ are strongly system-dependent, the ratio A^{-}/A^{+} was found to be about 2, suggesting that it might also be universal for all fluids. At the same time, the universality of the critical exponent and of the critical amplitude ratio was predicted by the renormalization-group theory (RG).[23–25] According to the latest RG calculations,[25,26] $\alpha \simeq 0.109$ and $A^{-}/A^{+} \simeq 1.91$. The theory also predicts corrections to the asymptotic power law given by Equation (5) if experimental data are considered in an extended critical region ($10^{-3} < \tau < 10^{-1}$).

Impressive measurements of the isochoric heat capacity in the critical region of SF_6 in microgravity conditions were performed in late 1990s by Haupt and Straub[27] with a scanning-radiation calorimeter during a German Spacelab Mission. For the slowest cooling runs of about $10^{-5}\,K\,s^{-1}$, undistorted data, shown in Figure 14.4, were obtained as close to the critical temperature as $10^{-3}\,K$. It was also found that the asymptotic power law given by Equation (5) was only valid at $\tau \le 1.6 \cdot 10^{-4}$.

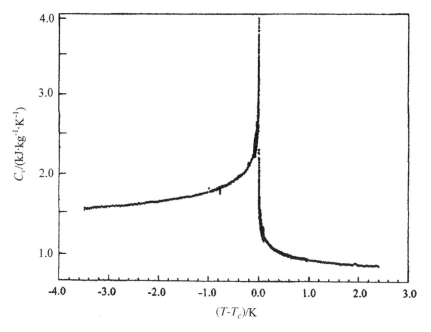

Figure 14.4 Isochoric heat capacity of SF_6 along the critical isochore performed with a scanning-radiation-calorimeter in microgravity during a German Spacelab Mission (adopted from ref. 27).

As shown in Figure 14.5, the experimentally obtained value of the critical exponent is strongly correlated with the value of the critical temperature. However, the best values of α and A^-/A^+, obtained from the fits of the SF_6 data in the asymptotic region, almost perfectly matched the theoretical universal values 0.109 and 1.91. More recently, Barmatz et al.[28] have reanalyzed the microgravity heat capacity measurements in SF_6 made by Haupt and Straub and concluded that the validity of the asymptotic power law given by Equation (5) may be extended to a reduced temperature up to $\tau \simeq 10^{-2}$. This result may indicate that non-asymptotic corrections, obtained in "ground" experiments, might be also affected by gravity.

14.3   The Nature of the Heat Capacity Anomaly in the Critical Region

In the classical, van der Waals-like, theory of critical phenomena there is no divergence of the isochoric heat capacity at the critical point. Instead, the classical theory predicts a finite discontinuity (a "jump") of the isochoric heat capacity at crossing of the phase boundary along the critical isochore. The modern theory of critical phenomena, RG, shows that the classical theory is an approximation, known as the mean-field approximation, which does not

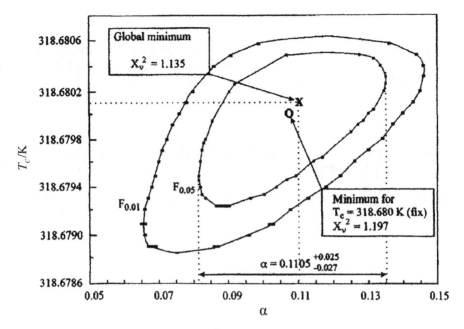

Figure 14.5 Contours of constant X_ν^2 in the T_c-α parameter space representing the 95 % (F0.05) and 99 % (F0.01) confidence level for the values of α and T_c determined in the asymptotic region of SF_6.[27]

properly account for the fluctuations of density. In the critical region, due to the anomalously large compressibility, the fluctuations of density become very large. The critical fluctuations significantly affect thermodynamic and transport properties of fluids, making them non-analytic functions of temperature and density. The size of fluctuation inhomogeneities ξ, known as the correlation length, diverges at the critical point. Along the critical isochore, asymptotically close to the critical point,

$$\xi = \xi_0^\pm |\tau|^{-\nu} \tag{6}$$

where $\nu = 0.630$ is a universal critical exponent[25,26] and $\xi_0^\pm$ is a system-dependent critical amplitude, which is on the order of the range of the inter-molecular forces. The power law given by Equation (6) is confirmed by accurate light scattering measurements. The RG theory predicts that the two universal critical exponents, α and ν, are interrelated as

$$2 - \alpha = 3\nu \tag{7}$$

While the system-dependent amplitudes, A^+ and ξ_0^+ form a universal relation,[26]

$$\frac{A^+}{\alpha} \rho_c \left(\xi_0^+ \right)^3 \simeq 0.171 \tag{8}$$

where ρ_c is the critical molar density. Equation (8), which is known as the "two-scale factor universality",[29] is confirmed by calorimetric and light-scattering experiments.[10] It follows from Equations (7) and (8) that the critical fluctuations are the only source of the divergence of the isochoric heat capacity at the critical point. If the range of the intermolecular forces, ξ_0^+, in fluids were infinite, the heat-capacity critical amplitude A^+ would vanish.

14.4 Isobaric Heat Capacity Near the Gas-Liquid Critical Point

From Section 14.2 it follows that not only C_p but also C_V diverges at the liquid-gas critical point. However, the divergence in C_p is much stronger than the one in C_V. This is clearly visible in Figure 14.6, where values of the specific heat

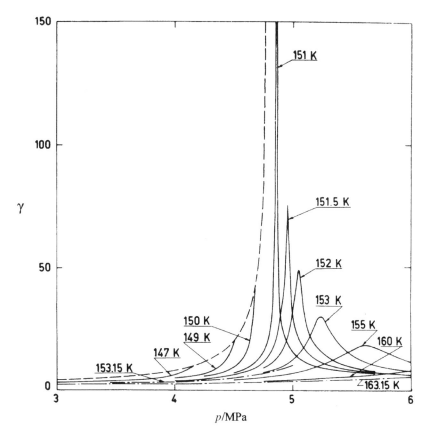

Figure 14.6 The specific heat capacity ratio γ for argon as a function of pressure for several isotherms above $T_c = 150.65$ K, and isotherms in the gaseous phase below the critical temperature.[30] The dashed line gives γ values for the saturated vapor. The dash-dotted lines are results obtained by Michels *et al.*[31]

capacity ratio

$$\gamma = \frac{C_p}{C_V} = \frac{\kappa_T}{\kappa_S} \tag{9}$$

are shown along isobars near the critical point of argon.[30] In Equation (9) κ_T is the isothermal compressibility and κ_S the adiabatic compressibility. The curves in Figure 14.6 have been calculated from a combination of experimental data for the sound velocity[30] and equation-of-state data.[31] While the ratio γ diverges at the critical point, it follows from Equation (1) that the divergence of C_p is the same as that of κ_T, because $(\partial p / \partial T)_V^2$ is finite at the critical point.

Because of the anomalously large isothermal compressibility, calorimetric measurements of the isobaric heat capacity near the gas-liquid critical point are not as accurate as those of the isochoric heat capacity. Measurements of the isobaric heat capacity are performed with a so-called flowing calorimeter.[32,33] In the critical region, even slight variations in pressure may result in large variations in the density. Along the critical isotherm,

$$p - p_c \propto |\rho - \rho_c|^\delta \tag{10}$$

where $\delta \simeq 4.8$ is a universal critical exponent.[10] Therefore, when heat is introduced into the calorimeter, the pressure is to be maintained as constant as possible.

Figure 14.7 demonstrates the comparison of experimental data on the isobaric heat capacity of methane at various pressures with the predictions of a crossover equation of state based on RG theory. In accordance with the thermodynamic relation given by Equation (1), the anomaly of the isobaric heat capacity indeed follows the anomaly of the isothermal compressibility.

A theoretical equation of state based on RG theory nicely correlates isobaric heat capacity and pVT data in an extended ("crossover") critical region.[34]

14.5 Isobaric and Isochoric Heat Capacities of Binary Fluids Near the Liquid-Liquid Critical Point

Binary fluid mixtures may exhibit complex phase behavior and, hence, various kinds of critical phenomena, gas-liquid, liquid-liquid, and gas-gas.[35] Heat capacities of weakly-compressible liquid mixtures near liquid-liquid critical points are best studied experimentally. Both the isobaric and isochoric heat capacities at constant overall concentration x, $C_{p,x}$ and $C_{V,x}$ demonstrate the anomalies apparently similar to that of C_V in single-component fluids. Both the isobaric and isochoric heat capacity can be measured in the same high-pressure calorimetric cell. In the presence of a vapor bubble (a non-critical phase), the measured heat capacity is very close to the isobaric one. Above the point where the vapor bubble disappeared, the measured heat capacity is isochoric. The temperature, where the vapor bubble disappears, can be controlled by the

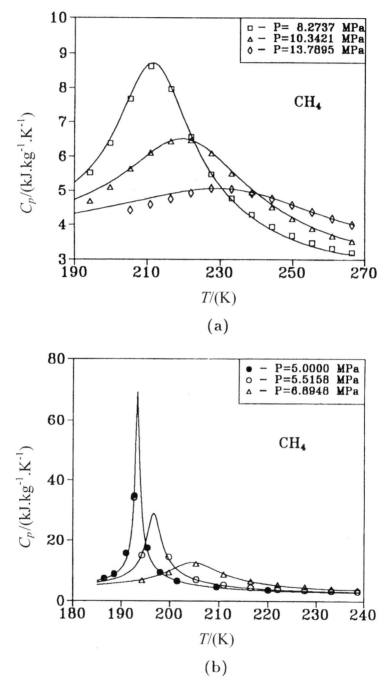

Figure 14.7 The isobaric specific heat capacity of methane at various pressures as a function of temperature. The symbols represent experimental data and the curves are calculated with an equation of state based on RG theory.[34] The critical pressure of methane is 4.599 MPa.

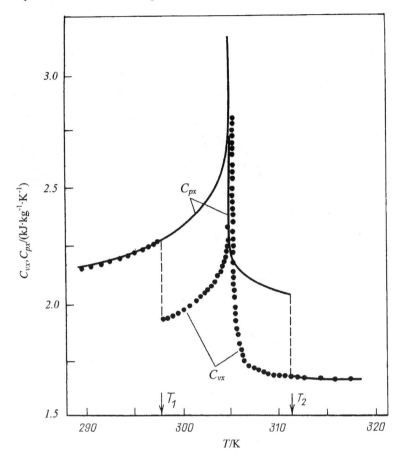

Figure 14.8 Isobaric and isochoric heat capacities of a nitroethane-isooctane mixture near the liquid-liquid critical points.[10] Dots indicate the measurements when the vapor bubble disappears below the critical temperature, at $T = T_1$, making the heat capacity near the critical temperature isochoric. Solid curve is an approximation for the measurements when the bubble disappeared above the critical temperature, at $T = T_2$, allowing the heat capacity to remain almost isobaric below T_2. The peaks are shifted with respect to each other since the liquid-liquid critical temperature slightly depends on pressure (~ 2 mK atm^{-1} for this particular mixture).

overall mass of the substance in the cell. An example of such an experiment is shown in Figure 14.8.

The corrections caused by the change in the saturation pressure (for $C_{p,x}$) and by the expansion of the calorimetric cell after the vapor bubble disappears (for $C_{V,x}$) were proven to be negligible.[10] The fact that both anomalies are apparently similar to each other is qualitatively explained by the thermodynamic relation given by Equation (1). The isothermal compressibility of binary fluids is particularly small for binary liquids far away from the gas-liquid critical points.

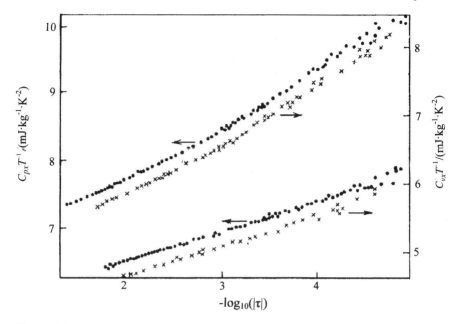

Figure 14.9 Isobaric and isochoric heat capacities of a nitroethane-isooctane mixture above (two lower data sets) and below (two upper data sets) the liquid-liquid critical point as a function of the distance from the critical temperature.[10] Dots indicate the isobaric heat-capacity measurements and crosses the isochoric heat capacity.

In Figure 14.9 the measurements of the heat capacities in the mixture of nitroethane and isooctane are presented in a semi-logarithmic scale.

The isobaric heat capacity was measured near the liquid-liquid critical points, both "upper" (the phase separation is induced by cooling) and "lower" (the phase separation is induced by heating), for many binary mixtures.[36–48] It was commonly observed that the isobaric heat capacity of binary mixtures at constant (critical) composition demonstrates the same divergence at the liquid-liquid critical point as the isochoric heat capacity of a single-component fluid at the gas-liquid critical point:

$$\frac{C_{p,x}}{R} = \frac{A^{\pm}}{\alpha} |\tau|^{-\alpha} + B \qquad (11)$$

with $\alpha \simeq 0.11$ and $A^{-}/A^{+} \approx 1.91$. Moreover, the isobaric heat capacities of binary liquid mixtures at constant composition obey the two-scale factor universality relation, Equation (8), established for the isochoric heat capacities of single-component fluids.

The size of the observed critical anomaly, as compared with that in pure fluids, is usually not very large and strongly depends on the correlation-length amplitude $\xi_0^{\pm}$ in accordance with Equation (8). The binary mixture of triethylamine-heavy water probably has the largest $C_{p,x}$ anomaly observed so far.[40] Data for this system (which exhibits a lower critical solution temperature) obtained by adiabatic slow-scanning calorimetry (ASC) are displayed in Figure 14.10 by the open symbols.

The fact that the direct experimental result in ASC is the enthalpy $H(T)$ as a function of temperature (see Chapter 13) allows one to introduce the quantity

$$C = \frac{H(T) - H_c}{T - T_c} \tag{12}$$

Data for this quantity are also given in Figure 14.10 by the solid dots. As pointed out in Chapter 13, the quantity C (with dimensions of a heat capacity) exhibits a critical divergences with the same critical exponent α as the heat capacity, but with a $1/(1-\alpha)$ times larger critical amplitude. In Figure 14.11, the C data are presented in a double logarithmic scale. The critical exponent $\alpha = 0.107 \pm 0.002$ (given by the slope of the data in Figure 14.11 in an undisturbed asymptotic region of $10^{-4} < |\tau| < 10^{-3}$ above and below T_c) is in excellent agreement with the theoretical RG value of 0.109.[26]

Therefore, high-resolution ASC experiments unambiguously confirm that binary fluids belong to the same class of critical-point universality as that of

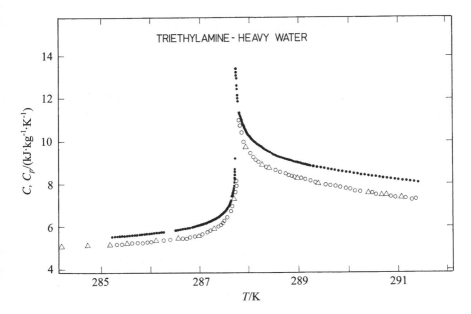

Figure 14.10 Adiabatic scanning calorimetric results for the isobaric heat capacity anomaly in the binary liquid mixture of triehylamine-heavy water at the critical concentration.[40] The open circles are heat capacity data obtained by numerical differentiation of the experimental temperature dependence of the enthalpy. Open triangles are results obtained by means of the heat-pulse adiabatic method. The solid dots are results for the quantity C defined by Equation (12).

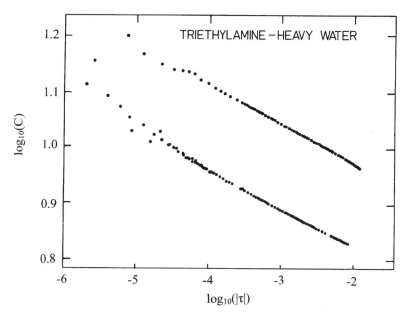

Figure 14.11   Double logarithmic plot of the C data below (lower curve) and above (upper curve) T_c of Figure 14.10. The slopes in the range of $10^{-4} < |\tau| < 10^{-3}$ are consistent with $\alpha = 0.107 \pm 0.002$.[40]

one-component fluids provided that only the proper ("isomorphic") properties are compared. In particular, the isobaric heat capacity of a binary solution at constant composition at the liquid-liquid critical point exhibits the same singularity as the isochoric heat capacity of a pure fluid at the gas-liquid critical point.

The isochoric heat capacity of binary mixtures at constant (critical) composition seemingly follows the divergence of the isobaric heat capacity, as demonstrated in Figure 14.9. The divergence of the isochoric heat capacity of mixtures at constant composition along the line of critical points does, generally, violate the stability criteria for mixtures and, hence, is thermodynamically impossible.[35] Therefore, $C_{V,x}$ must remain finite. However, it has been shown[10] that, due to the weak compressibility, in many liquid mixtures, the distance from the critical temperature, where $C_{V,x}$ saturates and tends to a finite limiting value, is too narrow to be experimentally detected. An extension of the principle of the critical-point universality to binary fluid mixtures,[49-54] often referred to a the "isomorphism of critical phenomena", specifically predicts which thermodynamic properties of binary fluids (and measured along which particular thermodynamic path) exhibit the same critical anomalies as those of single-component fluids. A characteristic parameter that controls this distance is the dependence of the critical temperature on pressure, dT_c/dp. If $dT_c/dp = 0$, $C_{V,x}$ diverges with the same exponent and with the same amplitude as $C_{p,x}$. The derivative dT_c/dp also controls the behavior of the isothermal compressibility at constant composition, which, in accordance with

Equation (1), diverges with the same exponent α as that of $C_{p,x}$ but with a very small critical amplitude, proportional to $(\mathrm{d}T_c/\mathrm{d}p)^2$. In particular, for the mixture of nitroethane-isooctane the amplitude of the divergence in κ_T is expected to be experimentally undetectable.[10]

14.6 Effects of Impurities and Confinement on the Heat Capacity Anomalies in the Critical Region

14.6.1 Renormalization of the Isochoric Heat-capacity Critical Exponent at the Liquid-gas Critical Point

The effects of (equilibrium) impurities on the heat-capacity anomalies are explained by the isomorphism theory. The isochoric heat capacity at constant (critical) composition is finite along the gas-liquid critical locus. However, the distance from the critical temperature, where the "ideal" (pure solvent) critical anomaly is suppressed, depends on the concentration and on the nature of the solute.

Figure 14.12 demonstrates the suppression of the critical isochoric-heat-capacity singularity in ethane by addition of small amount of heptane. In Figure 14.13, the data in the one-phase region are plotted in a semi-logarithmic

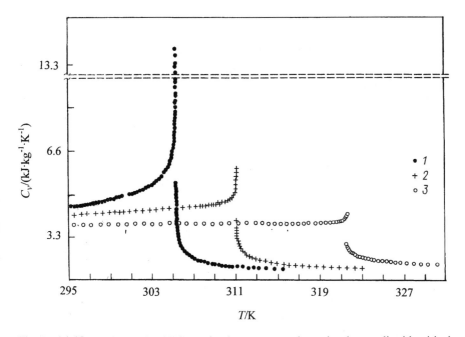

Figure 14.12 Effects of addition of n-heptane to ethane in the gas-liquid critical region.[55] Symbols represent experimental data: (1) pure ethane; (2) mole fraction 0.01 heptane; (3) mole fraction 0.03. heptane.

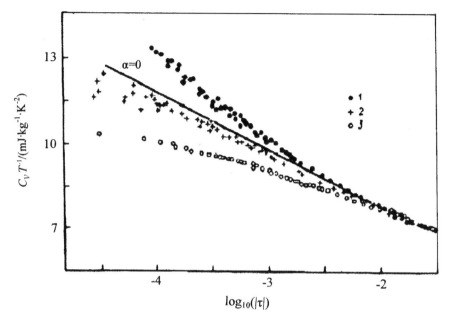

Figure 14.13 Renormalization of the isochoric-heat-capacity critical exponent in ethane-*n*-heptane mixtures in the one-phase region.[55] Symbols represent experimental data: (1) pure ethane; (2) mole fraction 0.01 heptane; (3) mole fraction 0.03 heptane.

scale as functions of the distance from the corresponding critical temperatures. It is clearly seen that the "effective" exponent α in the power law, Equation (5), already becomes negative for 1 % heptane; the effect is even more pronounced for 3 % heptane. The effect of changing the effective exponent value from positive to negative is known as "Fisher renormalization"[56] and is part of the isomorphism theory.

Let μ be the solute/solvent chemical-potential difference conjugate to the molar fraction x. According to isomorphism theory, the relation between the experimental scale $\tau \equiv \tau_x = |T-T_c(x)|/T_c(x)$ at constant x and the theoretical ("isomorphic") scale $\tau_\mu = |T-T_c(\mu)|T_c(\mu)$ is given by[34]

$$\tau \simeq \tau_\mu^{1-\alpha}\tau_0^\alpha\left[1 + \left(\frac{\tau_\mu}{\tau_0}\right)^\alpha\right] \qquad (13)$$

where τ_0 is a characteristic temperature scale, $\tau_0 \cong [A^{\pm}(1-\alpha)^{-1}x(1-x)(\mathrm{d}T/T_c\mathrm{d}x)]$.

Under the condition $\tau \ll \tau_0$, $\tau \simeq \tau_\mu^{1-\alpha}$ and the temperature scale is renormalized. In the van der Waals (mean-field) theory that neglects fluctuations, $\alpha = 0$ and the connection between τ and τ_μ is analytic. If the value of τ_0 is within the critical region, the isochoric heat capacity exhibits a crossover from the pure-fluid-like divergence (far away from the critical temperature) to a renormalized (cusp-like) behavior. The crossover expression for $C_{V,x}$ reads[34]

$$\frac{C_{V,x}}{R} \simeq \frac{A^{\pm}}{\alpha\left[1 + \left(\tau_{\mu}/\tau_0\right)^{\alpha}\right]\tau_0^{\alpha}} - \frac{B_{cr}\left(\tau_{\mu}/\tau_0\right)^{\alpha}}{1 + \left(\tau_{\mu}/\tau_0\right)^{\alpha}} + \text{const} \qquad (14)$$

where B_{cr} as a so-called "critical analytical background" induced by fluctuations,[34] which does not exist in the van der Waals mean-field theory. At the condition $\tau < \tau_0$, the isochoric heat capacity tends to a finite cusp at the critical point with an infinite slope and with a vanishing critical background. In the first approximation, expanding Equation (14) in terms of $(\tau_{\mu}/\tau_0)^{\alpha} < 1$ and taking into account for Equation (13), one obtains

$$\frac{C_{V,x}}{R} \simeq \frac{A^{\pm}}{\alpha\tau_0^{\alpha}}\left[1 - \left(\frac{\tau}{\tau_0}\right)^{\frac{\alpha}{1-\alpha}}\right] - B_{cr}\left(\frac{\tau}{\tau_0}\right)^{\frac{\alpha}{1-\alpha}} + \text{const} \qquad (15)$$

At $\tau \gg \tau_0$, but still in the critical domain $\tau \ll 1$, $C_{V,x}$ behaves as the isochoric heat capacity of single-component fluids in accordance with Equation (5). The predictions of isomorphism theory are also experimentally confirmed by more recent experiments on various binary and ternary mixtures in the gas-liquid critical region.[57–60]

14.6.2 Renormalization of the Isobaric Heat-capacity Critical Exponent at the Liquid-liquid Critical Point

Fisher critical-exponent renormalization[56] also can be observed for the isobaric-heat capacity at liquid-liquid critical points of binary mixtures with an added impurity. Bloemen et al.[41] investigated the effect of adding ethanol to the binary mixture of triethylamine–water for mole fractions of ethanol of $x_E = 0.07$ and $x_E = 0.01$. A graphical representation of these $C_{p,x}$ data is given in Figure 13.8 of Chapter 13. Complete Fisher renormalization from α to $-\alpha/(1-\alpha) = -0.122$ (with $\alpha = 0.109$) was not observed, but crossover ("effective") exponent values of 0.06 and 0.01 were obtained for $x_E = 0.07$ in the one-phase region and in the two phase region, respectively. For $x_E = 0.01$ the anomaly was very small but best fits gave near zero or negative effective critical exponent values.[41]

14.6.3 Near-critical Heat Capacities in Porous Media

The heat-capacity critical anomalies are suppressed in porous media. Voronov and co-workers investigated the isobaric and isochoric heat capacities of fluids in various porous media near the liquid-liquid critical points $(C_{p,x})$[43,61] and near the gas-liquid critical points $(C_{V,x})$.[62] The experimental data on the isobaric heat capacity of a 2,6-lutidine-water mixture at the critical composition near the liquid-liquid critical point are shown in Figure 14.14. While in the bulk sample the heat capacity obeys the power law given by Equation (11), in porous

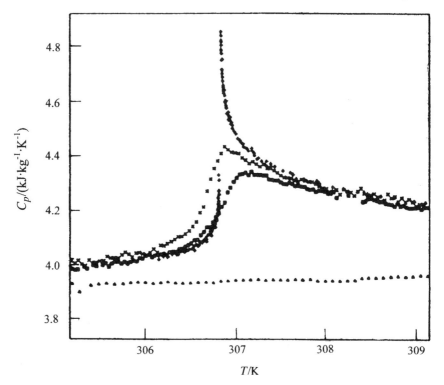

Figure 14.14 Isobaric heat capacity of a 2,6-lutidine-water mixture at the critical composition near the liquid-liquid critical point in porous media.[43,60] Vertical crosses indicate the results of measurements in a bulk sample (1 mm size micro-calorimeter); crosses 250 nm porous nickel; squares 100 nm porous glass, triangles 10 nm porous glass.

media the heat capacity remains finite and its maximum is shifted with respect to the critical temperature in the bulk sample. The magnitude of the anomaly depends on the pore size: in 10 nm porous glass the anomaly virtually vanishes.

The finite-size effects on the heat-capacity critical anomaly are explained by finite-size scaling theory.[63] The basic idea of the finite-size scaling is simple. The diverging correlation length ξ, which diverges at the critical point in accordance with Equation (6), competes with a characteristic size L of the system. In the range of temperatures, where $\xi \ll L$ ("bulk" regime), the heat capacities exhibit the divergences given by Equations (5) or (11). If $\xi \gg L$ (finite-size regime), the heat capacity reaches a limited value depending on the size L.

The crossover expression for the anomalous part $C(L, \xi)$ of the finite-size heat capacity near the critical point reads[64]

$$C(L, \xi) = C_\infty(\xi)\Psi(\xi/L) \tag{16}$$

where $C_\infty(\xi)$ is the bulk heat capacity anomaly and $\Psi(\xi/L)$ is a universal finite-size scaling function, At $(\xi/L) \ll 1$, $\Psi(\xi/L) \to 1$ and the heat capacity obeys the power laws (5) or (11). At $(\xi/L) \gg 1$, $\Psi(\xi/L) \to (\xi/L)^{-\alpha/\nu}$ and the heat capacity only depends on the size: $C(L) = (A^{\pm}/\alpha)(L/\xi_0^{\pm})^{\alpha/\nu}$.

14.6.5 Effects of Impurities on Heat-capacity Measurements in the Two-phase Region and the Yang-Yang Anomaly

The isochoric heat capacity of single-component fluids in the two-phase region can be expressed through the second temperature derivatives of pressure and chemical potential by the Yang-Yang thermodynamic relation:[65]

$$\frac{\rho C_V}{T} = \frac{d^2 p}{dT^2} - \rho \frac{d^2 \mu}{dT^2} \qquad (17)$$

Until recently, it was commonly believed that the heat-capacity anomaly at the critical point was solely associated with the divergence of the derivative of pressure and no singular contribution from the chemical potential variation. Moreover, all currently available equation-of-state models are based on this assumption.[34] However, this assumption was recently challenged by Fisher and co-workers[66,67] who argued that the pressure and the chemical potential derivatives, in general, both contribute to the specific heat singularity. The predicted divergence of the derivative $d^2\mu/dT^2$ at the critical point, called the "Yang-Yang anomaly", can, in principle, be experimentally verified by experimental studies of the isochoric heat capacity along different isochores in the two-phase critical region. Indeed, according to Equation (17), $d^2\mu/dT^2$ is a slope of the plot $\rho C_V/T$ vs. ρ. The Yang-Yang anomaly is caused by the gas-liquid asymmetry in phase coexistence and is expected to be strongly system-dependent.[54,68] A major obstacle to unambiguously verify the anomaly by isochoric heat capacity measurements in the two-phase region is an extremely high sensitivity of such measurements to small amounts of impurities.[69] Even a small impurity causes a mismatch in the chemical potentials of liquid and gas along the dew-bubble curve at constant temperature, strongly affecting the apparent slope in the Yang-Yang plot.[70,71] More heat-capacity experiments in the two-phase region of fluids with better purification are desirable.

Acknowledgements

M. A. Anisimov acknowledges support from Abu Dhabi National Oil and Gas Company (ADNOC), UAE during his sabbatical leave at The Petroleum Institute of ADNOC.

References

1. M. E. Bagatskii, A. V. Voronel and V. G. Gusak, *Sov. Phys. JETP*, 1963, **16**, 517.

2. A. V. Voronel, Yu. R. Chashkin, V. A. Popov and V. G. Simkin, *Sov. Phys. JETP*, 1964, **18**, 568.
3. A. V. Voronel, V. G. Snigirev and Yu. R. Chashkin, *Sov. Phys. JETP*, 1965, **21**, 653.
4. A. V. Voronel, V. G. Goburnova, Yu. Chashkin and V. V. Schekochikhina, *Sov. Phys. JETP*, 1966, **23**, 597.
5. L. D. Landau and E. M. Lifshitz, *Statistical Physics,* Pergamon, New York, 1958.
6. A. V. Voronel, Thermal measurements and critical phenomena in liquids, in *Phase Transitions and Critical Phenomena* eds. C. Domb and M. S. Green, **Vol. 5B**, Academic Press, London, 1976, pp. 343–391.
7. Jan Thoen, Chapter 13 in this book.
8. J. A. Lipa, C. Edwards and M. J. Buckingham, *Phys. Rev. Lett.*, 1970, **25**, 1086.
9. M. Barmatz, Inseob. Hahn, J. A. Lipa and R. V. Duncan, *Rev. Mod. Phys.*, 2007, **79**, 1.
10. M. A. Anisimov, *Critical Phenomena in Liquids and Liquid Crystals,* Gordon and Breach, Philadelphia, 1991.
11. M. A. Anisimov, A. T. Berestov, L. S. Veksler, L. S. Kovalchuk and V. A. Smirnov, *Sov. Phys. JETP*, 1974, **39**, 359.
12. M. R. Moldover and W. R. Little, *Phys. Rev. Lett.*, 1965, **15**, 54.
13. M. R. Moldover, *Phys. Rev.*, 1969, **182**, 342.
14. C. Edwards, J. A. Lipa and M. J. Buckingham, *Phys. Rev. Lett.*, 1968, **20**, 496.
15. A. V. Voronel, V. G. Gorbunova, V. A. Smirnov, N. G. Shmakov and V. V. Schekochihina, *Zh. Eksp. Teor. Fiz.*, 1972, **63**, 642.
16. Brown and Meyer, *Phys. Rev. A*, 1972, **6**, 364.
17. H. Schmidt, J. Opdycke and C. F. Gay, *Phys. Rev. Lett.*, 1976, **19**, 887.
18. I. M. Abdulagatov, L. N. Levina, Z. R. Zakaryaev and O. N. Mamchenkova, *Fluid Phase Equilib.*, 1997, **127**, 205.
19. I. M. Abdulagatov, V. A. Rabinovich and V.I. Dvoryanchikov, *Thermodynamic Properties of Fluids and Fluid Mixtures Near- and Supercritical Conditions,* Begell House, New York, 1998.
20. M. Barmatz, I. Han, F. Zhong, M. A. Anisimov and V. A. Agayan, *J. Low Temp. Phys.*, 2000, **121**, 633.
21. N. G. Polikhronidi, I. M. Abdulagatov, J. W. Magee and G. V. Stepanov, *Int. J. Thermophys.*, 2002, **23**, 745.
22. F. Zhong, M. Barmatz and I. Han, *Phys. Rev. E*, 2003, **67**, 021106.
23. Phase Transitions and Critical Phenomena, Vol. 6, 1976, C. Domb and M. S. Green, Eds., Academic Press, New York.
24. M. E. Fisher, *Rev. Mod. Phys.*, 1998, **70**, 653.
25. R. Guida and J. Zinn-Justin, *Phys. A: Math. Gen.*, 1998, **31**, 8103. M. Campostrini, A. Pelinetto, P. Rossi and E. Vicari, *Phys. Rev. E*, 1999, **60**, 3526; ibid. 2002, 65, 066127.
26. M. E. Fisher, S.-Y. Zinn, *J. Phys. A*, 1998, **31**, L629. M. E. Fisher, S.-Y. Zinn and P. J. Upton, *Phys. Rev.*, 1999, **B59**, 14533.
27. A. Haupt and J. Straub, *Phys. Rev. E*, 1999, **59**, 1795.

28. M. Barmatz, F. Zhong and A. Shih, *Int. J. Therm.*, 2004, **25**, 1667; ibid. 2005, 26, 921.
29. D. Stauffer, M. Ferer and M. Wortis, *Phys. Rev. Lett.*, 1972, **29**, 345.
30. J. Thoen, E. Vangeel and W. Van Dael, *Physica*, 1971, **52**, 205.
31. A. Michels, J. M. Levelt and G. J. Wolkers, *Physica*, 1958, **24**, 769.
32. A. M. Sirota and B. K. Maltsev, *Teploenergetica*, 1962, **9**, 52.
33. A. M. Sirota, B. K. Maltsev and P. E. Beljakova, *Teploenergetica*, 1963, **10**, 64.
34. M. A. Anisimov and J. V. Sengers, Critical region, in *Equations of State for Fluids and Fluid Mixtures* eds. J. V. Sengers, R. F. Kayser, C. J. Peters and H. J. White Jr, Elsevier, Amsterdam, 2000, pp. 381–434.
35. J. C. Rowlinson and F. L. Swinton, *Liquids and Liquid Mixtures,* Butterworth, London, 1982.
36. M. A. Anisimov, A. V. Voronel and T. M. Ovodova, *Sov. Phys. JETP*, 1972, **34**, 583.
37. M. A. Anisimov, A. V. Voronel and T. M. Ovodova, *Sov. Phys. JETP*, 1972, **35**, 536.
38. M. A. Anisimov, A. T. Berestov, V. P. Voronov, Y. F. Kiyachenko, B. A. Kovalchuk, V. M. Malyshev and V. A. Smirnov, *Sov. Phys. JETP*, 1979, **49**, 844.
39. J. Thoen, E. Bloemen and W. Van Dael, *J. Chem. Phys.*, 1978, **68**, 735.
40. E. Bloemen, J. Thoen and W. Van Dael, *J. Chem. Phys.*, 1980, **73**, 4628.
41. E. Bloemen, J. Thoen and W. Van Dael, *J. Chem. Phys.*, 1981, **75**, 1488.
42. J. Thoen, J. Hamelin and T. K. Bose, *Phys. Rev.E*, 1996, **53**, 6264.
43. L. V. Entov, V. A. Levchenko and V. P. Voronov, *Int. J. Thermophys.*, 1993, **14**, 221.
44. E. R. Oby and D. T. Jacobs, *J. Chem. Phys.*, 2001, **114**, 4918.
45. T. Heimburg, S. Z. Mirzaev and U. Kaatze, *Phys. Rev. E*, 2000, **62**, 4963.
46. B. Van Roie, G. Pitsi and J. Thoen, *J. Chem. Phys.*, 2003, **119**, 8047.
47. S. Pittois, B. Van Roie, C. Glorieux and J. Thoen, *J. Chem. Phys.*, 2004, **122**, 024504.
48. J. N. Utt, S. Y. Lehman and D. T. Jacobs, *J. Chem. Phys.*, 2007, **127**, 104505.
49. R. B. Griffiths and J. C. Wheeler, *Phys. Rev. A*, 1970, **2**, 1047.
50. W. F. Saam, *Phys. Rev. A*, 1970, **2**, 1461.
51. M. A. Anisimov, A. V. Voronel and E. E. Gorodetskii, *Sov. Phys. JETP*, 1971, **33**, 605.
52. M. A. Anisimov, E. E. Gorodetskii, V. D. Kulikov and J. V. Sengers, *Phys. Rev. E*, 1995, **51**, 1199.
53. M. A. Anisimov, E. E. Gorodetskii, V. D. Kulikov, A. A. Povodyrev and J. V. Sengers, *Physica A*, 1995, **220**, 277; ibid., 1996, **223**, 272.
54. C. A. Cerdeiriña, J. T. Wang, M. A. Anisimov and J. V. Sengers, *Phys. Rev. E*, 2008, **77**, 031127.
55. M. A. Anisimov, E. E. Gorodetskii and N. G. Shmakov, *Sov. Phys. JETP*, 1973, **36**, 1143.

56. M. A. Fisher, *Phys. Rev.*, 1968, **176**, 257.
57. M. A. Anisimov, S. B. Kiselev and S. Khalidov, *Int. J. Thermophys.*, 1988, **9**, 453.
58. V. P. Voronov and E. E. Gorodetskii, *JETP Lett.*, 2000, **72**, 516.
59. M. Y. Belyakov, V. P. Voronov, E. E. Gorodetskii and V. D. Kulikov, *JETP Lett.*, 2007, **86**, 20.
60. V. P. Voronov, E. E. Gorodetskii and S. S. Safonov, *J. Phys. Chem. B*, 2007, **111**, 11486.
61. V. P. Voronov and V. M. Buleĭko, *JETP*, 1998, **86**, 586.
62. V. P. Voronov, M. Yu. Belyakov, E. E. Gorodetskii, V. D. Kulikov, A. R. Muratov and V. B. Nagaev, *Transport in Porous Media*, 2003, **52**, 1573.
63. M. E. Fisher and M. N. Barber, *Phys. Rev. Lett.*, 1972, **28**, 1516.
64. R. Schmolke, A. Wacker, V. Dohm and D. Frank, *Physica B*, 1990, **165–166**, 575.
65. C. N. Yang and C. P. Yang, *Phys. Rev. Lett.*, 1964, **13**, 303.
66. M. E. Fisher and G. Orkoulas, *Phys. Rev. Lett.*, 2000, **85**, 696.
67. Y. C. Kim, M. E. Fisher and G. Orkoulas, *Phys. Rev. E*, 2003, **67**, 061506.
68. J. T. Wang and M. A. Anisimov, *Phys. Rev. E*, 2007, **75**, 051107.
69. A. K. Wyczalkowska, M. A. Anisimov, J. V. Sengers and Y. C. Kim, *J. Chem. Phys.*, 2002, **116**, 4202.
70. M. A. Anisimov, F. Zhong and M. Barmatz, *J. Low Temp. Phys.*, 2004, **137**, 69.
71. I. M. Abdulagatov, N. G. Polikhronidi, T. J. Bruno, R. G. Batyrova and G. V. Stepanov, *Fluid Phase Equilibria*, 2008, **263**, 71.

CHAPTER 15

Heat Capacity of Polymeric Systems

MAREK PYDA

Department of Chemistry, The University of Technology, 35-959, Rzeszow, Poland; and Department of Pharmacy, Poznan University of Medical Sciences, 61-701, Poznan, Poland; and ATHAS-MP Company, Knoxville, TN 37922, USA

15.1 Introduction

Heat capacity belongs to the fundamental quantities which are used in the evaluation of thermodynamic properties of polymeric materials derived from calorimetric measurements.[1] The thermodynamic heat capacity is defined as:

$$C_p = \lim \left(\frac{\Delta Q}{\Delta T} \right)_{p,n} = \left(\frac{\partial H}{\partial T} \right)_{p,n} \tag{1}$$

where Q is the heat required to increase the temperature, T, of the sample, of one mole (molar heat capacity, C_p) or one gram (specific heat capacity, $c_p = C_p/M$, where M is the molar mass) of a substance at constant pressure, p, by one kelvin in the absence of latent heat. The last term in Equation (1) is a partial differential of enthalpy at constant pressure, p, and composition, n. In general, the term apparent heat capacity, C_p^*, applies to a heat capacity that includes latent heat effects, such as heats of fusion, crystallization, reorganization, reaction and

Heat Capacities: Liquids, Solutions and Vapours
Edited by Emmerich Wilhelm and Trevor M. Letcher
© The Royal Society of Chemistry 2010
Published by the Royal Society of Chemistry, www.rsc.org

others. The change in enthalpy (H) can be written as:[2]

$$dH = \left(\frac{\partial H}{\partial T}\right)_{p,n} dT + \left(\frac{\partial H}{\partial n}\right)_{p,T} dn \qquad (2)$$

where the first part of Equation (2) represents thermodynamic heat capacity, C_p from Equation (1) and the second part $(\partial H/\partial n)_{p,T}$ is the latent heat effect, and dT and dn are the changes in temperature and composition, respectively.

The experimental apparent heat capacity of polymeric materials from calorimetric measurements can be represented only by thermodynamic C_p outside the transition region, which is below glass transition or above melting transition for semi-crystalline polymers. In other regions, such as between the glass and melting temperatures, the measured heat capacity contains latent heat effects. The transition parameters of first-order and glass transitions can be measured and separated from apparent C_p^* of polymeric materials.[3]

The thermodynamic heat capacity provides direct information about molecular motion. At low temperatures, the main contributions to the experimental heat capacity of flexible polymers come from vibrations. At high temperatures, additional contributions come from large-amplitude molecular motions – mainly conformational and anharmonic.[1,4]

Knowledge of heat capacity and transition parameters allows one to calculate the thermodynamic functions including enthalpy, entropy and the Gibbs function of crystalline, amorphous and semi-crystalline polymers over the whole temperature range.[1,5]

Macromolecules, so-called polymers, usually present a non-equilibrium semicrystalline system, which does not crystallize completely. On cooling from the melt, the random and entangled macromolecules of the liquid state can be transformed into a crystalline, mobile amorphous or rigid amorphous phase depending on kinetics and on the chemical structure of the macromolecules. Polymeric systems such as pure, one component macromolecules, and their copolymers and blends, can remain amorphous, super-cooled liquids or become a partial chain-folding crystal. These polymeric systems are not able to become complete extended-chain crystals in equilibrium and remain in a metastable state.[6-8]

Semi-crystalline polymers include polyethylene, polypropylene, polyamides (nylons), polyesters and biodegradable poly(lactic acid). Many polymers are completely amorphous, such as polystyrene, poly(methyl methacrylate), and its co-polymers, and poly(L-lactic acid) with a high percent content of D- isomer.[1,6,9]

In order to analyze and understand the thermodynamic properties of polymers based on heat capacity, quantitative thermal analysis needs to be performed. The experimental, apparent heat capacity, C_p^* should be analyzed in reference to the equilibrium baselines of the solid and liquid heat capacity, which arise from molecular motions and give the possibility of separating the thermodynamic heat capacity from melting, crystallization, annealing, reorganization, cold crystallization or enthalpy relaxation effects contained in the C_p^*. For semi-crystalline polymeric materials, the mobile-amorphous, rigid-amorphous and crystalline fractions are estimated from thermal analysis based on heat capacity.[1,10]

In this chapter, the heat capacity of semi-crystalline and amorphous polymeric systems is presented as a quantifiable thermal characteristic and is interpreted in terms of microscopic molecular motion. This is the basis of advanced thermal analysis. The experimental, apparent heat capacity results from measurements by adiabatic calorimetry, standard differential scanning calorimetry, and temperature-modulated differential scanning calorimetry, are linked to motion of macromolecules over the whole temperature range. The low-temperature, below the glass transition, experimental heat capacity of solid states is linked to vibration motion. The heat capacity of the liquid state of polymers is estimated from contributions of vibrational, conformational and anharmonic motions or from an empirical addition scheme based on contributions of the constituent chain-segments of polymers. Once the calculated, solid and liquid heat capacities are established, they can serve as two reference baselines for the quantitative thermal analysis of non-equilibrium semi-crystalline polymeric materials. Knowing heat capacities and transitions parameters, the integral functions such as the enthalpy (H), entropy (S) and free enthalpy (Gibbs function) (G) for equilibrium conditions are calculated and used as a reference for analysis. All recommended results, for over 200 polymers, have been collected and organized as part of the Advanced Thermal Analysis System (ATHAS) Data Bank and are available online (at http://www.athas.prz.rzeszow.pl). Examples of the quantitative thermal analysis of several amorphous and semi-crystalline polymer systems such as polyethylene, poly(trimethylene terephthalate), poly(butylene terephthalate), poly(lactic acid) and silk are presented here.

15.2 Instrumentation and Measurements

Apparent heat capacity ($C_p^* = C_p + (\partial H / \partial n)_{p,T}$) can be determined with a number of calorimetric techniques: adiabatic calorimetry, standard differential scanning calorimetry (DSC), temperature-modulated differential scanning calorimetry (TMDSC), and fast scanning calorimetry (FSC).[1,11–14]

15.2.1 Adiabatic Calorimetry

The low temperature heat capacity of polymeric materials, usually from 5 to 300 K, is measured by adiabatic calorimetry directly from experimental heat input and the corresponding temperature change of the polymeric sample ($C_p = (\Delta Q / \Delta T)$), which has been previously described.[11] Briefly, the heat capacity of the sample is 60–70 % of the total heat capacity of the calorimeter and substance over the whole temperature range. The calorimetric ampule is a cylindrical platinum vessel with a volume of *ca.* $15 \times 10^{-6}\,\mathrm{m}^3$. The heat capacity of an unloaded calorimetric ampule increased gradually from $0.0045\,\mathrm{J\,K^{-1}}$ to $1.440\,\mathrm{J\,K^{-1}}$ with increasing temperature from $T = 5\,\mathrm{K}$ to $T = 350\,\mathrm{K}$. The temperature is measured with a platinum resistance thermometer. Liquid helium and nitrogen are used to obtain low temperatures. Heat capacities are usually calibrated with a benzoic acid standard before the measurement of polymeric materials. The precision is estimated to be $\pm0.5\,\%$, usually from $T = 5\,\mathrm{K}$ to $T = 350\,\mathrm{K}$.[11]

The heat capacities of an amorphous or semi-crystalline polymer determined by adiabatic calorimetry, usually from $T = 5\,\mathrm{K}$ to $T = 200\,\mathrm{K}$, are used to establish the vibrational heat capacity of the solid state. At the lowest temperatures, the amorphous glass usually has a somewhat higher C_p than the crystalline solid heat capacity. The resulting vibrational heat capacities are then extrapolated from the measured range to higher temperature for the quantitative thermal analysis of the investigated polymer.

15.2.2 Differential Scanning Calorimetry

For the measurement of the heat capacities and the transition behaviour of polymeric materials between 200 and 550 K, differential scanning calorimetry or temperature modulated differential scanning calorimetry are used.[1,5,12,13] Both DSC and TMDSC can be either a heat-flux type or power-compensating calorimeter. Before any measurements, the standard calibration procedure should be performed for temperature and enthalpy calibration based on one or more melting material. For example, indium with a melting temperature $T_m^{\circ} = 330.75\,\mathrm{K}$ $(156.6\,^{\circ}\mathrm{C})$ and enthalpy $\Delta H\ (T_m^{\circ}) = 28.4\,\mathrm{J\,g^{-1}}$ and lead, with melting temperature $T_m^{\circ} = 600.55\,\mathrm{K}$ $(327.4\,^{\circ}\mathrm{C})$, are used very often for calibration for higher temperature ranges.

In the standard DSC experiment, the thermodynamic heat capacity is measured only in the steady state. The heat capacities of polymeric systems between 200 and 550 K are measured by standard DSC mostly at a heating rate, q, of $10\,\mathrm{K\,min^{-1}}$, but also at different heating rates from 0.1 to $100\,\mathrm{K\,min^{-1}}$ to identify changes of the samples during measurements. In calibrating the heat capacity, three runs are carried out: one with empty reference and empty sample pans, to correct for asymmetry of the DSC, one with an empty reference pan and a pan filled with sapphire for calibration, and one with an empty pan and a pan filled with the sample. Figure 15.1 shows an example of results of heat-flow rate versus time of these three runs. After the steady state is reached, the heat capacity was determined from the following equation[1,12]:

$$ mC_p = K\frac{\Delta T}{q} + C_s\frac{\mathrm{d}\Delta T}{\mathrm{d}T_s} \tag{3} $$

where K is determined as a function of temperature from the sapphire calibration, m is the sample mass, C_p is the heat capacity, ΔT is the temperature difference between reference and sample, proportional to the heat-flow rate

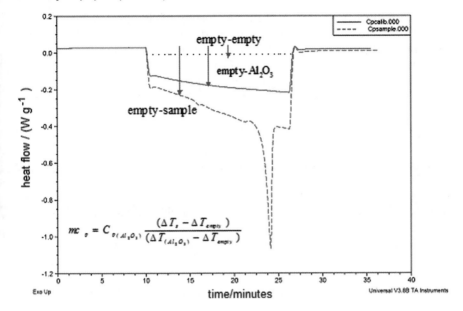

Figure 15.1 Example of three runs of heat flow-rate versus time for heat capacity calibration: empty-empty, empty-sapphire, and empty-sample.

$\Phi = K\Delta T$, C_s is the heat capacity of the sample calorimeter including sample and aluminium pan, and T_s is the sample temperature. The second part of Equation (3) is a correction of the heat capacity due to the changing heat rate of the sample and accounts for only about 1 % of the total heat capacity. The relative uncertainly of the measured C_p is usually estimated to be ± 3 % or better.

15.2.3   Temperature Modulated Calorimetry

In temperature modulated DSC, the programmed temperature is the result of the superimposition of the constant underlying heating rate and a sinusoidal, sawtooth-like, or other temperature modulation.[13]

The TMDSC is usually employed to obtain the total C_p, an apparent, reversing C_p and a non-reversing C_p in two modes of operation: standard and quasi-isothermal. In the standard TMDSC, the sinusoidal modulation of the sample temperature, $T_s(t)$, is employed with an amplitude A_{Ts} and period p ($\omega = 2\pi/p$) and in the steady state is given by:[13]

$$T_s(t) = T_0 + \langle q \rangle t - \frac{C_s}{K} + A_{Ts}\sin(\omega t - \varepsilon) \qquad (4)$$

where T_0 is the starting temperature of the experiment; $\langle q \rangle$, the underlying heating rate, obtained by forming a sliding average over a fixed number of complete modulation cycles. The phase shift is ε, taken relative to some internal reference frequency, and C_s is the heat capacity of the sample. The range of interest is covered by successive runs at different temperatures T_0. The apparent reversing heat capacity C_p (in $J\,K^{-1}\,mol^{-1}$) is given by:

$$C_p = \sqrt{C_{rel}^2 + C_{im}^2} = \frac{\langle A_\Phi \rangle}{\langle A_{Ts} \rangle \omega} K(\omega) \tag{5}$$

where

$$K(\omega) = \sqrt{1 + \tau^2 \omega^2} \tag{5a}$$

where C_p is the modulus of the complex heat capacity, the reversing C_p, with its real, C_{rel}, and imaginary, C_{im}, parts; K is the Newton's law of heat flow calibration constant; $\langle A_\Phi \rangle$ is the heat-flow rate amplitude, smoothed over one modulation cycle; $\langle A_{Ts} \rangle$ is the similarly smoothed modulation amplitude of the sample temperature; and $K(\omega)$ is the frequency-dependent calibration factor K. Under the usual conditions of continued steady state, linearity, and stationarity of response, τ is a constant $\tau = C_r/K$, where C_r is the heat capacity of the empty reference pan. Details about data de-convolution are described in the literature.[1,13]

In the quasi-isothermal TMDSC, measurements are performed with the underlying heating rate $\langle q \rangle = 0$ and modulation around constant temperature T_0 and in this case the apparent, reversing C_p are given also by Equations (5) and (5a). Typically, runs of 20–30 minute duration are carried out at each temperature and of which approximately the last 10 minutes are used for collection and analysis. Using quasi-isothermal TMDSC techniques, it is possible to determine the frequency-dependent reversing heat capacity using the sinusoidal or the complex sawtooth modulations. In the case of the complex sawtooth modulation, only one single measurement for a sample with the same thermal history gives similar results as produced by multiple measurements for the sinusoidal modulation. These quasi-isothermal techniques allow the separation of the thermodynamic from the kinetic processes, such as the glass or melting transitions in polymeric systems by measuring their reversing and non-reversing heat capacities.[1,15,16]

Specifically, in the so-called simplified complex sawtooth, in order to improve the evaluation of heat capacity from higher harmonic components, 14 segments of heating and cooling changes of temperature $T_x(t)$, as given by:

$$
\begin{aligned}
T_x(t) = T_0 \\
+ A[0.378\,\sin\omega t + 0.251\,\sin 3\omega t + 0.217\,\sin 5\omega t + 0.348\,\sin 7\omega t + \ldots]
\end{aligned}
\tag{6}
$$

are used, and have close to equal amplitudes for the four harmonics of the Fourier series and are thus easily generated by scanning calorimetry. More descriptions regarding complex modulation with sawtooth and different profiles can be found in literature.[17]

Fast scanning calorimetry (FSC), which is based on the thermal conductivity in chip calorimetry, can perform cooling and heating of micro-gram or nanogram film samples with rates of $10–30\,\text{kK s}^{-1}$ to avoid any crystallization or reorganization. This allows us to obtain full amorphous compositions for such polymers as polypropylene, poly(butylene terephthalate) and others. An application example of so-called superfast chip calorimetry, SFCC, with gauge TCG-3880 of the Xensor Intergration Co.,[18] has been described earlier and employed in investigating the heat capacity of superquenched polymers.[1,19]

15.3 Experimental Heat Capacity

Figure 15.2 shows the experimental heat capacities for one of the most popular polymers. The values of the experimental heat capacities for polyethylene (PE) have been extrapolated to heat capacities for the completely amorphous and crystalline states and as they are collected at ATHAS Data Bank[20] based on data from different laboratories.[1,21] Heat capacity of glassy amorphous and crystal PE below 150 K is almost the same. The C_p (amorphous) is a little higher than C_p (crystalline). The larger deviation between both heat capacities occurs continually due to the glass transition at 237 K. Above the glass transition temperature, the heat capacity of PE in the liquid state is a linear function of temperature. For crystalline PE between 150 K and the melting temperature of 414.6 K, the C_p is much lower than for amorphous PE.[1,20,21]

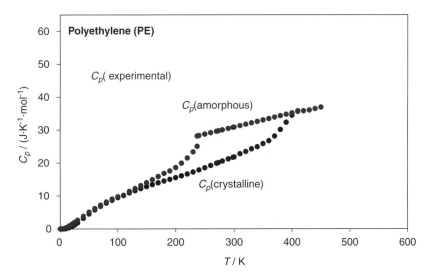

Figure 15.2   Experimental heat capacities of crystalline and amorphous polyethylene from ATHAS Data Bank.[20]

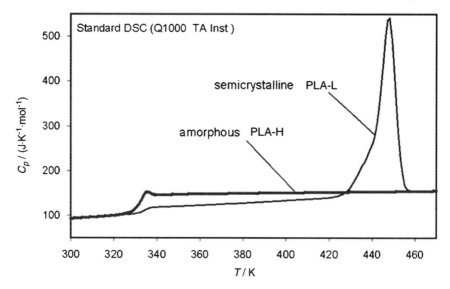

Figure 15.3 Molar heat capacity of a linear polymers, amorphous and semi-crystalline biodegradable poly(lactic acid). (PLA-H contains 16 % D-isomer, PLA-L contains 1.5 % D-isomer).

Figure 15.3 shows an example of the experimental heat capacity of a linear polymer such as biodegradable poly(lactic acid) (PLA) by standard differential scanning calorimetry (DSC) as a function of temperature for different physical states of semi-crystalline and full amorphous. Comparison of apparent heat capacities of both samples PLA-H and PLA-L is shown after being isothermally crystallized at 418.15 K (145 °C) for 15 hours after cooling from the melt and then reheating at 10 K min⁻¹ (reference 22). Both samples have the same C_p below glass transition and the same glass transition temperature, T_g, around of 330–333 K. Above T_g, semi-crystalline sample PLA-L shows much faster changing heat capacity than amorphous PLA-H with increasing temperature. Around 448 K, an endothermic peak related to the melting process of crystalline phase of the sample is presented. In contrast to PLA-L, the amorphous PLA-H behaves very differently and shows a linearly changing heat capacity with increasing temperature in the whole range of melt until reaching the degradation temperature of 580 K. Above approximately 460 K, the heat capacity of both samples are again the same and linear as is typical in the liquid state of polymeric materials.[22,23]

Figure 15.4 illustrates an example of the experimental apparent, total heat capacity measured by adiabatic calorimetry and standard differential scanning calorimetry for semi-crystalline poly(trimethylene terephthalate), PTT.[24] The apparent heat capacity, C_p, given in $J K^{-1} mol^{-1}$, changes with temperature and depends on the state of macromolecules. At low temperatures, below the glass temperature T_g of 331 K, heat capacity corresponds to the solid state of glass and crystal of PTT. The changes of the experimental C_p at around 331 K are due to the glass transition of the amorphous phase of PTT.

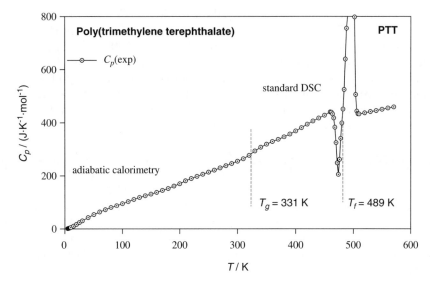

Figure 15.4 Experimental, apparent heat capacity of semi-crystalline poly(trimethylene terephthalate).

Above T_g the amorphous part of semi-crystalline PTT is in the high viscoelastic state or liquid state, while the crystalline phase is still in the solid state. Between the glass and melting transition, changes of heat capacity occur due to reorganization, pre-melting and exothermic crystallization. During the melting process, the apparent C_p shows an endothermic peak and the linear changes of C_p are in the isotropic liquid state over the whole temperature range.

15.4 Interpretation of the Heat Capacity

In order to understand the experimental heat capacity of polymeric materials, quantitative thermal analysis needs to be interpreted and the equilibrium functions of state and transition parameters need to be established for reference purposes. Heat capacity is a macroscopic quantity that should be analyzed in terms of microscopic molecular motion. The proper thermodynamic baselines of the solid and liquid heat capacity should be established and applied for discussion of phase transitions, such as the glass transition and melting, as well as for crystallization, cold crystallization or re-organization and enthalpy relaxation as is done to analyze apparent heat capacity.[1,20–24]

15.4.1 Calculation of the Heat Capacity for Solid State Polymers

The low-temperature experimental heat capacities of polymers from adiabatic calorimetry are used to calculate the solid heat capacity based on the common

and acceptable assumption that only the vibration motion of a polymer contributes to the $C_p(\exp)$. The computation of the vibrational heat capacity of
each analyzed polymer is based on the advanced method using the well-
established ATHAS scheme.[1,25,26] Briefly, a general scheme of this computation
is presented below. In order to link the heat capacity of the solid with the
vibrational spectrum, first, the experimental heat capacities at constant pressure $C_p(\exp)$, should be converted into heat capacities at constant volume,
$C_v(\exp)$, using the standard thermodynamic relationship:[26]

$$C_p(\exp) = C_v(\exp) + TV\frac{\alpha^2}{\beta} \tag{7}$$

where T is temperature, V is volume, and α and β are the coefficients of thermal
expansion and compressibility, respectively. The quantities α and β should be
considered as functions of temperature. In the case where α and β are not
available, the heat capacity can be estimated using the Nernst-Lindemann
approximation:[27]

$$C_p(\exp) - C_v(\exp) = 3RA_0\frac{C_p^2}{C_v}T/T_m \tag{8}$$

where R is the universal gas constant, A_0 is an approximately universal constant
equals 3.9×10^{-3} (mol K kJ^{-1}), and T_m is the equilibrium melting temperature.
With the assumption that below glass transition temperatures the experimental
heat capacity for sufficiently low temperatures contains only vibrational contributions $C_v(\exp)$, and can be separated into the heat capacities coming from
group, $C_v(\text{group})$, and skeletal, $C_v(\text{skeletal})$, vibrations:

$$C_v(\exp) = C_v(\text{group}) + C_v(\text{skeletal}) \tag{9}$$

The vibrational spectra of the solid state of each polymer, consisting of $3N$
vibrators, with N describing the total numbers of atoms in the repeating unit of
the polymer, can be separated into group and skeletal vibrations
($3N = N_{gr} + N_{sk}$). The numbers and types of group vibrations (N_{gr}) are derived
from the chemical structure of the sample as a series of single frequencies and
box frequencies over narrow frequency ranges. These frequencies can be
evaluated from normal-mode calculations on repeating units of the polymers
based on a fit to experimental infrared and Raman frequencies, or from suitable
low molar mass analogues. All approximate group vibrational frequencies of
the polymers were collected from the ATHAS Data Bank and from literature.[1,20,25,26] The skeletal vibrations, (N_{sk}) are not represented very well by
normal-mode calculations, but can be approximated by Debye's approach.[28]

The heat capacity due to the skeletal vibrations, N_{sk}, can be approximated by
fitting the experimental, low-temperature heat capacities to a Tarasov function
which consists of an appropriate combination of Debye functions, as will be
presented in more detail below.

The heat capacity from the group vibrations, C_v(group), of the polymers is estimated by the sum of the heat capacity from a series of single and box frequencies and is written as:

$$C_v(\text{group}) = C_v(\text{Einstein}) + C_v(\text{box}) \tag{10}$$

To evaluate C_v(group), first the heat capacity from single frequencies arising from normal modes (Einstein modes) has to be calculated and is given by a sum of the Einstein function:

$$C_v(\text{Einstein})/NR = \sum_i E(\Theta_{Ei}/T) = \sum_i \frac{(\Theta_{Ei}/T)^2 \exp(\Theta_{Ei}/T)}{[\exp(\Theta_{Ei}/T) - 1]^2} \tag{11}$$

where the summation takes place on the individual modes, and $\Theta_{Ei} = h\nu_i/k$, is the Einstein frequency in kelvin, and h and k are the Planck and Boltzmann constants, respectively. The heat capacity of a box-distribution, C_v(box), is given by a box-like spectrum and each box is represented by a sum of one-dimensional Debye functions, $\mathbf{D_1}$, for the sets of vibrations within the frequency interval from Θ_L to Θ_U:[28,29]

$$
\begin{aligned}
C_v(\text{box})/NR &= B(\Theta_U/T, \Theta_L/T) \\
&= \frac{\Theta_U}{\Theta_U - \Theta_L} [\mathbf{D_1}(\Theta_U/T) - (\Theta_L/\Theta_U)\mathbf{D_1}(\Theta_L/T)]
\end{aligned} \tag{12}
$$

where $\Theta_L = h\nu_L/k$ is the lower frequency and $\Theta_U = h\nu_U/k$ is the upper frequency in kelvin, respectively in the box-like spectrum.

After subtracting the contributions of all group vibrations from C_v(exp), the remaining portion of the experimental C_v(skeletal) is fitted at low temperatures to the general Tarasov function $\mathbf{T}$:[30]

$$
\begin{aligned}
C_v(\text{skeletal})/N_{sk}R &= \mathbf{T}(\Theta_1/T, \Theta_2/T, \Theta_3/T) \\
&= \mathbf{D_1}(\Theta_1/T) - (\Theta_2/\Theta_1)[\mathbf{D_1}(\Theta_2/T) - \mathbf{D_2}(\Theta_2/T)] \\
&\quad - (\Theta_3^2/\Theta_1\Theta_2)[\mathbf{D_2}(\Theta_3/T) - \mathbf{D_3}(\Theta_3/T)]
\end{aligned} \tag{13}
$$

to obtain the three characteristic theta-temperatures, $\Theta = h\nu/k$. The parameters Θ_1, Θ_2, and Θ_3 represent the maximum frequencies of the corresponding distributions in kelvin. The functions $\mathbf{D_1}$, $\mathbf{D_2}$ and $\mathbf{D_3}$ are the one-, two-, and three-dimensional Debye functions, respectively:[28,29]

$$C_v/NR = \mathbf{D_i}(\Theta_i/T) = i(T/\Theta_i)^i \int_0^{\Theta_i/T} \frac{(\Theta/T)^{i+1} \exp(\Theta/T)}{[\exp(\Theta/T) - 1]^2} d(\Theta/T) \tag{14}$$

where i is equal to 1, 2, or 3. In the Debye functions $\mathbf{D_i}(\Theta_i/T)$, N denotes the number of the vibrational modes for the frequency distribution to be described and R, the gas constant. The temperature Θ_3 describes skeletal contributions with a quadratic frequency distribution, as is usually found in solids for the lowest frequencies.

The temperatures Θ_2 and Θ_1 correspond to linear and constant distribution of density of states with frequency, respectively.[1,26]

For linear macromolecules, values of Θ_3 are usually less than 150 K. For planar molecular structures, an additional contribution with Θ_2 from 50 to 250 K is common. It yields a linearly increasing number of vibrational states with increasing frequency.[26] Finally, linear macromolecules commonly have values of Θ_1 in the 200 to 900 K range with a constant density of states as a function of frequency (box distribution).

Knowing Θ_1, Θ_2 and Θ_3 from a best fit of the experimental, skeletal heat capacities and with a list of group vibrations, one can calculate the heat capacity for the solid state at constant volume, $C_v(\text{total})$ as represented by Equation (9). Next $C_v(\text{total})$ is converted with Equations (7) or (8) to the calculated heat capacity at constant pressure $C_p(\text{vibration})$. This calculated heat capacity can be extended to higher temperatures and serves as a baseline of the vibrational heat capacity. The tables generated for the Data Bank range up to 1000 K.[20] It should be noted that the experimental and calculated heat capacity are reported per mole of the repeating unit.

Examples of calculations of vibrational heat capacities for different polymeric materials such as PE, PTT, PLA, and silk are presented in Figures 15.5, 15.6, 15.7, and 15.13 respectively.[22,24,31,32] For the evaluation of the vibrational heat capacity of PE $(\text{CH}_2)_x$ the 9 degrees of freedom resulting from the 3 atoms of the repeating unit, were separated into 7 group vibrations ($N_{gr} = 7$) and 2 skeletal vibrations ($N_{sk} = 2$) as shown in references 1 and 31. With the best fit of the

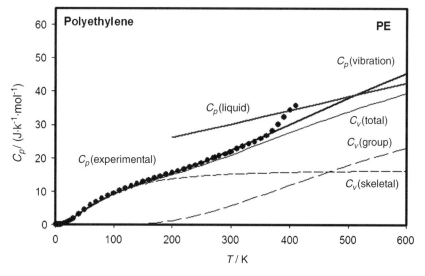

Figure 15.5   Evaluation of the solid, vibrational heat capacity of crystalline poly-ethylene.[20]

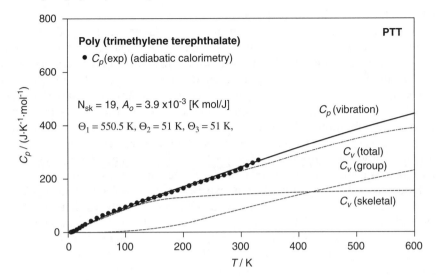

Figure 15.6 Experimental and calculated heat capacity of vibrational heat capacity of solid PTT.

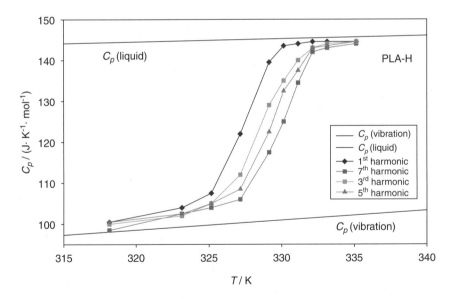

Figure 15.7 The vibration and liquid heat capacity of PLA with the experimental reversing heat capacity by complex temperature modulated DSC.[22,33]

experimental, skeletal heat capacity at low-temperatures to the Tarasov equation, Equation (13) the parameters $\Theta_1 = 519\,\mathrm{K}$ and $\Theta_2 = \Theta_3 = 158\,\mathrm{K}$ were calculated for crystalline polyethylene. With these parameters, the total, vibrational heat capacity for crystalline PE was estimated, C_p(vibration), and was extended from 0.1 to 1000 K. Figure 15.5 displays the evaluations of all contributions to vibrational C_p of crystalline PE and shows that the major contribution to the C_p(vibration), is from the skeletal heat capacity, C_p(skeletal) below 150 K which gets saturated around 250 K. The contribution from C_p(group) starts around 150 K and continues to increase into high temperatures but group vibrations are not excited even at 600 K. The sum of C_p(group) and C_p(skeletal) gives the total heat capacity at constant volume, C_v(total), that after conversion to constant pressure gives a baseline only from total vibrational motion, C_p(vibration). Figure 15.5 shows a comparison of experimental and calculated C_p for crystalline polyethylene. Good agreement between both is found at low temperatures with a precision greater than ± 3 %, and with deviations starting at around 300 K due to large-amplitude motions.

Figure 15.6 shows a comparison of experimental C_p(exp) by adiabatic calorimetry with calculated heat capacity, C_p(vibration) of poly(trimethylene terephthalate), PTT from 5 to 330 K. As was done for PE, in order to evaluate the vibrational heat capacity of PTT, the 75 degrees of freedom resulting from the 25 atoms of the repeating unit ($-CH_2-CH_2-CH_2-OCO-C_6H_4-COO-)_x$ were separated into 56 group vibrations ($N_{gr} = 18$) and 19 skeletal vibrations ($N_{sk} = 19$). All approximate group vibrational frequencies that are relevant to the current study of PTT were taken from normal–mode calculations based on IR and Raman spectroscopy.[24] The fit in the skeletal heat capacity shows a unique minimum in the chi-square statistical function at $\Theta_1 = 550.5$ K and $\Theta_2 = \Theta_3 = 51$ K. Within these parameters, the heat capacity for solid PTT attributed to vibrations only C_p(vibration), was collected from 0.1 to 1000 K in the ATHAS Data Bank.[1,24] Also, Figure 15.6 shows the evaluation of group, skeletal, and total heat capacities contribution for PTT. The calculated C_p(vibration) agrees with experimental C_p(exp) with a relative uncertainty of better than ± 3 % below 200 K.

Figure 15.7 shows an example of the calculated vibrational heat capacity for biodegradable polymeric material such as poly(lactic acid), PLA[22] together with results of reversing heat capacity from the complex saw-tooth modulation.[33] Again, as with PE, the 27 degrees of freedom resulting from the 9 atoms of the repeating unit ($-O-CHCH_3-CO-)_x$ were separated into 18 group vibrations ($N_{gr} = 56$) and 9 skeletal vibrations ($N_{sk} = 9$) for the evaluation of the group and skeletal vibrational heat capacity of PLA. The result of all contributions to calculated vibrational C_p(vibration) including skeletal heat capacity, C_v(skeletal) obtained from the best fit of the Tarasow equation ($\Theta_1 = 574$ K and $\Theta_2 = \Theta_3 = 52$ K) are presented in the reference 22 and only the final result of the total vibrational C_p(vibration) is presented in Figure 15.7 together with the experimental reversing heat capacity from the complex modulation. Within the range of measurement, the experimental data agree with calculated C_p within a few percent.

15.4.2 Calculation of the Heat Capacity for Liquid State Polymers

Liquid heat capacity of polymeric materials is more difficult to calculate and understand. The measured heat capacities for the liquid state for amorphous and semi-crystalline polymers are linear functions of temperature. Approximately, they can be described from an empirical addition scheme based on contributions of the constituent chain segments of polymers. For example, for aromatic polyesters the addition scheme of liquid heat capacity can be calculated from:[24]

$$C_p^{calc}(\text{liquid}) = N_{CH2}(17.91 + 0.0411T) + N_{COO}(64.32 + 0.002441T) + N_{C6H4}(73.13 + 0.1460T) \tag{15}$$

where N_{CH2}, N_{COO}, and N_{C6H4} are numbers of respective groups in the repeating unit of the polymer. For PTT with $N_{CH2} = 3$, $N_{COO} = 2$, and $N_{C6H4} = 1$ groups the value of calculated liquid heat capacity according to Equation (15) is: $C_p^{calc}(\text{liquid}) = 205.19 + 0.445T$ and agrees with experimental data[24] $C_p(\text{liquid}) = 211.6 + 0.434T$, as is shown in Figure 15.9.

In the case of macromolecules, the microscopic motion involved in liquid or high elastic rubbery states results from large-amplitude motion, such as conformational and anharmonic, in addition to the normal vibrations.

Generally, the heat capacity of polymers in the liquid state can be calculated from the standard thermodynamic relationship:[23,34]

$$C_p(\text{exp}) = C_v(\text{exp}) + TV\frac{\alpha^2}{\beta} \approx C_{vib}(\text{poly}) + C_{conf}(\text{poly}) + C_{ext}(\text{poly}) \tag{16}$$

where all quantities must be known as a function of temperature over the full range of the calculation. The experimental heat capacity $C_p(\text{exp})$, can be thus separated into the vibrational heat capacity $C_{vib}(\text{poly})$, the conformational heat capacity $C_{conf}(\text{poly})$, and the so-called external (anharmonic) contribution $C_{ext}(\text{poly})$. For the heat capacities of solids, Equation (16) is reduced to the two parts [$C_{vib}(\text{poly})$ and $C_{ext}(\text{poly})$] since the conformation of the chain is mostly fixed. The major contribution to the total heat capacity $C_p(\text{exp})$, comes from vibrational motion for both the solid and the liquid and $C_{vib}(\text{poly})$ is calculated according to Equation (9). The $C_{ext}(\text{poly})$ in Equation (16), is calculated either from Equation (7) or (8) for the liquid state.

The conformational contribution $C_{conf}(\text{poly})$ is calculated by making use of an previously derived equation for flexible macromolecules.[4] It is based on the one-dimensional Ising Model[35] with the following simplifying assumption: The conformational states of the bonds or flexible segments of the polymer can occur in only two discrete states, a ground state and an excited state, with an energy-difference between the two states of B. The energy B is modified by the parameter A, describing the interaction of the nearest conformational neighbours. The parameters B and A, have the meaning of stiffness and cooperativity, respectively. The conformations of the chain of a macromolecule with a total of N rotatable bonds can be described by the one-dimensional Ising-type model with the total energy:[4,35]

$$E_I = A\sum_{j=1}^{N} m_j m_{j+1} + B\sum_{j=1}^{N} m_j \tag{17}$$

where the conformation number $m_j = 0$ applies to the ground state with energy zero and degeneracy g_o. The conformation number $m_j = 1$ in Equation (17) corresponds to the excited state with energy B and degeneracy g_1. In the present analysis, the ratio of the degeneracies of the conformational states, $\Gamma = g_1/g_o$, is determined by a fit to the experimental heat capacity. Each conformation state can be modified by the energy A, depending on the conformation state of the next neighbour, and may be positive or negative.

Knowing E_1, using the transfer-matrix method one can calculate the partition function, the free energy per bond and then the conformational heat capacity in closed form:[41]

$$C_{\text{conf}}(\text{poly}) = R \frac{\Gamma[B/(k_B T)]^2 e^{-B/(k_B T)}}{[\Gamma e^{-B/(k_B T)} + 1]^2} [1 + \vartheta(B, A, \Gamma, T)] \qquad (18)$$

where R is the gas constant, k_B, the Boltzmann constant, and T, the temperature in kelvin. The first part in Equation (18) is the heat capacity identical to the rotational isomers model,[4] the second expression, ϑ (A, B, Γ, T), is too extensive to be shown in detail, but gives contribution to $C_{\text{conf}}(\text{poly})$ from the interaction of the nearest conformational neighbours. A full description of this model and calculation is given in the literature.[4,34,36,37]

An example of the evaluation of all contributions [$C_{\text{vib}}(\text{poly})$, $C_{\text{conf}}(\text{poly})$, $C_{\text{ext}}(\text{poly})$,] to the total liquid heat capacity of the amorphous PE above the glass transition is shown in Figure 15.8 with good agreement of experimental and calculated C_p. The experimental heat capacity data $C_p(\text{exp})$, in the temperature range from 250 to 600 K were separated as indicated by Equation (16). The

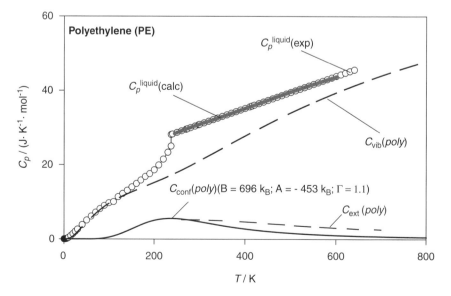

Figure 15.8   Evaluation of calculated liquid heat capacity of amorphous polyethylene.

vibrational heat capacity was calculated as was done previously for the solid state.[26] For the external heat capacity contribution, $C_{ext}(poly) = TV\frac{\alpha^2}{\beta_c}$, Equation (16) was used with the expansivity $\alpha_p = V^{-1}\left(\frac{\partial \ln V}{\partial T}\right)_p$, and the compressibility[†] $\beta_T = -V^{-1}\left(\frac{\partial \ln V}{\partial P}\right)_T$, derived from the experimental P-V-T diagram for the liquid-like state, collected from literature.[4,38] The conformational contribution to the total heat capacity of amorphous PE was then fitted to Equation (18) after subtracting the vibrational and external portions of heat capacity to find the three characteristic parameters, B, A, and Γ which are listed in Figure 15.8. Adding calculated conformational, vibrational, and external heat capacities, the total, calculated heat capacity, $C_p^{liquid}(calc)$, was computed, as is illustrated in Figure 15.8. Note that the parameters A and B are given in terms of Θ-temperatures in kelvin, so that the energy B of 696 K is obtained in units of J mol^{-1} by multiplication with the gas constant $R = 8.314$ J K^{-1} mol^{-1}($= 5.8$ kJ mol^{-1}).

In the case of the present three-parameter fitting, the effective energy difference between the two conformational states *gauche* and *trans* $B + A = (696 - 453) \times R = 2.0$ kJ mol^{-1} for PE, and is rather low compared to the value of 23 kJ mol^{-1} for glucose units in starch.[34,37] Since the degeneracies g_1 and g_0 must be an integer for each bond, the $\Gamma = 1.1$ from the fitting with three parameters must be interpreted such that only every eleventh *trans* state has easy availability of two *gauche* states. All others have only one.

Figure 15.8 shows a comparison of the experimental and calculated heat capacity with all contributions in the liquid-like state of amorphous PE. More details on the liquid heat capacity for different polymers and polymer-water systems can be found in the references.[4,34,36,37]

15.5 Quantitative Thermal Analysis of Polymeric Systems Based on the Heat Capacity

Knowing two equilibrium baselines of the solid heat capacity, C_p(vibration) and the liquid heat capacity, C_p(liquid), the apparent heat capacities of amorphous and semi-crystalline polymers as are obtained from DSC and TMDSC experiments can be analyzed quantitatively over the whole temperature range and time domain.

Figure 15.9 shows a comparison of experimental, semi-crystalline heat capacity of poly(trimethylene terephthalate) from adiabatic calorimetry and DSC (as shown in Figure 15.3) with the vibrational C_p(solid) and liquid heat capacity, C_p(liquid).[24] Agreement with the experimental C_p(exp) of PTT below glass transition temperature, T_g, with equilibrium C_p(solid) and above melting transition T_f with equilibrium C_p(liquid) is observed. Deviation of the apparent C_p(exp) from both equilibrium baselines between T_g and T_f shows endothermic and exothermic effects related with time-dependent, irreversible processes. Next, Figure 15.10 provides the details of the quantitative thermal analysis of semi-crystalline PTT from 250 to 550 K including these non-equilibrium effects

---

[†] In this chapter the isothermal compressibility is represented by the symbol β_T and not by κ_T as was recently recommended by IUPAC.

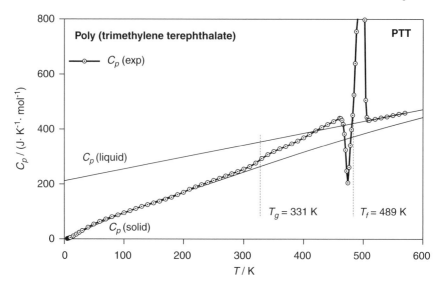

Figure 15.9 Experimental, semi-crystalline heat capacity of poly(trimethylene terephthalate) from adiabatic calorimetry and DSC (as shown in Figure
15.4) framed by the equilibrium of the vibrational C_p(solid) and liquid
C_p(liquid) heat capacities.

between T_g and T_f. The DSC trace of experimental C_p(exp) shows a broad glass
transition at 331 K before a broad small pre-melting peak around 450 K and
followed by a small ordering exotherm between 460 and 480 K. Finally, the
major melting endotherm is present with an onset temperature of $T_f = 489$ K.

Using these two baselines, the changes of heat capacity, ΔC_p, at $T_g = 331$ K was
estimated for the fully amorphous sample to be 88.8 J K^{-1} mol^{-1} and for the semi-
crystalline sample, to be 40 J K^{-1} mol^{-1}. The ratio of changes of heat capacity at
T_g for the semi-crystalline sample to changes of heat capacity at T_g of the fully
amorphous sample is 45 % of the mobile amorphous fraction, w_a, of PTT.

Quantitative thermal analysis allows one to separate all latent heats from
thermodynamic heat capacity in the apparent C_p(exp) between T_g and T_f using
the semi-crystalline baseline, $C_p^{\#}$ as is given by the solid line in Figure 15.10.
The semi-crystalline heat capacity, $C_p^{\#}$ was calculated from the change of
crystallinity with temperature and was made for a two-phase model as follows:

$$C_p(\exp) = w_c(T)C_p(\text{solid}) + (1 - w_c(T))C_p(\text{liquid}) - \frac{dw_c}{dT}\Delta_{\text{fus}}H(T) \quad (19)$$

where C_p(exp) is the experimental, apparent molar heat capacity in the melting
range, $w_c(T)$ is the temperature-dependent crystallinity, C_p(solid) and C_p(liquid)
are the calculated vibrational and liquid heat capacities, and $\Delta_{\text{fus}}H = H(\text{melt}) -$
$H(\text{crystal})$ is also temperature-dependent and is needed for the evaluation of the
temperature-dependence of the heat of fusion $[d\Delta_{\text{fus}}H/dT = C_p(\text{liquid}) -$
$C_p(\text{solid})]$.

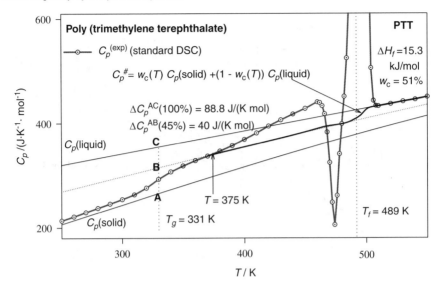

Figure 15.10 Quantitative thermal analysis of apparent heat capacity $C_p(\exp)$ of semi-crystalline PTT with details from 200 to 550 K.[24]

Using this function of $w_c,(T)$, the expected heat capacity $C_p(\text{expected})$ (in Figure 15.10 is marked as $C_p^{\#}$) was estimated according to expression:

$$C_p(\text{expected}) = w_c(T)C_p(\text{solid}) + (1 - w_c(T))C_p(\text{liquid}) \qquad (20)$$

which separate truly thermodynamic heat capacity from latent heat in the apparent heat capacity.

Next, with calculated crystallinity, $w_c(T)$ and an established $C_p(\text{expected})$ baseline, a corrected heat of fusion value, $\Delta_{\text{fus}}H$ for semi-crystalline PTT can be estimated. Integration of the area between the experimental, apparent heat capacity, and the expected heat capacity $C_p^{\#}$ gives the corrected value of heat of fusion, which is 15.3 kJ mol^{-1}. This corrected integration is possible due to the multi-steps scheme of the quantitative thermal analysis as is used in the ATHAS approach to polymeric materials.[1,24]

A similar approach, using quantitative thermal analysis can be presented for semi-crystalline PBT.[39,40] Figure 15.11 illustrates the details of the quantitative thermal analysis of the apparent $C_p(\exp)$, semi-crystalline PBT from the standard DSC and the quasi-isothermal TMDSC. Total and reversing $C_p(\exp)$ are analyzed using these two equilibrium baselines of vibrational $C_p(\text{solid})$ and $C_p(\text{liquid})$ in the range of temperatures from 250 to 550 K. The DSC traces of the heat capacity shows a very broad glass transition at 314 K, between **A** and **B** marked at the half-height of the change in heat capacity, and also a very broad double melting peak with a small peak around 480 K and a big peak around 497 K between **C** and **D**. Using these two baselines, the changes of heat capacity, ΔC_p, at $T_g = 314$ K were estimated for the fully amorphous sample to

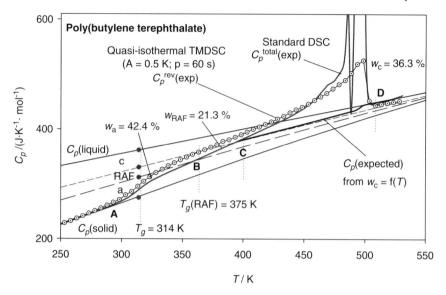

Figure 15.11 Quantitative thermal analysis of the experimental, total and reversing heat capacity of semi-crystalline PBT.

be $87.0 \, \text{J K}^{-1} \text{mol}^{-1}$ and for the semi-crystalline sample, to be $36.9 \, \text{J K}^{-1} \text{mol}^{-1}$. The ratio of changes of heat capacity at T_g for semi-crystalline sample to changes of heat capacity of the fully amorphous sample gives 42.4 % of the mobile amorphous fraction, w_a, of PBT. According to the two-phase model, the crystal starts melting at about 400 K and ends around 500 K, as is shown in Figures 15.11 and 15.12 respectively. The calculation of crystallinity was made according to Equation (13) and results of $w_c = f(T)$ are presented in Figure 15.12 with the highest (maximum) value of 36.3 %.

The expected C_p was established using this function of degree of crystallinity versus temperature and Equation (20) and then applying it to calculate the corrected value of the heat of fusion from the integration between the experimental and this C_p(expected) from C to D. The integration gives the value $\Delta_{\text{fus}}H = 11.6 \, \text{kJ mol}^{-1}$ for standard DSC measurement for semi-crystalline PBT as is associated with the crystalline fraction. The remaining fraction of the phase in the examined semi-crystalline PBT should be related with the rigid amorphous fraction, w_{RAF} with the value of 21.3 % ($w_{\text{RAF}} = 1 - w_a - w_c$). The change of the experimental, apparent C_p^{total}(exp) between B and C in Figure 15.11 corresponds to a relaxation of large-amplitude motion in the RAF mesophase of the sample. The different mobility of the two amorphous phases is reflected by the changes of C_p,(exp) for the more mobile fraction at a T_g of 314 K and the less mobile fraction of the rigid amorphous fraction at a T_g (RAF) of 375 K. In early papers,[41] it was shown that the major contribution to the changes of heat capacity, ΔC_p(exp) in the glass transition regions of amorphous polymers such polyethylene, PE, polypropylene, PP, polystyrene, PS, poly(methyl methacrylate), PMMA,[4] and

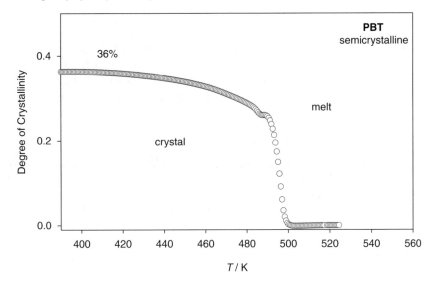

Figure 15.12 Change of the fraction degree of crystallinity $w_c = f(T)$ with temperature for the semi-crystalline PBT resulting from solution of Equation (15).

other aliphatic polyesters[23,42] are from the large-amplitude, conformational motion. Even more than $\Delta C_p(\exp)$ in the melting region are related to conformational motion of the crystal and appears during melting.[43] The apparent melting peaks of semi-crystalline PBT were studied by TMDSC using different frequencies in this publication.[44] In Figure 15.11, the reversing heat capacity, $C_p^{\text{rev}}(\exp)$ from the quasi-isothermal TMDSC is compared with total, $C_p^{\text{total}}(\exp)$ over the whole range of presented temperatures. The difference between the total and reversing $C_p(\exp)$ is the non-reversing heat capacity and provides information regarding irreversible processes. Agreement between both $C_p(\exp)$ with an error of (3 to 4) % can be observed in the solid and liquid states and a slightly higher value of reversing $C_p^{\text{rev}}(\exp)$ between (300 and 400) K can be related with the different thermal histories of the samples. The major differences between the apparent, total, and reversing $C_p(\exp)$ are found in the melting double peaks and described in the non-reversing portion, and this suggests that the major process of melting is practically an irreversible process as is expected for polymer crystals. Only the small reversing $C_p^{\text{rev}}(\exp)$ in the melting region contributes to the total endotherms beyond the expected C_p of semi-crystalline PBT and can be associated with the local ordering/disordering of small portions of the crystal. The local process of melting/re-crystallization, which follows heating and cooling during temperature modulation of polymeric materials, such as PBT, can be observed in time-domain and was studied earlier.[16,40,44]

Similar behaviour of the experimental $C_p(\exp)$ has been observed for PET, PTT and other synthetic and biological polymeric materials in the melting and overall, the temperature range between glass and melting transition where it shows many complicated overlapping processes.[10,23,24,37,45,46]

Interesting results of the quantitative thermal analysis of PLA are presented in Figure 15.7. The glass transition of amorphous poly(lactic acid), PLA is studied with temperature modulated differential scanning calorimetry. The reversing, non-reversing and total heat capacities were evaluated.[23] Using a multi-frequency, complex saw-tooth modulation in the quasi-isothermal mode, the apparent, reversing heat capacity in the glass transition region was obtained in a single experiment for the sample of poly(lactic acid) with the same thermal history. Figure 15.7 shows the reversing heat capacity from the first, third, fifth, and seventh harmonic component. For higher harmonics, the glass transition occurs in higher temperatures and below and above T_g heat capacities are practically the same and independent from frequency of modulation. The experimental heat capacities of amorphous PLA were analyzed in reference to the solid and liquid heat capacities of poly(lactic acid) found in the ATHAS Data Bank.[20,22] More information about this spectroscopy of heat capacity of PLA at glass transition region can be found in reference 33.

A final example of the quantitative thermal analysis, based on heat capacity in polymeric material, is given in Figure 15.13 for a more complex material, biological polymer dry silk fibroin.

Figure 15.13 shows a comparison of calculated vibrational heat capacity of silk fibroin, C_p^{Silk}(vibration) (solid curve) with the experimental heat capacity of dry amorphous *Bombyx mori* silk fibroin.[32] The experimental reversing heat capacity C_p^{rev}(experimental) was measured by TMDSC and the experimental, total heat capacity of dry silk, C_p^{total}(experimental) from standard DSC as was

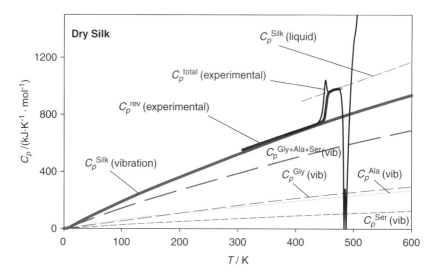

Figure 15.13   Comparison of the total and reversing experimental heat capacity of dry silk fibroin to calculated vibrational C_p. Also, the vibrational heat capacity of the three major amino acids components and their sum, are presented.[32]

presented earlier in the literature.[32,47,48] The data show a step change in heat capacity as silk fibroin undergoes its glass transition at $T_g = 451.15\,\text{K}$ (178 °C).[48] The experimental, total heat capacity curve shows a small enthalpy peak on the high temperature side of the glass transition, related to non-reversing processes such as physical aging. Immediately following the step at T_g, there is a strong exotherm related with a cold crystallization peak. Above this temperature, an exothermic increase of the total C_p^{total}(experimental) relating to processes of sample degradation, is observed. All non-reversing processes such as enthalpy relaxation, cold crystallization and degradation were separated by using quasi-isothermal TMDSC and did not show up in reversing heat capacity, C_p^{rev}(experimental). The quantitative thermal analysis of silk fibroin is possible after establishing baselines. First the solid heat capacity C_p^{Silk}(vibration), from vibrational motion, and next the liquid heat capacity is derived empirically from fitting the experimental data to the linear function of temperature C_p^{Silk}(liquid) as is presented in Figure 15.13.

Vibrational heat capacity of silk, C_p^{Silk}(vibration), was evaluated using the vibrational motion spectra of individual amino acids in silk fibroin.[32] The vibrational heat capacity was constructed according to:

$$C_p^{\text{Silk}} = N_{\text{Gly}}C_p(\text{Gly}) + N_{\text{Ala}}C_p(\text{Ala}) + \ldots + N_{\text{Met}}C_p(\text{Met}) = \sum_i N_i C_p(i) \quad (21)$$

using the vibrational C_p of the poly(amino acid)s from the ATHAS Data Bank.[20] Equation (21) shows the sum-over contributions from glycine, alanine, and so on through methionine, *i.e.*, over the individual amino acids comprising silk fibroin.[32] Knowing the numbers N_i of each type of amino acid in the silk fibroin and their corresponding vibrational heat capacity as function of temperature, the vibrational heat capacity of silk, C_p^{Silk}(vibration), was calculated over the whole range of temperatures from (0.1 to 600) K. The computation has been carried out on the complete heavy chain sequence of 5263 amino acids and the light chain sequence of 263 amino acids. The numbers N_i and compositions of the individual amino acids contained in *Bombyx mori* silk fibroin were taken from reference 49. For example, the 1593 molecules of alanine (Ala) in the heavy chain and the 37 molecules of Ala in the light chain of silk fibroin, were taken into account to determine the alanine contribution to the total vibrational heat capacity of dry silk, C_p^{Silk}(vibration). The molar mass of dry silk is $M_w(\text{silk}) = 419262\,\text{g mol}^{-1}$ and the molar masses of each amino acid, $M_w(i)$ contained in silk fibroin are listed in reference 32. In Figure 15.13, the vibrational heat capacities of the three amino acids, glycine, alanine, and serine, and their sum in relation to the total C_p^{Silk}(vibration), are also shown (by dashed or dotted lines lying below the experimental data curve). These three amino acids account for approximately 85 mol % of the silk fibroin molecule. We observe that the sum of the vibrational heat capacities of glycine, alanine, and serine, gives approximately 75 % of the total contribution to C_p^{Silk}(vibration). Good agreement between the experimental data and the calculated vibrational heat capacity of silk fibroin is observed below T_g with an a relative uncertainty of ± 1 % between (380 and 435) K, and with a relative uncertainty of ± 3 % below 380 K. This agreement supports the conclusion that the only contribution to the heat capacity below T_g for dry silk comes from the vibrational motion of the amino acid components.[32]

The expression for the liquid heat capacity was found empirically as a linear function of temperature: C_p^{Silk} (liquid) $= 297.676 + 1.455\, T$.

The thermal properties of the experimental data can be quantitatively analyzed knowing the calculated vibration and liquid heat capacity of dry silk as baselines. For example, the change in heat capacity, $\Delta C_p(\text{exp})$ at the temperature $T_g = 451.15$ K was estimated as $200.407\, \text{kJ K}^{-1}\text{mol}^{-1}$ between the vibrational and liquid baselines. According to Wunderlich,[1] each mobile unit in the chain of a polymer molecule contributes $11\, \text{J K}^{-1}\text{mol}^{-1}$ to the change of heat capacity at T_g. The ratio of our measured $\Delta C_p(\text{exp})$ to the contribution of a single mobile unit allows an estimate to be made of the total number of mobile units in non-crystalline silk fibroin. This ratio gives a total of 18219 units of silk fibroin, which start to become mobile at the glass transition.[32]

Next, the estimation of the number of mobile units from the mol. % of alanine, glycine, and serine, counting three rotating bonds for Ala, three rotating bonds for Gly, and four rotating bonds for Ser gives a sum of 14841 rotating units for these three amino acids in one silk fibroin molecule. This number compares very favorably to 85 % of 18219 (85 % of the estimated total number of units contributing to the heat capacity increment at T_g), which is 15486. We can conclude that the change in heat capacity at T_g of dry silk fibroin corresponds to the change anticipated for the fully non-crystalline material in our silk. The full description of this kind of thermal analysis can be found in the literature.[32]

The full data of the heat capacity of dry silk and silk with water has already been presented and discussed in references 32 and 48.

15.6 Conclusions

The heat capacity is an important quantity for the characterization of thermal properties of polymeric materials. Modern calorimetry and the quantitative thermal analysis based on heat capacity and interpretation on a molecular motion allows the study of the many processes in metastable polymers. Thermal analyses are often complicated by irreversible effects, such as the glass transition, enthalpy relaxation, partial crystallinity, reorganization, re-ordering and broad melting transition. All these non-equilibrium processes and states need the equilibrium thermodynamic references such as in the case of heat capacity: the solid, vibrational heat capacity and the liquid heat capacity. Temperature modulated, standard and adiabatic calorimetry together with advanced thermal analyses approaches permit the separation of reversible from irreversible processes that are part of apparent experimental heat capacities of polymeric systems.

Acknowledgments

This work was supported by European Union Grant (MIRG-CT-2006036558).

References

1. B. Wunderlich, *Thermal Analysis of Polymeric Materials,* Springer-Verlag, Berlin, 2005.
2. B. Wunderlich, *J. Thermal Anal.*, 2008, **93**, 7.
3. W. Qiu, M. Pyda, E. Nowak-Pyda, A. Habenschuss and B. Wunderlich, *Macromolecules*, 2005, **38**, 8454.
4. M. Pyda and B. Wunderlich, *Macromolecules*, 1999, **32**, 2044.
5. V. B. F. Mathot, (ed.) *Calorimetry and Thermal Analysis of Polymers,* Hanser Publishers, München, 1994.
6. B. Wunderlich, *Macromolecular Physics, Volume 1,2,3,* Academic Press, New York, London, 1973, 1976, 1980.
7. E. Turi, ed. *Thermal Characterization of Polymeric Materials,* Academic Press, New York, 1997.
8. S. Z. D. Cheng, ed. *Handbook of Thermal Analysis and Calorimetry,* **Vol.3**, *Applications to Polymers and Plastics,* Elsevier Science, Amsterdam, 2002.
9. D. R. Witzke, *Introduction to Properties Engineering and Prospects of Polylactide Polymers,* UMI, Dissertation, Michigan State University, East Lansing, MI, 1997.
10. B. Wunderlich, *Prog. Poly. Sci.*, 2003, **28**, 383.
11. B. V. Lebedev, T. G. Kulagina and N. N. Smirnova, *J. Chem. Thermodynam.*, 1988, **20**, 1383.
12. R. L. Danley, *Thermochim. Acta*, 2003, **395**, 201.
13. M. Reading, *Trends Polym. Sci.*, 1993, **8**, 248.
14. S. A. Adamowsky, A. A. Minakov and C. Schick, *Thermochim. Acta*, 2003, **52**, 403.
15. W. Qiu, A. Habenschuss and B. Wunderlich, *Polymer*, 2007, **48**, 1641.
16. M. Pyda, M. L. Di Lorenzo, J. Pak, P. Kamasa, A. Buzin, J. Grebowicz and B. Wunderlich, *J. Polymer Sci.: Part B: Polymer Phys.*, 2001, **39**, 1565.
17. M. Pyda, Y. K. Kwon and B. Wunderlich, *Thermochim. Acta*, 2001, **367/368**, 217.
18. A.W. Van Herwoarden, Thermochim. Acta, 2005, 432, 129, also: http:www.xensor.nl/.
19. A. A. Minakov, D. A. Moroivintsev and C. Schick, *Polymer*, 2004, **45**, 3755.
20. M. Pyda, ed. The ATHAS Data Bank, 2007: http://athas.prz.rzeszow.pl/.
21. U. Gaur, S.-F. Lau, H.-C. Shu, B. B.Wunderlich, M. Varma-Nair and B. Wunderlich, *J. Phys. Chem. Ref. Data*, 1981, 10, 89, 119, 1001, 1051; 1982, 11, 313, 1065; 1983, 12, 29, 65, 91; and 1991, 20, 349.
22. M. Pyda, R. C. Bopp and B. Wunderlich, *J. Chem. Thermodynamics*, 2004, **36**, 731.

23. M. Pyda and B. Wunderlich, *Macromolecules*, 2005, **38**, 10472.
24. M. Pyda, A. Boller, J. Grebowicz, H. Chuah, B. V. Lebedev and B. Wunderlich, *J. Polymer Sci., Part B: Polymer Phys.*, 1998, **36**, 2499.
25. B. Wunderlich, *Pure Applied Chem.*, 1995, **67**, 1019.
26. M. Pyda, M. Bartkowiak and B. Wunderlich, *J. Thermal Analysis*, 1998, **51**, 631.
27. W. Nernst and F. A. Lindemann, *Z. Electrochem.*, 1911, **17**, 817.
28. P. Debye, *Ann. Physik*, 1912, **39**, 789.
29. B. Wunderlich and M. Dole, *J. Polymer Sci.*, 1957, **24**, 201.
30. V. V. Tarasov, *Zh. Fiz. Khim.*, 1950, **24**, 111.
31. S.-F. Lau and B. Wunderlich, *J. Thermal Analysis*, 1983, **28**, 59.
32. M. Pyda, X. Hu and P. Cebe, *Macromolecules*, 2008, **41**, 4786.
33. B. Lisowska, E. Nowak-Pyda and M. Pyda, Proc. 35th NATAS Conf. in East Lasing, MI, 2007, August 26-29, edt. CD edition, 35, 10.
34. M. Pyda, in *The Nature of Biological Systems as Revealed by Thermal Methods* D. Lorinczy, ed. Kluver Academic Publisher, Amsterdam, 2004.
35. K. Huang, *K. Statistical Mechanics* ed. Wiley, New York, 1963.
36. M. Pyda, *J. Polymer Sci., Part B: Polymer Phys.*, 2001, **39**, 3038.
37. M. Pyda, *Macromolecules*, 2002, **35**, 4009.
38. P. Zoller, *J. Appl. Polym. Sci.*, 1979, **23**, 1051, and L. D. Loomis and P. Zoller, *J. Polym. Sci., Polym. Phys.*, 1983, **21**, 241, and O. Olabishi and R. Simha, *Macromolecules*, 1975, 8, 206.
39. M. Pyda E. Nowak-Pyda, J. Mays and B. Wunderlich, *J. Polymer Sci., Part B: Polymer Phys.*, 2004, **42**, 4401.
40. M. Pyda, E. Nowak-Pyda, J. Heeg, H. Huth, A. A. Minakov, M. L. Di Lorenzo, C. Schick and B. Wunderlich, *J. Polymer Sci., Part B: Polymer Phys.*, 2006, **44**, 1364.
41. M. Pyda and B. Wunderlich, *J. Polymer Sci., Part B: Polymer Phys.*, 2002, **40**, 1245.
42. M. Pyda, *Proc 3rd International Seminar on Modern Polymeric Materials For Environmental Applications*, 2008, **Vol. 3**, 201.
43. W. Qiu, M. Pyda, E. Nowak-Pyda, A. Habenschuss and B. Wunderlich, *J. Polymer Sci., Part B: Polymer Phys.*, 2007, **45**, 475.
44. M. C. Righetti and M. L. Di. Lorenzo, *J. Polymer Sci., Part B: Polymer Phys.*, 2004, **42**, 2191.
45. B. Wunderlich, Heat Capacity of Polymers in *Handbook of Thermal Analysis and Calorimetry, Vol.3, Applications to Polymers and Plastics*; S. Z. D. Cheng, ed., Elsevier Science, Amsterdam, 2002.
46. M. Pyda and B. Wunderlich, *J. Polymer Sci., Part B: Polymer Phys.*, 2000, **38**, 622.
47. X. Hu, D. Kaplan and P. Cebe, *Macromolecules*, 2006, **39**, 6161.
48. M. Pyda, X. Hu and P. Cebe, *NATAS Notes*, 2008, **40**(4), 20.
49. Swiss_Prot, Protein, data base:http://www.expasy.org/uniport/P05790FIBH_BOMMO and http://www.expasy.org/uniport/P21828FIBL_BOMMO.

CHAPTER 16
Protein Heat Capacity

WERNER W. STREICHER AND GEORGE I. MAKHATADZE

Department of Biology and Department of Chemistry and Chemical Biology, Center for Biotechnology and Interdisciplinary Studies, Rensselaer Polytechnic Institute, 110 8th Street, Troy, NY 12180, USA

16.1 Introduction

Heat capacity is one of the basic thermodynamic parameters specifying the state of a protein and is commonly measured in protein research. By definition, the specific heat capacity is equal to the amount of heat required to raise the temperature of a unit mass of a substance by 1 K. However, temperature is also commonly used as a perturbation to the conformational equilibrium of proteins. As the temperature increases, the population of protein molecules changes from the native state into the denatured state. This process of protein denaturation is accompanied by changes in the heat capacity: the heat capacity of the unfolded state is always higher than that of the native state.[1,2] There are several different factors that can contribute to the measured heat capacity; these are related to the global structural properties of proteins in the native and unfolded states and how these states interact with the solvent.

Proteins are heteropolymers consisting of 20 different building blocks, the amino acids. When covalently attached, the amino acid residues share a common backbone structure, the peptide unit, but differ in the chemical nature of their respective side chains. The physicochemical properties of side chains are vastly different: some are ionizable, *i.e.* carry positive or negative charge, some are polar and thus can act as hydrogen bonding donors and / or acceptors, and some are non-polar, consisting of different aliphatic or aromatic groups. The

Heat Capacities: Liquids, Solutions and Vapours
Edited by Emmerich Wilhelm and Trevor M. Letcher
© The Royal Society of Chemistry 2010
Published by the Royal Society of Chemistry, www.rsc.org

non-covalent interactions between various backbone and side chains define the native structure of the protein. The common feature is that a large fraction of the non-polar side chains are buried in the core of the protein, while a large fraction of polar and charged residues are located on the protein surface and are largely exposed to the solvent water. The unfolded state is an ensemble of conformations in which, on average, all groups become largely solvent exposed. Thus, upon unfolding, the groups that were previously buried in the interior of the protein become exposed to solvent and this exposure is believed to define the change in heat capacity upon protein unfolding. It has been shown that the heat capacity of the denatured state can be calculated, based on an additivity scheme according to Makhatadze and Privalov,[2] and is within experimental error of measured denatured state heat capacities. This not only allows direct comparison between measured heat capacities and calculated unfolded state heat capacities but has also allowed the calculated heat capacities to be used in the analysis of differential scanning calorimetric data (see for example references 3 and 4).

16.2 Definition of Protein Heat Capacity

The most commonly used definition of protein heat capacity at constant pressure, C_p, is based on the temperature derivative of the enthalpy, $H(T)_p$:

$$C_p = \frac{\mathrm{d}H(T)_p}{\mathrm{d}T} \tag{1}$$

From an experimental point of view, the specific heat capacity, c_p, is equal to the heat energy, ΔQ, required to raise the temperature of a unit mass by 1 K. The values of c_p, according to IUPAC recommendation, are expressed in joules per kelvin per gram ($J\,K^{-1}\,g^{-1}$).

16.3 Experimental Determination of Partial Heat Capacity

Differential scanning calorimetry (DSC) is used to measure the heat capacity of proteins in solution at constant pressure, C_p, as a function of temperature. The DSC operates in differential mode, which means that the heat capacity of the protein in an aqueous buffer solution in the sample cell is measured relative to the heat capacity of buffer in reference cell of equal volume.[3,5,6] The apparent difference in the heat capacities of sample and reference cells, $\Delta C_p^{\mathrm{app}}(T)$, is related to the partial heat capacity of the protein at temperature T, $\Delta C_{p,\mathrm{pr}}^{\mathrm{exp}}(T)$ as:

$$C_{p,\mathrm{pr}}^{\mathrm{exp}}(T) = \left[\frac{c_{p,\mathrm{H_2O}}}{\bar{v}_{\mathrm{H_2O}}} \cdot \bar{v}_{\mathrm{pr}} - \frac{\Delta C_p^{\mathrm{app}}(T)}{m_{\mathrm{pr}}} \right] \cdot M \tag{2}$$

where $c_{p,\mathrm{H_2O}}$ is the specific heat capacity of aqueous buffer, $\bar{v}_{\mathrm{H_2O}}(T)$ is the partial specific volume of the aqueous buffer (the $c_{p,\mathrm{H_2O}}/\bar{v}_{\mathrm{H_2O}}(T)$ term can be considered independent of temperature and equal to $4.184\ \mathrm{J\cdot(K\ cm^{-3})^{-1}}$), $\bar{v}_{\mathrm{pr}}$ is the partial specific volume of the protein (which can be determined by the additivity scheme shown by Equation (3)), m_{pr} is the mass of the protein in the calorimetric cell, and M is the molecular mass of protein.

The partial molar volume of the protein can be calculated from the additivity scheme described by Makhatadze *et al.*[7] as follows:

$$\bar{V}(\mathrm{calc}) = (N-1)\cdot V^0_\phi(-\mathrm{CHCONH}-) + \sum_{i=1}^{N} V^0_\phi(-R_j) \qquad (3)$$

where N is the number of amino acid residues in the protein sequence, $V^0_\phi(-\mathrm{CHCONH}-)$ is the contribution of the peptide unit, and $V^0_\phi(-R_i)$ is the contribution of the ith amino acid side chain.

A DSC profile, representing the temperature dependence of the partial molar heat capacity, $C_p^{\mathrm{exp}}(T)$, for a hypothetical protein is shown in Figure 16.1. The experimentally measured $C_p^{\mathrm{exp}}(T)$ consists of two terms,[6] the so-called progress

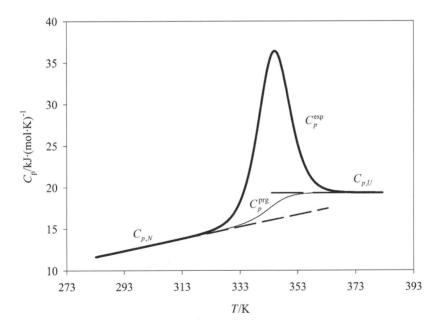

Figure 16.1 An example of a typical thermal unfolding profile of a protein monitored by DSC (solid line; C_p^{exp}). The long dashed line represents the heat capacity of the unfolded state $C_{p,\mathrm{U}}$, the short dashed line represents the heat capacity of the native state, $C_{p,\mathrm{N}}$, and the dotted line represents the progress heat capacity (C_p^{prog}).

heat capacity, $C_p^{\mathrm{prg}}(T)$, and the excess heat capacity, $\langle C_p(T)\rangle^{\mathrm{exc}}$:

$$C_p^{\mathrm{exp}}(T) = C_p^{\mathrm{prg}}(T) + \langle C_p(T)\rangle^{\mathrm{exc}} \tag{4}$$

The excess heat capacity is caused exclusively by the heat released, or absorbed, as the unfolding reaction proceeds.[8,9] The progress heat capacity is defined by the temperature dependencies of the heat capacities of various states. For a two state system it will be defined as:

$$C_p^{\mathrm{prg}}(T) = F_{\mathrm{N}}(T) \cdot C_{p,\mathrm{N}}(T) + F_{\mathrm{U}}(T) \cdot C_{p,\mathrm{U}}(T) \tag{5}$$

where $F_{\mathrm{N}}(T)$ and $F_{\mathrm{U}}(T)$ are the temperature dependencies of fractions of proteins in the native and unfolded states, respectively, and $C_{p,\mathrm{N}}(T)$ and $C_{p,\mathrm{U}}(T)$ are the temperature dependencies of the partial heat capacities of the native and unfolded states, respectively. By subtracting the progress heat capacity from the experimentally measured heat capacity and integrating the area under the excess heat capacity curve, the enthalpy of the reaction can be determined:

$$\Delta Q_{\mathrm{cal}} = \int_{T1}^{T2} \langle C_p(T)\rangle^{\mathrm{exc}} dT \tag{6}$$

For a two state unfolding process the ΔQ_{cal} is the enthalpy change upon protein unfolding. This parameter has important implications for the validation of the two-state behavior.[10] In addition, taken together with the transition temperature and the heat capacity change upon unfolding, it defines the temperature dependence of the Gibbs energy, $\Delta G(T)$, defined by the Gibbs-Helmholz equation:

$$\Delta G(T) = \Delta H_{\mathrm{cal}}(T_{\mathrm{m}}) + \Delta C_p \cdot (T - T_{\mathrm{m}}) - T$$
$$\cdot \left[\frac{\Delta H_{\mathrm{cal}}(T_{\mathrm{m}})}{T_{\mathrm{m}}} + \Delta C_p \cdot \ln\left(\frac{T}{T_{\mathrm{m}}}\right) \right] \tag{7}$$

where T_{m} is the so-called transition temperature, defined as a temperature at which the populations of the native and unfolded states are equal, ΔH_{cal} is the enthalpy of unfolding, and ΔC_p is the heat capacity change upon unfolding, usually assumed to be independent of temperature.[11]

16.4   Heat Capacity of Anhydrous Proteins

To determine the intrinsic heat capacity of proteins in the absence of solvent effects, the heat capacity of proteins in the anhydrous state have been measured.[12–15] Unfortunately, the number of examples where the anhydrous state heat capacities have been measured is rather limited relative to solution

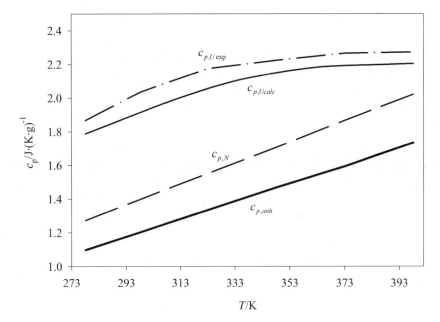

Figure 16.2 Temperature dependence of the calculated unfolded state specific heat capacity ($c_{p,U\text{calc}}$; solid line; (reference 17)), the measured specific heat capacity ($c_{p,U\exp}$; line dot), native state specific heat capacity ($c_{p,N}$; dashed line) and the specific heat capacity of the anhydrous state ($c_{p,\text{anh}}$; dotted line) of lysozyme.

state heat capacities. For the illustration of the effects of various factors on the heat capacity on proteins, lysozyme is used as an example. The reason for this is that lysozyme is one of the most well characterized proteins particularly when it comes to calorimetric measurements.[16,17] The anhydrous specific heat capacity, $c_{p,\text{anh}}$, of lysozyme was found to change in a linear fashion as the temperature was increased, within the temperature range of interest (see Figure 16.2). This trend has been shown to be true for several other proteins.[12,18] The major contributions to the anhydrous state heat capacity has been shown to be due to bending and stretching of each valence bond as well as internal rotations and is dominated by the covalent structure of the protein.[19] The good correlation between the additivity of the different amino acids constituting the protein, as well as the peptide bonds, and the experimentally determined heat capacity for the anhydrous state, shows that the covalent interactions dominate the heat capacity of proteins. By comparing the specific heat capacities for various anhydrous proteins, it has been shown that they are very similar at 298.15 K, (1.25 ± 0.013) J (K g)$^{-1}$, and have a similar temperature dependence. The difference in the heat capacities in the hydrated and anhydrous states of a protein provides some insight into the effects of hydration.

16.5   Heat Capacity of Hydrated Proteins

The heat capacity of proteins is usually measured in an aqueous environment. As mentioned previously, proteins undergo a conformational transition from the native state to the unfolded state upon an increase in temperature. For many small proteins, the unfolding can be described by a two state unfolding mechanism, where only native and/or unfolded proteins are significantly populated at different temperatures. Depending on the stability of the protein, the heat capacity of the native state is experimentally accessible in a relatively narrow temperature range where the native state is highly populated. As the temperature is increased, the denatured state becomes more populated until all protein molecules fully populate this state. Again, the experimentally accessible range of temperatures for measuring heat capacity of the unfolded state would depend on the stability of the protein and is also relatively narrow.

The general features of the heat capacities for the native or unfolded states of a protein can be observed in Figure 16.1. The temperature dependence of the native state heat capacity in this case is observed between 283.15 and 318.15 K, whereas the temperature dependence of the denatured state heat capacity is observed between 363.15 and 383.15 K.

16.5.1   Heat Capacity of the Native State

It is believed that the major difference in interactions between the solvated and anhydrous states of a protein is that in the solvated native state there are additional interactions of the protein surface with the solvent water. There is ample evidence that the native state heat capacities of proteins have linear temperature dependencies.[2,17,20] Figure 16.2 shows the temperature dependence for the specific heat capacity of the native state, $c_{p,N}$, of lysozyme. The same figure also shows the temperature dependence of the anhydrous specific heat capacity, $c_{p,anh}$, of lysozyme. It can be seen from Figure 16.2 that the $c_{p,N}$ is slightly increased relative to $c_{p,anh}$, which is due to solvation. Even though the difference in the case of lysozyme is relatively small, the $c_{p,N}$ can be quite different from that of the anhydrous state depending on the amino acid composition on the surface of the protein.

Depending on the sequence of the protein and thus the amino acids on the surface of the protein, the $c_{p,N}$ at 298.15 K ranges from 1.25 to 1.80 J $(K\ g)^{-1}$ and its dependence on temperature appears to be a linear function with slopes ranging from 0.005 to 0.008 J $(K^2\ g)^{-1}$.[21] Comparing these values with those for the anhydrous proteins, 1.0 to 1.5 J $(K\ g)^{-1}$ at 298.15 K, it is clear that ~ 70–80 % of the heat capacity of the native state in aqueous solution originates from the protein itself, but the major contribution to the variation is from the hydration of protein surfaces in the native state.

16.5.2   Heat Capacity of the Unfolded State

The unfolded state heat capacity is only accessible, experimentally, if the protein is unfolded which is at a much higher temperature range relative to the native state. Indirect estimates have suggested that the increase of the configurational freedom upon unfolding contributes approximately 30 % of the heat capacity increment[19] whereas hydration of groups that are newly exposed after unfolding contribute approximately 70 % of the heat capacity increment.[2] On average, at 298.15 K the values for the partial specific heat capacity for the unfolded state, $c_{p,U}$, range from 1.85 to 2.2 J (K g)$^{-1}$. These values increase to 2.1 to 2.4 J (K g)$^{-1}$ for the specific heat capacities of unfolded proteins at 373.15 K.[21]

There are two major differences between the heat capacities of the native and unfolded states, as can be seen in Figure 16.1 and Figure 16.2. The first major difference is the higher magnitude of the heat capacity of the unfolded state, relative to the heat capacity of the native state, which is caused by the difference in the hydration of protein internal groups, which are exposed to water upon unfolding, and by the rise of the configurational freedom of protein groups resulting from the disruption of its solid native structure.[2] Another major difference is the non-linearity of the temperature dependence of the unfolded state heat capacity. It has been shown that the non-linearity originates from the fact that increasing the temperature decreases the heat capacity for non-polar residues in a linear fashion, whereas increasing the temperature for polar residues increases their heat capacity in a non-linear fashion, which are illustrated in Figure 16.3.[2,22]

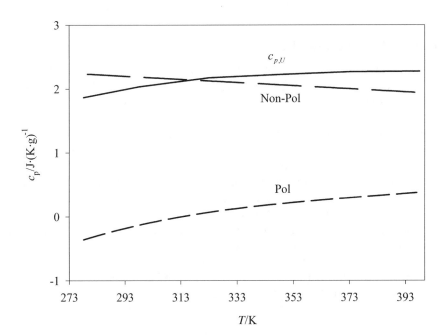

Figure 16.3 Temperature dependence of the unfolded state specific heat capacity, $c_{p,U}$, and the temperature dependence of the heat capacity contribution of the polar (Pol) and non-polar (Non-pol) groups of lysozyme (2).

Privalov and Makhatadze[2] have shown that the heat capacity of the unfolded state can be calculated based on the additive contribution of the partial molar heat capacities of the individual amino acids making up the sequence. The partial heat capacity of unfolded polypeptide chain can thus be presented as:[2]

$$C_{p,U}^{\text{calc}}(T) = \sum n_i \hat{c}_{p,i}(T) + (N-1)C_{p,\text{CHCONH}}(T) \tag{8}$$

where N is the total number of amino acid residues in the protein sequence, n_i is the number of the i-th type of amino acid residue, $\hat{c}_{p,i}(T)$ is the partial molar heat capacity for the side chain of the i-th type and $C_{p,\text{CHCONH}}(T)$ is the partial molar heat capacity of the peptide unit. The calculated heat capacity for the unfolded state has been compared with the experimentally determined heat capacity of the unfolded state of proteins and corresponds well, within 5 %,[2] not only in reproducing the temperature dependence, but also the absolute values. This is important as in some cases, it may be difficult to significantly populate the unfolded state of a protein if, for example, the protein is very stable which may lead to ambiguity in defining the unfolded state baseline. Experimentally, it may not be possible to alter the conditions to destabilize the protein, for example by the addition of denaturants as they alter the hydration properties of the unfolded state and consequently the heat capacity of the unfolded state.[23] Being able to calculate the unfolded state heat capacity in these instances is critical for data analysis.

16.6   Advancements

Measuring the heat capacities of proteins as a function of temperature has been invaluable in understanding the molecular mechanisms underlying protein folding and stability, and in particular the role of hydration.[1] However, the range of these investigations has been extended to include small peptides, representing in some cases, secondary structural elements of proteins or, in other cases, representing mini-proteins. Underpinning these differences are their unique structural features which have been studied using DSC.

The first example of this is the thermal unfolding of helical peptides. For brevity, only two specific peptides, namely A6 and G6, as described by Richardson and Makhatadze,[4] will be discussed. It was shown that the A6 peptide forms a helix, which is highly temperature dependent. In contrast, the G6 peptide is unstructured at all temperatures. Figure 16.4a shows the temperature dependence of the heat capacity for A6 as determined by DSC. It is evident that the thermal unfolding transition of A6 is rather broad, which is a general feature of helical peptides.[24] The broadness of the thermal transition makes it difficult to define the native state baseline, which in turn, makes it difficult to determine the progress heat capacity, which represents the intrinsic heat capacity of the polypeptide in solution. To remedy this, Richardson and Makhatadze,[4] as a first approximation, used the heat capacity of the unfolded (coiled) state, $c_{p,U}(T)$, to represent the progress heat capacity, $c_p^{\text{prg}}(T)$, where the

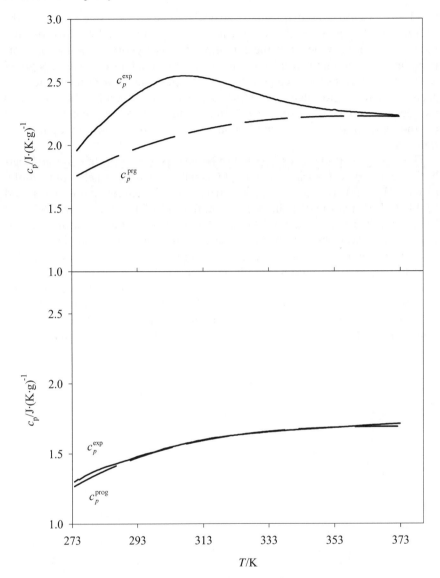

Figure 16.4 Temperature dependence of the experimentally determined specific heat capacity (c_p^{exp}; solid line) and progress heat capacity (c_p^{prg}; dashed line) of the A6 (panel A) and G6 peptides (panel B; (4)).

$c_{p,U}(T)$ can be calculated by Equation 8. To compare the calculated heat capacity for the unfolded (coiled) state, Richardson and Makhatadze compared the calculated heat capacity of the unfolded (coiled) state of G6 to the experimental partial molar heat capacity of G6 (a peptide that is always unstructured and thus behaves as a coiled state). The remarkable agreement (see Figure 16.4b) validated using the calculated heat capacity to represent the $c_p^{prg}(T)$ for A6 peptide. Such a treatment of a DSC profile, allowed for the first time, direct measurements of the enthalpy of helix to coil unfolding for short peptides. Combining the results for enthalpy measurements for several peptides with different transition temperature it was possible to determine the heat capacity change upon helix unfolding. The value was found to be very small and negative (-0.0032 kJ (mol K)$^{-1}$), in agreement with the predictions based on the type of the groups that become exposed to the solvent upon helix unfolding.

The trpcage peptide was suggested to behave as a mini-protein in that it has both secondary structural elements (an α-helix, a short 3_{10}-helix and a poly-proline II helix) as well as tertiary structural characteristics. Streicher and Makhatadze[3] investigated the thermal unfolding of trpcage using DSC and found that the trpcage was not 100 % folded at the lowest temperature used for DSC. In addition, the unfolded state baseline was difficult to obtain from the DSC data due to the broad thermal unfolding transition (see Figure 16.5). To remedy this, Streicher and Makhatadze[3] used the heat capacity calculated for the unfolded state, based on the amino acid composition of trpcage, using Equation (8). By using the calculated unfolded state baseline, and the difference in the heat capacity of the native and unfolded states, which defines the ΔC_p, Streicher and Makhatadze[3] were able to obtain a baseline for the native state

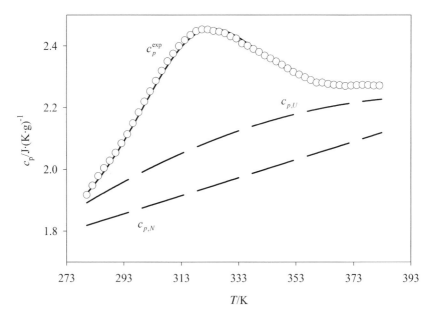

Figure 16.5   Temperature dependence of the experimentally determined specific heat capacity (c_p^{exp}; open symbols), the native state heat capacity ($c_{p,N}$) and unfolded state heat capacity ($c_{p,U}$) for trpcage as well as the data fitted to a two-state unfolding model (solid line) (reference 3).

through global fitting. This allowed a complete thermodynamic characterization of trpcage. In particular, it was found that thermodynamically, trpcage indeed behaves as a mini-protein as the unfolding transition closely follows a two-state model and there is a large positive heat capacity change upon unfolding.

A very different example of a peptide studied using DSC is the β-hairpin forming trpzip4 peptide (Figure 16.6). The temperature induced unfolding of this peptide suggests that it unfolds at relatively high temperature, and that even at 120 °C, the unfolding transition is not complete.[25] Thus the thermodynamic characterization of trpzip4 requires the knowledge of the temperature dependence of the heat capacity of the unfolded protein. This can be done by again calculating $c_{p,U}$ using the additivity scheme described by the Equation (8). This approach to the analysis of the DSC profiles provided calorimetric evidence that the unfolding of trpzip4 follows a two-state unfolding mode. What is truly remarkable is that the β-hairpin structure formed by trpzip4 peptide has a heat capacity change upon unfolding, ΔC_p, on a per residue basis, similar to that of larger proteins. It seems that this originates from the "core" formed in the peptide by the interdigitation of the four Trp residues. The behavior of trpzip4 is very different from the α-helical peptides (such as A6 discussed above) which show non two-state behavior upon thermal unfolding. It is possible that the differences in hydrogen bonding between these two

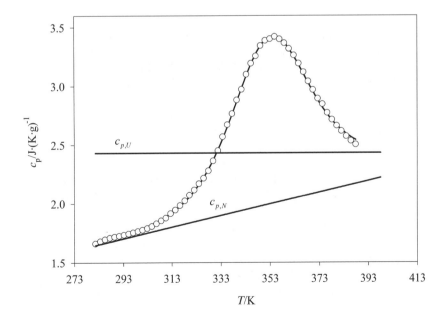

Figure 16.6 Temperature dependence of the experimentally determined specific heat capacity (c_p^{exp}; open symbols), the native state heat capacity ($c_{p,N}$) and unfolded state heat capacity ($c_{p,U}$) for trpzip as well as the data fitted to a two-state unfolding model (solid line) (reference 25).

different secondary structural elements could be responsible, as helices have relatively local hydrogen bonds, whereas β-hairpins have hydrogen bonds between regions separated in sequence space. Despite the large differences in size between proteins and peptides, it is clear that the fundamentals governing the heat capacities of proteins also hold true for peptides.

References

1. G. I. Makhatadze and P. L. Privalov, *Adv. Protein Chem.*, 1995, **47**, 307.
2. P. L. Privalov and G. I. Makhatadze, *J. Mol. Biol.*, 1990, **213**, 385.
3. W. W. Streicher and G. I. Makhatadze, *Biochemistry*, 2007, **46**, 2876.
4. J. M. Richardson and G. I. Makhatadze, *J. Mol. Biol.*, 2004, **335**, 1029.
5. M. M. Lopez and G. I. Makhatadze, *Methods Mol. Biol.*, 2002, **173**, 113.
6. G. I. Makhatadze, Current Protocols in Protein Science, 2001, Chapter 7, Unit 7. 9.
7. G. I. Makhatadze, V. N. Medvedkin and P. L. Privalov, *Biopolymers*, 1990, **30**, 1001.
8. R. L. Biltonen and E. Freire, *CRC Crit. Rev. Biochem.*, 1978, **5**, 85.
9. E. Freire, *Methods Enzymol.*, 1994, **240**, 502.
10. Y. Yu, G. I. Makhatadze, C. N. Pace and P. L. Privalov, *Biochemistry*, 1994, **33**, 3312.
11. P. L. Wintrode, G. I. Makhatadze and P. L. Privalov, *Proteins*, 1994, **18**, 246.
12. J. O. Hutchens, A. G. Cole and J. W. Stout, *J. Biol. Chem.*, 1969, **244**, 26.
13. G. M. Mrevlishvili, *Thermodynamic Data for Biochemistry and Biotechnology*, ed. H.-J. Hinz, Springer-Verlag, New York, 1986, 148.
14. J. Suurkuusk, *Acta Chem. Scand. B.*, 1974, **28**, 409.
15. P. H. Yang and J. A. Rupley, *Biochemistry*, 1979, **18**, 2654.
16. W. Pfeil and P. L. Privalov, *Biophys. Chem.*, 1976, **4**, 41.
17. P. L. Privalov, E. I. Tiktopulo, S. Venyaminov, V. Griko Yu, G. I. Makhatadze and N. N. Khechinashvili, *J. Mol. Biol.*, 1989, **205**, 737.
18. P. L. Privalov, in *Biological Microcalorimetry*, ed. E. A. Beezer, Academic Press, London, New York, Toronto, Sydney, San Francisco, 1980, p. 413.
19. G. Velicelebi and J. M. Sturtevant, *Biochemistry*, 1979, **18**, 1180.
20. P. L. Privalov and N. N. Khechinashvili, *J. Mol. Biol.*, 1974, **86**, 665.
21. G. I. Makhatadze, *Biophys. Chem.*, 1998, **71**, 133.
22. G. I. Makhatadze and P. L. Privalov, *J. Mol. Biol.*, 1990, **213**, 375.
23. W. Pfeil and P. L. Privalov, *Biophys. Chem.*, 1976, **4**, 33.
24. J. M. Richardson, K. W. McMahon, C. C. MacDonald and G. I. Makhatadze, *Biochemistry*, 1999, **38**, 12869.
25. W. W. Streicher and G. I. Makhatadze, *J. Am. Chem. Soc.*, 2006, **128**, 30.

CHAPTER 17
Heat Capacity in Liquid Crystals

M. MARINELLI, F. MERCURI AND U. ZAMMIT

Dipartimento di Ingegneria Meccanica, Università di Roma "Tor Vergata", Roma, Italia

17.1 Brief Introduction to Liquid Crystals

What characterizes the crystal phase of a material with respect to the liquid phase is the presence of long-range, three-dimensional (3D), orientational and positional order. Liquid crystals (LC) are materials that have some intermediate phases in which there is an intermediate order between a liquid and a crystal. For this reason these phases are called mesophases.

Most mesogenic compounds (which have mesophases) are made of rigid molecules, with a rod-like shape, in which the shape anisotropy gives rise to an intermolecular potential larger along the short axis of the molecule than in any other direction. Such an interaction can produce a phase called the nematic (N) phase (Figure 17.1b), in which there is a long-range orientational order due to the fact that, on average, the molecules are aligned along a certain direction $\hat{n}$, called the director.

A quantitative estimation of the nematic order is given by the nematic order parameter (see also Chapter 12), whose simplest and more useful formulation is given by

$$S = \left\langle \frac{2}{3}\cos^2\varphi - \frac{1}{2} \right\rangle \tag{1}$$

Heat Capacities: Liquids, Solutions and Vapours
Edited by Emmerich Wilhelm and Trevor M. Letcher
© The Royal Society of Chemistry 2010
Published by the Royal Society of Chemistry, www.rsc.org

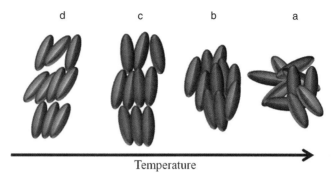

Temperature

Figure 17.1   Typical phase sequence of a rod-like LC with decreasing temperature: (a) isotropic, (b) nematic, (c) smectic A and (d) smectic C phases.

where the brackets indicate an average on the position of many molecules at the same time or the average on a single molecule in time, and φ is the angle between the molecular long axis and $\hat{n}$. Note that $S = 0$ corresponds to an isotropic liquid (Figure 17.1a) while $S = 1$ means that all the molecules are perfectly aligned along $\hat{n}$. Typical values of S, well within the nematic phase, range between 0.4 and 0.6.

Another phase that is usually encountered in mesogenic compounds is the smectic A (S_A) phase, in which, besides having orientational order, the molecules are, on average, disposed within parallel planes which form a 1D periodic array (Figure 17.1c). There is, however, no correlation among the position of the different molecules within each plane. The S_A phase can then be seen as a stack of two dimensional (2D) liquid layers. The smectic order parameter takes into account the periodicity of the density of the molecular centre of mass along $\hat{n}$ and it is generally defined as

$$\rho(z) = \rho_0(1 + \psi\cos(\varphi z)) \tag{2}$$

where ρ_0 is the average sample density, ψ is the amplitude of the density modulation, $\varphi = \frac{2\pi}{d}$, and d is the smectic layer spacing.

The smectic C (S_C) phase (Figure 17.1d) is similar to the S_A phase, but the molecules in this case are on average tilted by an angle θ with respect to the layer normal. As a consequence, the S_C order parameter must depend on this angle and is generally written as

$$\Theta = \theta e^{i\varphi} \tag{3}$$

where φ is the molecular azimuth.

Particularly interesting is the hexatic B (Hex B) phase (Figure 17.2), which has a structure similar to the S_A phase. While the latter is characterized by the absence of any in-plane order, in the Hex B phase there is a long-range bond orientational order (BOO) with no long-range positional order.

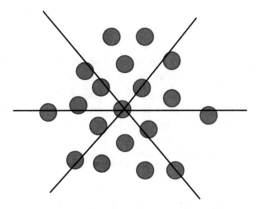

Figure 17.2   In-plane distribution of the LC molecules in the hexatic B phase.

There is a sixfold symmetry in the bond orientation, like in a 2D hexagonally packed crystals, but the distances between the molecules are randomly distributed. For these reasons, the order parameter can be written as

$$\Psi = |\Psi| \exp(i6\phi) \tag{4}$$

where ϕ is the average intermolecular bond-orientational angle.

It is clear from Figure 17.1 that, for decreasing temperature at constant pressure, the typical phase sequence of a LC is associated with a symmetry breaking that drives the system toward states where more and more order is present. There are also other symmetry-breaking transitions that can progressively transform an isotropic liquid into a crystal and there are, therefore, mesophases other than the ones depicted in the Figure 17.1. For a comprehensive review of the various mesophases, refer to Reference 1.

Of particular interest among rod-like mesogenic compounds are the ones whose molecules have a large permanent electric dipole.[2] Their importance relates to LC displays and other technological applications. In these LCs, which are called ferroelectrics, there are smectic phases that differ for the arrangement of the dipoles in the smectic layers.[3]

Since, as stated above, the main requirement for the formation of a mesophase is a particular anisotropy in the molecule shape and/or in the intermolecular interaction potential, there are other materials that can satisfy this requirement without having a rigid rod-like molecule. Example are LC polymers,[4] and also discotic (disk-like) mesogens[5] in which the interaction potential along the normal to the disks is larger than the one along any other direction. While in the former case the mesophases are similar to the ones found in rod-like molecules, in the latter, columnar mesophases[6] are usually found.

All the types of LCs mentioned above are called thermotropics since phase transitions are driven by a temperature change at constant pressure. However, LC mesophases can be also obtained, for example, by changing the concentration of hydrophobic molecules in a hydrophilic solvent. The repulsive interaction can form, by varying the concentration of the solute, mesophases that are completely different from the ones obtained in thermotropics. These LCs are called lyotropics and have found interesting applications in biology and medicine.[7,8]

In the following, we will focus our attention on the behaviour of the specific heat capacity c_p at some of the phase transitions among the different mesophases of the thermotropics described above. This has been a very active research field in the last three decades and has significantly contributed to the understanding of critical phenomena associated to phase transitions. The relevance of LCs in this field is mainly due to the very rich phase diagram and to the variety of phase transitions of these materials.

17.2 The Nematic–Isotropic Phase Transition

The nematic-isotropic (NI) transition has a weakly first-order character, as shown by the relatively small value of the latent heat which is usually of the order of few $J g^{-1}$.[9] The weakness of the first-order character is confirmed by the smallness of the parameter $\Delta = \frac{T_{NI} - T^*}{T_{NI}} \approx 10^{-2}$ to 10^{-3}, where T^* is the temperature corresponding to the stability limit of the isotropic phase in the nematic one. The study of the critical phenomena associated with this transition is very difficult since the critical region is confined to such a small temperature interval that it makes the critical point not easily accessible, even though high-resolution calorimetric techniques allow for measurements very close to the transition temperature (see Chapters 12 and 13). Moreover, at the transition, impurities act as nucleation centres and therefore the system is not monophasic (the so-called two-phase coexistence region). This region, which is very narrow for high purity samples, is usually not smaller than (5 to 10) mK[9] and represents an intrinsic limitation to the lowest experimentally accessible value of $t = \frac{T - T^*}{T_{NI}}$. For these reasons, it has been impossible, up to now, to determine the universality class to which the transition belongs.

One of the first attempts to find a theoretical description of the NI transition is based on the Landau-de Gennes (LdG) Mean Field (MF) theory, illustrated in Chapter 14. It uses the phenomenological expansion of the excess free energy at the transition

$$\Delta G = \frac{1}{2} a \left(\frac{T - T^*}{T_{NI}} \right) S^2 - \frac{1}{3} b S^3 + \frac{1}{4} c S^4 + \cdots \tag{5}$$

where a, b and c are constants. It turns out that, on the nematic side of the transition, the leading term in the c_p anomaly is given by

$$c_p = c_0 + A \left(\frac{T^{**} - T}{T_{NI}} \right)^{-\alpha} \tag{6}$$

where $T^{**} = T^* + T_{NI} \frac{b^2}{4ac}$ is the stability limit of the nematic phase in the isotropic one, c_0 is the background value of the specific heat capacity in the

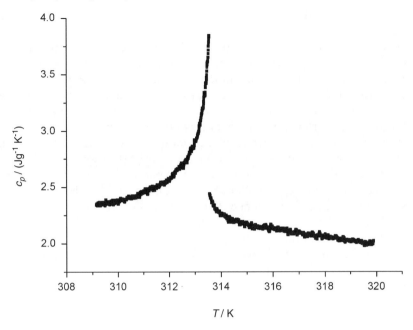

Figure 17.3 Temperature dependence of the specific heat capacity at the NI transition.

nematic phase, A is the amplitude of the critical part and $\alpha = 1/2$ is the value of the critical exponent.

A typical example of the temperature dependence of c_p close to T_{NI} is shown in Figure 17.3 obtained for octylcyanobiphenyl (8CB) LC.

The data reported in Figure 17.3 can be satisfactory fitted in the nematic phase with Equation (6). The prediction of the LdG theory, however, conflicts with the experimental results for the order parameter exponent β that for many different compounds has a value of approximately 0.25, far from the predicted LdG value of 0.5. This observation, together with the small values of the b and c coefficients in Equation (5) derived from the experiments, have suggested the so-called "tricritical hypothesis,"[10–12] analogous to the mean field tricritical behaviour for second-order phase transitions. The expected critical exponents in this case are $\beta = 0.25$ and $\alpha = 0.5$ and therefore in agreement with the experimental results. It should be noted, however, that the value for $T_{NI} - T^*$ obtained from the data of c_p is much smaller than the one obtained from susceptibility measurements and also that the experimental value of $\Delta c_p = c_0 - c_i$ (c_i being the value of c_p in the isotropic phase far from the transition) is much smaller than the one predicted by the LdG theory.[12] Finally, it has been shown that an improvement in the agreement between the experimental data and the theoretical prediction can be obtained by including in the ΔG expansion a sixth-order term and/or higher-order terms; this obviously increases the empirical nature of the model.[12] Improvements have also been obtained using a renormalization group approach but in all these cases the results are not fully satisfactory and open questions still remain.[13]

There are some other aspects that the LdG theory cannot explain; in particular the pre-transitional behaviour of the specific heat capacity observed in the isotropic phase as the transition temperature is approached (Figure 17.3). The specific heat capacity, in fact, is not flat approaching T_{NI} from above, but increases slightly. This increase is due to the presence of fluctuations that produces locally a transient nematic order in the isotropic phase and therefore transient nematic volumes, whose dimensions increase with decreasing temperature. Fluctuations are not included in standard MF theories, but they can be easily introduced in the free energy expansions using the Gaussian approximation.[14] The approximation is based on the fact that the susceptibility does not diverge at T_{NI} and implies that the interaction among fluctuations is relatively weak. As a result, a gradient term of the form $\frac{1}{2}K(\vec{\nabla}S)^2$, where K is the elastic constant in the isotropic phase, needs to be added to Equation (5). Neglecting the cubic term in the free energy expansion (second-order phase transition) in the Gaussian approximation, the result of the fluctuation theory is an additional term in the specific heat capacity given by[12]

$$\delta c_p \propto t^{-\frac{1}{2}} \tag{7}$$

This result does not change even when the cubic term is included in the ΔG expansion in the Gaussian approximation, since it only leads, in the framework of the de Gennes continuum theory, to a renormalization of the coefficients in Equation (5).[15]

The success of the attempts to reach the NI critical point depends on the possibility to eliminate the first-order character of the transition, which in turn depends on the presence of fluctuations. If the fluctuations could somehow be quenched, it would make the critical point experimentally accessible. However, calculations on the strength of a magnetic field capable of fulfilling this task give values of H in the range of 10^3 T,[16] which are very difficult to achieve. Experimentally, different approaches have been attempted to quench the fluctuations and are reported in the literature. Yokoyama[17] has used the surface-induced alignment in very thin film, but the results could not be considered conclusive due to the non-homogeneity of the aligning field applied to the sample across its thickness. More recently Taketoshi *et al.*[18] have investigated a mixture of bent shaped and rod-like LCs and they speculated that the caging of the latter in the former matrix can provide sufficiently strong a field to reach the NI critical point. Preliminary results have yielded a critical exponent value for the specific heat capacity which ranges from 0.5 (MF) to 2/3 (MF critical). It is interesting to note that in both cases the supposed quenching of the fluctuations produces an increase of the transition temperature.

Some recent experimental results suggest a glass-like behaviour in the isotropic phase as the NI transition temperature is approached. It has been shown, in fact, by means of optical heterodyne detection – optical Kerr effect measurements that nematogens have orientational relaxation characteristics similar to those of strongly supercooled molecular liquids. The exponential decay of the relaxation predicted by the LdG theory appears on a time scale of tens of ns while at shorter times, power law behaviour has been found.[19] Similar results have also been obtained from dielectric measurements.[20] Though the deviation from the LdG predictions takes place at very short times (very high frequencies), a regime, which is not easy to explore in standard calorimetric measurements, the possibility of glassy behaviour sheds new light on the possible interpretation of the NI transition.

17.3 The Smectic A–Nematic Phase Transition

The Smectic A–Nematic (AN) transition has been most extensively studied since de Gennes first pointed out its analogy with the superfluid helium transition and the normal–superconductor transition in metals.[21,22] In fact, similar to the last two phase transitions, the order parameter of the S_A phase is characterized by the two components, ψ and φ, thus placing the AN transition in the three-dimensional (3D) XY universality class, for second-order phase transitions.

The predicted values of the critical exponents for the specific heat capacity, the correlation length and the susceptibility, are $\alpha_{XY} = -0.013$, $\nu_{XY} = 0.669$ and $\gamma_{XY} = 1.316$ respectively.[23,24] However, the values of the critical exponents obtained for the AN transition of various LC compounds have been found to depend on the temperature range of the N phase, characterized by the value of the so called McMillan ratio $R_M = T_{AN}/T_{NI}$. Measurements have been performed in compounds with R_M ranging between 0.6 and 0.993 (reference 24) and values of the exponents α and γ consistent with the XY like predictions have only been obtained in compounds with a large-enough N range. For example, Figure 17.4 shows the behaviour of c_p at the AN transition for the compound octyloxythiolbenzoate ($8\bar{S}5$) characterized by $R_M = 0.936$. The c_p behaviour can be represented by the expression:

$$c_p = A^{\pm}|t|^{-\alpha}(1 + D|t|^{-0.5}) + B + Et \tag{8}$$

where $t = (T - T_{AN})/T_{AN}$ is the reduced temperature; the signs + and − refer to the data above and below the transition temperature, respectively; and the term in parenthesis represents the correction to scaling term.[23] The continuous line in Figure 17.4 corresponds to the best fit which yields the values $\alpha = -0.022 \pm 0.008$ and $A^-/A^+ = 0.97 \pm 0.02$ which are similar to those predicted for a 3D XY-like behaviour.[25]

The deviations of the values of α from α_{XY} for the LC compounds with a "small" nematic range has been explained in terms of the interaction between the nematic and smectic-order parameters, which tends to drive the transition towards a first-order character. In fact, as the temperature is lowered below the NI transition, the average molecular alignment improves, inducing a progressive increase of the orientational order parameter S until saturation is

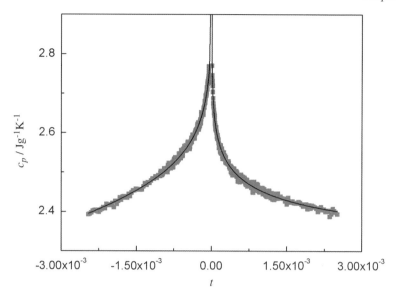

Figure 17.4   Specific heat behaviour over the smectic A–nematic phase transition of
8$\bar{S}$5. The continuous line represents the best fit result obtained with
Equation (8). The defination of t is given after equation (8).

achieved. The interaction between S and ψ arises because the onset of the
molecular "packing" in smectic planes at the AN transition, enhances the
orientational order, therefore affecting the value of S, if not yet saturated. This
effect is more relevant the closer the AN and NI transitions are to each other
and therefore the more S differs from the saturation value.

The second- to first-order crossover can be qualitatively justified on the basis
of a mean field approximation of the excess free energy of the S_A phase, with
respect to that of the N phase, containing a term representing the coupling
between S and ψ, as proposed by de Gennes:[21]

$$\Delta G = at\psi^2 + b\psi^4 + c\psi^6 + \frac{(\delta S)^2}{2\chi_N} - C\psi^2\delta S \qquad (9)$$

where a, b, c and C are all positive constants; $\delta S = S - S_N$ is the change in the
nematic order parameter induced by the formation of the smectic layers; χ_N is a
temperature-dependent nematic susceptibility whose value becomes smaller the
farther the AN transition is from the NI transition, and therefore the larger is
the value of S before the onset of the AN transition. Minimization of ΔG with
respect to δS yields :

$$\Delta G = at\psi^2 + b_0\psi^4 + c\psi^6 \qquad (10)$$

where $b_0 = b - C^2\chi_N/2$. Thus large χ_N values (AN transition close to NI transition) can induce negative values of b_0, causing the transition to become first

order.[21] The de Gennes crossover between second and first-order passes through a tricritical point (TCP) corresponding to $b_o = 0$, where the mean field value of the critical exponent is $\alpha_{TCP} = 1/2$. The effective values of α can then range between α_{XY} and α_{TCP} for varying range of the nematic phase of the investigated compound. Such a trend has indeed been verified in many samples,[24] among which are mixtures of compounds of the cyanobiphenyl (nCB) homologous series, giving rise to samples with different nematic range.[26]

However, even in compounds where the interaction between ψ and S is unimportant, and where good agreement with α_{XY} is obtained, deviations of the value of v from that predicted for the 3D XY universality class, were still observed. In particular, the exponents measured along a direction parallel ($v_{\parallel}$) and perpendicular ($v_{\perp}$) to the molecular director proved to be different from each other and from v_{XY}.[24] The anisotropy in the values of v were ascribed to the fact that the fluctuations δn of the molecular director, compete with the establishment of the smectic order by exciting anisotropic elastic deformations. So δn is coupled with the phase ϕ of the smectic order parameter, through an intrinsically anisotropic interaction. The implications of this interaction are still not well defined. A self-consistent one-loop theory[27] has been put forward according to which the anisotropy in the critical behaviour of the correlation length is related to the value of the splay elastic constant K_1.[27] Small values of K_1 at the AN transition (which occurs for small N range, since K_1 scales as S^2) yield strongly anisotropic behaviour for the correlation length, with a maximum anisotropy $v_{\parallel} = 2\,v_{\perp}$. In contrast, compounds with large N ranges should straddle the weak anisotropic and isotropic regimes. However, none of the mentioned extreme behaviours have been experimentally observed. The values of $v_{\parallel}$ and $v_{\perp}$ obtained even in a compound with the so far greatest accessible N range ($R_M = 0.66$) are close to those predicted by the theory in the crossover region between the strongly anisotropic and isotropic regimes. This indicates that isotropic behaviour could possibly require an even larger N range. The value $v_{\parallel}/v_{\perp} = 2$, on the other hand, would require so small a N range that the onset of the effects of the coupling between ψ and S could give rise to effective values for $v_{\parallel}$ and $v_{\perp}$ different from the ones predicted by the theory.

The study of the critical behaviour over phase transitions requires also the determination of the critical exponents of the order parameters as, for example, the orientational order parameter S. This can be achieved simultaneously with the study of the specific heat capacity when the photopyroelectric (PPE) calorimetric technique is adopted, since it is then possible to measure also the thermal conductivity k in a straightforward fashion and with a high-temperature resolution (see Chapter 12).[25,28] The thermal conductivity value in a LC measured parallel ($k_{\parallel}$) to the molecular long axis in homeotropically aligned samples has been shown to be larger than the one measured perpendicular ($k_{\perp}$) to the axis in planar samples. Moreover, the difference in the two values increases the more effective is the average alignment of the molecular director along the two perpendicular directions, and it is directly proportional to S. It is therefore possible to monitor the temperature dependence of S by detecting the temperature dependence of the thermal conductivity anisotropy, $(k_{\parallel} - k_{\perp})$,

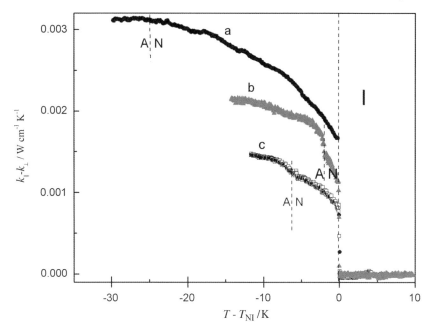

Figure 17.5   Thermal conductivity anisotropy over the nematic–isotropic and the smecticA–nematic phase transitions in: (a) $8\bar{S}5$, (b) 9CB; (c) 8CB liquid crystals. I indicates the isotropic region.

similarly to what is commonly obtained with the anisotropy of other tensorial quantities like the magnetic susceptibility and refractive index.[29] Figure 17.5 shows the behaviour of $(k_{\parallel} - k_{\perp})$, over the NI and AN (indicated by the arrows) transitions in LC compounds with different nematic ranges.[30] The abrupt changes over the AN transitions in the data corresponding to 8CB and 9CB are related to the previously mentioned (δS) induced by the onset of the smectic A phase in compounds where the nematic range is not large. As indicated earlier on, δS is more substantial in 9CB where the N range is smallest, and absent in $8\bar{S}5$ where it is largest. This capability offered by the thermal conductivity has also been successfully employed for the study of the critical exponent of S in the vicinity of the NI transition in several compounds of the nCB series.[31]

17.4   The Smectic A–Smectic C Phase Transition

The relevant order parameter for the smectic C–smectic A (AC) phase transition, as shown in paragraph 17.1, is characterized by the two components θ and φ. Therefore, as for the AN transition, the analogy with superfluid helium based on the structure of the order parameter, leads to the conclusion that it should be second order and it should belong to the 3D XY universality class.[32] Accordingly, one should expect an anomaly in the specific heat capacity, which is governed by the fluctuations, described by Equation (8), where the reduced temperature is, in this case, defined as $t = (T - T_c)/T_c$; T_c representing the AC transition temperature. However, the experimental results reported in literature show a mean-field-like behaviour in contrast to the predicted 3D XY one.[33] In particular it has been shown that the specific heat capacity can be described by the following equations:

$$\begin{cases} c_p = c_0 + c_g^+ & (T > T_c) \\ c_p = c_0 + c_g^- + A|T_m - T|^{-0.5} & (T < T_c) \end{cases} \qquad (11)$$

derived from the so called extended mean-field model, where the sixth-order term is included in the Landau excess free energy expansion[34,35]

$$\Delta G = at\theta^2 + b\theta^4 + c\theta^6 \qquad (12)$$

As in the case of the Equation (10), and also in Equation (12) the expansion coefficients a and c are positive while b can be positive or negative depending on the order of the transition. For second-order continuous transitions, $b \geq 0$ and the coefficients in the Equations (11) are defined as $A = \sqrt{a^3/12cT_c^3}$ and $T_m = T_c(1 + b^2/3ac)$ while $c_g^\pm$ represent the so called Gaussian terms, that account for a small contribution due to fluctuations. In the proximity of a tricritical point, where most of the studied AC transitions can be placed, $b \cong 0$ and the sixth-order term becomes particularly important with respect to the fourth-order one. A quantitative estimation of the tricritical character of the AC transitions is given by the parameter $t_0 = 3(T_m - T_c)/T_c = b^2/ac$ which vanishes at the tricritical point ($b = 0$) and can be evaluated by fitting the data with Equation (11). As an example, Figure 17.6 reports the specific heat capacity behaviour of the 4-(3-methyl-2-chlorobutanoyloxy)-4'-heptyloxybiphenyl (A7) liquid crystal compound at the AC transition, obtained by means of PPE measurements.[35] The graph shows the excellent fit of the experimental data with Equations (11). Similar results have been reported in the literature for other compounds indicating the general validity of the mean-field-like behaviour at the AC transition.

In order to explain the contradiction between the 3D XY theoretical prediction and the observed mean-field character, it has been proposed that the t range where the critical behaviour of the specific heat is governed by the fluctuations, is so narrow ($|t| \leqslant 10^{-5}$) as to become experimentally inaccessible.[36] Later on it had been suggested that such an inaccessibility concerns the specific heat capacity but not other quantities like the elastic constants obtained by ultrasound measurements which were proved to be more sensitive to the fluctuations. This could explain, for instance, why the c_p results are consistent with a Landau mean-field model whereas those associated with the ultrasound measurements are not. In the latter case it has been shown that the coupling between the order parameter and strain could be responsible for the near tricritical behaviour observed in most of the compounds exhibiting AC transition.[37,38] Actually the AC transition line has been shown to have a

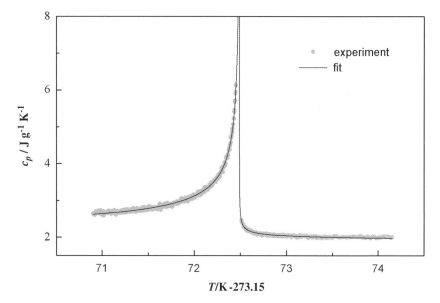

Figure 17.6 Specific heat capacity at the smectic A–smectic C transition of the A7
liquid crystal compound. The experimental data have been fitted by the
Equations (11).

mean-field tricritical point[39] but the reason for this is not yet clear, even though
it has been recently proposed that the tricritical nature could originate from the
coupling between θ and the smectic layer compression.[21]

17.5   The Smectic A–Hexatic B Phase Transition

A further transition between the smectic phases is the smecticA–hexaticB (AB)
transition. The Hex B phase shows a six-fold bond-orientational ordering in the
smectic planes and it can be considered as an intermediate phase between a
crystal and a S_A phase.[40–42] As in the case of the other two previously discussed
smectic phases, the Hex B phase can also be described by a two components
order parameter, $\Psi = |\Psi|e^{i6\varphi}$. It is therefore, once again, expected to belong to
the 3D XY universality class, with physical quantities governed by critical
fluctuations.[32] Nevertheless, based on the experimental results so far reported
in literature, a theoretical model actually describing the observed behaviour
over the AB transition is still lacking, with 3D XY-like behaviour being ruled
out by non-expected values of the measured critical quantities. It has been
shown, in fact, that the critical exponent of the specific heat capacity can vary
from $\alpha = 0.48$ to $\alpha = 0.68$, values that are very different from α_{XY}.[35,43–47]

It has been suggested, for this transition, that the large values of α, could be
an indication of tricritical behaviour,[48,49] but neither a continuous crossover
from the 3D XY to tricritical values of α, nor a correlation between such α
values and the temperature width of the Hex B and S_A phases, could be

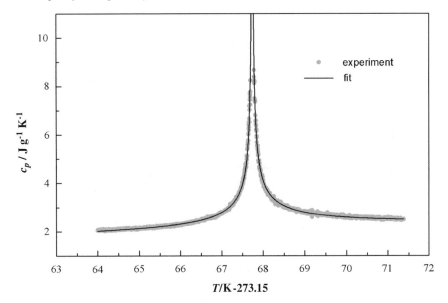

Figure 17.7   Specific heat capacity at the smectic A–hexatic B transition of the 65OBC liquid crystal compound. The experimental data have been fitted by Equation (8).

experimentally found.[50] More recently, it has been shown that the AB transition could have a very weak first-order character,[35,44–47,50] as in the case of the n-hexyl-4'-n-pentyloxybiphenyl-4-carboxylate (65OBC) LC, whose specific heat capacity behaviour is reported in Figure 17.7.[35]

This first-order nature of the AB transition was suggested by different theoretical predictions. It was initially assumed that the coupling between the hexatic order and other kinds of order could be responsible for the presence of a tricritical point. In particular, Bruinsma and Aeppli proposed, in the early 1980s, that this could have originated from the coupling between the long-range hexatic order and short-range fluctuations of the so-called herringbone order.[48] That is a kind of in-plane orientational order found in the crystal-E phase which is exhibited by some compounds below the Hex B phase. The existence of fluctuations of the herringbone order in the Hex B phase was confirmed by X-ray measurements on several compounds[41,43] but the hypothesis that their coupling with Ψ were responsible for *tricriticality* was later ruled out since tricriticality could be observed even in compounds without such an order.[51] Based on a different argument, the tricritical like behaviour was also proposed by Aharony *et al.* who suggested that the generic phase diagram of such systems should always have a structure similar to the one reported in Figure 17.8 [49,52] with a tricritical and a triple point. While the existence of the triple point was experimentally observed,[43] the evidence of a TCP was never revealed by experiments in mixtures of different compounds.[50]

Haga *et al.* have later proposed that the coupling of the Hex B order parameter with the in-plane density fluctuations or strain,[44] in all cases, induces a

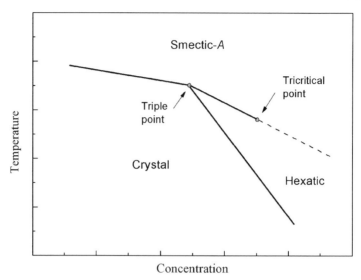

Figure 17.8   Generic temperature-concentration phase diagram near the crystal–Hexatic–Smectic triple point (TP). Full and dashed lines indicate first- and second-order transitions respectively.

first-order character. If the coupling is absent, the latent heat associated with the AB transition should become zero when the TCP is reached from the first-order side of the transition line and should obviously remain zero along the second-order segment of the line. If, on the contrary, the above-mentioned coupling is present, one should expect a so called *quasitricritical* behaviour,[53] since the latent heat decreases to a small but nonzero value in the second-order part of the transition line. Nevertheless, even if the results reported in the literature confirm the weak first-order nature of the AB transition,[45] the unusually large value observed for the specific heat capacity critical exponent does not conform either to 3D XY predictions or to the tricritical ones, leaving unsolved the puzzle of the universality class of this transition.

Recently, it has been shown that the disorder present in the sample plays a major role at the transition. It has been observed, in particular, that, when the sample was first cooled down from the isotropic phase, the specific heat capacity can show some unexpected anomalies at the AB transition which are strongly reduced and eventually suppressed during the subsequent heating and cooling cycling over the transition temperature. This behaviour has been ascribed to the presence of defects that are progressively annealed during the thermal cycling of the sample over the transition.[47]

Such evolution of the defects, that can affect the critical behaviour of the thermal parameters, eventually masking the real nature of the transition, is represented by the curves of Figure 17.9. The quantity I_L accounts for heat released by defect annealing processes and/or latent heat exchanged over the AB transition of the 65OBC LC and has been obtained by PPE measurements.[47]

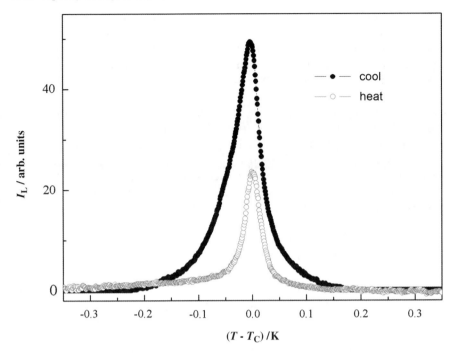

Figure 17.9   Internal heat source amplitude at the smectic A–hexatic B transition of the 65OBC liquid crystal compound under cooling and subsequent heating of the sample.

The large I_L peak shown at T_c (the AB transition temperature) during the first cooling from the isotropic phase has been basically correlated to the heat released by the defect annealing phenomena while the smaller peak obtained on subsequent heating measurement was possibly ascribed to the latent heat associated with the first-order character of the transition. The fact that the specific heat behaviour at the AB transition can be affected by the disorder which is present in the system, makes the study of samples with significantly different defect concentrations particularly interesting.

17.6   Phase Transitions in Restricted Volumes

The study of the influence of confinement and of substrate induced random disorder on the physical properties of LCs is of great interest because their weak translational and orientational order may be influenced by the presence of boundaries. The main effects of confinement on the LC phase transitions are generally those caused by finite size and random fields. Both such mechanisms induce an attenuation of the anomalies associated with the critical behaviour over the transitions. In the former case it is caused by the truncation of the maximum achievable correlation length of the fluctuation because of the limited available volumes, while in the latter case it is caused by the presence of random fields which weaken the average molecular ordering field. The random fields are associated with the local random pinning of the LC molecules to the confining walls which affect both nematic orientational order and the smectic layer positional order. However, when the local molecular orientation is strongly influenced by the interaction with the confining surfaces, substantial elastic strain may also be induced in the hosted LC, destroying long-range order and thus further altering the critical behaviour over the phase transition.

Effects caused by finite size have been observed in 5CB confined in Vycor glass which consists of an SiO_2 matrix with a 3D randomly interconnected pore segments with relatively uniform diameter $d \sim 70\,\text{Å}$ and average length $l \sim 300\,\text{Å}$.[54] The severe constraints imposed by the pores gives rise to an inhomogeneous and reduced nematic order, causing the NI transition to be severely smeared. A gradual increase of the local orientational order, with decreasing temperature, then appears over the NI transition.[55]

Aerogel networks constitute another kind of SiO_2-based network of interconnected pores similar to Vycor glass but with larger pore size and with the length to diameter aspect ratio close to one. Because of the larger pore size and consequently increased inter-porous interaction, effects associated with other than the mere finite size also affect the behaviour of the phase transition. Aerogels with pore size ranging between (120 to 1800)$\,\text{Å}$ were employed, corresponding to values of the concentration, ρ, of the matrix SiO_2 in the LC (mass of SiO_2 per unit volume of LC) ranging between $0.08\,\text{g\,cm}^{-3}$ and $0.6\,\text{g\,cm}^{-3}$ respectively.[56] In 8CB confined in aerogels, the excess specific heat, with respect to the background, associated with the NI transition was found to get progressively rounded, to broaden and to shift to a lower temperature with decreasing pore size. Similar results were also obtained for the AN transition except that the peak features proved to be more severely rounded even in correspondence with the largest average pore size. In both cases the correlation lengths in the smectic and nematic phases, determined by X ray scattering, are shown to extend beyond the average pore size, thus confirming the importance of the inter-porous interaction of the confined LC. Finite size effects were ruled out by the authors[56] as the main mechanism responsible for the induced changes because the reduction of the specific heat capacity peak value and the temperature shift of the transitions did not behave according to finite size scaling. Instead, random field effects had been at first invoked as the primary cause of the changes induced over the phase transitions of the LC confined in aerogels.[56] However, considerable strain is also induced in the confined LC because the director in the pore volume has to conform to the local constraint imposed by the SiO_2 surface which forces the molecular long axis to align perpendicular to the surface. Smearing effects by quenched elastic strain were therefore later considered as the dominant causes of the induced changes.[57] In fact, the results in LCs confined in aerogels were similar to those observed in 8CB LC confined in the cylindrical pores of Anopore membranes where substantial distortion of the molecular director also occurs, and induces quenched elastic strain effects. Anopopre membranes consist of a pretty uniform distribution of non-interconnected cylindrical pores in an alumina matrix with a radius much smaller than the axial length. After treating the pore surface with lecithin,[58] the LC molecules tend to align radially in the cylinders and to progressively bend along the cylinder axis as its position is approached, therefore causing molecular directed distortion.

A further SiO_2-based confining network is the aerosil gel which is obtained when silica nanoparticles are dispersed in the LC. For a silica particle concentration exceeding $\rho \sim 0.01 \, g \, cm^{-3}$, a hydrogen-bonded network of particles forms. The difference with respect to the previous networks is that as long as the particle concentration does not exceed a limiting value, the network is known to remain "soft",[57] with the hydrogen bonds being able to break and reform on a relatively short time scale, giving the possibility for the network to rearrange and partially relieve the strain induced by the molecular director distortion at the particle surface. For concentrations exceeding the limiting value, the network becomes "stiff" as it induces quenched elastic strain effects in the hosted LC similar to those which occur in a rigid aerogel network. Thus the random disorder and strain can be tuned by adjusting the particle concentration in the LC.

Figure 17.10 shows the behaviour of the specific heat capacity over the AN transition for 8CB LC in aerosil with various concentration of silica particles. For $\rho < 0.1 \, g \, cm^{-3}$, the peak of the c_p is shifted to a lower temperature (not evident in Figure 17.10), but, unlike the case of aerogels, it maintains a sharp critical-like profile, with a progressively lower peak value with increasing concentration. For $\rho > 0.1 \, g \, cm^{-3}$, the peak for the AN transition was dramatically rounded as in the case of aerogels.

This led to the conclusion that $\rho = 0.1 \, g \, cm^{-3}$ can be considered approximately as the threshold value above which the network becomes rigid, as in aerogels since similar effects are induced. It should nevertheless be pointed out that with similar average pore size, the effects induced by the aerogel have proved to be more severe than those in aerosil. When the LC was confined in "soft" aerosil gels the behaviour of the specific heat capacity could be characterized by values of the critical exponent which decreased with increasing particle concentration, approaching α_{XY}. It has been explained that the disorder induced random fields cause a reduction of the coupling between the smectic and nematic order parameters, discussed in Section 17.3, therefore reducing the effective value of α. The depression of the specific heat capacity peak value, the temperature rounding of the transition and the behaviour of the correlation length, determined by X-ray scattering measurements, could be reasonably well accounted for in terms of "finite size scaling" and "random field scaling" models described in Reference 59.

A similar trend, for varying particle concentration, was observed over the NI transition shown in Figure 17.11.[57] For $\rho < 0.1 \, g \, cm^{-3}$, c_p showed a double peaked feature. The sharper higher temperature peak (HTP) was associated with the bulk-like, first-order transition subjected to random field effects, while the rounded lower temperature peak (LTP) was correlated to the elastic strain coarsening occurring due to the progressive LC director "stiffening" as the

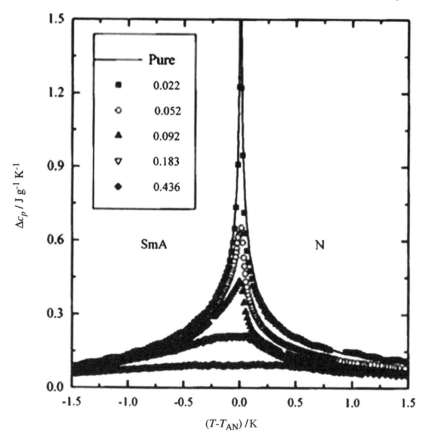

Figure 17.10   Specific heat capacity behaviour over the smecti A–nematic phase transition of 8CB in aerosil with different particle silica concentration ($g\,cm^{-3}$) (from Reference 57).

temperature is lowered below the NI transition.[57] As ρ approached the value of $0.1\,g\,cm^{-3}$, the HTP became progressively depressed while the LTP progressively grew and for $\rho > 0.1\,g\,cm^{-3}$, only the broadened rounded LTP survived, similar to the one which had been obtained with aerogel networks. In this case, the progressive lowering of the transition temperature and of the latent heat with particle concentration could be described in terms of mean field elastic strain model.[57]

The properties of the nematic material nucleated over the two peaks in "soft" aerosil gels were further analysed using complementary techniques. From polarizing microscopy[60] measurements and light scattering measurements,[61] performed simultaneously with the specific heat capacity measurements in a PPE set up, it was established that the nematic texture undergoes fragmentation, causing a reduction of its grain size, as the material goes through the LTP transition region, and that the nematic correlation length correspondingly assumes a minimum value.[62] It was suggested that the nematic material over the HTP

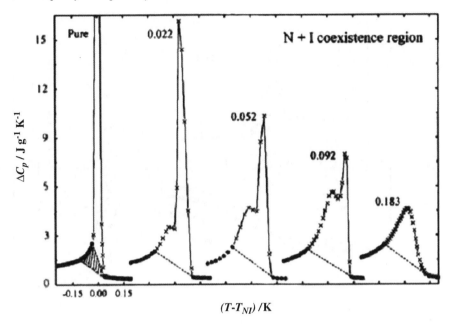

Figure 17.11 Specific heat capacity behaviour over the nematic–isotropic phase transition of 8CB in aerosil with different particle silica concentration (Reference 57).

should correspond to the one nucleating sufficiently distant from the particle surface, basically maintaining the bulk like characteristics. The material nucleating close to the particle surface is considerably more distorted and therefore nucleates at a lower temperature than the LTP. The texture fragmentation presumably occurs to relieve the strain accumulated during the nucleation of such a nematic phase.

Frequency dependent measurements of the specific heat capacity, as shown in Figure 17.12 for 8CB in a $0.05\,\mathrm{g\,cm^{-3}}$ sample of aerosil, enabled one to also investigate dynamics aspects related to the nematic material nucleated over each peak.[63] It was found that the decay with frequency of the peak values over the LTP is less rapid than the one observed over the HTP where it is very similar to the one observed for bulk 8CB (inset of Figure 17.12). The analogy of the results found at the HTP and for the bulk material was observed also in the hysteretic behaviour detected between heating and cooling measurements and in the ageing effects.[63] This supported the above mentioned scenario concerning the nematic material nucleating over the two peaks. The frequency dependence characteristics of the nematic over the HTP should then correspond to the partial phase conversion in the NI coexistence during the heating cycle, characteristic of bulk material, which gives rise to dynamical effects.[57] The different time scale involved in the frequency dependence of the LTP should then be associated with the presence of considerable strain which could eventually lead to glassy behaviour.[63,64] Slow dynamics and glassy behaviour had also been

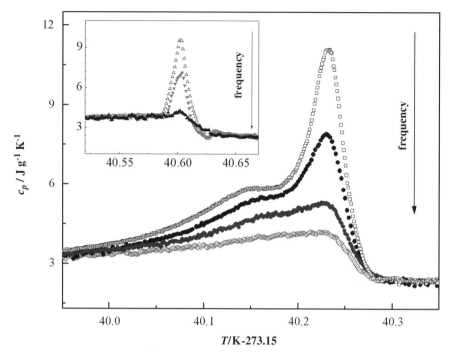

Figure 17.12 Frequency dependence of specific heat capacity behaviour over the nematic–isotropic phase transition of 8CB with $0.05\,g\,cm^{-3}$ particle silica concentration. Inset refers to bulk 8CB.

observed in other self-ordering systems, where the effect of the random field acted in opposition to the mean molecular ordering field, such being the case, for example, of diluted antiferromagnets (DAFF) in external fields,[65] superfluid ^{4}He in restricted geometry,[66] superconductors.[67]

The use of the PPE technique has also allowed to study the latent heat exchanged over each of the two peaks.[61] It was possible to discriminate between the latent heat involved over the HTP and LTP. The latent heat over the HTP was found to be considerably larger than the one over the LTP, again presumably because of the larger strain involved in the latter case. It is known that the first-order character features over phase transitions may get smeared with increasing disorder in the system[68] and eventually disappear. In fact it has been reported that latent heat over the NI transition is reduced in LC confined in spherical droplets,[69] and by the increase of the average particle concentration in aerosil networks,[57,70] consistently with what also predicted by a theoretical treatment.[71]

References

1. S. Kumar, *Liquid Crystals,* Cambridge University Press, Cambridge, 2001.
2. T. Ikeda, T. Sasaki and K. Ichimura, *Nature*, 1993, **361**, 428.

3. I. Muševic, R. Blinc and B. Zekš, *The Physics of Ferroelectric and Anti-ferroelectric Liquid Crystals,* World Scientific, Singapore, 2000.
4. A. M. Donald, A. H. Windle and S. Hanna, *Liquid Crystalline Polymers,* Cambridge University Press, Cambridge, 2005.
5. S. Kumar, *Chem. Soc. Rev.*, 2006, **35**, 83.
6. D. Adam, P. Schuhmacher, J. Simmerer, L. Häussling, K. Siemensmeyer, K. H. Etzbachi, H. Ringsdorf and D. Haarer, *Nature*, 2002, **371**, 141.
7. G. T. Steward, *Liquid Crystals*, 2003, **30**, 541.
8. G. T. Steward, *Liquid Crystals*, 2004, **31**, 443.
9. J. Thoen, H. Marynissen and W. Van Dael, *Phys. Rev. A*, 1982, **26**, 2886.
10. P. H. Keyes, *Phys. Lett. A*, 1978, **67A**, 132.
11. P. H. Keyes and J. R. Shane, *Phys. Rev. Lett.*, 1979, **42**, 722.
12. M. A. Anisimov, *Critical Phenomena in Liquids and Liquid Crystals,* Gordon & Breach Science Publishers, Philadelphia, 1991.
13. P. K. Mukherjee, *J. Phys. Condens. Matter*, 1998, **10**, 9191.
14. S. K. Ma, *Modern Theory of Critical Phenomena,* The Benjamin/Cummings Publishing Company, Reading, Massachusetts, 1976.
15. P. K. Mukherjee, *J. Chem. Phys.*, 1998, **109**, 3701.
16. P. J. Wojtowicz and P. Sheng, *Phys. Lett. A*, 1974, **48A**, 235.
17. H. Yokoyama, *J. Chem. Soc., Faraday Trans. 2*, 1988, **84**, 1023.
18. K. Taketoshi, K. Ema, H. Yao, Y. Takenishi, J. Watanabe and H. Takezoe, *Phys. Rev. Lett.*, 2006, **97**, 197801.
19. J. L. Hu Cang, H. C. Andersen and M. D. Fayer, *J. Chem. Phys.*, 2006, **124**, 14902.
20. S. J. Rzoska, M. Paluch, A. Drozd-Rzoska, J. Ziolo, P. Janik and K. Czuprynski, *Eur. Phys. J. E*, 2002, **7**, 387.
21. P. de Gennes and J. Prost, *The Physiscs of Liquid Crystals*, 2nd edn. Clarendon Press, Oxford, 1993.
22. P. de Gennes, *Sol. State Comm.*, 1972, **10**, 753.
23. C. Bagnulus and C. Berviller, *Phys. Rev. B*, 1985, **32**, 7209.
24. C. W. Garland and G. Nounesis, *Phys. Rev. E*, 1994, **49**, 2964.
25. M. Marinelli, F. Mercuri, U. Zammit and F. Scudieri, *Phys. Rev. E*, 1996, **53**, 701.
26. J. Thoen, H. Marynissen and W. Van Dael, *Phys. Rev. Lett.*, 1984, **52**, 204.
27. B. S. Andereck and B. R. Patton, *Phys. Rev. E*, 1994, 49, 1393 and references therein.
28. M. Marinelli, U. Zammit, F. Mercuri and R. Pizzoferrato, *J. Appl. Phys.*, 1992, **72**, 1096.
29. M. Marinelli, F. Mercuri, U. Zammit and F. Scudieri, *Phys. Rev. E*, 1998, **58**, 5860.
30. F. Mercuri, M. Marinelli and U. Zammit, *Phys. Rev. E*, 1998, **57**, 596.
31. M. Marinelli and F. Mercuri, *Phys. Rev. E*, 2000, **61**, 1616.
32. P. G. de Gennes, *Mol. Cryst. Liq. Cryst.*, 1973, **21**, 49.
33. C. C. Huang and J. M. Viner, *Phys. Rev. A*, 1982, **25**, 3385.
34. L. Reed, T. Stoebe and C. C. Huang, *Phys. Rev. E*, 1995, **52**, R2157.
35. F. Mercuri, M. Marinelli and U. Zammit, *Phys. Rev. E*, 2003, **68**, 051705.

36. C. R. Safiyna, M. Kaplan, J. Als-Nielsen, R. J. Birgeneau, D. Davidov, J. D. Litster, D. L. Johnson and M. Neubert, *Phys. Rev. B*, 1980, **21**, 4149.
37. L. Benguigui and P. Martinoty, *Phys. Rev. Lett.*, 1989, **63**, 774.
38. L. Benguigui and P. Martinoty, *Mol. Cryst. Liq. Cryst. Sci. Technol., Sect A*, 1997, **301**, 355.
39. T. Brauminger and B. M. Fung, *J. Chem. Phys.*, 1995, **102**, 7714.
40. R. J. Birgenau and J. D. Litster, *J. Phys. (Paris) Lett.*, 1978, **39**, L399.
41. R. Pindak, D. E. Moncton, S. C. Davey and J. W. Goodby, *Phys. Rev. Lett.*, 1981, **46**, 1135.
42. R. W. Goodby and R. Pindak, *Mol. Cryst. Liq. Cryst.*, 1981, **75**, 233.
43. C. C. Huang and T. Stoebe, *Adv. Phys.*, 1993, **42**, 343.
44. H. Haga, Z. Kutnjak, G. S. Iannacchione, S. Qian, D. Finotello and C. W. Garland, *Phys. Rev. E*, 1997, **56**, 1808.
45. H. Haga and C. W. Garland, *Phys. Rev. E*, 1998, **57**, 603.
46. B. V. Roie, K. Denolf, G. Pitsi and J. Thoen, *Eur. Phys. J. E*, 2005, **16**, 361.
47. F. Mercuri, S. Paoloni, U. Zammit, F. Scudieri and M. Marinelli, *Phys. Rev. E*, 2006, **74**, 041707.
48. R. Bruinsma and G. Aeppli, *Phys. Rev. Lett.*, 1982, **48**, 1625.
49. A. Aharony, R. J. Birgeneau, J. D. Brock and J. D. Litster, *Phys. Rev. Lett.*, 1986, **57**, 1012.
50. C. C. Huang, G. Nounesis, R. Geer, J. W. Goodby and D. Guillon, *Phys. Rev. A*, 1989, **39**, 3741.
51. E. Gorecka, L. Chen, W. Pyzuk, A. Krowczynski and S. Kumar, *Phys. Rev. E*, 1994, **50**, 2863.
52. T. Stoebe and C. C. Huang, *Int. J. Mol. Phys. B*, 1995, **9**, 2285.
53. D. J. Bergman and B. I. Halperin, *Phys. Rev. B*, 1976, **13**, 2145.
54. D. Finotello, G. S. Iannacchione and S. Qian, *Liquid Crystals in Complex Geometries*, ed. G. P. Crawford and S. Zumer, Taylor and Francis, London, p. 325 (1996), and references therein.
55. G. S. Iannacchione, G. P. Crawford, S. Zumer, J. W. Doane and D. Finotello, *Phys. Rev. Lett.*, 1993, **71**, 2595.
56. L. Wu, B. Zhou, C. W. Garland, T. Bellini and D. W. Shaefer, *Phys. Rev. E*, 1995, **51**, 2157.
57. G. S. Iannacchione, C. W. Garland, J. T. Mang and T. P. Rieker, *Phys. Rev. E*, 1998, **58**, 5966.
58. G. P. Crawford, R. Stannarius and J. W. Doane, *Phys. Rev. A*, 1991, **44**, 2558.
59. G. S. Iannacchione, S. Park, C. W. Garland, R. J. Birgeneau and R. L. Leheny, *Phys. Rev. E*, 2003, **67**, 011709.
60. F. Mercuri, S. Paoloni, U. Zammit and M. Marinelli, *Phys. Rev. Lett.*, 2005, **94**, 2478801.
61. F. Mercuri, M. Marinelli, S. Paoloni, U. Zammit and F. Scudieri, *Appl. Phys. Lett.*, 2008, **92**, 251911.
62. S. Paoloni, F. Mercuri, M. Marinelli, U. Zammit, C. Neamtu and D. Dadarlat, *Phys. Rev. E.*, 2008, **78**, 042701.

63. M. Marinelli, F. Mercuri, S. Paoloni and U. Zammit, *Phys. Rev. Lett.*, 2005, **95**, 2378801.
64. T. Bellini, M. Buscaglia, C. Chioccoli, F. Mantegazza, P. Pasini and C. Zannoni, *Phys. Rev. Lett.*, 2002, **88**, 245506.
65. R. J. Birgeneau, *J. Magn. Mater.*, 1998, **177–181**, 1.
66. M. Yamashita, A. Matsubara, R. Ishiguro, Y. Sasaki, Y. Kataoka, M. Kubota, O. Ishikawa, Yu. M. Bunkov, T. Ohmi, T. Takagi and M. Mizusaki, *Phys. Rev. Lett.*, 2005, **94**, 075301.
67. G. I. Menon, *Phys. Rev. B*, 2002, **65**, 104527.
68. Y. Imry and M. Wortis, *Phys. Rev. B*, 1979, **19**, 3580.
69. A. Golemme, S. Zumer, D. W. Allender and J. W. Doane, *Phys. Rev. Lett.*, 1988, **26**, 2937.
70. P. J. Jamee, G. Pitsi and J. Thoen, *Phys. Rev. E*, 2002, **66**, 21707.
71. A. V. Zakharov and J. Thoen, *Phys. Rev. E*, 2004, **69**, 11704.

CHAPTER 18

Heat Capacities and Phase Transitions for the Dynamic Chemical Systems: Conformers, Tautomers, Plastic Crystals, and Ionic Liquids

GENNADY KABO,[a] EUGENE PAULECHKA[a] AND MICHAEL FRENKEL[b]

[a] Chemistry Department, Belarusian State University, Minsk, Belarus;
[b] Thermodynamics Research Center, Thermophysical Properties Division, National Institute of Standards and Technology, Boulder, Colorado, USA

18.1 Introduction

18.1.1 Phenomenology of Heat Capacity Changes at Phase Transitions and Glass Transitions of Pure Compounds

Heat capacity of molecular liquids, gases, and glasses is a continuous function of temperature over all the range of existence of the corresponding state. The heat capacity of gas and glass monotonically increases with temperature. Normally, the heat capacity at constant pressure, C_p, for a liquid rises when the temperature is increased. The inverse dependence is sometimes observed, for example, for liquid water below 290 K[1] and liquid methanol below 190 K.[2]

Heat Capacities: Liquids, Solutions and Vapours
Edited by Emmerich Wilhelm and Trevor M. Letcher
© The Royal Society of Chemistry 2010
Published by the Royal Society of Chemistry, www.rsc.org

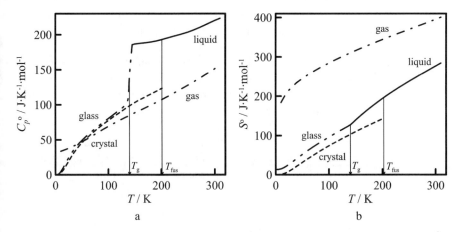

Figure 18.1 Heat capacity (a) and entropy (b) for different phases of cyclohexyl formate.[3,4]

The heat capacity of a liquid exceeds that of the corresponding gas for most compounds. Notable exceptions are hydrogen, helium, and mercury. The heat capacity of the crystal phases of molecular liquids (whose molecules do not have hydrogen bonds) at the temperature of fusion, $\Delta_{cr}^{liq}C_p\,(T_{fus})$, is about 30 % to 60 % of the heat capacity of the corresponding liquid phase, $C_{p,liq}$. The heat capacity of a liquid extrapolated to the temperature of the glass transition, T_g, is also substantially higher than the heat capacity of the glass at this temperature (Figure 18.1). For molecular liquids of that type, the glass heat capacity is about 40 % to 70 % of the liquid heat capacity.

The heat capacity of crystals without solid-phase transitions also rises monotonously with temperature to the melting point. As a rule, the crystal heat capacity extrapolated to T_{fus} is noticeably lower than the liquid heat capacity at that temperature. In the case of a molecular compound whose molecules contain no hydrogen bonds, the crystal heat capacity near T_{fus} is only 50 % to 80 % of the liquid heat capacity. Solid-phase transitions in crystals may be accompanied by heat capacity changes, but the heat capacities of the high-temperature phases are usually lower than that of the corresponding liquid.

It should also be noted that the relationships between the values in different phases for thermodynamic properties such as heat capacity (Figure 18.1a), entropy (Figure 18.1b) and thermal expansion coefficient at constant pressure, α_p, differ significantly. The thermal expansion coefficient is about $10^{-3}\,K^{-1}$ for most liquids. The value of α_p for rigid crystals near T_{fus} is less than that for liquids by a factor of 2 to 6, as calculated from the recommended densities of molecular crystals and liquids from Reference 5. A thermal expansion coefficient for an ideal gas is some times higher than that of a liquid and is equal to $1/T$.

All these well-known facts have not been interpreted, even qualitatively, due to the absence of generally accepted structural models of compounds in the condensed state. In order to understand heat capacity changes at phase transitions, it is very important to take into account that there are at least two types of equilibrium structural changes in pure compounds:

1. formation of equilibrium mixtures of different conformers of molecules in gases, liquids, and plastic crystals;
2. formation of systems of equilibrium intermolecular orientations (configurations) with minimal energy in liquids and plastic crystals.

In this context, an inhomogeneous compound can be considered as pure only as a physical approximation, despite its single triple point. Therefore, the refinement of a "pure compound" concept seems to be justifiable.

18.1.2   Concepts of Purity and Individual (Structurally Pure) Compounds

The problem of defining a "pure compound" was first discussed by van der Waals and Kohnstamm[6] and then by Kabo et al.[7] According to van der Waals, compounds are classified as follows.

1. A pure compound with respect to the Gibbs phase rule (definite values for the triple point, boiling temperature, fusion temperature) behaves as an individual compound, but reveals its complexity through complicated forms of the equation of state. A logarithmic term caused by mixing of different molecules is present in the entropy and Gibbs energy of this compound:

$$S^\circ = \sum_i x_i S_i^\circ - R \sum_{i=1}^{n} x_i \ln x_i \qquad (1)$$

$$\Delta G^\circ = \sum_i x_i \Delta G_i^\circ + RT \sum_{i=1}^{n} x_i \ln x_i \qquad (2)$$

where x_i is a mole fraction of molecules of the ith type, and S_i° and ΔG_i° are their entropy and Gibbs energy. Nitrogen dioxide is an example of such a compound where the following equilibrium takes place:

$$2NO_2 \rightleftharpoons N_2O_4$$

2. "Thermodynamic individual" behaves as one compound both with regard to the Gibbs phase rule and in all thermodynamic applications. Entropy and Gibbs energy for such "individuals" do not contain a logarithmic term. This group of compounds includes, for example, alkanes C_nH_{2n+2}, and their derivatives.
3. "Molecular individual" is a compound composed of molecules of one kind. Examples are benzene, cyclopropane, etc.

Kabo, Roganov, and Frenkel[7] suggested a different classification based on the molecular structure of compounds.

1. "Isomerically pure compound" consists of molecules of the same chemical formula. Mixtures of isomers such as isomer families within compounds of a particular elemental formula, C_5H_{10}, C_5H_{12}, C_6H_{12}, etc., belong to this group.
2. "Structurally pure compound" consists of molecules of the same structure, for example, butan-2-ol, halobutanes, etc., which form distinguishable chiral forms of molecules.
3. "Configurationally pure compound" consists of achiral or configurationally uniform molecules. This group includes *n*-alkanes, 2-methylalkanes, etc.
4. "Conformationally pure compound" consists of molecules that either do not form energetically and geometrically distinguishable conformers at internal rotation or are individual conformers due to high potential barriers to internal rotation "forcing" separation of individual conformers at a given temperature or due to thermodynamic limitations. Examples are ethane, toluene, etc.
5. "Individual compound" consists of molecules, for which only one conformation exists, such as benzene or cyclopropane.

Therefore, the use of terms "pure compound" or "purity of a compound" should be problem-specific with respect to the experimental and theoretical issues under consideration.

18.1.3 Kinetic and Thermodynamic Aspects of Conformational Transformations in Liquids and Gases

The equilibria between different *i* conformers of a compound *A* that exist due to the possibility of rotation of the molecule around single bonds can be presented as:

$$A_0 = \sum_i x_i A_i$$

These equilibria are defined by thermodynamic and kinetic properties of these transformations. Potential barriers to internal rotation and inversion $\Delta G^{\neq}$ are often close in gases and liquids. Normally they are below $50 \, kJ \, mol^{-1}$.[8,9] Conformational transformations are 1st order processes, for which the rate constant *k* can be evaluated from the equation:

$$k = (k_B T/h) \exp(-\Delta G^{\neq}/RT) \tag{3}$$

where $\Delta G^{\neq}$ is the Gibbs activation energy of a process, k_B is Boltzmann's constant, *h* is Planck's constant, *R* is the gas constant, and *T* is temperature. The mean lifetime for this transformation is equal to $\tau = 1/k$. The interrelation

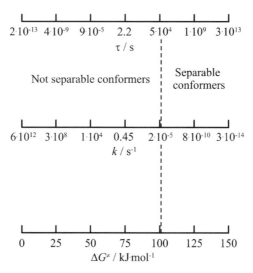

Figure 18.2   Activation scale of conformational transformations at $T = 298$ K. k is a rate constant of the reaction, calculated as $k = (k_B T/h) \exp(-\Delta G^{\neq}/RT)$, τ is an average lifetime $\tau = 1/k$.

between $\Delta G^{\neq}$, k, and τ is presented in Figure 18.2. One may assume that the conformers with $\tau > 24$ hours, which corresponds to $\Delta G^{\neq} = 101$ kJ mol^{-1} at $T = 298$ K, can be separated. This assumed border is shown in Figure 18.2 by a dashed line.

Average lifetimes of conformers determine possibilities of conformational equilibria down to low temperatures. For example, at $T = 300$ K and $\Delta G^{\neq} = 25$ kJ mol^{-1}, $\tau = 4 \times 10^{-9}$ s, and at 100 K, $\tau = 5.5$ s. That is why in gases and liquids conformational equilibria are easily achieved (Figure 18.2), and the composition is determined by the differences in thermodynamic properties of individual conformers. The conformers correspond to the Gibbs energy minima at the pathways of internal rotation.

Since in most cases the differences in entropy and in heat capacity between conformers are small, the differences between the enthalpy and Gibbs energy are negligible; $\Delta H^{\circ}(T) \approx \Delta G^{\circ}(T)$, and $\Delta H^{\circ}(T)$ is practically independent of temperature. The thermodynamic properties of conformationally inhomogeneous compounds can be presented by use of the relationships:[10]

$$C_p = \sum_{i=1}^{n} x_i C_{p,i} + C_{\text{conf}} \qquad (4)$$

$$S^{\circ} = \sum_{i=1}^{n} x_i S_i^{\circ} + S_{\text{conf}} \qquad (5)$$

$$\frac{H^{\circ} - H_0^{\circ}}{T} = \sum_{i=1}^{n} x_i \frac{H_i^{\circ} - H_{i,0}^{\circ}}{T} + \frac{\Delta_{\text{conf}} H}{T} \qquad (6)$$

where x_i is a mole fraction of the ith conformer, $C_{p,i}$, S_i°, H_i° are heat capacity, entropy, and enthalpy of the ith conformer; C_{conf}, S_{conf}, $\Delta_{\text{conf}}H$ are conformational contributions to heat capacity, entropy and enthalpy, respectively. These contributions can be calculated as:

$$C_{\text{conf}} = \frac{1}{RT^2} \sum_{i=1}^{n} x_i \sum_{j>i}^{n} x_j \left(\Delta H_{ij}\right)^2 \tag{7}$$

$$S_{\text{conf}} = -R \sum_{i=1}^{n} x_i \ln x_i \tag{8}$$

$$\frac{\Delta_{\text{conf}} H}{T} = \sum_{i=2}^{n} x_i \frac{\Delta H_{1i}}{T} \tag{9}$$

where ΔH_{ij} is the enthalpy difference between the ith and jth conformers.

Equations (4)–(9) can also be used for description of tautomeric equilibria, for example,

However, Equations (7) and (9) should be modified for description of equilibria for the reactions proceeding with the change of a total number of moles like $2NO_2 \rightleftharpoons N_2O_4$.

The differences in heat capacities of individual conformers are normally small, and it is reasonable to assume that

$$C_p = C_p^* + C_{\text{conf}} \tag{10}$$

where C_p^* is the heat capacity of one conformer. According to Equation (7), the conformational contribution to the heat capacity of a liquid and a gas is a continuous function of temperature. At a given temperature, the differences in conformational compositions of the liquid and gas are normally small.

The change of conformational composition of compounds on heating is accompanied by an increase of the fraction of energetically less stable conformers. The contribution to the heat capacity from a change in conformational composition, at a temperature change of 1 K, depends on the number of coexisting conformers and ΔH_{ij}. The dependences of C_{conf} and S_{conf} of the enthalpy difference between conformers, ΔH, for the systems formed by three and six conformers are presented in Figure 18.3.

It has been experimentally found that at the transition liquid → glass conformational composition of the liquid at the temperature close to T_g is frozen

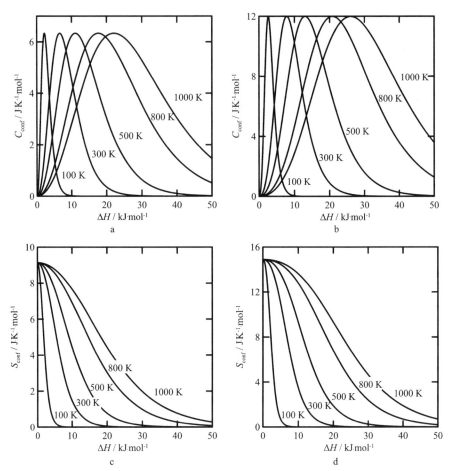

Figure 18.3   Conformational contributions to heat capacity and entropy for the systems: (a), (c) one most stable conformer and two conformers of relative enthalpy ΔH; (b), (d) one most stable conformer and five conformers of relative enthalpy ΔH.

due to kinetic restrictions. Therefore, the conformational contribution to heat capacity defined by Equation (7) is absent in glass.

At the transition liquid→plastic crystal, the conformational composition of liquid and plastic crystal does not differ much,[11] and there are no kinetic restrictions to conformational transformations. Vibrational spectra of 1-chloro-1-methylcyclohexane in various phases are presented in Figure 18.4. The band at $658\,\mathrm{cm}^{-1}$ is assigned to the C–Cl stretching of the chair conformer of the cyclohexane ring, with the chlorine atom in an equatorial position. The intensity of this band does not change much when going from the gas to the liquid and then to the plastic crystal (crI). The noted band completely disappears in the IR spectrum of the rigid crystal (crII).

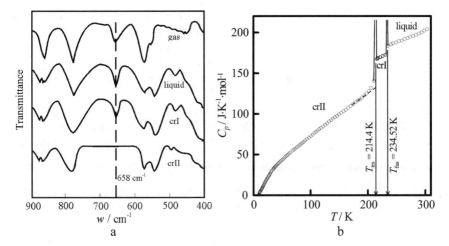

Figure 18.4   IR and heat capacity experimental data for 1-chloro-1-methyl-cyclo-hexane:[11] (a) IR spectra in different phase states: gas and liquid at $T = 293$ K; plastic crystal (crI) at $T = 223$ K; rigid crystal (crII) at $T = 205$ K; (b) temperature dependence of heat capacity.

Only one, the most stable conformer, normally remains in the lattice of rigid crystals at moderate crystallization rates.

By using only Equation (7) it is impossible to explain the experimentally observed changes in heat capacity at the transition of a liquid into other phases. That is why statistical thermodynamic relationships for various phases must be used in their interpretation.

18.2   Practical Calculations of Heat Capacity for Gases and Liquids by Statistical Thermodynamic Methods

18.2.1   Ideal Gas

The heat capacity of an ideal gas is given by

$$C_{p,\text{id.g}} = 4R + C_{\text{vib}} + C_{\text{int.rot}} \tag{11}$$

where $4R$ is a sum of translational and rotational contributions to heat capacity of nonlinear molecules $(3/2R + 3/2R)$ and $C_p - C_v = R$; C_{vib} is a contribution of $3n-6-n_r$ normal vibrations of molecules consisting of n atoms, and containing n_r tops or degrees of freedom for inversion,

$$C_{\text{vib}} = \sum_{i=1}^{3n-6-n_r} C_{\text{vib},i} \tag{12}$$

where $C_{\text{vib},i}$ is a vibrational contribution of the ith normal vibration with frequency ν_i to heat capacity calculated with the formula

$$C_{\text{vib},i} = R\left(\frac{h\nu_i}{kT}\right)^2 \exp\left(\frac{h\nu_i}{kT}\right)\left(\exp\left(\frac{h\nu_i}{kT}\right) - 1\right)^{-2} \tag{13}$$

If the potential function for internal rotation or inversion $V(\varphi)$, where φ is a phase angle, and the reduced moment of inertia are known, corresponding energy levels $(u_i)_{\text{int.rot}}$ can be calculated by solving the Schrödinger equation. The partition functions $Q_{\text{int.rot}} = \sum_{i=0}^{\infty} \exp\left(-\frac{u_{i,\text{int.rot}}}{kT}\right)$ and the corresponding contribution $C_{\text{int.rot}}$ can also be found. Often, spectral investigations of temperature dependences of vibrational spectra in combination with calculations by molecular mechanics and quantum chemistry allow one to evaluate a number of possible conformers and their energy differences. As was previously shown,[7] the calculation of the contribution of internal rotation can be efficiently performed from:

$$C_{\text{int.rot.}} = C'_{\text{vib}} + C_{p,\text{conf}} \tag{14}$$

Here C'_{vib} is a contribution of n_r torsional vibrations in a molecule calculated by Equation (13), and the conformational contribution C_{conf} is calculated by Equation (7).

The ideal-gas entropy can be found from:

$$S^{\circ}_{\text{id.g.}} = S^{\circ}_{\text{tr}} + S_{\text{rot}} + S_{\text{vib}} + S_{\text{int.rot}} \tag{15}$$

where S°_{tr}, S_{rot}, S_{vib}, $S_{\text{int.rot}}$ are contributions of translation, overall rotation, intramolecular vibrations, and internal rotation. These contributions are calculated from Equations (16)–(19):

$$S^{\circ}_{\text{tr}} = R\left(\frac{3}{2}\ln\frac{2\pi MkT}{h^2 N_A} + \ln\frac{RT}{p^{\circ}} + \frac{5}{2}\right) \tag{16}$$

$$S_{\text{rot}} = R\left(\ln\frac{8\pi^2}{\sigma} + \frac{3}{2}\ln\frac{2\pi kT}{h^2} + \frac{1}{2}\ln I_A I_B I_C + \frac{3}{2}\right) \tag{17}$$

$$S_{\text{vib}} = \sum_i S_{\text{vib},i} \tag{18}$$

$$S_{\text{vib},i} = R\left(\frac{h\nu_i}{kT}\left(e^{-\frac{h\nu_i}{kT}} - 1\right)^{-1} - \ln\left(1 - e^{-\frac{h\nu_i}{kT}}\right)\right) \tag{19}$$

where M is a molar mass, $I_A I_B I_C$ is a product of principal moments of inertia of the molecule, and σ is its symmetry number. The values of $S_{\text{int.rot}}$ are calculated either from the energy levels of a hindered rotator, as described above, or by use of Equation (20):

$$S_{\text{int.rot.}} = S'_{\text{vib}} + S_{\text{conf}} \tag{20}$$

where S'_{vib} is a contribution of n_r torsional vibrations in the molecule calculated by Equation (19). S_{conf} is determined from Equation (8).

18.2.2 Liquid

The heat capacity of liquid can be presented as:

$$C_{p,\text{liq}} = C_{v,\text{id.g}} + C_{p,\text{config}} \tag{21}$$

where $C_{v,\text{id.g}}$ is the heat capacity of the corresponding ideal gas and $C_{p,\text{config}}$ is the configurational heat capacity of the liquid. The $C_{v,\text{id.g}}$ value depends on temperature only, and $C_{p,\text{config}}$, related to the change of the energy of intermolecular interaction with temperature, depends both on temperature and pressure.

Configurational heat capacity for molecular liquids can be determined based on Equation (21) derived from the experimental data on liquid heat capacity and the calculated ideal-gas heat capacity, from molecular simulation, and from model of liquid state theories such as "hole" theories.

$C_{p,\text{config}}$ can be divided into two parts, $C_{v,\text{config}}$ and $(C_p - C_v)_{\text{config}}$:

$$
\begin{aligned}
C_{p,\text{config}} &= C_{v,\text{config}} + \left(C_p - C_v \right)_{\text{config}} \\
&= \left(\frac{\partial U_{\text{config}}}{\partial T} \right)_V + \left(p + \left(\frac{\partial U_{\text{config}}}{\partial V} \right)_T \right) \left(\frac{\partial V}{\partial T} \right)_p
\end{aligned} \tag{22}
$$

The values of the configurational contributions to the heat capacity for some liquids are presented in Table 18.1. As is shown in Table 18.1, $C_{v,\text{config}}$ is about $3R$ for many nonpolar liquids, and it decreases with increasing temperature (Figure 18.5). These facts may be explained by assuming that at low temperatures the molecules of liquid oscillate near certain stationary sites. The corresponding contributions to C_v of liquid will be close to $6R$ for a nonlinear molecule. For an ideal gas, the contributions of translation and overall rotation are equal to $3R$. The difference of the mentioned values close to $3R$ is a configurational contribution $C_{v,\text{config}}$. When temperature rises, the molecular motion becomes less hindered, which results in the decrease in $C_{v,\text{config}}$. However, $C_{v,\text{config}} > 0$ for liquid even at high temperatures.

$C_{v,\text{config}}$ can be higher than $3R$ in associated liquids. When the temperature is increased, even at $V = \text{const}$, the fraction of structures with higher relative energy increases, resulting in an additional contribution to the heat capacity. Formation of those structures may be caused, for example, by breaking hydrogen bonds.

Table 18.1   Contributions to configurational heat capacity for molecular liquids at 298.15 K

Compound	$C_{p,\text{liq}}{}^a$	$C_{v,\text{id.gas}}{}^a$	$C_{p,\text{config}}$ $\text{J K}^{-1}\text{mol}^{-1}$	$C_{v,\text{config}}$	$(C_p\text{-}C_v)_{\text{config}}$
Pentane	167.0±0.5	111.7±0.5	55.3±0.7	13.3±1.1	42.0±0.9
Hexane	194.1±3.0	134.3±3.0	59.8±4.2	8.0±4.9	51.8±2.5
Benzene	135.7±0.5	73.5±1.0	62.2±1.1	22.1±2.4	40.1±2.2
Toluene	156.7±1.7	95.5±1.0	61.2±2.0	20.8±2.1	40.5±0.8
Acetone	124.5±1.8	66.3±1.0	58.2±2.1	23.7±2.7	34.5±1.7
Butanone	158.8±0.4	95.0±2.0	63.8±2.0	23.8±2.8	40.0±2.0
Cyclohexyl acetate	245.1±1.6	161.4±1.6[4]	83.7±2.3	26.1±2.9[14]	57.6±1.7[14]
Fluorobenzene	146.3±0.4	86.0±3.0	60.3±3.0	20.8±3.3	39.5±1.3
Hexafluorobenzene	221.6±0.5	147.1±1.0	74.5±1.1	23.9±4.2	50.5±4.1
Methanol	81.2±0.7	35.7±0.2	45.5±0.7	31.8±0.9	13.7±0.5

[a]Evaluated using NIST Standard Reference Databases REFPROP[12] and ThermoData Engine[13]

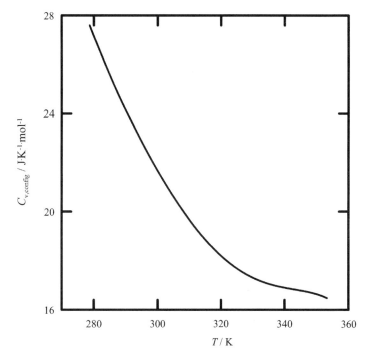

Figure 18.5   Temperature dependence of $C_{v,\text{conf.}}$ for benzene (calculated using critically evaluated data generated by NIST Standard Reference Databases.[12,13]

The $(C_p - C_v)_{\text{config}}$ value is equal to $(C_p - C_v)$ of a liquid and can be found analytically if the thermal expansion coefficient, α_p, the isothermal compressibility coefficient, κ_T, and the molar volume, V_m, of the liquid are known:

$$C_p - C_v = \frac{\alpha_p^2}{\kappa_T} V_m T \tag{23}$$

It was found[14] that at 298 K $(C_p - C_v)$ can be correlated to the vaporization enthalpy of liquid $\Delta_{vap}H$ and its thermal expansion coefficient:

$$(C_p - C_v) = n\Delta_{vap}H\alpha_p \tag{24}$$

where $n = 1.10\pm0.12$ for liquids whose molecules have no hydrogen bonds.

One possible approach for the calculation of the configurational contribution to the thermodynamic functions of liquid is through the use of hole models. In these models the liquid volume is divided into cells of equal volume v_0. To provide a correct description of thermal expansion and compressibility of liquids it is assumed that a part of the cells is occupied by molecules, while another part remains empty. These empty cells are called "holes". Thermal expansion of a liquid is related to the increase in v_0 and the increase of the number of unoccupied cells. The liquid volume is:

$$V = v_0\left(\sum_i r_i N_i + N_0\right) \tag{25}$$

where r_i is the number of cells occupied by the ith functional group, N_i is the number of those groups, N_0 is the number of empty cells (holes). In the simplest case, $i = 1$, and a molecule occupies r cells. The intermolecular interaction energy is calculated either in a random-mixing approximation or in a quasi-chemical approximation.[15] In the random-mixing approximation, the molar cohesion energy is:

$$E_{coh} = -\Delta_{vap}U^\circ = -u_{11}\frac{z}{2}\frac{(qN_A)^2}{qN_A + N_0} \tag{26}$$

where $\Delta_{vap}U^\circ$ is the change in standard internal energy for the vaporization of one mole of liquid; u_{11} is the interaction energy per one contact (molecule – molecule); $z = 10$; $q = r - \frac{2(l+(r-1))}{z}$ is the square of the molecule expressed as the ratio to the square of the hole;[16] $l = 0$ for open chain molecules; and $l = 1$ for the molecules containing a ring.

If the cohesion energy is described by Equation (26), and r does not depend on temperature, $C_p - C_v$ of liquid can be expressed as:[14]

$$\begin{aligned}(C_p - C_v)_{liq} &\approx \left(\frac{\partial \Delta_{vap}U^\circ}{\partial V}\right)_T \left(\frac{\partial V}{\partial T}\right)_p \\ &= \frac{\Delta_{vap}U^\circ}{qN_A + N_0}\left(\frac{\partial N_0}{\partial T}\right)_p + \Delta_{vap}U\left(\frac{\partial \ln u_{11}}{\partial v_0}\right)_T\left(\frac{\partial v_0}{\partial T}\right)_p\end{aligned} \tag{27}$$

The first term in Equation (27) is a hole contribution related to the changes of free volume of liquid. The second term is related to the dependence of the interaction parameter u_{11} on the cell volume.

18.2.3 Glass

The structure of a glass near T_g is similar to that of a liquid. This is indirectly supported by the fact that entropies of liquid and glass are equal at $T \to T_g$. However, in a glass, unlike a liquid, conformational transformations and changes of free volume are frozen in. These two circumstances result in substantial differences (sometimes greater than $100 \, \text{J} \, \text{K}^{-1} \, \text{mol}^{-1}$) between the heat capacities of the glass and the liquid.

The heat capacity of a glass can be considered as a sum of three terms:

$$C_{gl} = C_{lat,gl} + C_{vib} + (C_p - C_v)_{gl} \tag{28}$$

where $C_{lat,gl}$ is the contribution of pseudo-lattice vibrations. The contribution of intramolecular vibrations C_{vib} includes all $3n - 6$ vibrations. The $(C_p - C_v)_{gl}$ value is related to the thermal expansion of the glass caused mainly by the anharmonicity of pseudo-lattice vibrations.

Applying the model, described in Section 18.2.2, to glass, the increase of the cell volume v_0 should only be considered. A number of empty cells in glass below T_g remains unchanged:

$$\left(C_p - C_v\right)_{gl} = \Delta_{vap} U^\circ \left(\frac{\partial \ln u_{11}}{\partial v_0}\right)_T \left(\frac{\partial v_0}{\partial T}\right)_p \tag{29}$$

18.2.4 Rigid Crystals

The heat capacity of rigid molecular crystals can be defined as:

$$C_p = C_{lat} + C_{vib} + (C_p - C_v)_{cr} \tag{30}$$

where C_{lat} is the contribution of lattice vibrations and C_{vib} is the contribution of intramolecular vibrations. Often C_{lat} is presented as a sum:

$$C_{lat} = C_D(\theta_D/T) + \sum_{i=1}^{3} C_E(\theta_{E,i}/T) \tag{31}$$

where C_D is the Debye heat-capacity function for three degrees of freedom, θ_D is the characteristic Debye temperature, C_E is the Einstein heat-capacity function and $\theta_{E,i}$ is the characteristic temperature of the ith librational vibration. Near the melting point, the contribution of lattice vibrations are close to saturation, and one may assume $C_{lat} = 6R$.

The empirical Lord relationship,

$$(C_p - C_v)_{cr} = ATC_v^2 \tag{32}$$

assumes that an empirical parameter A is determined from the experimental data and calculations for C_p and C_v.

Near $T = 5$ K the differences of heat capacity between crystal and glass are caused mainly by differences in frequencies of lattice vibrations. Near T_g these differences are related to differences in $(C_p - C_v)$ and may reach 10–15 % of the crystal C_p.

18.3 Heat Capacity Changes at Transitions for Pure Organic Liquids

18.3.1 Gas-Liquid

Heat capacity changes at the transition gas→liquid for some compounds are presented in Table 18.2. As follows from Equation (21),

$$\Delta_g^{liq} C_p = C_{config} - R \tag{33}$$

if temperature $T < T_b$, then $C_{v,config} \approx 3R$ and

$$\Delta_g^{liq} C_p = 2R + \left(C_p - C_v \right)_{liq} \tag{34}$$

Then $\Delta_g^{liq} C_p > 2R$, and is determined mainly by $(C_p - C_v)_{liq}$

18.3.2 Glass-Liquid

Vitrification of organic liquids is accompanied by a significant change in heat capacity (Table 18.2). There are models for describing the heat capacity changes at vitrification for polymers, but the heat capacity changes at glass transitions of low-molecular compounds have attracted less attention. One of the most developed models for polymers is the DiMarzio-Dowell lattice model.[33] Below T_g, the thermal expansion of glass is caused only by the temperature dependence of the cell volume v_0, and the number and occupancy of the cells remain unchanged; conformational transformations are frozen in. Above T_g, thermal expansion is related to the increase in both the cell volume and the number of cells. Conformational transformations may occur as well.

The change in heat capacity at the transition glass→liquid[33] is equal to

$$\Delta_{gl}^{liq} C_p = C_{conf} + \Delta C_{vib} + C_h \tag{35}$$

where C_{conf} is the conformational contribution to the liquid heat capacity, C_h is the hole contribution caused by the change in the number of empty cells in the liquid with temperature, and ΔC_{vib} is a contribution related to changes in temperature dependence of vibrational frequencies while transforming from glass to liquid. In Equation (35), $\Delta_{gl}^{liq} C_v \approx C_{conf}$ and $\Delta_{gl}^{liq} (C_p - C_v) \approx \Delta_{vib} C + C_h$.

The ΔC_{vib} contribution[33] is given by:

$$\Delta C_{\text{vib}} = 0.5 C_{v\,\text{gl}} T \left(\alpha_{p,\text{liq}} - \alpha_{p,\text{gl}} \right) \tag{36}$$

The conformational contribution to the heat capacity change accompanying the transformation of glass to liquid, can be evaluated by Equation (7). Applying the model described in Section 18.2.2 for the calculation of the hole contributions, and taking into account that $N_0 \ll r N_A$ near T_g,

$$C_h \approx \Delta_{\text{vap}} U^{\circ} \left(\alpha_{p,\text{liq}} - \alpha_{p,\text{gl}} \right) \frac{r}{q} \tag{37}$$

Since the experimental data for the density of liquids from Table 18.2, near T_g, are not available, we assumed that $\alpha_{p,\text{liq}}(T_g) = \alpha_{p,\text{liq}}(T_{\text{fus}})$. $\alpha_{p,\text{liq}}(T_{\text{fus}})$ values were obtained using critically evaluated data generated by NIST Standard Reference Databases.[12,13] Based on the data reported in the Reference 4, we found that $\alpha_{p,\text{cr}}/\alpha_{p,\text{liq}} \approx 0.35$ at T_{fus}. We also assumed that in Equation (36), $\alpha_{p,\text{cr}}(T_{\text{fus}})/\alpha_{p,\text{liq}}(T_{\text{fus}}) = \alpha_{p,\text{gl}}(T_g)/\alpha_{p,\text{liq}}(T_g)$. Finally, we used the approximation that $C_{v,\text{gl}} \approx C_{p,\text{gl}}$. The differences in enthalpy for conformers were found by the molecular mechanics method (MM3 force field). The r parameter was assumed to be equal to the number of heavy atoms in a molecule.

The results of the calculations are presented in Table 18.2. The calculated $\Delta_{\text{gl}}^{\text{liq}} C_p$ values are on average 63 % of the experimental values. The hole contribution is the main contribution (>80 %) to the heat capacity change in glass → liquid transitions.

18.3.3 Crystal-Liquid

The contribution of lattice vibrations to C_v is close to $6R$ for a crystal near T_{fus}. Equations similar to Equations (35)–(37) can be used to determine the change in heat capacity at melting. The heat capacity change at melting is:

$$\Delta_{\text{cr}}^{\text{liq}} C_p = C_{\text{conf}} + \Delta C_{\text{vib}} + C_h \tag{38}$$

The obtained results are presented in Table 18.2. As in case of $\Delta_{\text{gl}}^{\text{liq}} C_p$, the experimental values are underestimated by about 18 % on average. The hole contribution is about 80 % of the total heat capacity change. Better agreement with experiment can probably be obtained using the quasi-chemical approximation. However, in that case, C_h cannot be explicitly related to the experimentally determined quantities. The calculated heat capacity change for 2-methylbutane is significantly higher than the experimental value. It is expected that partial disordering occurs in the corresponding crystal near T_{fus}, as in plastic crystals.

Table 18.2 Thermodynamic properties of molecular glass-formers[a]

Formula	Compound name	T_{fus} (K)	T_g (K)	T_b (K)	Experimental $\Delta_g^{liq}C_P°(T_g)$	Calculated $\Delta_g^{liq}C_P°(T_g)$	Experimental $\Delta_{cr}^{liq}C_P°(T_{fus})$	Calculated $\Delta_{cr}^{liq}C_P°(T_{fus})$	$\Delta_g^{liq}C_P°(T_{fus})$
					J K^{-1} mol^{-1}				
$C_4H_6O_2$	Vinyl acetate[17]	180.64±0.01	120±1	345.74±0.05	66±1	42	50±1	40	65±1
C_4H_8	1-Butene[18,19]	87.82±0.02	60±1	266.8±0.2	66±1	37	55±1	36	
$C_4H_{10}S$	2-Butanethiol[20]	133.01±0.06	93±1	358.10±0.01	83±1	40	71±2	36	
$C_4H_{10}S$	2-Methyl-1-propanethiol[21]	128.31±0.06	93.4±1.0	361.62±0.01	77±1	34	64±2	33	
C_5H_{10}	1-Pentene[22]	107.91±0.01	70±1	303.10±0.02	71±1	36	54±2	37	
$C_5H_{11}Br$	3-Bromopentane[23]	167.3±0.1	107.4±1.0	386±5	76±1		48±1		
C_5H_{12}	2-Methyl-butane[24]	113.33±0.18	65±1	301.0±0.3	68±1	41	34±2	40	70±1
C_6H_{14}	2-Methyl-pentane[25]	119.55±0.12	78.9±1.0	333.36±0.02	67±1	48	54±2	48	76±2
C_6H_{14}	3-Methyl-pentane[25]	110.26±0.02	75±1	336.40±0.02	72±1	51		53	
$C_7H_{12}O_2$	Cyclohexyl formate[3]	201.33±0.05	140±1	429±2	81±1	49	70±1	47	86±1
$C_8H_{14}O_2$	Butyl methacrylate[26]	197.78±0.01	131±1	437.0±0.9	111±1	66	70±1	65	
$C_8D_{14}O_2$	Perdeuteriobutyl methacrylate[26]	198.06±0.01	130±1		112±1		77±1		
C_7H_8	Toluene[27]	178.02±0.20	117±1	383.75±0.05	64±2	45	45±2	42	72±2
C_8H_{10}	Ethylbenzene[27]	178.20±0.02	115±1	409.32±0.02	76±2	43	47±3	41	77±2
C_9H_{12}	Isopropyl-benzene[28]	177.14±0.01	126±1	425.52±0.06	75±1	46	48±1	44	
$C_9H_{10}O_2$	3-Methyl toluate[29]	270.62±0.05	166.5±1.0	497.3±0.4	78±1	50	56±2	47	85±15
$C_9H_{16}O_2$	Cyclohexyl propanoate[30]	230.36±0.04	152.3±1.0	467±2	91±1	65	65±1	61	90±1
$C_{10}H_{18}O_2$	Cyclohexyl butanoate[3]	219.6±0.10	145.2±1.0	484±2	102±2	69	78±1	68	89±1
$C_{11}H_{20}O_2$	Cyclohexyl pentanoate[3]	222.4±0.08	145.8±1.0	505±3	112±2	80	75±2	78	93±2
$C_{10}H_{10}O_4$	Dimethyl phthalate[31]	274.18±0.02	190.3±1.0	555.88±0.03	92±2	67	81±2	61	116±18
$C_{12}H_{14}O$	Diethyl phthalate[32]	269.92±0.02		571.1±0.2				67	125±20
	Quenched		180.8±1.0		115 ± 2	86			
	Annealed		176.5±1.0		115 ± 2	86			

[a]The property information presented in this table represents a combination of the critically evaluated data presented in reference 4 or generated by the NIST Standard Reference Databases[12,13] and original experimental data from the listed sources.

18.4   Heat Capacity of Plastic Crystals as an Intermediate Phase at the Crystal-Liquid Transition

There are a large number of molecular crystals possessing very low enthalpy and entropy of fusion: $\Delta_{fus}H < 3RT_{fus}$ and $\Delta_{fus}S < 3R$, respectively. For these compounds, solid-phase transitions with $\Delta_{trs}H \gg 3RT$ and $\Delta_{trs}S \gg 3R$ are observed. These transitions are accompanied by changes in physical properties such as symmetry, lattice parameters, density, plasticity, and dielectric permeability. If in the high-temperature crystalline phase, orientational disorder of molecules occurs and internal rotation and ring inversion become possible, then this phase is called a "plastic crystal". The temperature range of existence of plastic crystals for different compounds depends strongly on the nature of the molecules (Figure 18.6).

At the transition rigid crystal→plastic crystal, an increase of entropy and heat capacity occurs (Table 18.3). This can be explained only by the existence of discrete energetically non-equivalent orientations. The changes of heat capacity and entropy at the considered solid-phase transition can be expressed as:[34]

$$\Delta_{trs}S = \Delta_v S + S_{conf} + S_{or} \tag{39}$$

$$\Delta_{trs}C_p = \Delta_v C_p + C_{conf} + C_{or} \tag{40}$$

where C_{or} and S_{or} are contributions to the heat capacity and entropy related to existence of various orientations, $\Delta_v S$ and $\Delta_v C_p$ are contributions caused by volume changes. It was previously demonstrated for cyclohexane derivatives[34] that $\Delta_v C_p \approx 0$ and $\Delta_v S \approx 0.4\, \Delta_{trs}S$

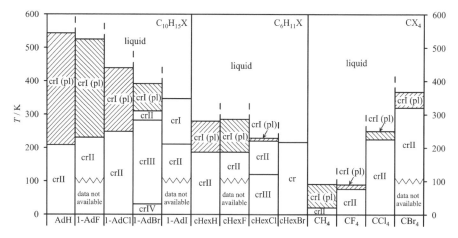

Figure 18.6   Phase behavior of halogen derivatives of adamantane (AdH), cyclohexane (cHexH), and methane. "pl" designates plastic crystal phase.

To connect the S_{or} and C_{or} contributions with the characteristics of orientational disorder the following assumptions were made:

1. in a plastic crystal there exists a number of energetically non-equivalent rotational orientations;
2. the lowest energy level is not degenerate, i.e., there is only one (basic) orientation in the crystal;
3. the entropy of ensembles consisting of molecules in one type of rotational orientation is postulated to be the same;
4. occupation of the energy levels of orientation does not depend on molecule conformation.

There are only two thermodynamic characteristics of orientational disorder obtained from experiment by Equations (39) and (40): S_{or} and C_{or}. Therefore, the calculations can be reduced to the search of two parameters of disorder: the number of non-equivalent orientations n_{or}, and the energy characteristic $\Delta_{or}E$. This allows one to calculate all the energy levels based on a particular algorithm. The values of n_{or} and $\Delta_{or}E$ are found either by direct solution of Equations (39) and (40) or as an intersection point of iso-entropic $S_{or} = \text{const}$ and iso-heat-capacity $C_{or} = \text{const}$ curves plotted from Equations (39) and (40) in the $n_{or} - \Delta_{or}H$ coordinates (Figure 18.7).

The difference in energy between the two lowest energy levels can be designated as $\Delta_{or}E$. The difference between every preceding k and following $(k+1)$ energy levels can be presented as:

$$\Delta_k^{k+1}E = f(k) \cdot \Delta_{or}E \qquad (41)$$

where $f(k)$ is a function.

When the $f(k)$ function is selected, the orientational contributions are calculated by use of Equations (7) and (8). Here x_i is the mole fraction of the ith orientation, and $\Delta H_{ij} \approx \Delta E_{ij}$ is the energy difference between two ensembles composed of the molecules in the ith and jth orientations, respectively. The enthalpy difference can be substituted by the corresponding energy difference:

$$\Delta_i^j H \approx \Delta_i^j E = \sum_{k=i}^{j-1} f(k) \cdot \Delta_{or}E \qquad (42)$$

The following expressions were obtained[51] for orientational contributions to changes of entropy and heat capacity at the phase transition; rigid crystal $\rightarrow$ plastic crystal:

$$S_{or} = -R \ln x_1 + x_1 \frac{\Delta_{or}E}{T} \cdot \sum_{i=2}^{n_{or}} \left[\left(\sum_{j=1}^{i-1} f(j) \right) \cdot \exp\left(-\frac{\Delta_{or}E}{RT} \cdot \sum_{j=1}^{i-1} f(j) \right) \right] \qquad (43)$$

$$C_{or} = R \cdot \left(x_1 \frac{\Delta_{or}E}{RT} \right)^2 \cdot \sum_{i=1}^{n_{or}-1} \left[\sum_{j=i}^{n_{or}-1} \left(\left\{ \sum_{k=i}^{j} f(k) \right\}^2 \cdot \exp\left\{ -\frac{\Delta_{or}E}{RT} \cdot \left(2 \cdot \sum_{k=1}^{j} f(k) - \sum_{k=i}^{j} f(k) \right) \right\} \right) \right]$$

$$(44)$$

Table 18.3　Temperatures, enthalpies, entropies, and heat capacity changes for solid-phase transitions and fusion of some organic compounds forming plastic crystals

Compound name	Transition[a]	T_{trs}/K	$\Delta_{trs}H°$/kJ mol^{-1}	$\Delta_{trs}S$ J K^{-1} mol^{-1}	$\Delta_{trs}C_p$ J K^{-1} mol^{-1}	Method[b], ref.
Tetrafluoromethane	crII–(crI)	76.23±0.04	1.709±0.003	22.42±0.04	4.2±0.2	AC[35]
	(crI)–liq	89.56±0.04	0.712±0.003	7.95±0.04	8.2±0.4	
Tetrachloromethane	crII–(crI)	225.6	4.60±0.21	20.4±0.9	8±2	HC[36]
	(crI)–liq	250.4	2.43±0.08	9.71±0.3	7±2	
	crII–(crI)	225.35±0.05	4.58±0.03	20.3±0.1	4.2±0.2	AC[37]
	(crI)–liq	250.3±0.2	2.51±0.05	10.1±0.2	5.5±0.4	
Neopentane	crII–(crI)	140.0±0.5	2.58±0.04	18.4±0.3	13.1±1.8	AC[38]
	(crI)–liq	256.5±0.2	3.26±0.05	12.7±0.2	15.9±3.6	
	crII–(crI)	140.5±0.5	2.630±0.002	18.72±0.01	13.8±0.4	AC[39]
	(crI)–liq	256.76±0.05	3.096±0.006	12.06±0.02	18.3±0.5	
2,3-Dimethylbutane	crII–(crI)	136.1±0.1	6.49±0.01	47.7±0.1	35.7±0.5	AC[25]
	(crI)–liq	145.19±0.06	0.801±0.001	5.52±0.01	7.2±0.5	
	crIII–(crI)	107±2	2.37±0.01	22.2±0.1	31.0±0.3	AC[40]
	crII–(crI)	136.02±0.02	6.427±0.006	47.24±0.04	36.2±0.3	
	(crI)–liq	145.04±0.02	0.794±0.004	5.47±0.03	6.3±0.4	
2,2-Dimethylbutane	crIII–(crII)	126.81±0.07	5.41±0.01	42.7±0.1	29.9±0.6	AC[25]
	(crII)–(crI)	140.79±0.07	0.285±0.004	2.03±0.03	3.0±0.8	
	(crI)–liq	174.28±0.06	0.579±0.004	3.32±0.02	11.3±0.4	
	crIII–(crII)	126.81±0.07	5.39±0.03	42.5±0.2	30.1±0.5	AC[41]
	(crII)–(crI)	140.88±0.02	0.283±0.008	2.01±0.06	3.3±0.5	
	(crI)–liq	174.16±0.07	0.579±0.006	3.33±0.03	10.7±0.5	

2,2,3-Trimethylbutane	crIV–crIII	86.8				
	crIII–crII	108.0				
	crII–(crI)	121.4±0.3	2.45±0.03	20.2±0.2	7.9±0.4	AC[42]
	(crI)–liq	248.57±0.07	2.261±0.008	9.10±0.03	0.5±0.5	
Cyclohexane	crII–(crI)	186.09±0.02	6.69±0.02	35.9±0.1	14.7±0.4	AC[43]
	(crI)–liq	279.84±0.02	2.628±0.005	9.38±0.02	17.1±0.6	AC[44]
	crII–(crI)	186.1±0.1	6.740±0.004	36.2±0.02	14.4±0.4	
	(crI)–liq	279.82±0.05	2.677±0.004	9.56±0.01	14.9±0.6	
Cyclohexanone	crII–(crI)	220.83±0.01	8.66±0.01	39.22±0.05	23.8±0.4	AC[45]
	(crI)–liq	245.21±0.01	1.328±0.002	5.41±0.01	10.6±0.4	
Adamantane	crII–(crI)	208.62±0.02	3.38±0.01	16.20±0.05	8.6±0.2	AC[46]
	(crI)–liq	543.2±0.2	14.0±0.3	25.7±0.5	0	DSC[47]
2-Adamantanone	crII–(crI)	216.4±0.1	7.63±0.01	35.24±0.06	11.4±0.8	AC[48]
	(crI)–liq	557.5±0.2	11.8±0.2	21.1±0.4	0	DSC[48]
1-Chloroadamantane	crII–(crI)	248.7±0.1	4.30±0.02	17.3±0.1	16.5±0.5	AC[49]
	(crI)–liq	439.7±0.1	5.53±0.01	12.6±0.1	6.7±0.9	
1-Bromoadamantane	crIV–crIII	31.0±0.1	0.0014±0.0003	0.05±0.01	0	AC[50]
	crIII–crII	282.2±0.1	1.39±0.01	4.92±0.05	0	
	crII–(crI)	309.9±0.1	7.416±0.007	23.93±0.02	24.5±1.3	
	(crI)–liq	391.8±0.2	3.97±0.08	10.1±0.2	5±2	DSC[50]

[a]Plastic crystal phases are in parentheses;
[b]AC, adiabatic calorimetry; HC, heat-conduction calorimeter; DSC, differential scanning calorimetry;

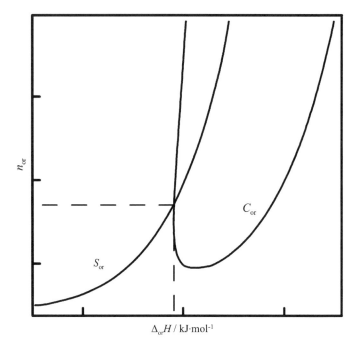

Figure 18.7   Typical iso-entropic $S_{or} = \text{const}$ and iso-heat-capacity $C_{or} = \text{const}$ curves in coordinates $n_{or}-\Delta_{or}H$ for the "rigid crystal $\rightarrow$ plastic crystal" phase transition.

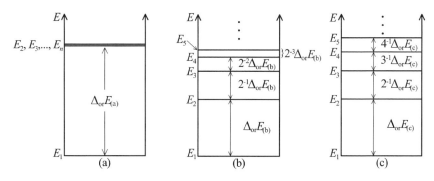

Figure 18.8   Various cases of energy levels distribution in plastic crystal.

where

$$x_1 = \left[1 + \sum_{i=2}^{n_{or}} \exp\left(-\frac{\Delta_{or}E}{RT} \cdot \sum_{j=1}^{i-1} f(j) \right) \right]^{-1} \quad (45)$$

Equations (43) and (44) are applicable for any $f(k)$ function. This approach for description of the energy levels in plastic crystals of organic compounds can quantitatively describe the changes both in entropy and heat capacity at the transition from the ordered to the plastic phase.

Various cases of energy level distributions in a plastic crystal are illustrated on Figure 18.8:

Case (a)

There is one (basic) orientation and $(n_{or} - 1)$ (plastic) orientations. All plastic orientations are geometrically distinguishable, but equal in energy. In this case, the $f(k)$ function can be written as:

$$f(k) = \begin{cases} 1, & \text{if } k = 1 \\ 0, & \text{if } k > 1 \end{cases}$$

Case (b)

The difference between neighboring orientational energy levels decreases with energy increase: $f(k) = 2^{1-k}$

Case (c)

Similar to Case (b), but $f(k)$ is defined as: $f(k) = k^{-1}$

In the selected cases of the energy level distribution in a plastic crystal, discussed above, the distributions are either limited at $n_{or} \to \infty$ (Cases (a) and Cases (b)) or characterized by a gradual energy level density increase (Case (c)).

The dependencies of orientational contributions S_{or} and C_{or} of $\Delta_{or}E$ for various functions $f(k)$ and numbers of non-equivalent orientations are demonstrated in Figures 18.9 and 18.10.

The results of calculations for the characteristics of orientational disorder $\Delta_{or}E$ and n_{or} are presented in Table 18.4. For symmetrical molecules (T_d point group), as can be expected, n_{or} and $\Delta_{or}E$ are small. For molecules of lower symmetry the numbers of orientations are sometimes very large, and can be related to the interference of the rotating molecules. The increase in polarity and size of molecules results in an increase of a number of orientations, as well. Similar trends are observed for the energy differences between orientations for all $f(k)$.

The data of Table 18.4 do not allow selection of the preferable set of orientational energy levels. Other similar sets corresponding to the experimental data can be proposed. However, the trends in the character of the orientational disorder as a function of the molecular structure were revealed. The $f(k)$ functions with unlimited increase in energy, such as $f(k) = 1, f(k) = k$, $f(k) = 2^{k-1}$, etc., do not provide agreement between S_{or} and C_{or} obtained from the experimental data.

It should also be noted that for the compounds presented in Table 18.3, the sum of heat capacity changes at solid-phase transitions and fusion $\sum_i \Delta_{trs} C_P =$ 10 to $45\,\mathrm{J\,K^{-1}\,mol^{-1}}$, which is significantly lower than $\Delta_{cr}^{liq} C_P^\circ$ for the compounds presented in Table 18.2.

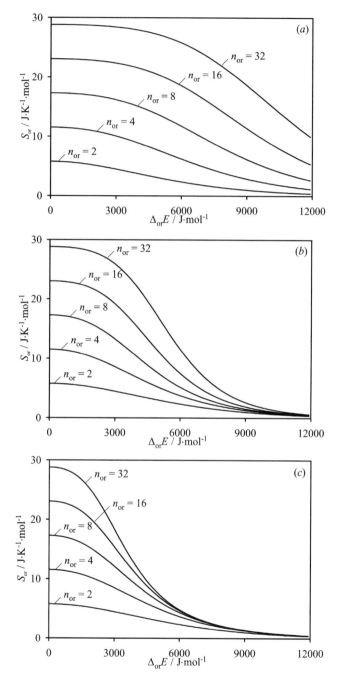

Figure 18.9   Dependences of orientational contribution S_{or} of the $\Delta_{or}E$ for various numbers of non-equivalent orientations of molecules in plastic crystal at $T = 298.15$ K.

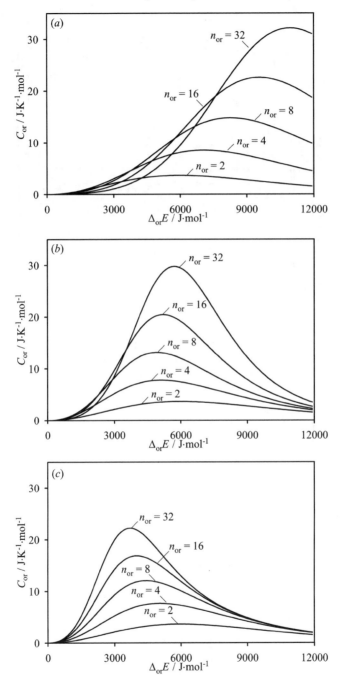

Figure 18.10   Dependences of orientational contribution C_{or} of the $\Delta_{or}E$ for various numbers of non-equivalent orientations of molecules in plastic crystal at $T = 298.15$ K.

Table 18.4 The number of non-equivalent orientations n_{or} and the difference in energy levels $\Delta_{or}E$ in plastic crystals

Compound name		Point group, symmetry number	$\dfrac{T_{trs}}{K}$	Model					
				(a)		*(b)*		*(c)*	
				n_{or}	$\dfrac{\Delta_{or}E}{\text{kJ mol}^{-1}}$	n_{or}	$\dfrac{\Delta_{or}E}{\text{kJ mol}^{-1}}$	n_{or}	$\dfrac{\Delta_{or}E}{\text{kJ mol}^{-1}}$
Tetrafluoromethane		T_d, 12	76.2[35]	5	0	5	0	5	0
Tetrachloromethane		T_d, 12	225.4[37]	10	4.8	11	2.8	13	2.4
Neopentane		T_d, 12	140.0[38]	9	2.6	9	1.6	10	1.4
2,3-Dimethylbutane	*trans*	C_{2h}, 2	136.0[40]	58	4.8	62	2.5	156	1.5
	gauche	C_2, 2							
2,2-Dimethylbutane		C_s, 1	140.9[41]	100	4.9	105	2.5	190	1.3
2,2,3-Trimethylbutane		C_s, 1	121.4[42]	41	4.1	45	2.2	108	1.4
Cyclohexane	*chair*	D_{3d}, 6	186.1[43,44]	19	4.4	20	2.4	25	1.8
	twist	D_2, 4							
Cyclohexanone	*chair*	C_s, 1	220.8[45]	49	6.6	52	3.4	78	2.0
	twist1	C_2, 2							
	twist2	C_1, 1							
Adamantane		T_d, 12	208.6[46]	6	4.1	7	2.7	7	2.5
2-Adamantanone		C_{2v}, 2	216.4[48]	24	4.9	25	2.6	30	1.8
1-Chloroadamantane		C_{3v}, 3	248.7[49]	20	8.1	23	4.6	67	3.6
1-Bromoadamantane		C_{3v}, 3	309.9[50]	82	12.0	89	6.3	373	3.5

18.5 Ionic Liquids

Room temperature ionic liquids (ILs) have attracted significant attention from scientists and engineers primarily because of their extremely low vapor pressures, and therefore, their potential for replacing traditional solvents. ILs are salts that remain liquid below 100 °C. Both van der Waals and Coulombic interactions have significant contributions to the physical properties of the ILs. The molar heat capacity of ILs is much higher than that of molecular liquids (Figure 18.11). The heat capacity changes glass→liquid at T_g; crystal→liquid and gas→liquid at T_{fus} for ILs presented in Table 18.5, are on average 29 %, 16 % and 25 % respectively, of $C_{p,liq}$; that is, substantially lower than for molecular liquids presented in Table 18.2 (45 %, 32 %, and 44 %, respectively).

The absolute values of the heat capacity changes for ILs not containing the bis(trifluoromethylsulfonyl)imide anion (NTf_2^-) are comparable with those for molecular liquids (Table 18.5). The conformational contributions to the heat capacity change at the glass transition and melting of these compounds are caused by conformational transformations in an alkyl chain of the cation and do not exceed 15 % of the $\Delta_{gl}^{liq}C_{p,m}^o(T_g)$ or 18 % of the $\Delta_{cr}^{liq}C_{p,m}^o(T_{fus})$.

NTf_2^--containing ILs possess much higher heat capacities (Figure 18.11) than the other ILs presented in Table 18.5 or molecular liquids. Heat capacity changes for these ILs (Table 18.5) are also significantly higher than those for the other ILs and molecular liquids. The NTf_2^- anion, unlike the other anions presented in Table 18.5, forms conformers distinguishable in energy. However, the main contribution to such a significant increase in heat capacity change is the configurational, but not conformational contribution. At the same time, the increase in heat capacity changes in the homologous series of 1-alkyl-3-methylimidazolium NTf_2^- ([$C_n$mim]NTf_2, where n is a number of carbon atoms

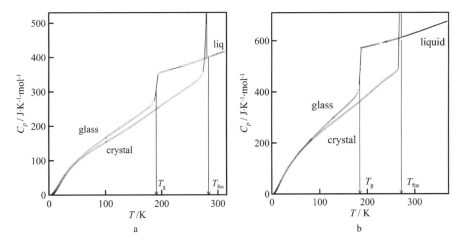

Figure 18.11   Heat capacity of ionic liquids: (a) 1-butyl-3-methylimidazolium hexafluorophosphate[54] and (b) 1-hexyl-3-methylimidazolium bis(trifluoromethylsulfonyl)imide.[57]

Table 18.5 Thermodynamic properties of ionic liquid glass-formers

Formula	Compound name	T_{fus}	T_g	$\Delta_{gl}^{liq} C_P^o (T_g)$	$\Delta_{cr}^{liq} C_P^o (T_{fus})$	$\Delta_g^{liq} C_P^o (T_{fus})$
		K			J K^{-1} mol^{-1}	
$C_8H_{15}BrN_2$	1-butyl-3-methylimidazolium bromide[52]	351.35±0.20	218.9±1.0	84±1	52±2	
$C_{10}H_{18}N_2O_2$	1-butyl-3-methylimidazolium acetate[53]		203.6±1.0	110±2		
$C_{10}H_{15}F_3N_2O_2$	1-butyl-3-methylimidazolium trifluoroacetate[53]	296.42±0.09	187.1±1.0	107±2	47±2	
$C_8H_{15}F_6N_2P$	1-butyl-3-methylimidazolium hexafluorophosphate[54]	283.51±0.12	190.6±1.0	82±2	45±2	95±10
$C_{10}H_{15}F_6N_3O_4S_2$	1-butyl-3-methylimidazolium bis(trifluoromethylsulfonyl) imide[55,56]	270.22±0.05	181.5±1.0	155±3	104±3	143±14
$C_{12}H_{19}F_6N_3O_4S_2$	1-hexyl-3-methylimidazolium bis(trifluoromethylsulfonyl) imide[57,58]	272.13±0.05[a]	184.3±1.0	171±3	111±2[a]	156±16
$C_{14}H_{23}F_6N_3O_4S_2$	1-octyl-3-methylimidazolium bis(trifluoromethylsulfonyl) imide[59]	263.96±0.05[a]	185.0±1.0	198±3	134±3[a]	168±17

[a]for crystalline modification with highest T_{fus}

in the alkyl substituent) ILs is caused mainly by the increase of conformational contribution due to a higher number of tops in the cation.

The formation of a large number of polymorphic modifications with different T_{fus} is characteristic for the [C$_n$mim]NTf$_2$ series. For example, for [C$_2$mim]NTf$_2$, four crystalline modifications were reported,[59] for [C$_6$mim]NTf$_2$ – one stable modification and one metastable sequence of crystalline phases with a solid-phase transition,[58,60] for [C$_8$mim]NTf$_2$ – three sequences of crystalline phases with a solid-phase transition in each sequence (Figure 18.12).[59] The differences in Gibbs energies between these polymorphs are within a few kJ mol^{-1}, and the crystalline modification stable at one temperature may become metastable when the temperature is changed.[59] It was shown[61] that the NTf$_2^-$ anion forms different conformations in various crystalline modifications of [C$_2$mim]NTf$_2$. Therefore, it might be assumed that polymorphism in [C$_n$mim]NTf$_2$ is related to the conformational flexibility of the anion.

The empirical relationship $T_g/T_{fus} = 2/3$, holding well for molecular liquids, is also valid for ILs (Table 18.5). It is slightly lower (0.62–0.66) for hydrophilic ILs ([C$_4$mim]Br, [C$_4$mim]CF$_3$COO) and higher (0.67–0.70) for hydrophobic ILs ([C$_4$mim]PF$_6$, [C$_n$mim]NTf$_2$).

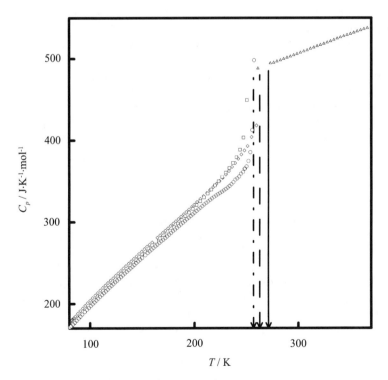

Figure 18.12 Experimental heat capacities for [C$_2$mim][NTf$_2$]: $\triangle$, liquid; $\diamondsuit$, crystal I; $\bigcirc$, crystal III; $\square$, crystal IV. The fusion temperatures designated by narrowed lines: ___, crystal I; _ _ _, crystal III; _._, crystal IV.[59]

Acknowledgment

For M. F. this work represents an official contribution of the U.S. National Institute of Standards and Technology and is not subject to copyright in the United States.

References

1. C. A. Angell, M. Oguni and W. J. Sichina, *J. Phys. Chem.*, 1982, **86**, 998.
2. H. G. Carlson and E. F. Westrum, *J. Chem. Phys.*, 1971, **54**, 1464.
3. A. A. Kozyro, A. V. Blokhin, G. J. Kabo and Y. U. Paulechka, *J. Chem. Thermodyn.*, 2001, **33**, 305.
4. Y. U. Paulechka, Dz. H. Zaitsau, G. J. Kabo and A. V. Blokhin, *Zh. Fiz. Khim.*, 2003, **77**, 791.
5. *TRC Thermodynamic Tables – Hydrocarbons and – Non-Hydrocarbons*, NIST, Boulder, Colorado, USA, 2008.
6. J. D. van der Waals and P. Kohnstamm, *Lehrbuch der Thermostatik Teil 1*, Verlag von Johann Ambrosius Barth, Leipzig, 1927.
7. G. Ya. Kabo, G. N. Roganov and M. L. Frenkel, Thermodynamics and Equilibria of Isomers, in *Thermochemistry and Equilibria of Organic Compounds,* ed. M. Frenkel, VCH, New York, 1993.
8. E. L. Eliel, N. L. Allinger, S. J. Angyal and G. A. Morrison, *Conformational Analysis,* Interscience, New York, 1965.
9. *Internal Rotation in Molecules*, ed. W. J. Orville-Thomas, John Wiley & Sons, New York, 1974.
10. P. A. Poleshchuk, G. Y. Kabo and M. L. Frenkel, *Zh. Fiz. Khim.*, 1988, **62**, 1105; G. J. Kabo, I. A. Yursha, M. L. Frenkel, P. A. Poleschuk, V. I. Fedoseenko and A. I. Ladutko, *J. Chem. Thermodyn.*, 1988, **20**, 429.
11. G. J. Kabo, A. V. Blokhin, A. A. Kozyro, V. V. Diky, L. S. Ivashkevich, A. P. Krasulin, V. M. Sevruk and M. Frenkel, *Thermochimica Acta*, 1998, **313**, 111.
12. E. W. Lemmon, M. L. Huber and M. O. McLinden. *REFPROP-Reference Fluid Thermodynamic and Transport Properties. NIST Standard Reference Database 23, Version 8*. NIST, Boulder, Colorado, USA, 2007. http://www.nist.gov/data/nist23.htm.
13. M. Frenkel, R. D. Chirico, V. Diky, C. D. Muzny, E. W. Lemmon and A. Kazakov. *NIST ThermoData Engine, NIST Standard Reference Database 103a-Pure compounds, Version 2.1*. NIST, Boulder, Colorado, USA, 2008. http://www.nist.gov/srd/nist103a.htm.
14. Y. U. Paulechka, Dz. H. Zaitsau and G. J. Kabo, *J. Mol. Liq.*, 2004, **115**, 105.
15. E. A. Guggenheim, *Mixtures,* Oxford University Press, Oxford, 1952.
16. A. J. Staverman, *Rec. Trav. Chim. Pays-Bas*, 1950, **69**, 163.
17. B. V. Lebedev and T. G. Kulagina, *J. Therm. Anal.*, 1998, **54**, 731.

18. J. G. Aston, H. L. Finke, A. B. Bestul, E. L. Pace and G. J. Szasz, *J. Am. Chem. Soc.*, 1946, **68**, 52.

19. K. Takeda, O. Yamamuro and H. Suga, *J. Phys. Chem. Solids*, 1991, **52**, 607.

20. J. P. McCullough, H. L. Finke, D. W. Scott, R. E. Pennington, M. E. Gross, J. F. Messerly and G. Waddington, *J. Am. Chem. Soc.*, 1958, **80**, 4786.

21. D. W. Scott, J. P. McCullough, J. F. Messerly, R. E. Pennington, I. A. Hossenlopp, H. L. Finke and G. Waddington, *J. Am. Chem. Soc.*, 1958, **80**, 55.

22. K. Takeda, O. Yamamuro and H. Suga, *J. Phys. Chem.*, 1995, **99**, 1602.

23. S. Takahara, O. Yamamuro and T. Matsuo, *J. Phys. Chem.*, 1995, **99**, 9589.

24. M. Sugisaki, K. Adachi, H. Suga and S. Seki, *Bull. Chem. Soc. Jpn*, 1968, **41**, 593.

25. D. R. Douslin and H. M. Huffman, *J. Am. Chem. Soc.*, 1946, **68**, 1704.

26. B. V. Lebedev, T. G. Kulagina and N. N. Smirnova, *J. Chem. Thermodyn.*, 1994, **26**, 941.

27. O. Yamamuro, I. Tsukushi, A. Lyndqvist, S. Takahara, M. Ishikawa and T. Matsuo, *J. Phys. Chem. B*, 1998, **102**, 1605.

28. K. Kishimoto, H. Suga and S. Seki, *Bull. Chem. Soc. Jpn*, 1973, **46**, 3020.

29. A. V. Blokhin, Y. U. Paulechka, G. J. Kabo and A. A. Kozyro, *J. Chem. Thermodyn.*, 2002, **34**, 29.

30. Y. U. Paulechka, Thermodynamics of Esters of Monocarboxylic Acids in the Condensed State, Ph.D. Thesis, Minsk, 2004.

31. I. B. Rabinovich, L. Ya. Martynenko and V. A. Maslova, *Tr. Khim. Khim. Tekhnol.*, 1969, No. 2, 10.

32. S. S. Chang, J. A. Horman and A. B. Bestul, *J. Res. NBS, Ser. A*, 1967, **71A**, 293.

33. E. A. DiMarzio and F. Dowell, *J. Appl. Phys.*, 1979, **50**, 6061.

34. G. J. Kabo, A. A. Kozyro, M. Frenkel and A. V. Blokhin, *Mol. Cryst. Liq. Cryst.*, 1999, **326**, 333.

35. J. H. Smith and E. L. Pace, *J. Phys. Chem.*, 1969, **73**, 4232.

36. D. R. Stull, *J. Am. Chem. Soc.*, 1937, **59**, 2726.

37. J. F. G. Hicks, J. G. Hooley and C. C. Stephenson, *J. Am. Chem. Soc.*, 1944, **66**, 1064.

38. J. G. Aston and G. H. Messerly, *J. Am. Chem. Soc.*, 1936, **58**, 2354.

39. H. Enokida, T. Shinoda and Y. Mashiko, *Bull. Chem. Soc. Jpn.*, 1969, **42**, 84.

40. K. Adachi, H. Suga and S. Seki, *Bull. Chem. Soc. Jpn.*, 1971, **44**, 78.

41. J. E. Kilpatrick and K. S. Pitzer, *J. Am. Chem. Soc.*, 1946, **68**, 1066.

42. H. M. Huffman, M. E. Gross, D. W. Scott and J. P. McCullough, *J. Phys. Chem.*, 1961, **65**, 495.

43. J. G. Aston, G. J. Szasz and H. L. Finke, *J. Am. Chem. Soc.*, 1943, **65**, 1135.

44. R. A. Ruehrwein and H. M. Huffman, *J. Am. Chem. Soc.*, 1943, **65**, 1620.

45. N. Nakamura, H. Suga and S. Seki, *Bull. Chem. Soc. Jpn.*, 1980, **53**, 2755.
46. S.-S. Chang and E. F. Westrum Jr., *J. Phys. Chem.*, 1960, **64**, 1547.
47. G. J. Kabo, A. V. Blokhin, M. B. Charapennikau, A. G. Kabo and V. M. Sevruk, *Thermochim. Acta*, 2000, **345**, 125.
48. A. B. Bazyleva, A. V. Blokhin, G. J. Kabo, A. G. Kabo and V. M. Sevruk, *Thermochim. Acta*, 2006, **451**, 65.
49. K. Kobashi, T. Kyomen and M. Oguni, *J. Phys. Chem. Solids*, 1998, **59**, 667.
50. A. B. Bazyleva, A. V. Blokhin, G. J. Kabo, A. G. Kabo and Y. U. Paulechka, *J. Chem. Thermodyn.*, 2005, **37**, 643.
51. A. B. Bazyleva, G. J. Kabo and A. V. Blokhin, *Physica B: Condens. Matter*, 2006, **383**, 243.
52. Y. U. Paulechka, G. J. Kabo, A. V. Blokhin, A. S. Shaplov, E. I. Lozinskaya and Ya. S. Vygodskii, *J. Chem. Thermodyn.*, 2007, **39**, 158.
53. A. A. Strechan, Y. U. Paulechka, A. V. Blokhin and G. J. Kabo, *J. Chem. Thermodyn.*, 2008, **40**, 632.
54. G. J. Kabo, A. V. Blokhin, Y. U. Paulechka, A. G. Kabo, M. P. Shymanovich and J. W. Magee, *J. Chem. Eng. Data*, 2004, **49**, 453.
55. A. V. Blokhin, Y. U. Paulechka, A. A. Strechan and G. J. Kabo, *J. Phys. Chem. B*, 2008, **112**, 4357.
56. Y. Shimizu, Y. Ohte, Y. Yamamura and K. Saito, *Chem. Lett.*, 2007, **36**, 1484.
57. A. V. Blokhin, Y. U. Paulechka and G. J. Kabo, *J. Chem. Eng. Data*, 2006, **51**, 1377.
58. Y. Shimizu, Y. Ohte, Y. Yamamura, K. Saito and T. Atake, *J. Phys. Chem. B*, 2008, **110**, 13970.
59. Y. U. Paulechka, A. V. Blokhin, G. J. Kabo and A. A. Strechan, *J. Chem. Thermodyn.*, 2007, **39**, 866.
60. D. G. Archer, National Institute of Standards and Technology, Investigation Report-6645, Gaithersburg, MD, 2006.
61. Y. U. Paulechka, G. J. Kabo, A. V. Blokhin, A. S. Shaplov, E. I. Lozinskaya, D. G. Golovanov, K. A. Lyssenko A. A. Korlynkov and Ya. S. Vygodskii, *J. Phys. Chem. B*, 2009, **113**, 9538.

CHAPTER 19

The Estimation of Heat Capacities of Pure Liquids

MILAN ZÁBRANSKÝ,[a] ZDEŇKA KOLSKÁ,[b] VLASTIMIL RŮŽIČKA[a] AND ANATOL MALIJEVSKÝ[a]

[a] Department of Physical Chemistry, Institute of Chemical Technology, Technická 5, 166 28 Prague 6, Czech Republic; [b] Department of Chemistry, Faculty of Science, J. E. Purkinje University, České Mládeže 8, 400 96 Ústí nad Labem, Czech Republic

19.1 Introduction

The heat capacities of liquids are important thermodynamic quantities and are necessary in many industrial and theoretical calculations. Heat capacities must be known in order to evaluate heat balances and the temperature dependence of many thermodynamic properties. It is also needed for calculations related to the molecular structure of liquids, *etc.*[1]

Heat capacity, C, can be measured directly and the heat capacities of liquids obtained by calorimetric measurement are known for almost 2000 organic and inorganic pure liquids having melting points below 573 K.[2-4] Unfortunately, most of the data has been determined at only one temperature (mostly at 298.15 K), or in a narrow temperature range, and about 10 % of the data presented in the compilations[2-4] are either old values (data obtained before 1920) or inaccurate values. Furthermore, for many compounds, some which are important for industrial applications and environmental calculations, heat capacity data are missing.[5] However, unavailable heat capacity data can be determined by using estimation methods and there are more than 50 methods,

Heat Capacities: Liquids, Solutions and Vapours
Edited by Emmerich Wilhelm and Trevor M. Letcher
© The Royal Society of Chemistry 2010
Published by the Royal Society of Chemistry, www.rsc.org

described in the open literature. Most of them provide values of isobaric heat capacity C_p, which is useful in many industrial calculations.

19.2 Survey of Estimation Methods for Liquid Heat Capacity

All estimation methods can be divided into two groups depending on the required input data.[6] QPPR methods (Quantity-Property-Property-Relationship) require the values of some physicochemical properties in order to estimate heat capacities. Methods based on the corresponding states theorem (CST), thermodynamic methods and empirical methods belong to this group. These first two methods have been largely developed to calculate the difference between the liquid heat capacity C_p^l and the heat capacity of an ideal gas $C_p^{g,id}$:

$$\Delta C_p = C_p^l - C_p^{g,id} \tag{1}$$

which can be described exactly by the thermodynamic equation:

$$\Delta C_p = -R - T \int_V^{\infty} \left[(\partial^2 p / \partial T^2)_V \right]_T dV - T(\partial p / \partial T)_V^2 / (\partial p / \partial V)_T \tag{2}$$

where R is universal gas constant. The isobaric heat capacity of an ideal gas can be found either from the literature[7] or estimated using Benson's group contribution method.[8,9]

The second group involve the QSPR[6,10] methods (Quantity-Structure-Property-Relationship). These methods need only a knowledge of the chemical structure of a compound to predict the heat capacity of a liquid. QSPR methods use some structural characteristics, such as number of fragments (atoms, bonds or group of atoms in a molecule), topological indices (see e.g. Reference 11) or other structural information to express the relation between the molecular property and the molecular structure of the compound (see Section 19.2.5).

19.2.1 Theoretical Methods

Estimation methods discussed in the next sections of this chapter are based on a generalization of experimental results and/or phenomenological thermodynamics. This section deals with the more fundamental approaches of statistical thermodynamics.[12] The aim of the statistical thermodynamics is to calculate thermodynamic quantities from intermolecular interactions among molecules forming the bulk of a system.

The simplest example is a model of an ideal gas where the particles do not mutually interact. In this case the energy of a molecule is the sum of translational, rotational, vibrational, *etc.*, energies and the heat capacity is the sums of related contributions. The contributions may be calculated from the knowledge of the geometry of a given molecule and its fundamental frequencies. These calculations are in excellent agreement with calorimetric data. The agreement is typically within 1 %. In simple cases the calculated values are even more accurate than the experimental data.[13]

A good agreement is also found at the opposite extreme to the low density, high temperature (ideal) gas; namely, the ideal crystal (high density, low temperature). Here, the models of the Einstein ideal crystal, the Debye ideal crystal and their extensions, give good results.[14]

It is much more difficult to estimate heat capacities from first principles for dense gases and liquids. Methods for their determination are the subject of intense research. The tools used can be divided into two groups; computer simulations and theoretical approaches. Computer simulations (Monte Carlo and molecular dynamics),[15] have some features which are related to real experiments (they are sometimes called the pseudo experiments). Their advantage is an absence of any approximations. On the other hand they are demanding on computer time and they do not provide an explanation of the bulk behaviour.

Presently, the mainstream theoretical approaches falls into two groups: perturbation methods and theories of integral equations.[16] In a perturbation method, a reference system (typically hard spheres) is chosen. Thermodynamic quantities are calculated as deviations from the reference system. The theories of integral equations (typically the Ornstein-Zernike equation approaches) are based on relations among the pair correlation functions describing the internal molecular structure of the macroscopic thermodynamic system.[17]

The theoretical methods involve approximations whose reliability must be tested. Besides, they involve models of intermolecular interactions whose reliability must also be tested. In this field, statistical thermodynamics together with quantum chemistry is used in order to calculate pair and three body intermolecular potentials from first principles.[18] Remarkable results in this field have been reached for the simplest real systems such as argon[18,19] and other rare gases,[20,21] nitrogen, oxygen, carbon monoxide, carbon dioxide and other simple systems. However, for the more challenging fluids such as those found in chemical engineering applications (*i.e.* gases at high densities and liquids), there is a lack of data at the molecular level, and these methods do not produce reliable and quantitative results, but only qualitative estimates.

19.2.2 Thermodynamic Methods

Thermodynamic methods stem from an exact relationship between the difference of heat capacities of the liquid and the vapour, along the saturation curve, and the temperature derivative of the enthalpy of vaporization, $\Delta_{vap}H$. The first derivation was presented by Watson:[22]

$$\Delta C_\sigma = -\left(\partial \Delta_{vap}H / \partial T\right)_\sigma - \left[\partial\left(H^{g0} - H^g_\sigma\right)/\partial T\right]_\sigma \quad (3)$$

The second term in Equation (3) is the temperature derivative of the difference of the enthalpy of the ideal and real vapour. The main uncertainty comes from the temperature derivative of enthalpy of vaporization, the first term in Equation (3). This is particularly true below the room temperature and up to $T_r = 0.85$, where $T_r = T/T_c$. The method is recommended for vapours above the normal boiling temperature, where experimental data are rare and all other estimation methods usually fail.

Several authors[23-27] have proposed modifications to Equation (3) involving different ways of calculating the first and second terms. The modificatibn by Tyagi[28] was widely criticised and it is therefore not used.

Coniglio *et al.*[29] based their estimation method on Equation (2). They used the modified Peng-Robinson[30] equation of state (EOS) to calculate the derivatives of pressure with respect to temperature and to volume. Two additional parameters, in addition to critical temperature and critical pressure, are required for the application of the modified Peng-Robinson EOS, namely a pseudocovolume and a shape parameter, both of which can be calculated by the Bondi group contribution method.[31] This group contribution method was used in the work on hydrocarbons occurring in crude oil.[29] The method was used to estimate the vapour pressure, saturated liquid density, enthalpy of vaporization, and saturated liquid heat capacity of 134 compounds. The relative uncertainty was 0.5 % for vapour pressure and for liquid density and was satisfactory for the enthalpy of vaporization (the estimation error was 0.7 %, but for only 51 of the compounds due to insufficient experimental data). However the relative uncertainty was between 1 and 10 % for liquid heat capacity (for only 51 compounds) due to the sensitivity of the second derivative in Equation (2) to the input data. Coniglio *et al.*[32] presented a modification of the method, that included more input data (69 compounds), which also extended the method to include other properties.

Illoukani and Nikoobakht[33] used a similar approach to Coniglio *et al.*[29] and tested five equations of state to estimate the heat capacity for hydrocarbons and heterocyclic compounds. They concluded that the Schmidt and Wenzel EOS was the best.[34]

Diedrichs *et al.*[35] estimated the liquid heat capacity of 33 pure compounds, from their melting temperature to their normal boiling temperature, by using the group contribution approach for the parameters of the VTPR equation of state (volume translated Peng-Robinson equation of state). To improve the estimation error the authors included the so-called α-parameter (a function of reduced temperature) proposed by Twu *et al.* for Peng-Robinson EOS.[36]

19.2.3 Methods Based on the Corresponding State Theorem

The corresponding state theorem (CST) was introduced by van der Waals and extended by Pitzer.[37] It was modified and extended to include a wide range of input parameters and can predict a wide range of properties. Sakiadis and Coates[38] are considered to be the first to propose a CST method for the estimation of C_p^l. Their method was developed for alkanes only, and the estimation error was about 1 %.

Rowlinson[39] derived a relationship based on CST from the liquid molecular theory. Its final form was designed by Bondi.[40] This equation is presented in a dimensionless form in the book by Poling *et al.*[10]

$$\Delta C_p/R = 1.586 + 0.49/\tau + \omega\left[4.2775 + 6.3\tau^{1/3}/T_r + 0.4355/\tau\right] \quad (4)$$

where ω is an acentric factor defined by Pitzer and Curl[37], $\tau = (1-T_r)$, T_r is the reduced temperature $T_r = T/T_c$ where T_c is the critical temperature.

Sterling and Brown presented a similar simple equation.[40] Both of these relations need only critical property values and the acentric factor to estimate heat capacities. They are reliable only for non-polar compounds in the temperature range of $0.4 \le T_r \le 1$ with an average estimation error of 4 %. Morad *et al.*[41] tested the equation by Bondi[40] for the higher members of the homologous series of triacylglycerols and found that the estimation error was of about 5 %.

Yuan and Stiel[42] suggested two equations, one for non-polar liquids with two terms:

$$\Delta C_\sigma = \Delta C_\sigma^{(0)} + \omega\Delta C_\sigma^{(1)} \quad (5)$$

and one for polar liquids with six terms:

$$\Delta C_\sigma = \Delta C_\sigma^{(0)} + \omega\Delta C_\sigma^{(1)} + X\Delta C_\sigma^{(2)} + X^2\Delta C_\sigma^{(3)} + \omega^2\Delta C_\sigma^{(4)} + \omega X\Delta C_\sigma^{(5)} \quad (6)$$

where X is the so-called Stiel polar factor, defined as:

$$X = (\log p_r)_{T_r=0.6} + 1.70\omega + 1.552 \quad (7)$$

In this work, the terms, $\Delta C_\sigma^{(i)}$, from Equations (5) and (6) are presented as functions of reduced temperature $T_r \in \langle 0.4 \div 0.96 \rangle$. Equation (6) is considered to be one of the best for polar compounds. The estimation error ranges from 1 to 4 %.

Lee and Kesler[43] suggested a three-parameter generalized thermodynamic relation where the third parameter is the acentric factor ω:

$$\Delta C_p/R = \Delta C_V/R - 1 - T_r(\partial p_r/\partial T_r)^2_{V_r}/(\partial p_r/\partial V_r)_{T_r} \quad (8)$$

They used the modified Benedict-Webb-Rubin (BWR) equation of state to evaluate the last term in Equation (8) in an analytical form. The method provides values of several properties of pure compounds and of mixtures including the difference of isobaric ΔC_p and isochoric ΔC_V heat capacities[43] as a function of temperature and pressure. The method is reliable for non-polar compounds in a wide range of reduced temperatures approaching 1 and for reduced pressure up to 6. Filippov[44] proposed a relation valid to $T_r = 0.95$, which in addition to reduced temperature and critical volume uses an interatomic distance. However, as this last parameter is not always known, the method cannot be applied to molecules with a complex molecular structure.

The four-parameter CST method for calculating the difference between the molar heat capacity along the saturation curve and the ideal gas molar heat capacity at constant pressure ΔC_σ, was suggested by Lyman and Danner.[45] They used, in addition to the two reduced quantities T_r and p_r, two additional parameters, the gyration radius R_g, which is determined from molecular structure, and the association factor κ determined from critical properties, normal boiling temperature and the gyration radius:

$$\Delta C_\sigma = A_1 + T_r\left(A_2 + A_3 R_g\right) + T_r^5\left(A_4 + A_5 R_g\right) + A_6 R_g^2/T_r^2 + A_7 R_g/T_r^3$$
$$+ A_8/T_r^5 + \kappa\left(B_1 + B_2 T_r^2 + B_3 T_r^5\right) + \kappa^2\left(B_4 + B_5 T_r^2\right) \tag{9}$$

The additional parameters, R_g and κ, are difficult to obtain; in the original work they were tabulated, together with parameters A_1–A_8 and B_1–B_5, for 250 compounds. The equation for ΔC_σ is valid for the interval $T_r \in \langle 0.35 \div 0.96 \rangle$. The estimation error for non-polar compounds is around 2 % and for polar ones, around 5 %. However, for polar compounds the method is unreliable for $T_r > 0.7$, because at the time the method was being developed, there was insufficient reliable data available for calculating the adjustable parameters. Tarakad and Danner[46] proposed to replace the parameter κ by the parameter Φ, obtained from the second virial coefficient. This modification makes the method more widely applicable because there is more input data (second virial coefficient) available. The resulting estimation error is similar to that found for the original method by Lyman and Danner.[45]

Over the last two decades a few new methods based on CST have been proposed, see for example, Prasad *et al.*,[47] Al-Shorachi *et al.*[48] However, these methods either use an obsolete concept (molar refraction as input para-meters[47]), or were applied to, and tested on a small number of compounds.[48]

19.2.4 Empirical Methods

Empirical methods are losing their importance and are being replaced mainly by semi-empirical or semi-theoretical methods which are more generally applicable and more readily extendable to other families of compounds. We present here just two empirical methods as examples.

Pachaiyappan *et al.*[49] proposed a two-parameter relation to estimate heat capacity as a function of molar mass M. They presented parameters for their estimation equation, for nine homologous series (including alcohols, ketones, acids, amines, *etc.*). The relative uncertainty was up to 5 %, with a few exceptions.

Riazi and Roomi[50] presented several four-parameter relationships to esti-mate thermodynamic and transport properties using the refractive index as an input parameter. They presented sets of 4 constants to account for the tem-perature dependence of C_p for several families of hydrocarbons and reported a relative uncertainty up to 2 %. Their method is applicable also to undefined mixtures occurring typically in crude oil.

19.2.5 Group Contribution Methods

Group contribution methods are presently the most frequently used methods for the estimation of liquid heat capacity and are based solely on the knowledge of the chemical structure of the compound in question. The compound is divided into fragments which are usually atoms, bonds or group of atoms. A fragment has a partial value called a contribution. A property of a compound is obtained by summing up the values of the contributions in the molecule. Several methods have been proposed for the estimation of heat capacities of organic compounds[51–80] and inorganic compounds.[81,82]

The simplest methods stem from the relationship between heat capacity and the number of carbon atoms n_C or the number of methylene groups n_{CH2} in the molecule. Briard et al.[83] proposed a relationship to represent all phase change properties and also liquid heat capacity as a function of the number of carbon atoms n_C and of temperature for 27 hydrocarbons, C_{18}–C_{60}.

More sophisticated methods are based on more complex fragments. Benson and Buss[54] suggested dividing group contribution methods into three sequentially increasing classes. Zero-order methods cater for the additivity of atomic fragments that do not depend upon mutual bonds. For cyclic and aromatic compounds, the ring fragment is considered to be an indivisible unit. First-order methods use bonds between neighbouring atoms as structural fragments. Again, cyclic fragments are considered to be unique units. Second-order methods are based on the additivity of groups. A group is defined as a polyvalent central atom surrounded by all its ligands.

Most group contribution methods permit an estimation of liquid heat capacity at a single temperature[8,54,55,57,58,60,63,65,73,75,78] (mostly at 293 or 298 K), at several discrete temperatures[66,71,72] or as a function of temperature.[51,59,62,64,66,69,74,76,77,79,80]

19.2.5.1 Estimation at One Temperature

The Kopp's rule published in 1844 is the oldest zero-order group contribution method. It is still used today to estimate the heat capacities of inorganic compounds. Other examples of the zero-order group contribution method include that by Gambill,[60] who developed contributions at 293 K for the most common elements, resulting in an average relative uncertainty of 30 to 40 %, and by Hurst and Harrison,[63] who developed new contributions at 298 K for 8 elements and one "universal" contribution for all other elements. Hurst and Harrison used 477 data points for their parameter development which resulted in an average relative uncertainty of 9 %. The latter method should be used only when the empirical formula of a compound is known and the values of higher order contributions are missing.

Several first-order group contribution methods have been suggested, see for example, Shaw[78] and Johnson and Huang,[65] the former being the more cited of the two. Shaw's work[78] contained values of 63 contributions at 298 K with an average difference between experimental and estimated values of 6.3 J K^{-1} mol^{-1}. Most of the methods provide an estimate of the isobaric liquid heat capacity, even though this is not always clearly stated. The work by Ogiwara et al.[73] may be considered as an isolated example of a method to provide an estimation of the isochoric liquid heat capacity of organic compounds; 12 structure contributions were published.

Chickos et al.[55] proposed a group additivity method which provides heat capacity estimates of the condensed phase. The method was developed using a database of experimental data that consisted of 810 liquids and 446 solids. The structural fragments were made up of carbon in various common substituted and hybridized states and 47 functional groups. The standard relative uncertainty was 19.5 J K^{-1} mol^{-1} (for liquids) and 26.9 J K^{-1} mol^{-1} (for solids). This method has been refined for solid organo-sulfur compounds and their oxides, by Roux et al.[75]

The second-order group contribution method that takes into account interactions with the closest neighbours of a central atom, was introduced by Benson and collaborators.[8,54] This approach, applied to hydrocarbons in the gas, liquid and solid state was adopted by Domalski and Hearing.[57] The authors developed 48 structural contributions to estimate liquid heat capacities at 298.15 K, resulting in a relative uncertainty between experimental and estimated values for 390 compounds of 1.9 J K^{-1} mol^{-1}. The method was later extended to cover also other organic compounds.[58]

19.2.5.2   Estimation over a Temperature Range

The first attempt to calculate contributions at different temperatures was by Missenard[71,72] who estimated heat capacities at six discrete temperatures from $-25\,°C$ to $100\,°C$ in steps of $25\,°C$. He used 24 first-order contributions and the resultant relative uncertainty was below 5 %.

Temperature-dependent second-order contributions developed originally by Benson and Buss for ideal gas heat capacities[54] was extended by Luria and Benson[69] to also include liquid hydrocarbons. Luria and Benson presented 33 group contribution, each expressed as a function of temperature by a cubic polynomial (the temperature interval the parameters cover, corresponds to experimental data), resulting in an average deviation of 3.3 J K^{-1} mol^{-1} for 117 hydrocarbons.

Dvorkin et al.[59] proposed a group contribution method, similar to that developed for other properties,[84] for the calculation of heat capacities of organic liquids. The temperature dependence of heat capacity was expressed through an empirical polynomial equation and the parameters of this equation were listed at various temperature intervals for the most common functional groups.

An extensive investigation was published by Růžička and Domalski,[76,77] using a second-order group contribution approach together with a quadratic polynomial to express the temperature dependence of C_p. They used recommended liquid heat capacities[2] for 333 liquid organic compounds containing elements C, H, N, O, S and halogens and calculated 149 contributions, resulting in an average relative uncertainty of 2.9 % over the whole temperature interval in which the parameters are valid. For comparison they selected the widely used Lee and Kesler method;[43] for 173 compounds the latter method gives an average relative uncertainty of 7.8 % whereas the Růžička and Domalski method[76,77] gave a relative uncertainty of 2.8 %.

The Růžička and Domalski method[76,77] was later extended and refined by Zábranský and Růžička[80] for estimating the heat capacity of pure organic liquids as a function of temperature in the range from the melting temperature to the normal boiling temperature. The group contribution parameters were developed from an extensive database of recommended heat capacity values of more than 1800 compounds[2-4] and provided estimates with an overall standard percent deviation, between experimental and estimated data, of 1.7 %.

It is important when developing a large number of temperature dependent group contributions, to cover a wide range of families of compounds and to have an extensive database of critically assessed experimental data (see Zábranský *et al.*[2-4]). As this database was not available prior to 1996, several authors published estimation methods covering only a limited number of families of compounds, mostly hydrocarbons (e.g. Luria and Benson,[69] Akhmedov[51]). The method by Hadden[62] and Sokolov[79] is also applicable only to hydrocarbons; for the temperature dependant contributions it uses reduced temperature rather than temperature. Other authors developed sets of contributions to cover other families of compounds, such as coal liquefaction products (Le and Allen[68]), esters of higher fatty acids with industrial applications (Phillips and Mattamal[74]), or high molecular-weight *n*-alkanes (Jin and Wunderlich[64]).

A novel and universal method for the estimation of physicochemical properties by a group contribution approach was proposed by Marrero and Gani.[70] This approach was first modified by Kolská *et al.*[66,67] for the estimation of enthalpy of vaporization and entropy of vaporization at 298.15 K and at the normal boiling temperature T_b,[67] and later for estimation of liquid heat capacity as a function of temperature.[66] The estimation of properties involves a three-level calculation procedure, covering structural fragments of the first, second and third-levels. In the first or primary level, contributions from simple groups are employed-this allows the representation of a wide variety of organic compounds. These fragments, however, are insufficient to capture the proximity effect and the differences between isomers and are able only to estimate, correctly, values for simple and mono-functional compounds. To accommodate this deficiency, Marrero and Gani,[70] and before them, Constantinou and Gani[56] introduced the second and third-level contributions,[70] involving poly-functional and structural groups that provide more information about the molecular structure of the more complex compounds. These involve poly-functional compounds with at least one ring in a molecule, or non-ring chains including more than four carbon atoms in a molecule, and multi-ring compounds with a fused or non-fused aromatic or non-aromatic rings. Using the second-level and/or the third-level contributions it is possible to distinguish between some isomers; this is a unique feature in comparison with other common group contribution methods.

The new model for the estimation of $C_p^l(T)$ has the form:

$$C_p^l(T) = C_{p0}^l(T) + \sum_i N_i C_{p1-i}^l(T) + w \sum_j M_j C_{p2-j}^l(T)$$
$$+ z \sum_k O_k C_{p3-k}^l(T) \tag{10}$$

The temperature dependence of the group contributions was expressed through an empirical polynomial equation:[2]

$$C_{pq^{th}level-i,j\,or\,k}^l(T) = a_{q-i,j\,or\,k} + b_{q-i,j\,or\,k} \cdot (T/100) + d_{q-i,j\,or\,k} \cdot (T/100)^2 \tag{11}$$

where $C_{p1-i}^l(T)$ is the contribution of the first-level group of type-i, $C_{p2-j}^l(T)$ the contribution of the second-level group of type-j, $C_{p3-k}^l(T)$ the contribution of the third-level group of type-k. N_i, M_j, and O_k is the number of occurrences of the individual groups (of type-i, -j or -k, respectively) in a compound. $C_{p0}^l(T)$ (which could be considered as the contribution of the zero-level group) is an additional adjustable parameter. Variables w and z are weighting factors that are assigned to 0 or 1, depending on whether the second-level and third-level contributions, respectively, are used or not. In Equation (11), $a_{q-i,j\,or\,k}$, $b_{q-i,j\,or}$ $_k$, and $d_{q-i,j\,or\,k}$ are adjustable parameters for the temperature dependence of $C_{p0}^l(T)$, $C_{p1-i}^l(T)$, $C_{p2-j}^l(T)$, and $C_{p3-k}^l(T)$. All compounds were described by the same set of groups as defined previously.[67] In total, 224 group contribution parameters were determined for estimating C_p^l. The average relative errors of liquid heat capacity estimation are as follows: 1.2 % for 549 compounds from the basic set covering data over the whole temperature range, 1.5 % for 404 data points at 298.15 K for compounds from the basic dataset, and 2.5 % for 149 compounds from an independent test set from data obtained at 298.15 K.

In Table 19.1 a comparison of estimation errors for liquid heat capacities is presented for the three group contribution methods[66,76,77,80] and for the corresponding-states-approach method by Rowlinson.[85]

Recently group contribution methods were published to estimate heat capacity of ionic liquids.[86,87] Whereas Waliszewski[87] presented parameters for estimation at 293.15 K with average estimation error of 12 %, Gardas et al.[86] proposed a method to estimate heat capacity as a function of temperature by using experimental data for 19 ionic liquids and reported estimation errors not exceeding 2.5 %. In their work 12 contributions were used (3 for cations, 6 for anions and 3 for groups of aliphatic chain). However, their estimated values were not compared with an independent set of experimental data not used in the parameter development.

Table 19.1 Comparison of relative uncertainties for liquid heat capacity.

Family of compounds	Rowlinson[85]		Růžička and Domalski[76,77]		Zábranský and Růžička[80]		Kolská et al.[66]	
	NC	STD/%	NC	STD/%	NC	ARE/%	NC	ARE/%
Alkanes	49	5.0	68	1.5	24	1.0	24	1.5
Cycloalkanes	23	3.8	90	2.8	43	1.0	43	1.0
Unsaturated aliphatic hydrocarbons	17	6.3	30	1.3	25	0.5	25	0.9
Unsaturated cyclic and aromatic hydrocarbons	25	3.2	74	1.2	65	0.8	65	0.6
Halogenated hydrocarbons	43	4.9	100	2.3	78	1.2	78	0.9
Amines	13	6.9	28	1.2	23	1.2	23	1.6
Nitriles	5	19.8	10	2.8	8	0.6	8	0.8
Heterocyclic N-compounds	11	3.8	26	2.2	20	1.0	20	0.6
Ethers	12	4.4	36	1.3	22	1.4	22	0.9
Alcohols/*Alcohols and phenols	32	19.2	74	6.1	41*	2.8*	41*	3.7*
Aldehydes and ketones	15	4.0	55	3.2	30	1.6	30	1.3
Acids	10	9.7	31	2.1	21	1.6	21	1.2
Esters	18	3.4	85	1.8	48	1.5	48	1.0
Heterocyclic O-compounds	6	2.3	13	3.8	9	1.5	9	1.6
Other O-compounds (except as above)	4	6.5	0		11	1.8	11	1.2
Sulfides	12	2.7	15	1.1	20	0.7	20	0.8
Thiols	14	2.3	18	1.2	17	0.8	17	1.0
Heterocyclic S-compounds	6	1.9	11	0.9	9	0.6	9	1.2
Other families of compounds					35	1.6	35	1.7
TOTAL	315	7.1	764	2.6	549	1.2	549	1.2

NC, number of compounds, ARE, average relative uncertainty: $ARE = 100|C_{p,\mathrm{exp}}^{\mathrm{l}} - C_{p,\mathrm{calc}}^{\mathrm{l}}|/C_{p,\mathrm{exp}}^{\mathrm{l}}$, STD, standard percent deviation:

$$STD = 100\sqrt{(1/(n-m))\sum_{i=1}^{n}(|C_{p,\mathrm{exp}}^{\mathrm{l}} - C_{p,\mathrm{calc}}^{\mathrm{l}}|/C_{p,\mathrm{exp}}^{\mathrm{l}})^2}$$

$C_{p,\mathrm{exp}}^{\mathrm{l}}$ is experimental value, $C_{p,\mathrm{calc}}^{\mathrm{l}}$ is estimated value, n is total number of data points, m is number of adjustable parameters in the model.

19.3 Conclusion

The survey of liquid heat capacity estimation methods presented here clearly indicates there is no universal method widely applicable to any family of compounds and for any particular temperature interval. Either the input data may be missing or fragments are not available for the application of some group contribution method. Methods based on the corresponding state theorem and the thermodynamic methods will be used, in particular, for the estimation of heat capacities at temperatures close to the critical region and for the first members of a homologous series, where group contribution methods usually fail.

Presently group contribution methods are most widely applicable for the estimation of heat capacities of liquids, as well as for gases and solids. Method selection for a particular application depends on the family of compounds being considered and on the required accuracy of estimated values. We may expect that group contribution methods will be improved and refined as more precise experimental data on new families and/or new compounds becomes available.

Generally, the best results are obtained when estimating liquid heat capacities for non-polar compounds. This corresponds to higher reported deviations between experimental and estimated data for polar compounds, such as alcohols, acids and other compounds. It is obvious that better results are reported for methods developed to cover only compounds of some homologous series or a limited range of compounds than for those widely and generally applicable.

Acknowledgement

This work was conducted within the Institutional Research Plans No. MSM 6046137307 and with the help of an Academy of Sciences of the Czech Republic Grant No. IAA 400720710.

References

1. K. Růžička and V. Majer, *AIChE J.*, 1996, **42**, 1723.
2. M. Zábranský, V. Růžička, V. Majer and E. S. Domalski, *Heat Capacity of Liquids. Critical Review and Recommended Values,* American Chemical Society, Washington, D.C., 1996, (ISBN 1-56396-600-X).
3. M. Zábranský, V. Růžička, V. Majer and E. S. Domalski, *J. Phys. Chem. Ref. Data*, 2001, **30**, 441.
4. M. Zábranský, V. Růžička and E. S. Domalski, *J. Phys. Chem. Ref. Data*, 2001, **30**, 1199.
5. M. Zábranský and V. Růžička, *Fluid Phase Equilib.*, 2002, **194–197**, 817.
6. E. J. Baum, *Chemical Property Estimation: Theory and Practice,* Lewis Publishers, Boca Raton, 1998, (ISBN 0-87371-938-7).
7. M. L. Frenkel, G. J. Kabo, K. N. Marsh, G. N. Roganov and R. C. Wilhoit, *Thermodynamics of Organic Compounds in the Gas State,* TRC, TAMUS, College Station, Texas, 1994, (ISBN 1-883400-03-1).
8. S. W. Benson, F. R. Cruickshank, D. M. Golden, G. R. Haugen, H. E. O'Neal, A. S. Rodgers, R. Shaw and R. Walsh, *Chem. Rev.*, 1969, **69**, 279.
9. M. Bureš, V. Majer and M. Zábranský, *Chem. Eng. Sci.*, 1981, **36**, 529.
10. B. E. Poling, J. M. Prausnitz and J. P. O'Connell, *Properties of Gases and Liquids,* McGraw-Hill Professional Publishing, 2000, (ISBN 0070116822/ 9780070116825).
11. K. Roy and A. Saha, *J. Ind. Chem. Soc.*, 2006, **83**, 351.
12. J. O. Hirschfelder, C. F. Curtiss and R. B. Bird, *Molecular Theory of Gases and Liquids,* J.Wiley, New York, 1966, (ISBN 978-0-471-40065-3).

13. K. Lucas, *Applied Statistical Thermodynamics,* Springer-Verlag, Berlin, 1991, (ISBN 0-387-52007-4/3-540-52007-4).

14. G. Grimvall, *Thermophysical Properties of Materials,* North Holland; 2Rev Ed edition (22 Sep 1999), 1999, (ISBN 0444827943/978-0444827944).

15. D. Frenkel and B. Smit, *Understanding Molecular Simulations,* Academic Press, San Diego, 2002, (ISBN 0122673700/978-0122673702).

16. J. P. Hansen and I. R. McDonald, *Theory of Simple Fluids,* Academic Press, Amsterdam, 2006, (ISBN 0123705355/978-0123705358).

17. A. Mulero, *Theory and Simulation of Hard-Sphere Fluids and Related Systems,* Springer, Berlin/Heidelberg, 2008, (ISBN 978-3-540-78766-2).

18. P. Slavíček, R. Kalus, P. Paška, I. Odvárková, P. Hobza and A. Malijevský, *J. Chem. Phys.,* 2003, **119**, 2102.

19. A. Malijevský, F. Karlický, R. Kalus and A. Malijevský, *J. Phys. Chem. C,* 2007, **111**, 15565.

20. R. A. Aziz, *Chapter 2. Interatomic Potentials for Rare-Gases: Pure and Mixed Interactions.,* Springer-Verlag, Berlin, 1984, (ISBN 3-540-13128-0/3-387-13128-0).

21. A. Malijevský and A. Malijevský, *Molec. Phys.,* 2003, **101**, 3335.

22. K. M. Watson, *Ind. Eng. Chem.,* 1943, **35**, 398.

23. R. C. Reid and J. E. Sobel, *Ind. Eng. Chem., Fundam.,* 1965, **4**, 328.

24. C. F. Chueh and A. C. Swanson, *Can. J. Chem. Eng.,* 1973, **51**, 596.

25. C. F. Chueh and A. C. Swanson, *Chem. Eng. Progr.,* 1973, **69**, 83.

26. V. Svoboda, Z. Wagner, P. Voňka and J. Pick, *Collect. Czech. Chem. Commun.,* 1981, **46**, 2446.

27. X. Zhao, Z. Liu, W. Wang and Z. Chen, *Gongcheng Rewuli Xuebao,* 1996, **17**, 17.

28. K. P. Tyagi, *Ind. Eng. Chem., Process Des. Develop.,* 1975, **14**, 484.

29. L. Coniglio, E. Rauzy and C. Berro, *Fluid Phase Equilib.,* 1993, **87**, 53.

30. M. Rogalski, B. Carrier, R. Solimando and A. Peneloux, *Ind. Eng. Chem. Res.,* 1990, **29**, 659.

31. A. Bondi, *J. Phys. Chem.,* 1964, **68**, 441.

32. L. Coniglio, L. Trassy and E. Rauzy, *Ind. Eng. Chem. Res.,* 2000, **39**, 5037.

33. H. Illoukani and B. Nikoobakht, *Ind. J. Chem., Sect. A,* 1999, **38**, 1264.

34. G. Schmidt and H. Wenzel, *Chem. Eng. Sci.,* 1980, **35**, 1503.

35. A. Diedrichs, J. Rarey and J. Gmehling, *Fluid Phase Equilib.,* 2006, **248**, 56.

36. C. H. Twu, J. E. Coon and J. R. Cunningham, *Fluid Phase Equilib.,* 1995, **105**, 49.

37. K. S. Pitzer and R. F. Curl, *J. Am. Chem. Soc.,* 1955, **77**, 3427.

38. B. C. Sakiadis and J. Coates, *AIChE J.,* 1956, **2**, 88.

39. J. S. Rowlinson, *Trans. Faraday Soc.,* 1955, **51**, 1317.

40. A. Bondi, *Ind. Eng. Chem., Fundam.,* 1966, **5**, 442.

41. N. A. Morad, A. A. M. Kamal, F. Panau and T. W. Yew, *J. Am. Oil Chem. Soc.,* 2000, **77**, 1001.

42. T. -F. Yuan and L. I. Stiel, *Ind. Eng. Chem., Fundam.,* 1970, **9**, 393.

43. B. I. Lee and M. G. Kesler, *AIChE J.,* 1975, **21**, 510.

44. L. P. Filipov, *Vestn. Mosk. Univ., Ser. 3: Fiz., Astron.,* 1979, **20**, 87.

45. T. J. Lyman and R. P. Danner, *AIChE J.*, 1976, **22**, 759.
46. R. R. Tarakad and R. P. Danner, *AIChE J.*, 1977, **23**, 944.
47. T. E. V. Prasad, A. Rajiah and D. H. L. Prasad, *Chem. Eng. J.*, 1993, **52**, 31.
48. H. N. Al-Shorachi and E. T. Hashim, *Petrol. Sci. Tech.*, 2007, **25**, 1513.
49. V. Pachaiyappan, S. H. Ibrahim and N. R. Kuloor, *Chem. Eng. Chem. Metal. Eng.*, 1967, **74**, 241.
50. M. R. Riazi and Y. A. Roomi, *Ind. Eng. Chem. Res.*, 2001, **40**, 1975.
51. A. G. Akhmedov, *Zh. Fiz. Khim.*, 1979, **53**, 2387.
52. L. Becker and J. Gmehling, *J. Chem. Eng. Data*, 2001, **46**, 1638.
53. N. Behmanesh and D. T. Allen, *Fluid Phase Equilib.*, 1989, **53**, 423.
54. S. W. Benson and J. H. Buss, *J. Chem. Phys.*, 1958, **29**, 546.
55. J. S. Chickos, D. G. Hesse and J. F. Liebman, *Struct. Chem.*, 1993, **4**, 261.
56. L. Constantinou and R. Gani, *AIChE J.*, 1994, **40**, 1697.
57. E. S. Domalski and E. D. Hearing, *J. Phys. Chem. Ref. Data.*, 1988, **17**, 1637.
58. E. S. Domalski and E. D. Hearing, *J. Phys. Chem. Ref. Data.*, 1993, **22**, 805.
59. P. L. Dvorkin, G. L. Ryzhova and Y. A. Lebedev, *Bull. Acad. Sci. USSR,Div. Chem. Sci.*, 1984, **33**, 982.
60. W. R. Gambill, *Chem. Eng. (N.Y.)*, 1957, **64**, 261.
61. J. P. Guthrie and K. F. Taylor, *Can. J. Chem.*, 1984, **62**, 363.
62. S. T. Hadden, *J. Chem. Eng. Data*, 1970, **15**, 92.
63. J. E. Hurst and B. K. Harrison, *Chem. Eng. Commun.*, 1992, **112**, 21.
64. Y. Jin and B. Wunderlich, *J. Phys. Chem.*, 1991, **95**, 9000.
65. A. I. Johnson and C. -J. Huang, *Can. J. Technol.*, 1955, **33**, 421.
66. Z. Kolská, J. Kukal, M. Zábranský and V. Růžička, *Ind. Eng. Chem. Res.*, 2008, **47**, 2075.
67. Z. Kolská, V. Růžička and R. Gani, *Ind. Eng. Chem. Res.*, 2005, **44**, 8436.
68. T. T. Le and D. T. Allen, *Fuel*, 1985, **64**, 1754.
69. M. Luria and S. W. Benson, *J. Chem. Eng. Data*, 1977, **22**, 90.
70. J. Marrero and R. Gani, *Fluid Phase Equilib.*, 2001, **183–184**, 183.
71. F. A. Missenard, *C. R. Acad. Sc. Paris*, 1965, **260**, 5521.
72. F. A. Missenard, *Rev. Gen. Therm.*, 1966, **5**, 337.
73. K. Ogiwara, Y. Arai and S. Saito, *J. Chem. Eng. Jpn.*, 1981, **14**, 156.
74. J. C. Phillips and M. M. Mattamal, *J. Chem. Eng. Data*, 1976, **21**, 228.
75. M. V. Roux, M. Temprado, P. Jimenez, R. Guzman-Mejia, E. Juaristi and J. S. Chickos, *Thermochim. Acta*, 2003, **406**, 9.
76. V. Růžička and E. S. Domalski, *J. Phys. Chem. Ref. Data*, 1993, **22**, 597.
77. V. Růžička and E. S. Domalski, *J. Phys. Chem. Ref. Data*, 1993, **22**, 619.
78. R. Shaw, *J. Chem. Eng. Data*, 1969, **14**, 461.
79. S. N. Sokolov, *Zh. Fiz. Khim.*, 1979, **53**, 2089.
80. M. Zábranský and V. Růžička, *J. Phys. Chem. Ref. Data*, 2004, **33**, 1071.
81. I. B. Sladkov and O. S. Neganova, *Russ. J. Appl. Chem.*, 1992, **65**, 1764.

82. J. D. Williams, Prediction of melting and heat capacity of inorganic liquids by the method of group contributions. Thesis, New Mexico State Univ., Las Cruces, NM, USA, 1997.

83. A. J. Briard, M. Bouroukba, D. Petitjean and M. Dirand, *J. Chem. Eng. Data*, 2003, **48**, 1508.

84. P. L. Dvorkin, G. L. Ryzhova and Y. A. Lebedev, *Izv. Akad. Nauk SSSR Ser. Khim.*, 1983, **5**, 1101.

85. J. S. Rowlinson, *Liquids and Liquid Mixtures*, Butterworths, London, 1969.

86. R. L. Gardas and J. A. P. Coutinho, *Ind. Eng. Chem. Res.*, 2008, **47**, 5751.

87. D. Waliszewski, I. Stepniak, H. Piekarski and A. Lewandowski, *Thermochim. Acta*, 2005, **433**, 149.

CHAPTER 20

Computer Simulation Studies of Heat Capacity Effects Associated with Hydrophobic Effects

DIETMAR PASCHEK,[a] RALF LUDWIG[b] AND JÖRG HOLZMANN[b]

[a] Department of Physics, Applied Physics, and Astronomy, Rensselaer Polytechnic Institute, 110 Eighth Street, Troy, NY 12180-3590 USA; [b] Institut für Chemie, Abteilung Physikalische Chemie, Universität Rostock, Dr. Lorenz Weg 1, D-18051 Rostock, Germany

20.1 Introduction

The solvation of small non-polar particles in water behaves in an anomalous way: their solubility increases strongly with decreasing temperature. This feature has been related to the tendency of water to maintain its hydrogen bond network and has been coined the "hydrophobic effect". Extensive reviews have appeared[1-3] focusing on different aspects of the subject. The hydrophobic effect is important when studying the biophysics of cells, as it is considered to be the dominant driving force in the initial steps of protein folding.[4] Furthermore, it is responsible for the structural organization of amphiphilic aggregates, such as micelles, lyotropic mesophases, and lipid membranes.

The "hydrophobic effect" is manifested thermodynamically, by a low solubility (large positive solvation free energy) for nonpolar molecules in water.

Heat Capacities: Liquids, Solutions and Vapours
Edited by Emmerich Wilhelm and Trevor M. Letcher
© The Royal Society of Chemistry 2010
Published by the Royal Society of Chemistry, www.rsc.org

Quite surprisingly, the low solubility of small sized apolar particles is not due to a particularly weak interaction with their surrounding environment.[1] The heat of solvation of methane in water at ambient temperature has roughly the same magnitude as the heat of vaporization of pure liquid methane.[5] The positive solvation free energy of small apolar particles is a consequence of a negative solvation entropy, which overcompensates the solvation enthalpy of the same sign. This "entropy penalty" has been related to the orientational ordering of the hydration shell of water molecules, which is supposed to maintain the hydrogen bond network.[1] In parallel to the entropy decrease at low temperatures, theoretical and experimental studies have reported a slowing down of the translational and reorientational dynamics of water in the hydration shell of apolar molecules.[6–9] An important thermodynamic signature of hydrophobic hydration is the large positive solvation heat capacity. The solvation heat capacity is attributed to the temperature-induced changing mutual interactions between the solvent molecules in the hydration shell.[10] It is considered to be caused by the progressing disintegration of the hydrogen bond network around the solute with increasing temperature.[1,10] Since the solvation of small apolar moieties is accompanied by an entropy decrease of the solvent, the formation of contact pairs of apolar particles is a way to reduce this "entropy penalty". The tendency to form apolar contact pairs in solution is termed "hydrophobic interaction" and is essentially determined by the solvent. Since the association of small apolar particles is entropically favorable, a temperature increase leads to more stable apolar contacts. Hence the "hydrophobic interaction" is a classical example of an "entropic force".

Contrasting the behavior of *small* apolar solutes, water behaves differently at an extended (planar) interface. Here the thermodynamic features are mostly governed by water's interfacial tension, being essentially enthalpic in nature (weakening with increasing temperature). As a consequence, at some size or length-scale a "crossover" occurs [11,12] from an entropy to an enthalpy dominated solvation behavior. In addition, at small length-scales the solvation free energy is proportional to the particle volume, whereas at large length-scales it scales with the surface area. Recent computer simulation studies (Figure 20.1) indicate that this transition appears at a length-scale significantly below 1 nm.[13,14]

The thermodynamic signatures of small apolar particle hydration have been modeled by simple two state models[16–19] solely focusing on water's hydrogen bonding as the supposed dominant effect. Stronger hydrogen bonds between water molecules close to an apolar particle are counter-balanced by fewer possible binding partners. Silverstein *et al.*[19] were able to consistently relate experimental spectroscopy data, describing water's hydrogen bond equilibrium, with hydrophobic solvation calorimetric data. Their model calculations suggests that at lower temperatures the hydrogen bonds are more intact than in the bulk, whereas at high temperatures more hydrogen bonds are broken. In some ways the model reflects the idea of Franks and Evans expressed in their so-called "iceberg" model.[20] Here hydrophobic particles were thought to be stabilizing ice-like water configurations. However, as the entropy change

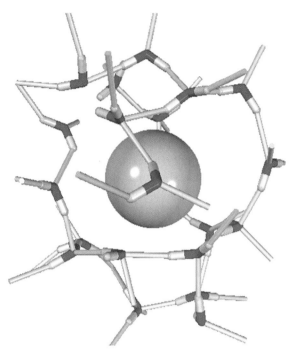

Figure 20.1   Water molecules forming a hydrophobic hydration shell around a Lennard-Jones particle of the size of a methane molecule. The snapshot was taken from a molecular dynamics simulation at 1 bar and 280 K. The molecules are trying to maintain an intact hydrogen bond network. However, the picture also reveals a significant amount of disorder in terms of transiently broken hydrogen bonds. A movie of the hydration shell dynamics of a closely related system of 5 ps duration is available on youtube.[15]

experienced by a water molecule in a hydrophobic hydration shell is about a factor of five smaller than the ice-crystal,[1] the "iceberg" model has been considered to strongly exaggerate the ordering found in a hydrophobic hydration shell.[1] In addition, two-state models seem to fail in reliably predicting absolute solvation free energies.[1] This has been recognized as being due to the fact that altering hydrogen bonding is not providing sufficient information to also determine the entropic "free volume contribution",[21,22] also named "cavitation free energy".

Realistic atomic detailed computer models for water were proposed over 30 years ago.[23] However, even relatively simple effective water model potentials based on point charges and Lennard-Jones interactions are computationally expensive. Hence, significant progress with respect to accurately describing water's thermodynamic structural and dynamical anomalous features has only quite recently been achieved.[24–26] Earlier studies, however, have shown that the water models were essentially able to capture the effects of hydrophobic hydration and interaction on a qualitative level.[6,7,27] Recent simulations suggest that the solvation entropy of hydrophobic molecules is tightly related to the water models ability to account for water's thermodynamic anomalous behavior.[28-31]

Simplifying concepts, such as the hydrogen-bond two-state models, strongly rely on the assumption that hydrophobic hydration is essentially determined by hydrogen bonding effects. The benefit of realistic atomic detail models, predicting sufficiently accurate solvation properties without any further fitting, might provide the opportunity to investigate to what extent these assumptions are justified. This is the scope of this chapter.

20.2 Methods

20.2.1 Molecular Dynamics Simulations

We present molecular dynamics simulations of liquid water using system sizes of 1000 TIP4P-Ew model water molecules.[26] The electrostatic interactions are treated in the "full potential" approach by the smooth particle mesh Ewald summation[32] with a real space cutoff of 0.9 nm and a mesh spacing of approximately 0.12 nm and 4th order interpolation. The Ewald convergence factor α was set to 3.38 nm^{-1} (corresponding to the relative accuracy of the Ewald sum of 10^{-5}). A 2.0 fs timestep was used for all simulations and the geometric constraints were solved using the SETTLE procedure.[33] All simulations were carried out by the GROMACS 3.2 simulation program.[34] The simulations were performed under isobaric/isothermal conditions for a pressure of 1 bar using a Nosè-Hoover[35,36] thermostat and a Rahman-Parrinello barostat[37,38] with coupling times of $\tau_T = 1.0$ ps, and $\tau_P = 2.0$ ps (assuming the isothermal compressibility to be $\chi_T = 4.5\,10^{-5}$ bar^{-1}). All properties were studied for 13 temperatures ranging from 250 to 370 K with a temperature interval of 10 K. In addition, for the same set of (P,T)-states we have also performed simulations of systems additionally containing one methane molecule. Here methane is treated as a single Lennard-Jones particle, employing the parameters of Hirschfelder *et al.*[39] with $\sigma = 3.730$ nm and $\varepsilon/k = 147.5$ K. The methane water cross parameters were obtained from Lorentz-Berthelot mixing rules. In all cases at least 2.4×10^3 configurations were stored for analysis.

20.2.2 Infinite Dilution Properties

The solubility of a gaseous solute is measured by the Ostwald coefficient $L^{l/g}$ = ρ_B^l/ρ_B^g, where ρ_B^l and ρ_B^g are the number densities of the solute in the liquid and the gas phase, respectively, when both phases are in equilibrium. Here A denotes the solvent and B indicates the solute. Equilibrium between both phases leads to a new expression for $L^{l/g}$, namely,

$$L^{l/g} = \exp\left[-\beta\left(\mu_B^{E,l} - \mu_B^{E,g}\right)\right] \tag{1}$$

where $\beta = 1/kT$ and $\mu_B^{E,l}$ and $\mu_B^{E,g}$ denote the excess chemical potentials of the solution, the liquid and the gas phase, respectively. When the gas phase has a sufficiently low density, then $\mu_B^{E,g} \approx 0$, hence $L^{l/g}$ becomes identical to the solubility parameter $\gamma_B^l = \exp[-\beta\mu_B^{E,l}]$. For our study the excess chemical potential of apolar solutes in the gas phase can practically be considered to be zero. The chemical potential of a solute can be obtained from a constant pressure simulation (NPT-Ensemble) of the pure solvent using the particle distribution theorem[40,41] according to

$$
\begin{aligned}
\mu_B^l = &- \beta^{-1} \ln \frac{\langle V \rangle}{\Lambda^3} \\
&- \beta^{-1} \ln \left\langle V \int d\vec{s}_{N+1} \exp(-\beta\Delta U) \right\rangle / \langle V \rangle \\
= &\, \mu_{id,B}^l(\langle \rho_B^l \rangle) + \mu_B^{E,l}
\end{aligned}
\tag{2}
$$

where $\Delta U = U(\vec{s}^{N+1};L) - U(\vec{s}^N;L)$ is the potential energy of a randomly inserted solute $(N+1)$-particle into a configuration containing N solvent molecules. The $\vec{s}_i = L^{-1}\vec{r}_i$ (with $L = V^{1/3}$ being the length of a hypothetical cubic box) are the scaled coordinates of the particle positions and $\int \vec{s}_{N+1}$ denotes an integration over the whole space. The brackets $\langle \ldots \rangle$ indicate isothermal-isobaric averaging over the configuration space of the N-particle system (the solvent). Λ represents the thermal wavelength of the solute particle. The first term $\mu_{id,B}^l$ is the ideal gas contribution of the solute chemical potential at an average solute number density $\langle \rho_B^l \rangle = 1/\langle V \rangle$ at the state (T, P). The entropic and enthalpic contributions to the excess chemical potential can are obtained as temperature derivatives according to

$$
s^E = -\left[\partial\mu^E/\partial T\right]_P
\tag{3}
$$

and

$$
h^E = -T^2\left[\partial(\mu^E/T)/\partial T\right]_P
\tag{4}
$$

and the isobaric heat capacity contribution according to

$$
c_P^E = -T\left[\partial^2\mu^E/\partial T^2\right]_P
\tag{5}
$$

As an alternative to the Ostwald coefficient, the solubility of gases is expressed in terms of Henry's constant k_H. The relationship between Henry's constant and the solubility parameter γ_B^l in the liquid phase is given by[42]

$$
k_H = \rho_A^l RT/\gamma_B^l
\tag{6}
$$

where ρ_A^l is the number density of the solvent. We use this relation in order to compare the experimental with the simulation data. The thermodynamic solvation properties discussed in this paper belong to the so called *number density scale*. In the experimental literature,[43–46] however, the properties are often discussed on the *mole fraction scale* with the solvation free energy being

$$\Delta G = -RT \ln k_{\mathrm{H}} \tag{7}$$

where the Henry's constant k_{H} is expressed in bars.[42,47] Care must be taken to which scale the discussed properties belong, since properties determined for the mole fraction scale contain additional terms depending on the thermal expansivity of the liquid.[47–49]

In order to perform the calculation efficiently, we have made use of the excluded volume map (EVM) technique[50,51] by mapping the occupied volume onto a grid of approximately 2×10^{-2} nm mesh-width. Distances smaller than $0.7 \times \sigma_{\mathrm{ij}}$ with respect to any solute molecule (oxygen site) were neglected and the term $\exp(-\beta \Delta U)$ taken to be zero. With this setup the systematic error was estimated to be less than $0.02 \,\mathrm{kJ \, mol}^{-1}$. This simple scheme improves the efficiency of the sampling by almost two orders of magnitude. For the calculation of the Lennard-Jones insertion energies ΔU we have used cut-off distances of 0.9 nm in combination with a proper cut-off correction. Each configuration has been probed by 10^3 *successful* insertions (*i.e.* insertions into the free volume contributing non-vanishing Boltzmann-factors).

In addition, we discuss the effect of having a polarizable methane solute. Therefore we add a polarization term to the insertion energy according to

$$\Delta U = \Delta U_{\mathrm{LJ}} + \Delta U_{\mathrm{pol}} \tag{8}$$

with

$$\Delta U_{\mathrm{pol}} = -\frac{1}{2} \alpha |\vec{F}|^2 \tag{9}$$

where $\alpha = 2.60 \times 10^{-3} \,\mathrm{nm}^3$ is the methane polarizability and $\vec{F}$ is the electric field created by all water molecules at the position where the particle is inserted. $\vec{F}$ is evaluated using the classical Ewald summation technique with a Ewald convergence factor of $2.98 \,\mathrm{nm}^{-1}$ (corresponding to a relative accuracy of the Ewald sum of $\approx 10^{-4}$) in combination with a real space cut-off of 0.9 nm and a reciprocal lattice cut-off of $|\vec{k}_{\mathrm{max}}|^2 = 25$.

20.2.3   "Computational Calorimetry"

In order to provide a spatial resolution of the water contribution to the solvation excess heat capacity, we calculate the individual potential energies of the water and solute molecules. This is done by a reaction field method based on the minimum image convention in combination with a minimum image "cubic" cutoff. This approach was originally proposed by Neumann[52] and was discussed by Roberts and Schnitker.[53,54] The reaction field approach in general is well suited for our purposes since it allows one to cleanly dissect the potential energy contributions according to individual molecules. For convenience we partition the potential energies in contributions assigned to individual molecules with

$$E = \sum_{i=1}^{M} E_i$$

$$E_i = \left(\frac{1}{2}\sum_{j=1}^{M} E_{ij}\right) + E_{i,\text{corr.}} \tag{10}$$

where E_i is the potential energy assigned to molecule i, M is the total number of molecules. The molecule-molecule pair energy

$$E_{ij} = \sum_{\alpha}\sum_{\beta} 4\varepsilon_{i\alpha j\beta}\left[\left(\frac{\sigma_{i\alpha j\beta}}{r_{i\alpha j\beta}}\right)^{12} - \left(\frac{\sigma_{i\alpha j\beta}}{r_{i\alpha j\beta}}\right)^{6}\right] + \frac{q_{i\alpha}q_{j\beta}}{r_{i\alpha j\beta}} \tag{11}$$

is then obtained as sum over discrete interaction sites α and β, with $r_{i\alpha j\beta} = |\vec{r}_{j\beta} - \vec{r}_{i\alpha}|$ based on the molecule/molecule center of mass minimum image separation. We employ long range corrections $E_{i,\text{corr.}} = E_{i,\text{corr.}}^{\text{el}} + E_{i,\text{corr.}}^{\text{LJ}}$ accounting for electrostatic, as well as Lennard-Jones interactions. The electrostatic correction

$$E_{i,\text{corr.}}^{\text{el}} = \frac{2\pi}{3V}\vec{D}\vec{d}_i \tag{12}$$

is a reaction field term, corresponding to the cubic cutoff, assuming an infinitely large dielectric dielectric constant. Here $\vec{d}_i = \sum_{\alpha} q_{i\alpha}\vec{r}_{i\alpha}$ is the dipole moment of molecule i, $\vec{D} = \sum_{i} \vec{d}_i$ is the total dipole moment of all molecules in the simulation cell and V is the instantaneous volume of the simulation box. $E_{i,\text{corr.}}^{\text{el}}$ has also been considered as the *extrinsic* potential and has been shown to provide configurational energies quite close to the values obtained by Ewald summation (with tin-foil boundary conditions).[53]

In order to be consistent with the applied cubic cutoff procedure for the electrostatic interactions, we also use a Lennard-Jones correction term for the cubic cutoff

$$E_{i,\text{corr}}^{\text{LJ}} = \frac{2}{V}\sum_{\alpha}\sum_{j}\sum_{\beta}\frac{\kappa_6}{b^3}\left(-\varepsilon_{i\alpha j\beta}\sigma_{i\alpha j\beta}^6\right)$$
$$+ \frac{\kappa_{12}}{b^9}\left(\varepsilon_{i\alpha j\beta}\sigma_{i\alpha j\beta}^{12}\right) \tag{13}$$

with $b = V^{1/3}/2$ denoting the half box length. $\kappa_6 \approx 2.5093827$ and $\kappa_{12} \approx 0.4106497$ are analytically integrated factors accounting for the cubic cutoff geometry.[10]

In addition to the procedure outlined in Section 20.2.2, we can also directly use the individual energies to calculate the solvation enthalpies and heat capacities according to

$$
\begin{aligned}
h^{E} =& \langle E_{\text{solute}} \rangle \\
&+ \langle n_{\text{shell}} \rangle \times (\langle E_{\text{shell}} \rangle - \langle E_{\text{bulk}} \rangle) \\
&+ \langle P \rangle \times \left(\langle V_{\text{shell}} \rangle - \langle V_{\text{bulk}} \rangle \frac{\langle n_{\text{shell}} \rangle}{\langle n_{\text{bulk}} \rangle} \right)
\end{aligned}
\tag{14}
$$

where $\langle E_{\text{solute}} \rangle$ is the average potential energy of the solute molecule, $\langle E_{\text{shell}} \rangle$ is the average potential energy of the water molecules in a sufficiently large solvation sphere of volume $\langle V_{\text{shell}} \rangle$ (here we use a radius of 1.0 nm) around the solute molecule, whereas $\langle E_{\text{bulk}} \rangle$ is the energy of the water molecules outside this sphere. $\langle n_{\text{shell}} \rangle$ and $\langle n_{\text{bulk}} \rangle$ represent the number of water molecules in the solvation sphere and the bulk. The total volume and total number of water molecules are accordingly $V = V_{\text{shell}} + V_{\text{bulk}}$ and $n = n_{\text{shell}} + n_{\text{bulk}}$. From the temperature dependence of h_{ex} we can easily obtain the corresponding heat capacities. We would like to point out that for a pressure $P = 0.1$ MPa as used in the present study the volume term in Equation (14) can be practically neglected.

20.2.4 Hydrophobic Interaction

The "hydrophobic interaction" is characterized by the potential of mean force (or profile of free energy) for separating two methane particles in aqueous solution. It is therefore related to the solvent mediated interaction between the methane particles. We calculate the profile of free energy $w(r)$ for the association process directly from the potential distribution theorem[40] according to

$$
w(r) = -kT \ln \langle V \exp(-\beta \Phi(\vec{r}_i)) \delta(|\vec{r}_{\text{ref}} - \vec{r}_i| - r) \rangle / \langle V \rangle - \mu^{E} \tag{15}
$$

Here $\Phi(\vec{r}_i)$ is the energy of randomly inserting a virtual methane gas particle into a simulation of a methane water-solution already containing one real methane particle, $\vec{r}_{ref}$ represents the position of the *real* reference particle in the simulation and μ^{E} is the excess chemical potential of methane in the aqueous solution, which has been calculated independently before. The "solvent mediated interaction" $c(r) = w(r) - v(r)$ is given as the profile of free energy for the association of two methane molecules, subtracting the methane-methane intermolecular potential $v(r)$. $w(r)$ is related to the gas-gas pair distribution function $g(r)$ according to $-kT \ln g(r) = w(r)$. As $c(r)$ is essentially determined by the molecules forming a solvation cage around the pair of methane particles, it has been referred to a "cavity potential".[55] $w(r)$ is calculated from configurations of the methane/water simulation. An excluded volume map is used mapping the occupied volume onto a grid of approximately 2×10^{-2} nm meshwidth. Here distances smaller than $0.75 \times \sigma_{ij}$ with respect to any solute molecule (oxygen site) were neglected and the term $\exp(-\beta \, \Delta U)$ taken to be zero. 8×10^3 *successful* (nonzero) insertions are sampled per configuration and a total of 2.4×10^4 configurations are analyzed per state point.

We use temperature derivatives of quadratic fits of $w(r,T)$ to calculate the enthalpic and entropic contributions at each methane-methane separation r. For the fits all 13 temperatures were taken into account. The entropy and enthalpy contributions are then obtained as

$$s(r) = -[\partial w(r, T)/\partial T]_P \tag{16}$$

and

$$h(r) = -T^2[\partial(w(r, T)/T)/\partial T]_P \tag{17}$$

In addition, the corresponding heat capacity change relative to the bulk liquid is available according to

$$c_P(r) = -T[\partial^2 w(r, T)/\partial T^2]_P \tag{18}$$

20.3 Discussion

20.3.1 Solvation of Methane and the Expansivity of Water

The solvation entropy of a small size particle is to a large extent determined by the available free volume. For a hard-sphere solute it is in fact the only contribution[2,56–58] to the solvation free energy. Hence it is quite obvious that thermophysical properties, such as density and thermal expansivity, play an important role in predicting the solvation free energy of small apolar particles. A comparison of various water models revealed that the temperature derivatives of the solvation free energy respond strongly to the varying thermal expansivities.[30,59]

Here we present thermal expansivity data for the TIP4P-Ew model[26] calculated from MD simulations. The MD data, as well as the experimental data according to Kell[60] are shown in Figure 20.2. Most prominent is the difference of location of the density maximum. The TIP4P-Ew model was actually parameterized to achieve a density maximum close to the experimental value.[26] Although this feature has been significantly improved compared to the original TIP4P-model,[24,61] the temperature of maximum density $T_{md} = 271$ K is still about 6 K below the experimental value of $T_{md} = 277$ K. The expansivity $\alpha_P = 1/V \; (\partial V/\partial T)_P$ for water and for the TIP4P-Ew model are shown in Figure 20.2b. The data suggest that the low-pressure part of the PVT-surface of TIP4P-Ew is shifted by about 6 to 8 K towards lower temperatures.

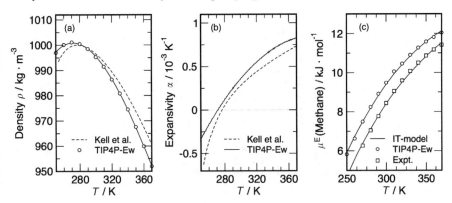

Figure 20.2 (a) Mass density ρ and (b) expansivity α_P of water at one bar pressure. The experimental data is according to Kell.[60] The density maximum of the TIP4P-Ew model for water is shifted by 6 K towards lower temperatures. (c) Excess chemical potential μ^E of methane in water. Circles: data obtained from the MD simulations. Squares: experimental data according to Fernandez-Prini *et al.*[63] Full lines: "IT model" prediction with $\mu^E = a\rho^2 T + b$ with $a = 7.15 \times 10^{-2}\,\mathrm{kJ\,mol^{-1}\,K^{-1}\,cm^6\,g^{-2}}$ and $b = -11.85$ kJ mol^{-1} (here we use $b = -12.9\,\mathrm{kJ\,mol^{-1}}$ for the experimental data).

From computer simulations of water, Hummer *et al.* have derived an information theory (IT) model for hydrophobic hydration,[58] proposing simple analytic expressions for the solvation free energy of small apolar particles as a function of temperature and density. The leading term in the IT model strongly suggests a quadratic relation between the excess chemical potential and the solvent number density ρ' according to $\mu^E/k \approx \rho'^2 T v^2/2\sigma_n^2$,[58] where v denotes the volume of a hydrophobic hard sphere particle, while $\sigma_n^2 = \langle n^2 \rangle - \langle n \rangle^2$ indicates the variance of the number of water molecules in a sphere of volume v. We have recently shown that these results are in line with the effect of pressure and temperature on the solubility of small apolar particles in TIP5PE model water over broad temperature and pressure ranges.[31] In addition, we could show that this simple relation also describes the behavior of the solvation of methane in aqueous salt solutions quite well.[62] Figure 20.2c indicates that the same scaling relation apparently describes the solvation free energy of methane in both water and TIP4P-Ew water quite successfully, when either experimental densities or TIP4P-Ew densities are employed. The observed shift of the experimental chemical potentials by about 1 kJ mol^{-1} to lower values is described here by a lower "offset" parameter b for the IT-model.

20.3.2 Temperature Dependence of the Solvation of Methane: Effect of a Polarizable Solute

The excess chemical potentials of methane μ^E as a function of temperature are given in Figure 20.3. Derivatives of the solvation free energy with respect to

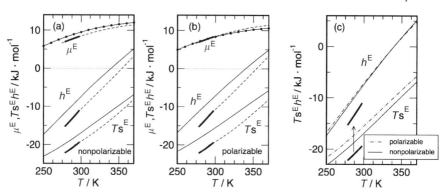

Figure 20.3 Thermodynamic parameters characterizing the solvation of methane in water. Given are the solvation free energy μ_{ex}, as well as its enthalpic and entropic contributions h^E, and Ts^E. Filled symbols: excess chemical potential μ^E, obtained from simulation data. Solid Lines: Third order polynomial fits of the solvation free energy obtained from simulation data.[64] All temperature derivatives are obtained from the fitted parameters. (a) The polarizability of methane has been neglected. (b) The data shown is explicitly considering the polarizability of methane. Heavy solid lines: experimental data according to Ben-Naim and Yaacobi.[65] Dashed lines: Experimental data according to Fernandez-Prini and Crovetto.[63] (c) Direct comparison of the differences caused by including the polarization of methane.

temperature were calculated from fitted third order polynomials.[64] The experimental data of the solvation of methane in water have been directly obtained from the Henry coefficients provided by Fernandez-Prini and Crovetto,[63] employing the pure water densities of Kell.[60] In addition, we also provide the solvation data published by Ben-Naim and Yaacobi. Those data were derived from the fitted Ostwald coefficients $\gamma = \exp[-\mu^E/kT]$ reported by Ben-Naim and Yaacobi.[65]

The excess chemical potential of methane in TIP4P-Ew water at 300 K of 9.48 kJ mol^{-1} is similar to the value reported by Krouskop et al.[59] and Holzmann et al.[62] and reasonably close to the simulated values of 9.79 kJ mol^{-1} and 9.78 kJ mol^{-1} obtained by Shimizu and Chan, as well as Paschek[10,28,30] for the original TIP4P model at 300 K and 1 atm. The differences with respect to the latter have to be attributed to the changes in the water model. However, the value is at least about 1 kJ mol^{-1} larger than the experimental value of 8.4 kJ mol^{-1}.[65] Dyer et al. have recently shown that this difference can be significantly reduced when the solute polarizability is explicitly considered.[66] It has been argued that polarizability would enhance the interaction with the solvent[30] and would thus strengthen the enthalpic part.

Figure 20.3 reveals that the temperature dependence of the simulated excess chemical potentials, as well as the derived enthalpic and entropic contributions behave qualitatively similar to the experimental data.[65] The excess chemical potential of methane in water is positive and increases with temperature, being consistent with a dominating negative entropy of hydrophobic solvation. However, both the solvation enthalpy h^E and the contribution from the solvation entropy Ts^E are found to be *less negative* than the corresponding experimental values. For pure water at 298 K we obtain $h^E = -7.7\,\text{kJ}\,\text{mol}^{-1}$ (Expt.: $h^E = -10.9\,\text{kJ}\,\text{mol}^{-1}$) and for the entropy we get $Ts^E = -17.2\,\text{kJ}\,\text{mol}^{-1}$ (Expt.: $Ts^E = -19.3\,\text{kJ}\,\text{mol}^{-1}$). Note that the *larger* μ^E of $9.5\,\text{kJ}\,\text{mol}^{-1}$ (Expt.: $\mu^E = 8.4\,\text{kJ}\,\text{mol}^{-1}$) is a consequence of a possibly overemphasized entropy effect and underestimation of the solvation enthalpy.

The systematic underestimation of the solvation enthalpy seems to be in line with the polarizability arguments raised by Dyer *et al.*[66] As shown in Figure 20.3b, the solute polarizability indeed lowers the excess chemical potential at 300 K to $8.35\,\text{kJ}\,\text{mol}^{-1}$, leading to a quantitative agreement with experimental data at this particular temperature. However, as indicated in Figure 20.3c, the apparently better agreement with experimental data is a consequence of an even stronger deviation of the solvation entropy from the experimental data, whereas the enthalpic contribution remains almost unchanged. Therefore the good agreement of the chemical potential might be considered a lucky coincidence, and being the consequence of compensating errors.

Having traced the temperature dependence of the hydrophobic solvation of methane over a broad temperature interval, the fitted data allows us to reliably determine the second derivative of the solvation free energy: the solvation heat capacity $c_P{}^E$. For 298 K we find a solvation heat capacity of $(195 \pm 10)\ \text{J}\,\text{K}^{-1}\,\text{mol}^{-1}$, compared to the $228\,\text{J}\,\text{K}^{-1}\,\text{mol}^{-1}$ according to the data set of Ben-Naim,[65] and the $234\,\text{J}\,\text{K}^{-1}\,\text{mol}^{-1}$ according to Rettich *et al.*[44] when transforming their data on a number density scale (see Reference 30 for a discussion of this issue). The effect of solute polarization is found to influence the heat capacity only little. It is found to consistently lower the heat capacity of solvation by about $(7\text{ to }8)\,\text{J}\,\text{K}^{-1}\,\text{mol}^{-1}$ over the entire temperature range, leading to a value of $(188 \pm 10)\ \text{J}\,\text{K}^{-1}\,\text{mol}^{-1}$ at 298 K. Again, considering the polarizability explicitly leads to a less perfect agreement with experimental data.

The presented simulation data clearly demonstrate the limitations of the currently available potential models. A combination of existing model potentials for methane does provide qualitative to semi-quantitative agreement with experimental data. Truly quantitative agreement will require further refinement of both the solvent model and the specific solvent-solute interaction. Paschek has shown that an increasing water solute Lennard-Jones σ will increase the heat capacity of solvation.[30] Modifications of this type will be required to overcompensate the effects introduced by the polarizability. However, the agreement with experimental data is sufficient to provide detailed insight into the molecular origin of the heat capacity of hydrophobic solvation.

20.3.3 Origin of the Individual Heat Capacity Contributions

Here we calculate the solvation enthalpy based on the "individual" potential energies of solute and solvent molecules using the reaction field method

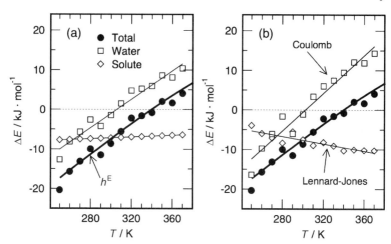

Figure 20.4   Contributions of the solvation enthalpy h^E. Heavy solid line: data according to the temperature derivative of the fitted solvation free energies obtained from the potential distribution theorem. Filled circles: Solvation energy enthalpies according to the differences between water molecules in the solvation region ($r \leq 1.0$ nm) and the bulk including the energy contribution of the methane particle. (a) Open squares: Energy contributions by the water molecules. Open diamonds: Energy contribution of the solute. (b) Open squares: Coulomb part of the solvation energy. Open diamonds: Lennard-Jones part of the solvation energy.

discussed in Section 20.2.3. We are employing exactly the same procedure discussed in Reference 10. We determine $\Delta E \approx h^E$ from the energies of the solute molecule and energies and numbers of the water molecules in the bulk and in the hydration sphere of radius 1.0 nm. As shown in Figure 20.4, the directly calculated data for ΔE are in good agreement with the solvation enthalpy h^E obtained from the temperature dependence of the solvation free energy. In addition, Figure 20.4a reports a partition of the solvation energy ΔE into contributions according to the solute $\langle E_{\text{solute}} \rangle$ (solute) and the solvent $\Delta E- \langle E_{\text{solute}} \rangle$ (water). The calculated excess heat capacity c_P^E of 191 J K^{-1} mol^{-1}, averaged over the entire temperature range, is dominated by the water contribution of 180 J K^{-1} mol^{-1}, whereas the solute part with 11.0 J K^{-1} mol^{-1} contributes only to about 5 %. The observed value for c_P^E of 191 J K^{-1} mol^{-1}, agrees reasonably well with the value of 228 J K^{-1} mol^{-1}, observed experimentally for $T = 300$ K.[65] Note that the change of sign of the methane solvation enthalpy from negative to positive at about 340 K has to be attributed essentially to the greatly enhanced heat capacity of the water molecules in the hydrophobic hydration shell.

Being able to semiquantitatively describe the solvation properties of methane from an atomic detail model, we might be able to compute to what extent the different intramolecular forces, such as "van der Waals" and "hydrogen bonding" contribute to the heat of solvation. Therefore we divide the solvation enthalpy data into their Coulomb and Lennard-Jones contributions, as shown in Figure 20.4b. It turns out that the large positive solvation heat capacity of methane is almost exclusively due to the Coulomb-interaction. In fact, the Coulomb-part of the heat capacity has a value of $237\,\mathrm{J\,K^{-1}\,mol^{-1}}$ which is even larger than the total heat capacity of solvation, and is balanced by a negative Lennard-Jones contribution of $-46\,\mathrm{J\,K^{-1}\,mol^{-1}}$. The strongly orientation dependent water-water Coulomb interactions of associated water molecules is what we basically consider as "hydrogen bonding" interaction. Consequently, our atomic detail simulations clearly support the "hydrogen bond view" of hydrophobic hydration.

In order to elucidate the radial dependence of the solvent contribution to the excess heat capacity, we also calculate the potential energy of the water molecules as a function of distance to the central carbon atom. Figure 20.5b shows the change of the potential energy of the water molecules around a methane particle with respect to the bulk value as a function of temperature. To give an impression of the hydration shell, we also show the methane-water pair distribution functions (Figure 20.5a). A rather strong temperature dependence of the potential energy of the water molecules in the first hydration shell is clearly evident. Similarly to what has been shown for other water models,[10,30] it can be seen that for lower temperatures the water molecules found in the distance interval between 0.35 nm and 0.5 nm exhibit a potential energy even lower than the bulk value. With increasing temperature this behavior is reversed and the location of the molecules in the hydration shell becomes more and more energetically unfavorable. The corresponding change in heat capacity as a function of distance to the methane particle is shown in Figure 20.5a.

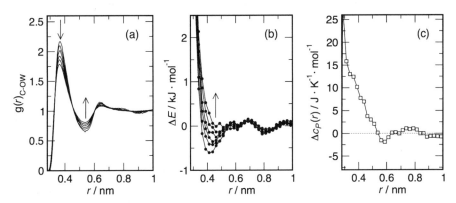

Figure 20.5 (a) Methane-carbon-water-oxygen radial pair distribution functions obtained for 260 K to 360 K (with steps of 20 K). The arrow indicates increasing temperatures. (b) Relative change of the water potential energy $\Delta E_{\mathrm{water}}(r) = E_{\mathrm{water}}(r) - E_{\mathrm{bulk\ water}}$ for all investigated temperatures for the same set of temperatures. (c) Change of the heat capacity of the water molecules around a methane particle $\Delta c_P(r) = c_P(r) - c_{P\mathrm{bulk}}$ as a function of distance to the methane particle. $\Delta c_P(r)$ is obtained as a linear regression of the datasets shown in (b).

Significant changes in the heat capacity are found to be restricted mostly to the first solvation shell. Those water molecules in the first hydration shell experience a roughly 20 % increased (configurational) heat capacity with respect to the bulk value of about $64 \, J \, K^{-1} \, mol^{-1}$.

20.3.4   On Hydrophobic Interactions

In order to quantify hydrophobic interactions, we calculate the profile of free energy for the association of two methane particles at all temperatures. Representative curves for selected temperatures are shown Figure 20.6a. Similar to our previous studies[30] we find that the free energy well at the contact state deepens with increasing temperature. This observation is in accordance with the interpretation that the association of two hydrophobic particles is stabilized by minimizing the solvation entropy penalty and has been reported by a large number of publications.[28,29,67–74] The pair-association is schematically depicted in Figure 20.7: the enhanced ordering of the solvent molecules in the hydration shell provides a negative solvation entropy. When two particles associate, the corresponding hydration shells overlap, hence leading to a positive net entropy. Consequently contact-configurations should become increasingly stabilized at higher temperatures. In parallel, the increased heat capacity of the water molecules in the hydrophobic hydration shell should lead to a negative heat capacity contribution for the association of two particles, thus weakening the entropy contribution at elevated temperatures.

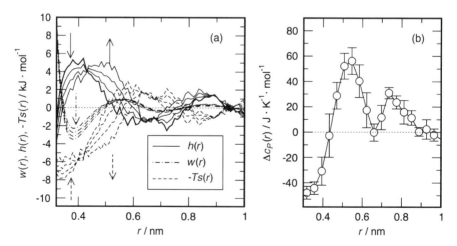

Figure 20.6   (a) Profile of free energy $w(r)$ for the association of two methane particles for selected temperatures (280 K, 300 K, 320 K, and 340 K), as well as the corresponding enthalpic and entropic contributions $h(r)$ and $-Ts(r)$. The arrows indicate increasing temperatures. (b) Relative heat capacity change $\Delta c_P(r) = c_P(r) - c_P^{bulk}$ as a function of distance to the two methane particles obtained for T = 300 K.

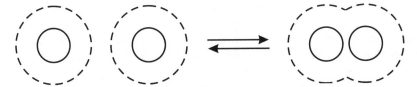

Figure 20.7 Simplifying schematic of the hydrophobic association process according to the surface area picture: The contact configuration is stabilized with increasing temperatures by minimizing the entropy penalty. The prominent peak of the heat capacity at the desolvation barrier, however, indicates that localized hydration phenomena might be also important.

From the temperature derivative of the $w(r,T)$ data set we obtain the entropy profiles for the association of two methane particles, also shown in Figure 20.6a. The entropic $-Ts(r)$ and enthalpic $h(r)$ contributions to the profile of free energy are given. The temperature variation of the entropy profiles is quantified by the corresponding heat capacity profiles $c_P(r)$ given in Figure 20.6b. As previously suggested by Smith and Haymet[67,68] and others, the hydrophobic association process is found to be entropically favored and enthalpically disfavored. The value of $-6.0\,\mathrm{kJ\,mol^{-1}}$ for the entropic contribution to the profile of free energy at $300\,\mathrm{K}$ at the contact distance for TIP4P-Ew water is larger than the $-4.14\,\mathrm{kJ\,mol^{-1}}$ observed for methane in the original TIP4P water reported by Shimizu *et al.*[28] Figure 20.6 indicates that there is a tendency for the entropy at very short distances to decrease with increasing temperature, whereas in the region around $6\,\mathrm{\mathring{A}}$, at the so called *desolvation barrier*, the entropy increases with temperature. The overall tendency to smaller contact-entropies is in accordance with the decrease of the absolute values of the solvation entropies. The deepening of the free energy well in the contact state with increasing temperature, however, is apparently a consequence of a stronger decrease of the enthalpic contribution.

The temperature dependence discussed here can be quantified by the association heat capacities $c_P(r)$ shown in Figure 20.6b. Shimizu and Chan[28,29,75] report for the association of methane particles in TIP4P water a change of the heat capacity for the contact state close to zero, in qualitative disagreement with expectation due to the overlapping hydration shells. In addition, they observe a maximum of the heat capacity of about $120\,\mathrm{J\,K^{-1}\,mol^{-1}}$ at the location of the desolvation barrier at a distance of $5.5\,\mathrm{\mathring{A}}$. In qualitative agreement with Shimizu *et al.*[29,75] and Southall and Dill[76] we find a maximum of the heat capacity at the desolvation barrier in the region around $5.5\,\mathrm{\mathring{A}}$ with $50\,\mathrm{J\,K^{-1}\,mol^{-1}}$, which is identical to the value reported recently for the TIP4Pmodel.[30] For the contact state, however, we observe a negative net heat capacity of $-40\,\mathrm{J\,K^{-1}\,mol^{-1}}$, significantly below zero. The observation that the heat capacity for the association of two hydrophobic particles exhibits a maximum located at the desolvation barrier has been recently explained.[10] MD simulations indicate that this behavior is the consequence of two compensating effects. On one hand, the decreasing solvent accessible surface leads to a decreasing heat capacity. In addition, however, an even more strongly increased heat capacity is observed for water molecules that are located in the joint hydration shell between the two hydrophobic particles. Their heat capacity increases by about 60 % compared to the bulk.[10] This is apparently even overcompensating the effect of a reduced solvent accessible surface just when the two methane particles are separated by about 5.5 Å.[10,77] A detailed analysis of the water-water pair interactions in the different states (bulk, hydration shell, joint hydration shell of two hydrophobic particles) has revealed that the heat capacity effects can be rationalized as a counterbalance of strengthened hydrogen bonds in a state of tension (low water density) and enhanced disintegration of the hydrogen bond network with increasing temperature. The reduced number of water neighbors in different parts of the hydrophobic hydration shell can be considered similar to the situation found in (metastable) water at negative pressures.[10] Experimental compressibility data for water are indicating that the observed heat capacity effects have the same order of magnitude, as it would be expected by extrapolating the data for pure water to a density range similar to the *local* densities observed in the different hydration shell states.[10]

Finally, we would like to discuss the apparent strength of hydrophobic effects and their temperature dependence in general. In the field of biophysics of proteins, it is considered essential that the interactions are *not too strong*, thus allowing for transient conformational diversity.[78,79] It is therefore required that hydrophobic interactions are in the range of kT. Recently Widom et al.[3] have proposed a relationship between the hydrophobic interaction strength of two methane particles in contact and their solvation free energy. Their conclusion was deduced from simplified lattice models, where the interaction parameters were tuned to quantitatively reproduce the (experimental) free energies of solvation. The solvation entropy part has been obviously represented by the form and connectivity of the lattice. Interestingly, their calculations predicted an almost linear relation between the hydrophobic interaction c/RT of two methane particles in the contact state and the free energy of solvation μ^E/RT. This property turned out to be almost independent of the type of the lattice that was employed, and was found for the temperature interval between 273 K and 333 K and possibly also even lower temperatures. Here we show the behavior of methane dissolved in TIP4P-Ew-model water. Figure 20.8 depicts the cavity potential for two methane particles in contact, as well as the solvation free energy for methane. All quantities are given in multiples of RT, as it has been suggested by Widom et al. The plot shown in Figure 20.8c indicates that at least for temperatures sufficiently below the maximum of $\mu^E(T)/RT$ the lattice model predictions are consistent with our data. For the TIP4P-Ew model an almost linear behavior in the interval 250 K and 310 K is observed. We would like to point out that the slope of 0.7 observed by Widom et al. for methane is reasonably close to the value of 0.6 obtained for TIP4P-Ew water. Note, however, that the hydrophobic interaction between two methane particles observed here is even weaker than the interaction predicted by Widom et al. for the limiting case according to the Bethe-Guggenheim approximation ($Z = \infty$).

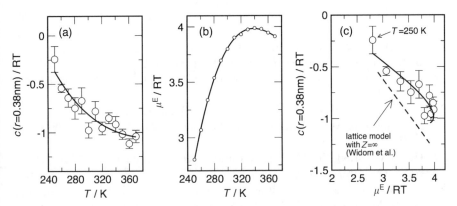

Figure 20.8 (a) Variation of the strength of the hydrophobic interaction for two methane particles in contact $c(0.38\,\text{nm}) = \mu^E(0.38\,\text{nm}) - \mu^E(\infty) - v(0.42\,\text{nm})$ in multiples of RT with temperature. (b) Solvation free energy of a single methane particle (in units of RT) at infinite dilution as a function of temperature. (c) Plot of hydrophobic interaction strength versus solvation free energy as suggested by Widom *et al.*[3] The dashed line represents the prediction according to the lattice solvent in the Bethe-Guggenheim approximation $Z = \infty$. Symbols are simulation data shown in a and b. The heavy line represents the fits shown in a and b.

References

1. N. T. Southall, K. A. Dill and A. D. J. Haymet, *J. Phys. Chem. B*, 2002, **106**, 521.
2. L. R. Pratt and A. Pohorille, *Chem. Rev.*, 2002, **102**, 2671.
3. B. Widom, P. Bhimalapuram and K. Koga, *Phys. Chem. Chem. Phys.*, 2003, **5**, 3085.
4. K. A. Dill, *Biochemistry*, 1990, **29**, 7133.
5. A. D. J. Haymet, K. A. T. Silverstein and K. A. Dill, *Faraday Discuss.*, 1996, **103**, 117.
6. A. Geiger, A. Rahman and F. H. Stillinger, *J. Chem. Phys*, 1979, **70**, 263.
7. D. A. Zichi and P. J. Rossky, *J. Chem. Phys.*, 1985, **83**, 797.
8. R. Haselmeier, M. Holz, W. Marbach and H. Weingärtner, *J. Phys. Chem.*, 1995, **99**, 2243.
9. Y. L. A. Rezus and H. J. Bakker, *Phys. Rev. Lett.*, 2007, **99**, 148301.
10. D. Paschek, *J. Chem. Phys.*, 2004, **120**, 10605.
11. K. Lum, D. Chandler and J. D. Weeks, *J. Phys. Chem. B*, 1999, **103**, 4570.
12. D. Chandler, *Nature (London)*, 2005, **437**, 640.
13. S. Rajamani, T. M. Truskett and S. Garde, *Proc. Natl. Acad. Sci. USA*, 2005, **102**, 9475.
14. M. V. Athawale, G. Goel, T. Ghosh, T. M. Truskett and S. Garde, *Proc. Natl. Acad. Sci.*, 2007, **104**, 733.
15. http://www.youtube.com/watch?v = ETMmH2trTpM.

16. S. J. Gill, S. F. Dec, G. Olofsson and I. Wadsö, *J. Phys. Chem.*, 1985, **89**, 3758.
17. N. Muller, *Acc. Chem. Res.*, 1990, **23**, 23.
18. B. Lee and G. Graziano, *J. Am. Chem. Soc.*, 1996, **118**, 5163.
19. K. A. T. Silverstein, A. D. J. Haymet and K. A. Dill, *J. Am. Chem. Soc.*, 2000, **122**, 8037.
20. H. S. Frank and M. W. Evans, *J. Chem. Phys.*, 1945, **13**, 507.
21. E. Gallicchio, M. M. Kubo and R. M. Levy, *J. Phys. Chem. B*, 2000, **104**, 6271.
22. H. S. Ashbaugh, T. M. Truskett and P. G. Debenedetti, *J. Chem. Phys.*, 2002, **116**, 2907.
23. A. Rahman and F. H. Stillinger, *J. Chem. Phys.*, 1971, **55**, 3336.
24. M. W. Mahoney and W. L. Jorgensen, *J. Chem. Phys.*, 2000, **112**, 8910.
25. S. W. Rick, *J. Chem. Phys.*, 2004, **120**, 6085.
26. H. W. Horn, W. C. Swope, J. W. Pitera, J. D. Madura, T. J. Dick, G. L. Hura and T. Head-Gordon, *J. Chem. Phys.*, 2004, **120**, 9665.
27. J. C. Owicki and H. A. Scheraga, *J. Am. Chem. Soc.*, 1977, **99**, 7413.
28. S. Shimizu and H. S. Chan, *J. Chem. Phys.*, 2000, **113**, 4683.
29. S. Shimizu and H. S. Chan, *J. Am. Chem. Soc.*, 2001, **123**, 2083.
30. D. Paschek, *J. Chem. Phys.*, 2004, **120**, 6674.
31. D. Paschek, *Phys. Rev. Lett.*, 2005, **94**, 217802.
32. U. Essmann, L. Perera, M. L. Berkowitz, T. A. Darden, H. Lee and L. G. Pedersen, *J. Chem. Phys.*, 1995, **103**, 8577.
33. S. Miyamoto and P. A. Kollman, *J. Comp. Chem.*, 1992, **13**, 952.
34. E. Lindahl, B. Hess and D. van der Spoel, *J. Mol. Model.*, 2001, **7**, 306.
35. S. Nosé, *Mol. Phys.*, 1984, **52**, 255.
36. W. G. Hoover, *Phys. Rev. A*, 1985, **31**, 1695.
37. M. Parrinello and A. Rahman, *J. Appl. Phys.*, 1981, **52**, 7182.
38. S. Nosé and M. L. Klein, *Mol. Phys.*, 1983, **50**, 1055.
39. J. O. Hirschfelder, C. F. Curtiss and R. B. Bird, *Molecular Theory of Gases and Liquids*, Wiley, New York, 1954.
40. B. Widom, *J. Chem. Phys.*, 1963, **39**, 2808.
41. D. Frenkel and B. Smit, *Understanding Molecular Simulation. From Algorithms to Applications*, 2nd Aufl., Academic Press, San Diego, 2002.
42. R. P. Kennan and G. L. Pollack, *J. Chem. Phys.*, 1990, **93**, 2724.
43. E. Wilhelm, R. Battino and R. J. Wilcox, *Chem. Rev.*, 1977, **77**, 219.
44. T. R. Rettich, Y. Handa, R. Battino and E. Wilhelm, *J. Phys. Chem.*, 1981, **85**, 3230.
45. H. Naghibi, S. F. Dec and S. J. Gill, *J. Phys. Chem.*, 1986, **90**, 4621.
46. A. Braibanti, E. Fisicaro, F. Dallavalle, J. D. Lamb, J. L. Oscarson and R. Sambasiva Rao, *J. Phys. Chem.*, 1994, **98**, 626.
47. A. Ben-Naim, *J. Phys. Chem.*, 1978, **82**, 792.
48. D. Ben-Amotz and B. Widom, *J. Phys. Chem. B*, 2006, **110**, 19839.
49. D. Ben-Amotz and B. Widom, *J. Chem. Phys.*, 2007, **126**, 104502.

50. G. L. Deitrick, L. E. Scriven and H. T. Davis, *J. Chem. Phys.*, 1989, **90**, 2370.

51. G. L. Deitrick, L. E. Scriven and H. T. Davis, *Molecular Simulation*, 1992, **8**, 239.

52. M. Neumann, *Mol. Phys.*, 1983, **50**, 841.

53. J. E. Roberts and J. Schnitker, *J. Chem. Phys.*, 1994, **101**, 5024.

54. J. E. Roberts and J. Schnitker, *J. Phys. Chem.*, 1995, **99**, 1322.

55. A. Ben-Naim, *Hydrophobic Interactions*, Plenum Press, New York, 1980.

56. G. Hummer, S. Garde, A. E. García, A. Pohorille and L. R. Pratt, *Proc. Natl. Acad. Sci. USA*, 1996, **93**, 8951.

57. G. Hummer, S. Garde, A. E. García, M. E. Paulaitis and L. R. Pratt, *J. Phys. Chem. B*, 1998, **102**, 10469.

58. G. Hummer, S. Garde, A. E. García and L. R. Pratt, *Chem. Phys.*, 2000, **258**, 349.

59. P. E. Krouskop, J. D. Madura, D. Paschek and A. Krukau, *J. Chem. Phys.*, 2006, **124**, 016102.

60. G. S. Kell, *J. Chem. Engineer. Data*, 1967, **12**, 66.

61. H. Tanaka, *J. Chem. Phys.*, 1996, **105**, 5099.

62. J. Holzmann, R. Ludwig, A. Geiger and D. Paschek, *Chem. Phys. Chem.*, 2008, **9**, 2722.

63. R. Fernandez-Prini and R. Crovetto, *J. Phys. Chem. Ref. Data.*, 1989, **18**, 1231.

64. Fitted excess chemical potential $\mu_{ex}(T) = \mu_0 + \mu_1 T + \mu_2 T^2 + \mu_3 T^3$ of methane in TIP4P-Ew water: $\mu_0 = -5.07 \times 10^1 \, \text{kJ} \, \text{mol}^{-1}$, $\mu_1 = 3.91 \times 10^{-1} \, \text{kJ} \, \text{mol}^{-1} \, \text{K}^{-1}$, $\mu_2 = -7.92 \times 10^{-4} \, \text{kJ} \, \text{mol}^{-1} \, \text{K}^{-2}$, $\mu_3 = 5.19 \times 10^{-7} \, \text{kJ} \, \text{mol}^{-1} \, \text{K}^{-3}$. Including polarization: $\mu_0 = -4.94 \times 10^1 \, \text{kJ} \, \text{mol}^{-1}$, $\mu_1 = 3.81 \times 10^{-1} \, \text{kJ} \, \text{mol}^{-1} \, \text{K}^{-1}$, $\mu_2 = -7.83 \times 10^{-4} \, \text{kJ} \, \text{mol}^{-1} \, \text{K}^{-2}$, $\mu_3 = 5.22 \times 10^{-7} \, \text{kJ} \, \text{mol}^{-1} \, \text{K}^{-3}$.

65. A. Ben-Naim and M. Yaacobi, *J. Phys. Chem.*, 1974, **78**, 170.

66. P. J. Dyer, H. Docherty and P. T. Cummings, *J. Chem. Phys.*, 2008, **129**, 024508.

67. D. E. Smith, L. Zhang and A. D. J. Haymet, *J. Am. Chem. Soc.*, 1992, **114**, 5875.

68. D. E. Smith and A. D. J. Haymet, *J. Chem. Phys.*, 1993, **98**, 6445.

69. L. X. Dang, *J. Chem. Phys.*, 1994, **100**, 9032.

70. S. Lüdemann, H. Schreiber, R. Abseher and O. Steinhauser, *J. Chem. Phys.*, 1996, **104**, 286.

71. S. Lüdemann, R. Abseher, H. Schreiber and O. Steinhauser, *J. Am. Chem. Soc.*, 1997, **119**, 4206.

72. S. W. Rick and B. J. Berne, *J. Phys. Chem. B*, 1997, **101**, 10488.

73. S. W. Rick, *J. Phys. Chem. B*, 2000, **104**, 6884.

74. T. Ghosh, A. E. García and S. Garde, *J. Chem. Phys*, 2002, **116**, 2480.

75. S. Shimizu and H. S. Chan, *Proteins: Struct., Funct., Genet.*, 2002, **49**, 560.

76. N. T. Southall and K. A. Dill, *Biophys. Chem.*, 2002, **101–102**, 295.
77. G. Graziano, *J. Chem. Phys.*, 2005, **123**, 034509.
78. H. Frauenfelder, N. A. Alberding, A. Ansari, D. Braunstein, B. R. Cowen, M. K. Hong, I. E. T. Iben, J. B. Johnson, S. Luck, M. C. Marden, J. R. Mourant, P. Ormos, L. Reinisch, R. Scholl, A. Schulte, E. Shyamsunder, L. B. Sorensen, P. J. Steinbach, A. H. Xie, R. D. Young and K. T. Yue, *J. Phys. Chem.*, 1990, **94**, 1024.
79. R. Jaenicke, *Eur. J. Biochem*, 1991, **202**, 715.

CHAPTER 21

Partial Molar Heat Capacity Changes of Gases Dissolved in Liquids

EMMERICH WILHELM[a] AND RUBIN BATTINO[b]

[a] Institute of Physical Chemistry, University of Wien, Währinger Strasse 42, A-1090, Wien (Vienna), Austria; [b] Department of Chemistry, Wright State University, Dayton, OH 45435, USA

21.1 Introduction

Quantitative experimental investigations of physical properties of solutions and of phase equilibria involving solutions have always held a prominent position in physical chemistry. The scientific insights gained in these studies can hardly be overrated and have been of immense value for the development of the general discipline of solution thermodynamics. An important sub-discipline of this field concerns the solubility of gases in liquids, which has a long and distinguished tradition by itself going back to Bunsen's work of 1855.[1] A few selected representative reviews are provided by references 2–14, documenting the impressive number of vapour-liquid equilibrium (VLE) measurements involving a supercritical solute, that is to say, of experiments at conditions where the experimental temperature exceeds the critical temperature of the solute. In addition to its profound theoretical interest,[15–22] this topic is of importance in many areas of the applied sciences, for instance in chemical engineering, geo-chemistry, the environmental sciences, and biomedical technology. Water is the most abundant liquid on earth, and because it sustains life as we know it, it is also the most important liquid solvent. Thus, studies of the solubility of gases in

Heat Capacities: Liquids, Solutions and Vapours
Edited by Emmerich Wilhelm and Trevor M. Letcher
© The Royal Society of Chemistry 2010
Published by the Royal Society of Chemistry, www.rsc.org

water, in particular of the rare gases and of simple hydrocarbons, hold a prominent position in biophysics. They are a major source of information on hydrophobic effects that are thought to be of pivotal importance in the formation and stability of higher order structures of biological substances, such as proteins, nucleic acids, and cell membranes.[23–36] The preponderance of scientific papers dealing with aqueous solutions is therefore not surprising.

The unusual partial molar thermodynamic properties of nonpolar solutes in water (compared to those of the same solute in other solvents) were discussed by Frank and Evans[37] in terms of some unspecified ordering of the water molecules around the solute, the famous "iceberg" formation. While this concept should not be taken literally, it has provided a convenient platform for discussing semi-quantitatively the large entropy losses and the large heat capacity changes[10,18,21,22,38,39] measured when dissolving a nonpolar solute in water. What has been indicated by computer simulations[40,41] and model calculations,[34] has been confirmed by structural studies using the method of neutron diffraction and isotopic substitution, NDIS, on aqueous solutions of argon[42] and methane.[43] The picture that emerges from this work is one in which the solute, say, methane is surrounded by a relatively strong first coordination shell containing about 19 water molecules which are oriented tangentially to the CH_4 molecule, with the pair distribution functions for methane in water showing no evidence of a second coordination shell (a similar behavior was observed by Broadbent and Neilson[42] in their NDIS study of the hydration structure around an argon atom). Apparently, the interaction between the apolar molecules and water is much shorter-ranged than suggested by the models used for computer simulations. The results also suggest that water molecules point their OH bonds towards the bulk. At elevated temperatures, these pseudo-clathrate cages should gradually disappear,[44] and concomitantly the partial molar heat capacity $C_{P,2}^{L\infty}$ at constant pressure P of solute 2 at infinite dilution in the liquid (L) solvent 1 (water) should also diminish, though it will increase again when the critical temperature $T_{c,1}$ of water is approached: $C_{P,2}^{L\infty}$ will diverge to $+\infty$ as $T_{c,1}$ is approached from *lower* temperatures (at $P = P_{c,1}$, the critical pressure of water). The effects of this divergence are felt relatively far from the critical point.[45]

Since the heat capacity at constant pressure is related to the second derivative of the Gibbs energy with respect to temperature, it is a particularly sensitive property to be used for comparison with results of model calculations. With the availability of reliable partial molar heat capacity changes on solution, determined either *via* van't Hoff analysis of high-precision gas solubility data or calorimetrically, this is indeed becoming an attractive possibility.

21.2   Experiment

Only very few laboratories have produced gas solubility data (VLE) of sufficient precision as a function of temperature to allow a van't Hoff-type analysis for obtaining reliable partial molar heat capacity changes on solution of gases in liquids. By high precision we mean results where the experimental imprecision does not exceed, say, 0.1 % (the imprecision of most gas solubility data in the literature is in the range 0.5 % to 3 %). The most important ones are the three laboratories of B. B. Benson and co-workers,[46,47] of Battino and co-workers,[48] and of Cook and co-workers.[49] The first two groups reported low-pressure gas solubility data of many aqueous systems, and the last group reported only on the low-pressure solubility of hydrogen and deuterium in eight nonpolar solvents from 248.15 K to 308.15 K for total pressures always close to 0.1 MPa. We note that gases reacting *chemically* with water, such as NH_3, CO_2, H_2S, and SO_2, are very soluble and will not be considered here.

Direct calorimetric determinations of the partial molar enthalpy change on solution of a gas in a liquid,

$$\Delta H_2^{\infty}(T, P_{s,1}) = H_2^{L\infty} - H_2^{pg*} \tag{1}$$

have been carried out also by only a limited number of researchers, primarily due to the very small heat effects involved. Here, $H_2^{L\infty}$ is the partial molar enthalpy of solute 2 at infinite dilution in the liquid solvent 1, H_2^{pg*} is the molar enthalpy of the pure (*) solute in the perfect gas (pg) state, T denotes the thermodynamic temperature, and $P_{s,1} = P_{s,1}(T)$ is the vapour pressure of the solvent. For aqueous solutions, the partial molar enthalpy changes at 298.15 K vary, for instance, from about $-0.6\,kJ\,mol^{-1}$ (for He) to $-19.5\,kJ\,mol^{-1}$ (for C_2H_6). The experimental difficulty is connected with the mole fraction solubility x_2 at *ca.* 0.1 MPa partial gas pressure (usual experimental conditions) being very small; it is about 10^{-4} for gas solubilities in organic solvents, and about 10^{-5} for gas solubilities in water. Thus the calorimetrically measured heat effects are also correspondingly small with the related energy changes amounting to fractions of 1 J. Besides calorimetric sensitivity, the major experimental problem is achieving the dissolution of an accurately known amount of gas in a time interval compatible with the stability of the calorimeter.

From the *temperature dependence* of $\Delta H_2^{\infty}(T,P_{s,1})$ one may subsequently obtain the *partial molar heat capacity change on solution* of gases in liquids (for details see below),

$$\Delta C_{P,2}^{\infty}(T, P_{s,1}) = C_{P,2}^{L\infty} - C_{P,2}^{pg*} \tag{2}$$

and thus $C_{P,2}^{L\infty}$, with $C_{P,2}^{pg*}$ being the molar heat capacity at constant pressure of the pure solute in the perfect gas state.

There were three early reports on direct calorimetric measurements of ΔH_2^{∞}. Two of them reported data at 298.15 K only,[50,51] while the third one, by Cone *et al.*,[52] presented values for methane, ethane, propane, ethene, and carbon dioxide dissolved in benzene, tetrachloromethane, and tetrahydrofuran at 288.15 K and 298.15 K. However, the derived $\Delta C_{P,2}^{\infty}$ values are in error by about an order of magnitude, as shown conclusively by Battino and Marsh.[53] Using a modified isothermal displacement calorimeter,[54] these latter authors measured ΔH_2^{∞} of argon and nitrogen in tetrachloromethane, cyclohexane, and benzene at 298.15 K, and ΔH_2^{∞} of carbon dioxide, methane, ethane, ethene, and propane in the same three solvents at 298.15 K and 318.15 K. Their derived $\Delta C_{P,2}^{\infty}$ values are much more reasonable. The *major* step forward for measuring ΔH_2^{∞} with a precision high enough to allow reliable determination, from its temperature dependence, of $\Delta C_{P,2}^{\infty}$ is connected with the development of microcalorimeters (batch or flow) by I. Wadsö at the Thermochemistry Laboratory in Lund, Sweden, and by S. J. Gill in the Chemistry Department of the University of Colorado in Boulder, Colorado, USA.[55,56] Regrettably, so far no other research group has continued work in this field.

There exist only *seven sets* of data of *directly* determined heat capacities of gases dissolved in water: the apparent molar heat capacities for aqueous argon,[57,58] xenon,[58] methane,[59] and ethene[58] (and CO_2, H_2S, and NH_3, see Reference 59) were determined over very large temperature ranges by Wood and co-workers with flow calorimeters described in detail elsewhere.[60–62] The concentrations of the dissolved gases used in these studies are dilute enough to make the apparent molar heat capacities *approximately equal* to the partial molar heat capacities at infinite dilution within experimental error (at temperatures below 500 K). These calorimetric measurements were all performed at elevated pressures between *ca.* 17 MPa and 32 MPa. In principle, the pressure dependence of $C_{P,2}^{L\infty}$ can be estimated from volumetric data,[57,63,64] but since its influence is less than the experimental error, the "raw" comparisons presented here in this chapter are justified.

21.3 Thermodynamics, Data Reduction, Data Correlation and Selected Results

The exact thermodynamic relations of relevance for a van't Hoff-type analysis of high-precision gas solubility data have been derived by Wilhelm,[17–19,65–68] and recently summarized.[21,22,38] Specifically, the Henry fugacity[†] of solute 2 dissolved in liquid solvent 1 at temperature T and pressure P is defined by

$$h_{2,1}(T,P) = \lim_{x_2 \to 0} \frac{f_2^{L}(T,P,x_2)}{x_2} \tag{3}$$

where $f_2^{L}(T, P, x_2)$ denotes the fugacity of solute 2 in the liquid phase at mole fraction x_2. This quantity is also known as Henry's law constant. For VLE, because of the phase equilibrium criterion

$$f_2^{L}(T,P,x_2) = f_2^{V}(T,P,y_2) \tag{4}$$

where f_2^{V} denotes the fugacity of solute 2 in the coexisting vapor (V) phase at

---

[†] In this chapter the Henry fugacity (Henry's law constant) is represented by the symbol $h_{2,1}$ as suggested by E. Wilhelm in *Experimental Thermodynamics, Vol. VII*, R. D. Weir and Th. W. De Loos, eds., Elsevier/IUPAC 2005, and not by $k_{H,2,1}$ as was recently recommended by IUPAC in the Green Book.

mole fraction y_2, f_2^L may conveniently be expressed as

$$f_2^L(T, P, x_2) = \phi_2^V(T, P, y_2) y_2 P \tag{5}$$

where ϕ_2^V (T, P, y_2) is the fugacity coefficient of component 2 in the coexisting vapor phase. Thus, at temperature T and at the vapor pressure $P_{s,1}(T)$ of the solvent, the Henry fugacity $h_{2,1}(T, P_{s,1})$ pertaining to the *liquid phase* is rigorously accessible from *isothermal* VLE measurements at decreasing total pressure $P \to P_{s,1}$ according to

$$h_{2,1}(T, P_{s,1}) = \lim_{x_2 \to 0}\left[\frac{\phi_2^V(T, P, y_2) y_2 P}{x_2}\right] \tag{6}$$

The vapor-phase fugacity coefficient must be calculated from a suitable equation of state, say the volume-explicit virial equation.

Once Henry fugacities $h_{2,1}(T, P_{s,1})$ have been experimentally determined over a reasonably large temperature range, then one obtains therefrom

$$\frac{\Delta H_2^\infty(T, P_{s,1})}{RT} = -T\frac{d\ln[h_{2,1}(T, P_{s,1})/\mathrm{Pa}]}{dT} + \frac{V_2^{L\infty}}{R}\frac{dP_{s,1}}{dT} \tag{7}$$

and

$$\frac{\Delta C_{P,2}^\infty(T, P_{s,1})}{R} = \frac{d\Delta H_2^\infty(T, P_{s,1})}{RdT} - \frac{1}{R}\left[V_2^{L\infty} - T\left(\frac{\partial V_2^{L\infty}}{\partial T}\right)_P\right]\frac{dP_{s,1}}{dT} \tag{8}$$

$$\begin{aligned} =&2T\frac{d\ln[h_{2,1}(T, P_{s,1})/\mathrm{Pa}]}{dT} - T^2\frac{d^2\ln[h_{2,1}(T, P_{s,1})/\mathrm{Pa}]}{dT^2} \\ &+2\frac{T}{R}\frac{dV_2^{L\infty}}{dT}\frac{dP_{s,1}}{dT} - \frac{T}{R}\left(\frac{\partial V_2^{L\infty}}{\partial P}\right)_T\left(\frac{dP_{s,1}}{dT}\right)^2 + \frac{TV_2^{L\infty}}{R}\frac{d^2P_{s,1}}{dT^2} \end{aligned} \tag{9}$$

These equations constitute the rigorous basis for any van't Hoff-type analysis of gas solubility data leading to $\Delta H_2^\infty(T, P_{s,1})$ and $\Delta C_{P,2}^\infty(T, P_{s,1})$, respectively. Here, R is the gas constant, and $V_2^{L\infty}$ is the partial molar volume of dissolved gas at infinite dilution. The ordinary differential quotients in the above equations indicate differentiation while maintaining orthobaric conditions. The first term on the right-hand side of Equation (7) as well as the first and the second term on the right-hand side of Equation (9) may all be obtained from any one of the selected fitting equations for $\ln[h_{2,1}(T, P_{s,1})/\mathrm{Pa}]$. Until recently, the remaining terms on the right-hand sides of Equations (7), (8), and (9), respectively, containing $V_2^{L\infty}$ and its derivatives with respect to T and P together with $dP_{s,1}/dT$ and $d^2P_{s,1}/dT^2$ – now referred to in the literature[69–71] as *Wilhelm terms* – have been overlooked. As pointed out repeatedly, at temperatures well below the critical temperature of the solvent, the magnitude of these terms will often be smaller than the experimental error of the measurements. However, their contributions increase rapidly with increasing temperature because of the increase of $dP_{s,1}/dT$ and $d^2P_{s,1}/dT^2$, and, of course, $V_2^{L\infty}$, $dV_2^{L\infty}/dT$ and $|(\partial V_2^{L\infty}/\partial P)_T|$.

For the mathematical representation of experimental Henry fugacities as a function of temperature, most frequently either the Clarke-Glew equation[72,73]

$$\ln[h_{2,1}(T, P_{s,1})/\text{Pa}] = A_0 + A_1(T/\text{K})^{-1} + A_2\ln(T/\text{K}) + \sum_{i=3}^{n} A_i(T/\text{K})^{i-2} \quad (10)$$

or the Benson-Krause equation[46,47]

$$\ln[h_{2,1}(T, P_{s,1})/\text{Pa}] = \sum_{i=0}^{m} a_i(T/\text{K})^{-i} \quad (11)$$

are used. On the basis of the ability to fit highly accurate Henry fugacities over reasonably large temperature ranges, and of simplicity, the Benson-Krause power series in T^{-1} appears to be superior. Whatever representation is selected, any correlation for $h_{2,1}(T,P_{s,1})$ extending up to the *critical region* must incorporate the thermodynamically correct *limiting behavior* of the Henry fugacity for $T \to T_{c,1}$ and $P_{s,1} \to P_{c,1}$ (with $T_{c,1}$ and $P_{c,1}$ denoting the critical temperature and the critical pressure of the solvent, respectively):[18,19,21,22,38,67,68,74]

$$\lim_{T \to T_{c,1}} h_{2,1}(T, P_{s,1}) = P_{c,1}\phi_2^{V\infty}(T_{c,1}, P_{c,1}) \quad (12)$$

and for *volatile* solutes, when the critical point of the solvent is approached along the orthobaric curve[69,75]

$$\lim_{T \to T_{c,1}} \{d\ln[h_{2,1}(T, P_{s,1})/\text{Pa}]/dT\} = -\infty \quad (13)$$

Here we note only the versatile semiempirical correlating equation suggested by Harvey[76] (where $T_r = T/T_{c,1}$ is the reduced temperature):

$$\ln[h_{2,1}(T, P_{s,1})/\text{Pa}] = \ln(P_{s,1}/\text{Pa}) + \frac{A}{T_r} + \frac{B(1 - T_r)^{0.355}}{T_r} + \frac{C\exp(1 - T_r)}{T_r^{0.41}} \quad (14)$$

For details we refer to references 21, 22, and 38, and the original papers quoted therein.

If the Benson-Krause equation is selected for the correlation of experimental Henry fugacities, van't Hoff determination of ΔH_2^∞ and $\Delta C_{P,2}^\infty$ according to Equations (7) and (9), respectively, is based on[21,22,38]

$$\frac{\Delta H_2^\infty(T, P_{s,1})}{RT} = \sum_{i=1}^{m} ia_i(T/\text{K})^{-i} + \frac{V_2^{L\infty}}{R}\frac{dP_{s,1}}{dT} \quad (15)$$

and

$$
\begin{aligned}
\frac{\Delta C_{P,2}^{\infty}(T, P_{s,1})}{R} = &- \sum_{i=2}^{m} i(i-1)a_i(T/K)^{-i} + 2\frac{T}{R}\frac{dV_2^{L\infty}}{dT}\frac{dP_{s,1}}{dT} \\
&- \frac{T}{R}\left(\frac{\partial V_2^{L\infty}}{\partial P}\right)_T \left(\frac{dP_{s,1}}{dT}\right)^2 + \frac{TV_2^{L\infty}}{R}\frac{d^2P_{s,1}}{dT^2}
\end{aligned}
\tag{16}
$$

The direct calorimetric determination of ΔH_2^{∞} for more than 20 gases dissolved in water was carried out by Gill and Wadsö and their co-workers, and their data were used to calculate $\Delta C_{P,2}^{\infty}$ values *via* an excellent approximation to Equation (8). Specifically, in the temperature region considered here, the second term on the right-hand side of Equation (8) is frequently smaller than the experimental uncertainty, whence the approximate Equation (17) is entirely satisfactory:

$$
\frac{\Delta C_{P,2}^{\infty}(T, P_{s,1})}{R} \approx \frac{d\Delta H_2^{\infty}(T, P_{s,1})}{RdT}
\tag{17}
$$

As already pointed out, only *very* few direct calorimetric determinations of molar heat capacities of gases dissolved in liquid water have been reported in the literature, all originating from Robert H. Wood's laboratory at the University of Delaware, Newark, DE, USA. Under the selected experimental conditions, these apparent molar heat capacities are approximately equal to the partial molar heat capacity at infinite dilution, $C_{P,2}^{L\infty}$. According to the defining Equation (2), the partial molar heat capacity *change* on solution (at infinite dilution) is obtained by subtracting the molar heat capacity at constant pressure of the pure solute in the perfect gas state, $C_{P,2}^{pg*}(T)$, at the temperature of interest.

For a large number of nonelectrolyte solutes in water, Table 21.1 contains *calorimetry*-based results on partial molar enthalpy and heat capacity changes at 298.15 K, as well as *van't Hoff* ΔH_2^{∞} values and $\Delta C_{P,2}^{\infty}$ values obtained from solubility measurements of gases in water according to the thermodynamically rigorous approaches outlined above (incidentally, in a few cases the solutes are subcritical vapours, *i.e.* $T < T_{c,2}$). Scanning the results presented in Table 21.1 for the rare gases as well as for O_2, CH_4, C_2H_6, C_2H_4, and C_3H_8 in water shows the remarkable agreement between these two data sets: it is usually within the combined experimental uncertainties reported by Benson and co-workers, and Battino and co-workers (gas solubility measurements), and those of Gill and Wadsö and co-workers, and Wood and co-workers (calorimetry).

For the sake of comparison, where it makes sense, we also included a few recent less precise van't Hoff results as well as van't Hoff values obtained from older measurements as published in the review article by Wilhelm *et al.*[10] more than 30 years ago. Although these values were the result of a critical evaluation

Table 21.1 Partial molar enthalpy changes on solution $\Delta H_2^\infty (T,P_{s,1})$ and partial molar heat capacity changes on solution $\Delta C_{P,2}^\infty (T, P_{s,1})$ of several gases dissolved in liquid water at $T = 298.15$ K and $P_{s,1} = 3.1691$ kPa: comparison of values obtained *via* van't Hoff analysis of precision gas solubility measurements [see Equations (7) and (9)] with values obtained by calorimetric methods. Usually, the quoted uncertainty is twice the standard deviation as calculated according to the method of Clarke and Glew,[72,73] or Benson and Krause.[46,47] Note that in some cases $T < T_{c,2}$.

	From solubility (van't Hoff analysis)		From calorimetry	
Dissolved Gas	$-\dfrac{\Delta H_2^\infty (T, P_{s,1})}{\text{kJ} \cdot \text{mol}^{-1}}$	$\dfrac{\Delta C_{P,2}^\infty (T, P_{s,1})}{\text{J} \cdot \text{K}^{-1} \cdot \text{mol}^{-1}}$	$-\dfrac{\Delta H_2^\infty (T, P_{s,1})}{\text{kJ} \cdot \text{mol}^{-1}}$	$\dfrac{\Delta C_{P,2}^\infty (T, P_{s,1})}{\text{J} \cdot \text{K}^{-1} \cdot \text{mol}^{-1}}$
He	0.67±0.07[46] 0.54±0.04[71] *0.76±0.20[10]*	117±2[46] 122±4[71] *105[10]*	0.65±0.04[95] 0.52±0.04[96]	135±7[95]
Ne	3.76±0.08[46] 3.64±0.01[71] *3.75±0.14[10]*	149±2[46] 143±2[71] *151[10]*	3.64±0.07[95] 3.95±0.09[96] 5.9±1.7[50]	145±1[95]
Ar	11.95±0.18[46,a] 11.92±1[71] 11.96[78] *12.27±0.12[10]*	186±5[46,a] 195±1[71] 192[78] *178[10]*	12.01±0.08[95] 11.94±0.05[96] 12.0±0.8[50] 12.07±0.16[97]	200±5[95] 185±20[57,e]
Kr	15.51±0.06[46] 15.34±0.01[71] *15.70±0.13[10]*	210±3[46] 218±1[71] *204[10]*	15.29±0.06[95] 15.28±0.04[96] 15.8±0.8[50] 16.06±0.33[97]	220±4[95]
Xe	19.18±0.08[46] 19.06±22[71] *18.42±0.40[10]*	251±4[46] 250±2[71] *197[10]*	18.87±0.12[95] 19.10±0.12[96] 17.2±0.8[50] 17.24±0.21[97]	250±9[95] 240±5[58,e]
Rn	*21.4[10]*	*293[10]*		
H₂	*4.04[10]*	*148[10]*		
N₂	10.77[46,b] 10.16[46,c] 10.45±0.04[81] *10.44[10]*	189[46,b] 232[46,c] 214±2[81] *221[10]*		
O₂	12.19[46] 12.09[80] 12.01[47] 11.97[82] *12.06[10]*	192[46] 138[80] 196[47] 200[82] *200[10]*	12.06±0.02[55] 12.03±0.04[56] 12.00±0.05[95]	205±2[95]
CO	10.78[83] 11.13[10]	215[83] 194[10]		
CH₄	13.19±0.04[48] *13.8[10]*	237±3[48] *207±49[10]*	13.06±0.15[95] 13.18±0.07[99] 13.12±0.07[100]	242±6[95] 218±5[101] 209±3[100] 212±5[59,e]
CD₄	13.03[84]	206[84]		
C₂H₆	19.50±0.08[48] *19.8[10]*	272±10[48] *301±37[10]*	19.30±0.12[95] 19.52±0.12[99] 19.43±0.10[102]	317±10[95] 284±10[101] 273±2[102]
C₂H₄	16.40±0.08[85] *15.30±0.07[10]*	239±10[85] *171±54[10]*	19.30±0.12[95]... 16.46±0.07[99]	237±5[101] 221±5[58,e]

Table 21.1 *(Continued)*

Dissolved Gas	From solubility (van't Hoff analysis)		From calorimetry	
	$-\dfrac{\Delta H_2^\infty(T,P_{s,1})}{kJ \cdot mol^{-1}}$	$\dfrac{\Delta C_{P,2}^\infty(T,P_{s,1})}{J \cdot K^{-1} \cdot mol^{-1}}$	$-\dfrac{\Delta H_2^\infty(T,P_{s,1})}{kJ \cdot mol^{-1}}$	$\dfrac{\Delta C_{P,2}^\infty(T,P_{s,1})}{J \cdot K^{-1} \cdot mol^{-1}}$
C_2H_2	14.8 ± 0.57^{10}	178 ± 39^{10}	14.62 ± 0.02^{99}	154 ± 7^{101}
C_3H_8	23.41 ± 0.22^{86}	370 ± 54^{86}	22.90 ± 0.08^{95}	389 ± 20^{95}
	22.5 ± 1.2^{10}	369 ± 70^{10}	23.27 ± 0.26^{99}	332 ± 38^{101}
			23.11 ± 0.13^{102}	319 ± 3^{102}
C_3H_6			21.64 ± 0.12^{99}	278 ± 21^{101}
$c\text{-}C_3H_6$			23.26 ± 0.06^{99}	303 ± 11
$n\text{-}C_4H_{10}$	26.0 ± 2.5^{10}	373 ± 160^{10}	25.93 ± 0.08^{95}	425 ± 17^{95}
			25.92 ± 0.17^{99}	390 ± 28^{101}
			25.70 ± 0.15^{103}	383 ± 9^{103}
Isobutane (2-methyl-propane)			24.19 ± 0.25^{99} 23.97 ± 0.19^{103}	377 ± 20^{101} 360 ± 23^{103}
1-Butene			24.88 ± 0.11^{99}	389 ± 28^{101}
$C(CH_3)_4$	27.8 ± 8.1^{10}	520 ± 270^{10}	25.11 ± 0.17^{99}	486 ± 31^{101}
CCl_2F_2	23.5 ± 1.7^{87}	197 ± 52^{87}	26.13 ± 0.18^{104}	315 ± 11^{104}
$CClF_3$	17.2 ± 1.5^{87}	375 ± 68^{87}	21.24 ± 0.38^{104}	278 ± 22^{104}
$CBrF_3$			23.72 ± 0.21^{104}	318 ± 17^{104}
CF_4	14.5 ± 2.1^{88}	637 ± 172^{88}	15.77 ± 0.14^{104}	343^{104}
	15.77 ± 0.33^{89}	410 ± 123^{89}	14.54 ± 0.22^{105}	268 ± 18^{105}
	15.06^{10}	380^{10}		
C_2F_6	17.82 ± 0.84^{89}	724 ± 217^{89}		
	25.24^{90}	1130^{90}		
C_2F_4	17.28^{10}	296^{10}		
C_3F_6	20.2^{10}			
$c\text{-}C_4F_8$	27.74^{87}	763^{87}		
	24.77 ± 0.50^{89}	791 ± 237^{89}		
NF_3	15.53^{10}	293^{10}		
$SF_6^{\,d}$	19.99^{10}	523^{10}	20.66 ± 0.44^{105}	311 ± 25^{105}

[a] In their paper, Benson and Krause[46] recalculated the original experimental data on argon dissolved in water of Murray and Riley.[77]

[b] In their paper, Benson and Krause[46] recalculated the original experimental data on nitrogen dissolved in water of Murray et al.[79]

[c] In their paper, Benson and Krause[46] recalculated the original experimental data on nitrogen dissolved in water of Klots and Benson.[80]

[d] The results quoted for SF_6 dissolved in water[10] are based on the solubility data of Ashton et al.,[91] which are in very close agreement with the more recent results of Cosgrove and Walkley.[92] Henry fugacities from 75 °C to 230 °C, though with distinctly greater imprecision, have recently been measured by Mroczek.[93] He combined them with the low-temperature data from literature and correlated this data set with the semiempirical large temperature range Harvey equation.[76] Additional thermodynamic properties of $x_1SF_6 + x_2H_2O$ have been measured by Strottmann et al.[94]

[e] After extrapolating the apparent molar heat capacity to 298.15 K and setting it approximately equal to $C_{P,2}^{L\infty}$ (see text), equation (2) was used with $C_{P,2}^{pg*}(Ar) = C_{P,2}^{pg*}(Xe) = 20.79 \, J \, K^{-1} \, mol^{-1}$, and with perfect-gas state heat capacities $C_{P,2}^{pg*}(CH_4, 298.15\,K) = 35.79 \, J \, K^{-1} \, mol^{-1}$ and $C_{P,2}^{pg*}(C_2H_4, 298.15\,K) = 43.63 \, J \, K^{-1} \, mol^{-1}$, respectively, taken from Landolt-Börnstein.[98]

of the extant literature at that time, agreement with calorimetric data is distinctly less satisfactory, simply because in general experimental accuracy was lower. In particular, this is the case for some of the heavier hydrocarbon gases. To facilitate comparison, the values from reference 10 are given in *italics*. When no data (or no reliable data) are available, the empty boxes in the table will serve as a guide to indicate where experiments would be highly desirable. We note in passing that we gave up our attempts to do high-precision gas solubility measurements on *n*-butane in water due to inconsistent results, which we ended up attributing to adsorption effects of this gas in our primarily glass apparatus.

To further illustrate the correspondence between the indirect van't Hoff approach and the direct calorimetric approach, Table 21.2 presents results for methane dissolved in water in the temperature range 273.15 K through 323.15 K. The van't Hoff results were obtained by Rettich *et al.*,[48] while the calorimetry-based results were taken from the work of Naghibi *et al.*[100] Agreement between the respective ΔH_2^∞ values is close to perfect at 298.15 K, and only towards the edges of the temperature range do the values drift apart slightly. The calorimetry-based $\Delta C_{P,2}^\infty$ values of Naghibi *et al.* are always lower than the corresponding van't Hoff values, yet the differences are quite small and are *comparable* with the differences observed between different calorimetric data sets (see also Table 21.1): for instance, at 298.15 K the van't Hoff value[48] for $\Delta C_{P,2}^\infty/\text{J K}^{-1}\text{mol}^{-1}$ is 237 ± 3, while the calorimetry-based values are $\Delta C_{P,2}^\infty/\text{J K}^{-1}\text{mol}^{-1} = 209\pm3$ (Naghibi *et al.*[100]), 218 ± 5 (Dec and Gill[101]), and 242 ± 6 (Olofsson *et al.*[95]). According to Equation (2), the directly obtained partial molar heat capacity (at high dilution) of Hnĕdkovský and Wood[59] yields,

Table 21.2 Partial molar enthalpy changes on solution $\Delta H_2^\infty(T,P_{s,1})$ and partial molar heat capacity changes on solution $\Delta C_{P,2}^\infty(T,P_{s,1})$ of methane, CH_4, dissolved in liquid water at selected temperatures T and corresponding vapour pressure $P_{s,1}(T)$ of pure water: comparison of our values obtained *via* van't Hoff analysis of high-precision gas solubility measurements[48] [see Equations (7) and (9)], with values obtained calorimetrically.[100]

	From solubility (van't Hoff analysis)		From calorimetry	
T/K	$-\dfrac{\Delta H_2^\infty(T,P_{s,1})}{\text{kJ}\cdot\text{mol}^{-1}}$	$\dfrac{\Delta C_{P,2}^\infty(T,P_{s,1})}{\text{J}\cdot\text{K}^{-1}\cdot\text{mol}^{-1}}$	$-\dfrac{\Delta H_2^\infty(T,P_{s,1})}{\text{kJ}\cdot\text{mol}^{-1}}$	$\dfrac{\Delta C_{P,2}^\infty(T,P_{s,1})}{\text{J}\cdot\text{K}^{-1}\cdot\text{mol}^{-1}}$
273.15	19.43±0.19	262.7±16.1	18.56±0.08	226.9±9.8
278.15	18.13±0.12	257.5±13.2	17.44±0.05	223.3±7.5
283.15	16.85±0.06	252.3±10.4	16.33±0.05	219.7±5.7
288.15	15.60±0.04	247.1±7.6	15.24±0.06	216.1±4.0
293.15	14.38±0.04	241.9±4.6	14.17±0.07	212.5±2.9
298.15	13.18±0.04	236.7±3.0	13.12±0.07	209.0±2.9
303.15	12.01±0.05	231.5±3.2	12.08±0.07	205.4±4.2
308.15	10.87±0.04	226.3±5.3	11.06±0.07	201.8±5.8
313.15	9.75±0.04	221.0±8.0	10.06±0.08	198.2±7.7
318.15	8.66±0.07	215.8±10.8	9.08±0.10	194.6±9.6
323.15	7.59±0.13	210.6±13.6	8.12±0.14	191.0±11.5

after extrapolation from 303.96 K down to 298.15 K and with the perfect-gas heat capacity $C_{P,2}^{pg*}(298.15\,\text{K})/\text{J K}^{-1}\text{mol}^{-1} = 35.79$,[98] for the partial molar heat capacity change on solution $\Delta C_{P,2}^{\infty}$ (298.15 K, $P_{s,1} = 3.1691\,\text{kPa}/\text{J K}^{-1}\text{mol}^{-1}) = 247.6-35.8 \approx 212$. This clearly illustrates the spread of results prevalent in this demanding area of experimental thermodynamics.

21.4 Concluding Remarks

Chemical thermodynamics of solutions in general, and of solutions of gases in liquids in particular, continue to be exciting, developing fields. The major driving forces are advances in instrumentation leading to increased precision, lower uncertainities and greater speed of measurements, and the need to increase the application range, that is to say, higher temperatures, higher pressures, smaller concentrations, new types of solvents (*e.g.* room temperature *ionic liquids*, see Chapter 4). This is paralleled by advances in (classical) chemical thermodynamics, in the statistical-mechanical treatment of solutions, and by increasingly sophisticated computer simulations. In this chapter we have concisely presented the thermodynamic formalism required for the determination of the partial molar heat capacity changes $\Delta C_{P,2}^{\infty}$ (and partial molar enthalpy changes ΔH_2^{∞}) on solution of gases in liquids at infinite dilution from VLE measurements, *i.e.* for a thermodynamically rigorous *van't Hoff*-type analysis of *high-precision* gas solubility measurements. Here, the thermodynamic quantity of central importance is the *Henry fugacity*, or Henry's law constant, $h_{2,1}(T,P_{s,1})$, and its temperature dependence. Whenever possible, the results obtained so far have been compared with *calorimetrically* determined quantities, and *vice versa*. Evidently, comparing van't Hoff derived enthalpy changes (*one* differentiation level with respect to T) and heat capacity changes (*two* differentiation levels with respect to T) with high-quality calorimetric results is a severe test of solubility data. In general, the agreement between these two approaches was found to be entirely satisfactory, that is, it was usually within the combined experimental errors. What a credit to the skills of solution thermodynamicists!

Acknowledgements

This chapter is based on the experimental gas solubility work we have done with T. R. Rettich and Y. P. Handa. Most of it was supported by the Public Health Service of the United States of America *via* grants from the National Institute of General Medical Sciences.

References

1. R. Bunsen, *Justus Liebigs Ann. Chem. Pharm.*, 1855, **93**, 1.
2. A. E. Markham and K. A. Kobe, *Chem. Rev.*, 1941, **28**, 519.
3. T. J. Morrison and N. B. Johnstone, *J. Chem. Soc.*, 1954, 3441.

4. D. M. Himmelblau, *J. Phys. Chem.*, 1959, **63**, 1803.
5. D. M. Himmelblau and E. Arends, *Chem.-Ing.-Techn.*, 1959, **31**, 791.
6. D. M. Himmelblau, *J. Chem. Eng. Data*, 1960, **5**, 10.
7. R. Battino and H. L. Clever, *Chem. Rev.*, 1966, **66**, 395.
8. K. W. Miller and J. H. Hildebrand, *J. Am. Chem. Soc.*, 1968, **90**, 3001.
9. E. Wilhelm and R. Battino, *Chem. Rev.*, 1973, **73**, 1.
10. E. Wilhelm, R. Battino and R. J. Wilcock, *Chem. Rev.*, 1977, **77**, 219.
11. R. Battino, *Rev. Anal. Chem.*, 1989, **9**, 131.
12. R. Fernández-Prini, J. L. Alvarez and A. H. Harvey, *J. Phys. Chem. Ref. Data*, 2003, **32**, 903.
13. (a) R. Battino and H. L. Clever, in *Development and Application in Solubility*, T. M. Letcher, ed., The Royal Society of Chemistry/IUPAC, Cambridge, UK, 2007, pp. 66–77; (b) H. L. Clever and R. Battino, in *The Experimental Determination of Solubilities*, G. T. Hefter and R. P. T. Tomkins, eds., *Wiley Series in Solution Chemistry, Vol. 6*, Wiley, Chichester, 2003, pp. 101–150.
14. J. L. Alvarez and R. Fernández-Prini, *J. Solution Chem.*, 2008, **37**, 1379.
15. J. H. Hildebrand, J. M. Prausnitz and R. L. Scott, *Regular and Related Solutions*, Van Nostrand-Reinhold, New York, 1970.
16. J. M. Prausnitz, R. N. Lichtenthaler and E. G. Azevedo, *Molecular Thermodynamics of Fluid Phase Equilibria*, Prentice-Hall PTR, Upper-Saddle River, NJ, 1999.
17. (a) E. Wilhelm, *CRC Crit. Rev. Analyt. Chem.*, 1985, **16**, 129; (b) E. Wilhelm, *Pure Appl. Chem.*, 1985, **57**, 303.
18. (a) E. Wilhelm, *Fluid Phase Equil.*, 1986, **27**, 233; (b) E. Wilhelm in *Interactions of Water in Ionic and Nonionic Hydrates*, H. Kleeberg, ed., Springer-Verlag, Berlin, BRD, 1987, pp. 117–123; (c) J.-P. E. Grolier and E. Wilhelm, *Pure Appl. Chem.*, 1991, **63**, 1427.
19. E. Wilhelm, in *Molecular Liquids: New Perspectives in Physics and Chemistry*, J.J.C. Teixeira-Dias, ed., Kluwer Academic Publishers, Amsterdam, The Netherlands, 1992, pp. 175–206.
20. E. Wilhelm, *High Temp.-High Press.*, 1997, **29**, 613.
21. E. Wilhelm, in *Experimental Thermodynamics, Vol VII: Measurement of the Thermodynamic Properties of Multiple Phases*, R. D. Weir and Th. W. de Loos, eds., Elsevier/IUPAC, Amsterdam, The Netherlands, 2005, pp. 137–176.
22. E. Wilhelm, in *Development and Applications in Solubility*, T. M. Letcher, ed., The Royal Society of Chemistry/IUPAC, Cambridge, UK, 2007, pp. 3–18.
23. W. Kauzmann, *Adv. Protein Chem.*, 1959, **14**, 1.
24. C. Tanford, *The Hydrophobic Effect: Formation of Micelles and Biological Membranes*, Wiley, New York, 1973.
25. F. H. Stillinger, *J. Solution Chem.*, 1973, **2**, 141.
26. L. R. Pratt and D. Chandler, *J. Chem. Phys.*, 1977, **67**, 3683.
27. A. Ben-Naim, *Hydrophobic Interactions*, Plenum Press, New York, 1980.
28. P. L. Privalov and S. J. Gill, *Adv. Protein Chem.*, 1988, **39**, 191.

29. K. P. Murphy, P. L. Privalov and S. J. Gill, *Science*, 1990, **247**, 559.
30. K. A. Dill, *Biochemistry*, 1990, **29**, 7133.
31. W. Blokzijl and J. B. F. N. Engberts, *Angew. Chem., Int. Ed. Engl.*, 1993, **32**, 1545.
32. G. Hummer, S. Garde, A. E. Garcia, M. E. Paulaitis and L. R. Pratt, *J. Phys. Chem. B*, 1998, **102**, 10469.
33. K. A. T. Silverstein, A. D. J. Haymet and K. A. Dill, *J. Am. Chem. Soc.*, 1998, **120**, 3166.
34. G. Hummer, S. Garde, A. E. Garcia and L. R. Pratt, *Chem. Phys.*, 2000, **258**, 349.
35. D. Chandler, *Nature*, 2005, **437**, 640.
36. D. Ben-Amotz, *J. Chem. Phys.*, 2005, **123**, 184504.
37. H. S. Frank and M. W. Evans, *J. Chem. Phys.*, 1945, **13**, 507.
38. E. Wilhelm, *Thermochim. Acta*, 1997, **300**, 159.
39. R. Battino, *J. Chem. Eng. Data*, 2009, **54**, 301.
40. B. Guillot, Y. Guissani and S. Bratos, *J. Chem. Phys.*, 1991, **95**, 3643.
41. R. L. Mancera and A. D. Buckingham, *J. Phys. Chem.*, 1995, **99**, 14632.
42. R. D. Broadbent and G. W. Neilson, *J. Chem. Phys.*, 1994, **100**, 7543.
43. P. H. K. De Jong, J. E. Wilson, G. W. Neilson and A. D. Buckingham, *Molec. Phys.*, 1997, **91**, 99.
44. K. Shinoda, *J. Phys. Chem.*, 1977, **81**, 1300.
45. J. M. H. Levelt Sengers, in *Supercritical Fluid Technology: Reviews in Modern Theory and Applications,* T. J. Bruno and J. F. Ely, eds., CRC Press, Boca Raton, 1991, pp. 1–56.
46. B. B. Benson and D. Krause Jr., *J. Chem. Phys.*, 1976, **64**, 689.
47. B. B. Benson, D. Krause Jr. and M. A. Peterson, *J. Solution Chem.*, 1979, **8**, 655.
48. T. R. Rettich, Y. P. Handa, R. Battino and E. Wilhelm, *J. Phys. Chem.*, 1981, **85**, 3230.
49. (a) M. W. Cook, D. N. Hanson and B. J. Alder, *J. Chem. Phys.*, 1957, **26**, 748; (b) M. W. Cook and D. N. Hanson, *Rev. Sci. Instrum.*, 1957, **28**, 370.
50. D. M. Alexander, *J. Phys. Chem.*, 1959, **63**, 994.
51. R. Jadot, *J. Chim. Phys. Chim. Biol.*, 1973, **70**, 352.
52. J. Cone, L. E. S. Smith and W. A. Van Hook, *J. Chem. Thermodyn.*, 1979, **11**, 277.
53. R. Battino and K. N. Marsh, *Aust. J. Chem.*, 1980, **33**, 1997.
54. M. J. Costigan, L. J. Hodges, K. N. Marsh, R. H. Stokes and C. W. Tuxford, *Aust. J. Chem.*, 1980, **33**, 2103.
55. S. J. Gill and I. Wadsö, *J. Chem. Thermodyn.*, 1982, **14**, 905.
56. S. F. Dec and S. J. Gill, *Rev. Sci. Instrum.*, 1984, **55**, 765.
57. D. R. Biggerstaff, D. E. White and R. H. Wood, *J. Phys. Chem.*, 1985, **89**, 4378.
58. D. R. Biggerstaff and R. H. Wood, *J. Phys. Chem.*, 1988, **92**, 1994.
59. L. Hnědkovský and R. H. Wood, *J. Chem. Thermodyn.*, 1997, **29**, 731.
60. D. E. White, R. H. Wood and D. R. Biggerstaff, *J. Chem. Thermodyn.*, 1988, **20**, 159.

61. R. W. Carter and R. H. Wood, *J. Chem. Thermodyn.*, 1991, **23**, 1037.
62. L. Hnědkovský, V. Majer and R. H. Wood, *J. Chem. Thermodyn.*, 1995, **27**, 801.
63. D. R. Biggerstaff and R. H. Wood, *J. Phys. Chem.*, 1988, **92**, 1988.
64. L. Hnědkovský, R. H. Wood and V. Majer, *J. Chem. Thermodyn.*, 1996, **28**, 125.
65. E. Wilhelm, *The Solubility of Gases in Liquids. Thermodynamic Considerations*, in *Solubility Data Series (IUPAC)*, *Vol. 10*, R. Battino, ed., Pergamon Press, Oxford, pp. XX–XXVIII, 1982.
66. E. Wilhelm, in *Proceedings of the Plenary Lectures, International Meetings on Phase Equilibrium Data* (*IUPAC, CODATA*), Paris, France, 5–13 September 1985, Vol. 1, pp. 73–83.
67. E. Wilhelm, *Thermochim. Acta*, 1987, **119**, 17.
68. E. Wilhelm, *Thermochim. Acta*, 1990, **162**, 43.
69. M. L. Japas and J. M. H. Levelt Sengers, *AIChE J.*, 1989, **35**, 705.
70. D. Krause Jr. and B. B. Benson, *J. Solution Chem.*, 1989, **18**, 803.
71. D. Krause Jr. and B. B. Benson, *J. Solution Chem.*, 1989, **18**, 823.
72. E. C. W. Clarke and D. N. Glew, *Trans. Faraday Soc.*, 1966, **62**, 539.
73. (a) P. D. Bolton, *J. Chem. Educ.*, 1970, **47**, 638; (b) D. M. Alexander, D. J. T. Hill and L. R. White, *Aust. J. Chem.*, 1971, **24**, 1143; (c) S. F. Dec and S. J. Gill, *J. Chem. Educ.*, 1985, **62**, 879.
74. D. Beutier and H. Renon, *AIChE J.*, 1978, **24**, 1122.
75. W. Schotte, *AIChE J.*, 1985, **31**, 154.
76. A. Harvey, *AIChE J.*, 1996, **42**, 1491.
77. C. N. Murray and J. P. Riley, *Deep-Sea Res.*, 1970, **17**, 203.
78. T. R. Rettich, R. Battino and E. Wilhelm, *J. Solution Chem.*, 1992, **21**, 987.
79. C. N. Murray, J. P. Riley and T. R. S. Wilson, *Deep-Sea Res.*, 1969, **16**, 297.
80. C. E. Klots and B. B. Benson, *J. Phys. Chem.*, 1963, **67**, 933.
81. T. R. Rettich, R. Battino and E. Wilhelm, *J. Solution Chem.*, 1984, **13**, 335.
82. T. R. Rettich, R. Battino and E. Wilhelm, *J. Chem. Thermodyn*, 2000, **32**, 1145.
83. T. R. Rettich, R. Battino and E. Wilhelm, *Ber. Bunsenges. Phys. Chem.*, 1982, **86**, 1128.
84. M. F. Costa Gomes and J.-P. E. Grolier, *Phys. Chem. Chem. Phys.*, 2001, **3**, 1047.
85. T. R. Rettich, R. Battino and E. Wilhelm, in preparation; communicated by E. W. at the 20th ICCT, Warsaw, Poland, August 3–8, 2008: p. 127, paper ES-RS-O-8.
86. T. R. Rettich, R. Battino and E. Wilhelm, in preparation; communicated by E. W. at the 19th ICCT (THERMO International 2006), Boulder, CO, USA, July 30–August 4, 2006: p. 47.
87. P. Scharlin and R. Battino, *Fluid Phase Equil.*, 1994, **94**, 137.
88. P. Scharlin and R. Battino, *J. Solution Chem.*, 1992, **21**, 67.

89. W.-Y. Wen and J. A. Muccitelli, *J. Solution Chem.*, 1979, **8**, 225.
90. R. P. Bonifácio, A. A. H. Pádua and M. F. Costa Gomes, *J. Phys. Chem. B*, 2001, **105**, 8403.
91. J. T. Ashton, R. A. Dawe, K. W. Miller, E. B. Smith and B. J. Stickings, *J. Chem. Soc. A*, 1968, **1793**.
92. B. A. Cosgrove and J. Walkley, *J. Chromatogr.*, 1981, **216**, 161.
93. E. K. Mroczek, *J. Chem. Eng. Data*, 1997, **42**, 116.
94. B. Strottmann, K. Fischer and J. Gmehling, *J. Chem. Eng. Data*, 1999, **44**, 388.
95. G. Olofsson, A. A. Oshodj, E. Qvarnström and I. Wadsö, *J. Chem. Thermodyn.*, 1984, **16**, 1041.
96. S. F. Dec and S. J. Gill, *J. Solution Chem.*, 1985, **14**, 417.
97. G. A. Krestov, V. N. Prorokov and V. V. Kolotov, *Russ. J. Phys. Chem.*, 1982, **56**, 152.
98. Landolt-Börnstein, 6. Auflage, II. Band, 4. Teil, *Kalorische Zustandsgrößen*, K. Schäfer and E. Lax, eds., Springer, Berlin, 1961.
99. S. F. Dec and S. J. Gill, *J. Solution Chem.*, 1984, **13**, 27.
100. H. Naghibi, S. F. Dec and S. J. Gill, *J. Phys. Chem.*, 1986, **90**, 4621.
101. S. F. Dec and S. J. Gill, *J. Solution Chem.*, 1985, **14**, 827.
102. H. Naghibi, S. F. Dec and S. J. Gill, *J. Phys. Chem.*, 1987, **91**, 245.
103. H. Naghibi, D. W. Ownby and S. J. Gill, *J. Chem. Eng. Data*, 1987, **32**, 422.
104. H. Naghibi, D. W. Ownby and S. J. Gill, *J. Solution Chem.*, 1987, **16**, 171.
105. D. Hallén and I. Wadsö, *J. Chem. Thermodyn.*, 1989, **21**, 519.

CHAPTER 22

Heat Capacities of Molten Salts

YIZHAK MARCUS

Institute of Chemistry, The Hebrew University of Jerusalem, 91904, Jerusalem, Israel

22.1 Introduction

The heat capacities of molten salts are important from both a theoretical and a practical aspect. The former depend on the insight that can be gained from examination of the heat capacity values in relation to the properties of the constituent ions of the salt and of other properties of the melts. The latter devolve around the use of molten salts as heat transfer agents and as media for sensible thermal energy storage. However, contrary to the availability of heat capacity data for many crystalline salts in the first half of twentieth century, accurate data of the molar heat capacity change on their melting, hence of the molar heat capacity of the molten salts, were scarce before the middle of the twentieth century. Thus, when Bockris and Richards[1] discussed the difference between the molar heat capacity at constant pressure, C_p, and that at constant volume, C_v, or their ratio, γ, they required estimates of the C_p of seven out of nine alkali metal halides. These estimates were based on the relation $C_p(l)/C_p(c) = 1.1$, where (l) and (c) denote the melt and the crystal.

The work of Dworkin and Bredig[2,3] among others in the early 1960s provided many required data on the heat capacity change on melting, $\Delta_m C_p$, hence of C_p [the (l) of $C_p(l)$ is dropped in the following unless required to set the value of the melt off against that of another phase]. Thus the *Molten Salts Handbook* by Janz[4] in 1967 had already a good collection of C_p values, as did the *Metallurgical Thermochemistry* of Kubachewski *et al.*[5a] of the same year. A rather complete collection of data is available in *Thermochemical Properties*

Heat Capacities: Liquids, Solutions and Vapours
Edited by Emmerich Wilhelm and Trevor M. Letcher
© The Royal Society of Chemistry 2010
Published by the Royal Society of Chemistry, www.rsc.org

of Inorganic Substances by Barin and Knacke,[6] published in 1973, and in its successor, *Materials Thermochemistry* by Kubachewski *et al.*[5b] of 1993. An interesting feature of these data, as well as those accumulated since then and shown in Table 22.1, is that the C_p values of the molten salts are practically all reported as invariant with the temperature. The only exceptions[5,6] are LiCl, NaCl, NaOH, $K_2B_4O_7$, CsCl, CsI, TlBr, $CaCl_2$, LuF_3, and $TiCl_4$, but for no obvious reasons.

This chapter discusses the heat capacities of molten salts constituted of inorganic ions (but excludes very high melting oxides, silicates, *etc.*) and leaves the discussion of room temperature ionic liquids, the cations of which are generally organic, for another review. Most of the salts of interest have melting temperatures $T_m \geq 470$ K; those that melt at a lower temperature may not be completely ionic in the liquid state, but are included anyway.

22.2 Methodology

The following is a very brief account of the methods employed for obtaining the molar heat capacities of molten salts. Most of the more reliable values of the molar heat capacities of molten salts at constant pressure (nominally 1 atm. or 0.1 MPa), C_p, have been obtained from calorimetric measurements at elevated temperatures. The usual arrangement was that of drop calorimetry,[2,7] in which the sample of mass w_s was preheated to a definite temperature, T_s, in a furnace, then dropped into the calorimeter of heat capacity C to which it imparted its heat, raising its temperature from T_1 to T_2. The heat content of the sample per unit mass at T_s is (on replacing the integral $\int C dT$ by the product):

$$h_s = C(T_2 - T_1)/w_s \tag{1}$$

The molar enthalpy change of the sample between T_s and the mean temperature in the calorimeter, $T_{mean} = (T_2 + T_1)/2$ is:

$$\Delta H(T) = H(T_s) - H(T_{mean}) = M_s h_s \tag{2}$$

where M_s is the molar mass of the sample. The values are generally normalized to $T_{mean} = 298.15$ K. The molar heat capacity of the sample is then the slope of $\Delta H(T)$ as a function of T_s:

$$C_p = d\Delta H(T)/dT_s \tag{3}$$

In practice $\Delta H(T)$ is generally a linear function, within experimental error, hence the invariance of C_p with the temperature noted above. This procedure is applicable both below and above the melting temperature, T_m. The values thus obtained were claimed in the older literature to be accurate to 2 % or $2 \, J \, K^{-1} \, mol^{-1}$, the probable error having been halved in the more reliable recent determinations.

Table 22.1 The heat capacities of molten salts at constant pressure, C_p. Reported are the melting temperature, T_m, the temperature range over which C_p was determined, the reported C_p and the reference, and the difference between C_p of the molten salt and the sum of the ideal gas heat capacities of its ions, ΔC_p. Probably incorrect values are in [ ].

Salt	T_m/K	T/K range	$C_p/J\,K^{-1}\,mol^{-1}$	Ref.	$\Delta C_p/J\,K^{-1}\,mol^{-1}$ from data	calculated
LiH	962	962–1223	58.6	5, 6	17.0	
LiF	1121	1121–1966	64.2	5, 6	22.6	23
LiCl	883	883–1653	64.0	5, 6	22.4	25
LiBr	823	823–1562	65.3	5, 6	23.8	26
LiI	742	742–1449	63.2	5, 6	21.7	27
LiOH	744	744–1803	87.1, 86.8	5, 6	36.7	32
LiClO₃	401	401–430	122.2	24	33.9	38
LiClO₄	509	509–1500	161.1	5, 6	53.8	40
LiNO₃	525		112.5, 152.7	25, 40	28.2, 58.9	37
Li₂CO₃	993	993–2000	142.8, 185.4	50, 5	25.5, 71.3	36
Li₂SiO₃	1474	near T_m	167.4	5		
Li₂SO₄	1130		207.9	26	62.9	59
Li₂CrO₄	755		200.0	26	57.3	61
Li₂MoO₄	974		215.1	26	69.5	62
Li₂WO₄	1013		205.0, 218.8	5, 26	59.1	64
Li₂B₄O₇	1193	near T_m	453.6	5		
NaF	1269	1269–1983	70.5	5, 6	28.9	24
NaCl	1074	1074–1738	67.8	5, 6	26.2	27
NaBr	1020	1020–1666	62.3	5, 6	20.8	28
NaI	933	933–1577	64.9	5, 6	23.4	29
NaOH	593	900–1663	87.1, 83.7	5, 6	37.1, 33.7	35
NaCN	835	835–1803	79.5	5, 6	25.9	39
NaNO₂	557	571–630	116.7	27	48.4	39
NaNO₃	579	579–700	136.8, 149.4	49, 40	50.0, 55.6	40
NaClO₃	533	533–570	133.9	28	39.9	40
NaBF₄	679	near T_m	165.4	5	47.9	42

NaHSO$_4$	455	458–543	246.1	41	*130.9*	42
NaHCO$_2$	531	550–560	142.3	46	*67.1*	42
NaCH$_3$CO$_2$	601	570–590	154.8, 174.5	46, 40	*32.6, 62.4*	42
NaC$_2$H$_5$CO$_2$		570–580	226.4	46		57
Na$_2$CO$_3$	1123	1123–2000	189.5	5, 6	*70.3*	57
Na$_2$SiO$_3$	1361	near T_m	179.1	5		63
Na$_2$SO$_4$	1155	1157–2000	197.4, 204.2	5, 6, 26	*52.0, 58.8*	63
Na$_2$CrO$_4$	1070		204.6, 212.5	5, 26	*58.3, 66.2*	64
Na$_2$MoO$_4$	962		213.0, 215.1	5, 26	*67.5, 69.6*	66
Na$_2$WO$_4$	967		209.2, 216.3	5, 26	*63.7, 70.8*	67
Na$_2$S$_2$O$_7$	674	680–720	244.8	42	*43.2*	69
Na$_2$B$_4$O$_7$	1016	1016–2000	444.9, 355	5, 6, 29		66
Na$_2$SiF$_6$	1120	near T_m	276.1	5	*79.6*	66
Na$_3$AlF$_6$	1279	1279–1600	396.2, 390.8	5, 6	*175.2, 169.8*	
KF	1130	1130–1783	66.9	5, 6	*25.4*	26
KCl	1044	1044–1710	73.6	5, 6	*32.0*	28
KBr	1007	1007–1671	69.9	5, 6	*28.4*	29
KI	954	954–1618	72.4	5, 6	*30.8*	30
KOH	673	673–1600	83.1	5, 6	*33.0*	38
KCN	895	895–1898	75.3	5, 6	*21.3*	42
KSCN	446	453–493	[429]	9		
KNO$_3$	607	607–700	123.4, 141.0	5, 6	*35.4, 53.0*	43
KHF$_2$	512	near T_m	104.6	5	*37.1*	40
KBF$_4$	843	near T_m	167.2	5	*46.1*	45
KHCO$_2$	441	443–459	[282]	9		
KCH$_3$CO$_2$	565	573–623	[380]	9		
KHSO$_4$	481	483–540	287.0	41	*169.2*	45
KMgCl$_3$	755	772–859	168.1	30		
K$_2$MgCl$_4$	705	706–776	282.9	30		
K$_2$S	1221	near T_m	101.0	5	*38.6*	48
K$_2$CO$_3$	1174	1174–2000	183.8, 209.2	50, 5	*63.8, 89.1*	61
K$_2$SO$_4$	1341	1342–1700	200.0, 204.2	5, 26	*50.1, 54.3*	67
K$_2$CrO$_4$	1250	near T_m	209.2	5	*60.3*	68

Table 22.1 (*continued*).

Salt	T_m/K	T/K range	$C_p/\text{J K}^{-1}\text{mol}^{-1}$	Ref.	$\Delta C_p/\text{J K}^{-1}\text{mol}^{-1}$ from data	calculated
K_2WO_4	1189		213.4, 218.0	5, 26	65.7, 70.3	71
$K_2S_2O_7$	692	692–717	267	42	64.3	73
$K_2Cr_2O_7$	671		415.9	40	*184.6*	77
$K_2B_4O_7$	1089	near T_m	464.6	5		
RbF	1048	1200–1663	59.4, 58.8	5, 6	*17.8, 17.2*	27
RbCl	988	988–1654	64.0	5, 6	22.4	29
RbBr	953	953–1625	66.9	5, 6	25.4	30
RbI	913	913–1577	66.5, 66.9	5, 6	25.0, 25.4	31
RbOH	658	near T_m	83.7	5	33.5	38
$RbNO_3$	583	near T_m	[102.9]	52	*15.9*	45
$RbMgCl_3$	825	834–890	164.7	30		
Rb_2MgCl_4	740	750–783	294.3	30		
Rb_2SO_4	1341	near T_m	207.1	5	57.3	68
Rb_2WO_4	1225		213.8	26	65.6	73
$Rb_2S_2O_7$	708	724–757	272.2	42	68.7	74
CsF	976	976–1504	74.1	5, 6	32.6	28
CsCl	918	918–1597	74.4, 76.6	5, 31	33.6, 35.8	30
CsBr	908	908–1573	77.4	5, 6	35.8	31
CsI	907	908–1172	72.4, 71.3	5, 31	30.8, 29.7	32
CsOH	616	near T_m	81.6	5	*31.5*	40
$CsNO_3$	680	near T_m	136.0	5	45.4	46
$CsMgCl_3$	883	898–949	167.4	30		
Cs_2MgCl_4	813	820–864	308.1	30		
Cs_2SO_4	1286	near T_m	207.1	5	*58.9*	71
Cs_2CrO_4	1234		210.9	26	62.3	72
Cs_2MoO_4	1220	near T_m	210.0	5, 26	*61.8*	73
Cs_2WO_4	1217		214.2	26	66.1	75
$Cs_2S_2O_7$	728	728–777	222.3	42	87.8	77

Salt	T	T range		Ref		Ref
CuCl	703	703–1485	66.9	5, 6	25.3	26
CuBr	761	761–1591	66.9	5, 6	25.3	26
CuI	861	861–1480	70.8, 66.9	5, 6	29.2, 25.3	27
AgCl	728	728–1837	61.2, 66.9	51, 5	25.3, 19.6	27
AgBr	703	703–1833	59.3, 62.3	51, 5	20.8, 17.7	28
AgI	831	831–1778	58.6	5, 6	17.0	30
AgNO$_3$	483	483–600	139.1, 128.0	49, 5	57.1, 46.0	41
Ag$_2$S	1103	near T_m	93.1	5	30.7	
Ag$_2$SO$_4$	937	933–1200	205.0, 182.0	5, 6, 26	62.9, 39.9	64
InCl	498	near T_m	62.8	5	21.2	27
InBr	558	near T_m	60.7	5	19.1	28
InI	638	near T_m	60.7	5	19.1	29
TlF	600	600–973	67.3	5, 6	25.7	27
TlCl	702	702–1089	59.4, 74.9	5, 47	17.8, 33.3	29
TlBr	733	733–1098	77.9, 71	5, 6	36.3, 29.4	30
TlI	713	713–1188	72.0	5, 6	30.4	31
TlNO$_3$	484	484–513	130.3	48	48.2	42
Tl$_2$SO$_4$	905	near T_m	205.0	5	63.4	68
NH$_4$SCN	423	423–440	[523]	32		
NH$_4$HCO$_2$	389	389–410	[417]	32		
NH$_4$CH$_3$CO$_2$	387	387–410	[574]	32		
NH$_4$NO$_3$	443	442–460	161.1, 168.1	33, 40	59.1, 66.1	41
BeF$_2$	815	815–1424	87.9	5, 6	25.5	25
BeCl$_2$	688	688–805	121.4	5, 6	59.0	36
BeBr$_2$	761	761–794	113.0	5, 6	50.6	38
BeI$_2$	758	758–755	113.0	5, 6	50.6	41
MgF$_2$	1536	1536–2605	94.4	5, 6	32.0	36
MgCl$_2$	987	987–1691	92.5	5, 6	30.1	40
MgBr$_2$	984	984–1503	104.6, 97.3	5, 6	42.2, 34.9	43
MgI$_2$	923		100.4	4, 5	38.0	46
MgSO$_4$	1400	1400–2000	159.0, 159.8	5, 6	40.9	
Mg$_3$(PO$_4$)$_2$	1621	1621–2000	462.1	5	176.0	

Table 22.1 (*continued*).

Salt	T_m/K	T/K range	$C_p/\text{J K}^{-1}\text{mol}^{-1}$	Ref.	$\Delta C_p/\text{J K}^{-1}\text{mol}^{-1}$	
					from data	calculated
CaH$_2$	1273	near T_m	75.3	5	12.9	
CaF$_2$	1691	1691–2783	99.2, 100.0	5, 6	36.8, 37.6	38
CaCl$_2$	1045	1045–2273	106.7, 103.3	5, 6	44.3, 40.9	44
CaBr$_2$	1015	1015–2073	113.0, 114.6	5, 6	50.6, 52.2	46
CaI$_2$	1052	1053–1373	103.3, 117.2	5, 6	40.9, 54.8	50
CaB$_4$O$_7$	1263	near T_m	444.8	5		
Ca$_2$P$_2$O$_7$	1626	near T_m	405.0	5		
SrF$_2$	1673	1673–2753	99.0, 100.4	5, 6	36.6, 38.0	40
SrCl$_2$	1146	1146–2273	104.6	5, 6	42.2	46
SrBr$_2$	916	916–2150	116.4, 108.8	5, 6	54.0. 46.4	48
SrI$_2$	811	811–1850	110.0, 113.0	5, 6	47.6, 50.6	52
BaF$_2$	1641	1563–2488	99.8	5	38.4	43
BaCl$_2$	1235	1235–2103	108.8	5, 6	46.4	49
BaBr$_2$	1130	1127–1300	104.9	5, 6	42.5	51
BaI$_2$	985	985–1300	113.0	5	50.6	55
CrCl$_2$	1088	1088–1573	100.4	5	38.0	42
MnF$_2$	1203	1129–2093	92.1, 98.3	5, 6	29.7, 35.9	36
MnCl$_2$	923	923–1504	94.6, 96.2	5, 6	31.4, 33.8	42
MnBr$_2$	971	971–1300	100.4	5, 6	42.0	44
MnI$_2$	911	911–1290	108.8	5, 6	46.4	47
FeF$_2$	1373	1373–2110	98.3	5, 6	35.9	35
FeCl$_2$	950	950–1285	102.1	5, 6	39.7	41
FeBr$_2$	964	964–1207	106.8	5, 6	44.4	43
FeI$_2$	860	860–1366	113.0	5, 6	50.6	47
CoCl$_2$	1013	1013–1298	99.2, 100.4	5, 6	36.8, 38.0	41
NiCl$_2$	1303	1303–1350	100.4	34	38.0	39
CuF$_2$	1043	1043–1722	100.4, 94.1	5, 6	38.0, 31.7	34
ZnF$_2$	1148	near T_m	94.1	5	31.7	34

Salt	T_m	Range					
$ZnCl_2$	591	591–1005	100.8, 92.5	5, 6, 40	38.4, *30.1*	41
$ZnBr_2$	675	675–923	113.8	5, 6	51.4	43
$PdCl_2$	952	near T_m	94.1		31.7	40
$CdCl_2$	841	841–1234	110.0, 104.6	5, 6	47.6, 42.2	44
$CdBr_2$	568	near T_m	101.7	5	39.3	46
CdI_2	661	near T_m	102.1	5	39.7	49
HgF_2	918	918–920	102.1	5, 6	39.7	38
$HgCl_2$	550	550–577	102.1	5, 6	39.7	45
$HgBr_2$	514	514–592	102.1	5, 6	39.7	47
HgI_2	525	near T_m	104.6	5	42.2	48
$SnCl_2$	520	520–925	96.2, 92.0	5, 6	33.8, *29.6*	41
$SnBr_2$	505	near T_m	99.6	5	37.2	43
SnI_2	593	near T_m	94.6	5	32.2	47
PbF_2	1091	1091–1576	109.2, 100.4	5, 6	46.8, 38.0	40
$PbCl_2$	768	768–1226	111.5, 104.2	5, 40	49.1, 41.8	47
$PbBr_2$	640	640–1187	112.1, 110.0	5, 40	49.7, 47.6	49
PbI_2	680	680–1135	108.6, 135.6	5, 6	46.2, 73.2	52
$PbSO_4$	1443	near T_m	179.9	5	61	
$EuBr_2$	1095	1096–1300	105.4	43	43.0	48
UO_2Cl_2	851	near T_m	159.8	5	58.6	
$AlCl_3$	465	466–1000	111.3	5, 6	28.2	46
$AlBr_3$	370	370–528	125.0	5, 6	41.9	53
AlI_3	464	464–658	121.3	5, 6	38.2	60
Al_2S_3	1370	near T_m	156.9	5	52.9	
ScF_3	1825	near T_m	88.9	5	5.7	40
$ScCl_3$	1240	near T_m	143.4	5	60.2	53
YF_3	1428	near T_m	133.7	5	50.5	44
YCl_3	954	near T_m	177.6	5	94.4	57
LaF_3	1766	near T_m	152.7	5	69.5	48
$LaCl_3$	1128	1128–2085	125.5, 157.7	3, 5, 6	42.4, *74.6*	61
$LaBr_3$	1061	near T_m	151.1	53	67.9	67
LaI_3	1062	near T_m	151.8	5	68.6	71

Table 22.1 (continued).

Salt	T_m/K	T/K range	C_p/J K^{-1} mol^{-1}	Ref.	ΔC_p/J K^{-1} mol^{-1}	
					from data	calculated
CeF₃	1705	near T_m	125.0	5	21.0	47
CeCl₃	1095	near T_m	145.2	5	62.0	60
CeBr₃	1006	near T_m	152.7	5	69.6	64
CeI₃	1034	near T_m	152.7	3	69.6	71
PrF₃	1672	near T_m	130.8	5	47.6	47
PrCl₃	1059	1000–1090	133.9, 155.3	4, 8	50.8, 72.2	59
PrBr₃	966	near T_m	154.8	4	71.7	66
PrI₃	1010	near T_m	143.1	3	60.0	73
NdF₃	1650	near T_m	133.9	5	50.7	47
NdCl₃	1031	1030–1090	146.4, 149.5	3, 8	63.3, 66.4	59
NdBr₃	955	near T_m	148.5	4	65.4	61
NdI₃	1057	near T_m	155.7, 151.9	5, 3	68.8, 72.6	72
SmCl₃	955	950–1300	145.3	35	62.2	58
EuCl₃	896	894–1300	142.3, 156.0	5, 35	58.3	58
GdF₃	1505	near T_m	127.8	5	44.6	46
GdCl₃	882	880–970	141.0, 139.9	4, 8	57.9, 56.8	58
GdBr₃	1043	near T_m	139.3, 135.1	5, 4	56.2	64
GdI₃	1204	near T_m	155.9	5	72.7	72
TbCl₃	855	866–965	139.3	54	56.1	58
TbBr₃	1103	1103–1300	145.0	36	61.9	64
DyF₃	1430	near T_m	156.9	5	73.7	46
DyCl₃	924	935–960	144.8, 159.4	5, 8	61.7, 76.2	57
HoF₃	1416	near T_m	96.0	5	12.8	45
HoCl₃	991	near T_m	147.7, 148.7	4, 5	64.6, 65.6	57
ErF₃	1419	near T_m	139.1	5	55.9	46
ErCl₃	1049	near T_m	148.5, 141.0	5, 4	65.4	56
TmF₃	1431	near T_m	140.3	5	57.1	46
TmCl₃	1092	1092–1300	148.5	35	65.4	56

Salt						
YbCl₃	1143	near T_m	121.3	5	38.1	56
LuF₃	1457	near T_m	214.1	5	130.9	46
FeCl₃	577	577–670	133.9, 134	5, 37	50.9	50
GaCl₃	351	near T_m	128.0	5	44.8	49
GaBr₃	395	near T_m	125.5	5	42.3	53
GaI₃	485	near T_m	128.5	5	45.3	60
InI₃	480	near T_m	136.0	5	52.8	64
AsF₃	260	276–288	126.8	38	42.5	
AsCl₃	257	298–403	133.9	5, 6	50.4	41
AsI₃	414	near T_m	133.9	5	50.4	55
SbF₃	564	near T_m	127.6	5, 6	44.4	60
SbCl₃	346	346–493	123.9, 123.4	5, 6	40.8, 40.3	64
SbBr₃	368	368–562	125.5	5, 6	42.4	58
SbI₃	443	near T_m	143.5	5	60.3	60
BiF₃	922	near T_m	184.6	5	101.4	67
BiCl₃	503	near T_m	143.5, 127.6	4, 5	60.4, 44.4	69
BiBr₃	492	500–560	157.7	39	74.6	
BiI₃	681	near T_m	157.7	5	74.6	
Bi₂S₃	1036	near T_m	188.3	5	84.3	
VOCl₃	194	298–1100	150.6	5, 6	54.7	
TiF₄	428	near T_m	156.5	5	52.5	17
TiCl₄	249	298–409	143.6	5, 6	39.7	39
TiBr₄	311	311–504	151.9	5, 6	48.0	45
VCl₄	248	298–425	161.7	5, 6	57.8	41
MoCl₄	600	near T_m	149.8	5	45.6	40
ThF₄	1383	near T_m	152.7	5	48.7	37
ThCl₄	1043	1043–1195	163.2, 159.0	5, 6	59.2, 55.1	58
ThBr₄	952	near T_m	171.5	5	67.5	65
ThI₄	839	near T_m	177.4	5	73.4	76
UF₄	1039	1039–1730	165.3, 160.2	5, 6	61.4, 56.3	36
UCl₄	863	863–1062	162.5	5, 6	58.6	57
UBr₄	792	792–1050	163.2	5, 6	59.3	64

Table 22.1 (*continued*).

Salt	T_m/K	T/K range	C_p/J K^{-1} mol^{-1}	Ref.	ΔC_p/J K^{-1} mol^{-1}	
					from data	*calculated*
UI$_4$	793	793–1030	165.7, 163.2	5, 6	61.8, 59.3	75
NbF$_5$	352	near T_m	177.8	5	53.0	
NbCl$_5$	477	477–523	184.1	5, 6	59.4	
NbBr$_5$	540	near T_m	184.1	5	59.4	
TaF$_5$	369	near T_m	177.8	5	53.0	
TaCl$_5$	490	490–513	180.8, 184.1	5, 6	56.0, 59.4	
TaBr$_5$	543	near T_m	184.1	5	59.4	

More common recently is the use of differential scanning calorimetry (DSC) that provides lower accuracy due to the small samples, a few mg, compared with the substantial samples of the order of 1 g in drop calorimetry. Measurements are generally made against a standard of well known specific heat, say sapphire.[8] The measured heat capacity depends somewhat on the heating rate of the DSC and the overall relative uncertainty is >2 % or 5 J K^{-1} mol^{-1}.

Another method that was recently introduced is the use of a heated wire technique[9] that yields the volumetric heat capacity, ρc, in J K^{-1} m^{-3}, ρ being the density of the molten salt. The product of the molar volume V (in m^3 mol^{-1}) and ρc then gives the molar heat capacity C_p. The claimed relative uncertainty of the method is 3 %, but its application to, say, KNO$_3$ produced[9] the value C_p ~ 335 J K^{-1} mol^{-1} that is more than twice the established values of alkali metal nitrates (see Table 22.1). No obvious reason was offered for this discrepancy, so that results from this method for other salts cannot be considered as valid.

The molar heat capacity of molten salts at constant volume, C_v, is useful for theoretical discussions and for comparison with some computer simulations. This value is obtained from calorimetrically measured C_p values and the thermodynamic relation:

$$C_v = C_p - \alpha_p^2 VT/\kappa_T = C_p[1 - (1 + C_p/u^2\rho\alpha_p^2 VT)^{-1}] \qquad (4)$$

where $\alpha_p = (\partial \ln V/\partial T)_P$ is the isobaric expansibility and $\kappa_T = -(\partial \ln V/\partial P)_T$ is the isothermal compressibility. The latter is conveniently obtained from the speed of ultrasound, u, in the melts that yields the adiabatic compressibility $\kappa_S = u^{-2}\rho^{-1}$. This is converted to the isothermal one using $\kappa_T = \kappa_S + \alpha_p^2 VT/C_p$.

22.3   Theory and Computer Simulations

No complete theory of molten salts exists (see the old chapter by Stillinger,[10] since which no fundamental progress has been made). However, model-depending approximate descriptions have been reported, from which the heat capacities can be deduced. The most useful description of a molten salt is in terms of a quasi-lattice in which cations and anions occupy alternate sites, so that each cation is surrounded mainly by anions (and some 'holes', unoccupied sites) and vice versa. The force between a pair of ions includes strong short range repulsion, medium range attraction due to London forces and multipolar interactions, and long range coulomb attraction between ions of opposite charges and repulsion between ions of the same charge. The configurational energy of the ensemble of ions is calculated as the sum of all the pair-wise interactions of the ions, and from this the partition function is obtained, yielding eventually the equation of state and the heat capacity.

A commonly used model from which the heat capacity is derived is the restrictive primitive model that regards the ions as hard spheres. They interact with infinitely large repulsion at distances $r \leq \sigma$, where σ is the mean diameter of the ions. At larger distances coulombic interaction prevails with a pair-wise potential energy of $z_+z_-e^2/\varepsilon r$, where z is the charge of the designated ion and ε the permittivity.[11] The latter is a rather elusive quantity, that at best is given the values 1 at $r \leq \sigma$ and n^2 at $r > \sigma$, where $n^2 \sim 2$ to 5 is the square of the refractive index.[10] The temperature and density variables employed in the statistical thermodynamic derivation by Larsen[11] are replaced by the dimensionless $q = (ze)^2/\varepsilon\sigma k_B T$ and the packing fraction $y = \pi\rho\sigma^3/6$, where $\rho = 2N_A/V$ is now the number density for a symmetrical salt CA. For such a model the heat capacity is given to a good approximation as:[12]

$$C_p = 3R + 2RT\alpha_p(1 + y + y^2)/(1 - y)^3 \tag{5}$$

The values thus calculated depend strongly on the choice of the mean diameter σ of the ions depicted as hard spheres. Yosim and Owen[12] took the C–A distances in the molten alkali halides (measured by X-ray diffraction) as representing σ, but the sum of the ionic radii $r_C + r_A$ prevailing in other condensed phases (crystals and solutions) could serve as well as these.

The improvement introduced by Itami and Shimoji[13] takes into account the temperature dependence of σ, and they arrived at an expression for the heat capacity at constant volume per mole of ions:

$$C_v = 1.5R - 4RT(2 - y)(\partial y/\partial T)_v/(1 - y)^3 - 2RT^2(5 - 2y)(\partial y/\partial T)_v^2/(1 - y)^4 \tag{6}$$

The derivative $(\partial y/\partial T)_v$ is negative (hence the second term on the right-hand side of Equation (6) is positive) and is estimated from equating the value of α_p/κ_T derived from the equation of state with the experimental values. Its inclusion is essential since otherwise the resulting $C_v = 1.5R$ per mole of ions for all molten salts is definitely wrong. Note that since the temperature range involved with molten salts is some hundreds of K, the temperature dependence of the radii of the ions, hence of σ, is of importance, although ignored in most discussions of electrolyte solutions (but see Krestov[14]).

A quite different approach was taken by Bagchi[15] who applied a cell model in the statistical thermodynamic calculation. The size of the cell is $(4\pi/3)y\sigma^3$, where y, the packing fraction, is taken as that of close-packed spheres (0.283 for KF, 0.203 for LiCl, NaCl, and KCl). Values of C_p for these four molten salts were given, but without an explicit expression of how they were obtained.

Computer simulations of molten salts can yield values of their heat capacities. Woodcock[16] applied both isothermal ionic dynamics (i.e. molecular dynamics) and Monte Carlo computations to LiCl, NaCl, and KCl at 1273 K to derive their $C_v/J K^{-1} mol^{-1}$ values, 47.4, 46.8, and 45.5. These were compared with nominally experimental values reported by Bockris and Richards,[1] 49.4, 47.3, and 45.2, derived from C_p data by means of Equation. (4), the C_p value for LiCl being actually an estimate. The Monte Carlo simulations by Larsen[17] did not involve actual molten salts but only hypothetical ones, with varying q and y values pertaining to the restricted primitive model presented above. The resulting C_v values ranged between $3.10R$ for low q and y (2 and 0.15) to $7.0R$ for high q and y (100 and 0.39). Typical values for the alkali halides were said to be $q = 70$ and $y = 0.35$.

The difficulties involved in the computer simulations with the long-range nature of charge-charge interactions is generally overcome by using the Ewald summation method, but recently Avendano and Gil-Villegas[18] proposed the use of the Wolf method instead. They also lifted the 'restriction' from the restricted primitive model by using a softened repulsion potential instead of the hard sphere one. They did not apply their calculations to actual molten salts, but as Larsen[17] did previously, to hypothetical ones over a range of q and y values. The Monte Carlo simulations yielded values of the excess constant volume heat capacity $C_v^{ex} = C_v - 3R$ similar to those that were obtained by the application of the mean spherical approximation (MSA) and the hypernetted chain integral equation (HNC).

In all these theoretical derivations and computer simulations of the heat capacities of molten salts only uni-univalent salts of monatomic ions, specifically molten alkali halides, were considered, and the derived C_p (or C_v) values were compared with experimental data. The agreements were generally only moderate, and moreover, the C_p values of the alkali halides reported by different authors vary considerably, Table 22.2. Neither asymmetrical salts (e.g.

Table 22.2 Heat capacities at constant pressure of alkali metal halides, $C_p/\mathrm{J\,K^{-1}\,mol^{-1}}$, reported by various authors.

Salt	Ref. 12	Ref. 4	Ref. 44	Ref. 45	Ref. 6	Ref. 5
LiF	64.9	64.9			64.2	64.2
LiCl	64.0	62.8	74.1		65^a	65.3^a
LiBr	65.3		66.5		65.3	65.3
LiI	63.2				63.1	63.2
NaF	68.6	68.6			70.6	70.5
NaCl	71.1	66.9	79.	69.0	67.8	69.7^a
NaBr	69.9		73.2	67.2	62.3	62.3
NaI	69.0		67.4	65.6	64.9	64.9
KF	66.9	66.9			66.9	66.9
KCl	69.9	66.9	74.5	73.2	73.6	73.6
KBr	69.9		72.8	69.2	69.9	69.9
KI	72.4		67.4	68.2	72.4	72.4
RbF	69.9				58.8	59.4
RbCl	77.0		73.5	73.5	64.0	64.0
RbBr	73.6		72.8	72.8	66.9	66.9
RbI	72.8		72.1	72.1	66.9	66.5
CsF	74.1				74.1	74.1
CsCl	77.4	75.3	75.9	75.9	77.4	74.4^a
CsBr	77.4	72.0	77.9	77.9	77.4	77.4
CsI	75.3		74.3	74.3	85^a	72.4

[a]The values vary with T, given is the mean value over the T range shown in Table 1.

CA_2, alkaline earth halides) nor salts with polyatomic ions (e.g., NO_3^-, except in[1]) were included in the considerations.

Cantor[19] applied the Debye theory of heat capacities of solids to yield a semi-empirical expression for the entropy of molten salts:

$$S = A(v) + 1.5vR \ln[(M/M_0)(V/V_0)^{2/3}] \qquad (7)$$

where v is the number of ions per formula unit of the salt and subscript $_0$ denotes unit mass and volume. The coefficient A is presumed to have a single value for salts with the same v but would be different for salts with different v. This approach could equally well be applied to the heat capacities using the thermodynamic relation $C_p = T(\partial S/\partial T)_P$ that yields:

$$C_p = 3vR + vRT\alpha_p - 1.5vRT(\partial \ln E/\partial T)_p \qquad (8)$$

Here E is a characteristic (quasi-lattice) vibration energy that is included in the coefficient A derived by Cantor from the Debye theory. Contrary to the theoretical and computer simulation treatments presented above, Cantor could calculate by means of Equation (7) the entropies of molten salts not only of the alkali halides but also of the alkaline earth halides and of salts with polyatomic ions with reasonable agreement with experimental values. It is, therefore, tempting to apply Equation (8) to the heat capacities.

22.4 Correlations

Table 22.1 lists the C_p values of molten salts and includes in its last two columns values of $\Delta C_p = C_p(l) - C_p(\text{i.g.})$, where $C_p(\text{i.g.})$ is the heat capacity of an ideal gas composed of the same ions as the molten salt. In the case of monatomic ions $C_p(\text{i.g.}) = 2.5vR = 20.8v$ J$\,$K^{-1}mol^{-1} and is temperature independent. Any electronic heat capacity (e.g. for $MnCl_2$ and $NdBr_3$) is ignored. When gaseous polyatomic ions are involved their heat capacity depends moderately on the temperature due to the vibrational and rotational modes being excited, and values for the ions included in Table 22.1 are taken from the compilation by Loewenschuss and Marcus[20] inter- or extrapolated to $1.1T_m$. It may be noted that $C_p(\text{i.g.})$ for MO_3^{z-} and MO_4^{z-} anions depend only slightly on the nature of the central atom M. It is assumed that the vibrational modes of the polyatomic ions in the salt melt are the same as in the ideal gas and that free rotation of the ions takes place in the melt.

The difference, ΔC_p, then reflects the effects of the electrostatic interactions in the molten salt that is taken as a quasi-lattice assemblage of hard spheres. The following correlations were obtained from the data in Table 22.1. The values of ΔC_p depend linearly on the cation–anion distance, d_{C-A}. For salts constituted of univalent monatomic ions:

$$(\Delta C_p/vz_+)/\text{J}\,\text{K}^{-1}\,\text{mol}^{-1} = (6.4 \pm 2.4) + (22.7 \pm 8.0)(d_{C-A}/\text{nm}) \qquad (9)$$

for CA_2-type salts with monatomic ions:

$$(\Delta C_p/vz_+)/\text{J K}^{-1}\text{mol}^{-1} = (0.6 \pm 1.1) + (22.9 \pm 4.0)(d_{C-A}/\text{nm}) \qquad (10)$$

for CA_3-type salts with monatomic ions:

$$(\Delta C_p/vz_+)/\text{J K}^{-1}\text{mol}^{-1} = (-1.3 \pm 2.1) + (22.3 \pm 7.5)(d_{C-A}/\text{nm}) \qquad (11)$$

and for CA_4-type salts with monatomic ions:

$$(\Delta C_p/vz_+)/\text{J K}^{-1}\text{mol}^{-1} = (-3.4 \pm 0.5) + (22.5 \pm 2.0)(d_{C-A}/\text{nm}) \qquad (12)$$

Although the scatter of the values is large, resulting in large standard errors of the linear fitting coefficients, it is noteworthy that the slopes for the different types of salts are the same, but that the intercepts vary. For salts with polyatomic ions the correlation expression is:

$$(\Delta C_p/vz_+)/\text{J K}^{-1}\text{mol}^{-1} = (8.8 \pm 4.6) + (36.8 \pm 14.1)(d_{C-A}/\text{nm}) \qquad (13)$$

with a slope different from that of the salts with monatomic ions.

Values of ΔC_p that differ by more than $3v$ J K^{-1} mol^{-1} from those obtained in Equations (9) to (13) are shown in the penultimate column of Table 22.1 in *italics*. This may or may not signify that the reported values of C_p are incorrect. There may certainly be reasons why the simple correlation with the cation-anion distance should not be valid. For instance, it has been argued[10] that the quasi-lattice model should not be operative for molten lithium salts in which the large anions are in contact and the small cations occupy interstitial positions. Another cause for non-adherence to the linear correlations could be a partial, even extensive, covalent bonding of cations and anions in the melts.

Still, the conformal liquids approximation of Reiss, Katz, and Mayer[21] that has been shown to be applicable to the molten alkali metal (except lithium) halides and some alkaline earth halides does specify a single variable, d_{C-A}, for describing the properties of molten salts.[22] It must, however, be realized that a positive slope of C_p with regard to d_{C-A} is not expected from simple electrostatic considerations of a Born-type expression (ref. 10 p. 23). The electrostatic attraction energy of unlike charges should become smaller with increasing mean distances between the charges, as also its temperature dependence and C_v or C_p.

References

1. J. O'M. Bockris and N. E. Richards, *Proc. Royal Soc. (London)*, 1957, **241**, 44.
2. A. S. Dworkin and M. A. Bredig, *J. Phys. Chem.*, 1960, **64**, 269.
3. A. S. Dworkin and M. A. Bredig, *J. Phys. Chem.*, 1963, **67**, 697.

4. G. J. Janz, *Molten Salts Handbook*, Academic Press, New York, 1967, pp. 200–202.
5. (a) O. Kubachewski, E. L. Evans and C. B. Alcock, *Metallurgical Thermochemistry*, Pergamon, London, 1967; (b) O. Kubachewski, C. B. Alcock and P. J. Spencer, *Material Thermochemistry*, Pergamon, Oxford, 1993.
6. I. Barin and O. Knacke, *Thermochemical Properties of Inorganic Substances*, Springer, Berlin. 1973.
7. K. K. Kelley, B. F. Naylor and C. H. Shomate, Bur. Mines Tech. Paper 686, 1946.
8. M. Gaune-Escard, A. Bogacz, L. Rycerz and W. Szczepaniak, *J. Alloys Comp.*, 1996, **235**, 176.
9. A. A. El-Sharkawi, M. T. Dessouky, M. B. S. Osman, A. Z. Dakroury and S. R. Atalla, *High Temp. High Press*, 1993, **25**, 63.
10. F. J. Stillinger Jr, in *Molten Salt Chemistry*, ed. M. Blander, Interscience, New York, 1964, 1.
11. B. Larsen, *J. Chem. Phys.*, 1976, **65**, 3431.
12. S. J. Yosim and B. B. Owen, *J. Chem. Phys.*, 1964, **41**, 2032.
13. T. Itami and M. Shimoji, *J. Chem. Soc., Faraday Trans. II*, 1980, **76**, 1347.
14. G. A. Krestov, *Russ. J. Phys. Chem.*, 1967, 41, 1272; *Thermodynamics of Solvation*, Ellis Horwood, New York, 1991, 16. .
15. S. N. Bagchi, *Phys. Lett.*, 1979, **74A**, 271.
16. L. V. Woodward, *Chem. Phys. Lett.*, 1971, **10**, 257.
17. B. Larsen, *Chem. Phys. Lett.*, 1974, **27**, 47.
18. C. Avendano and A. Gil-Villegas, *Mol. Phys.*, 2006, **104**, 1475.
19. S. Cantor, *Inorg. Nucl. Chem. Lett.*, 1973, **9**, 1275.
20. A. Loewenschuss and Y. Marcus, *J. Phys. Chem. Ref. Data*, 1987, **16**, 61.
21. H. Reiss, S. W. Mayer and J. L. Katz, *J. Chem. Phys.*, 1961, **35**, 820.
22. M. Harada, M. Tanigaki, M. Yao and M. Kinoshita, *J. Chem. Soc., Faraday Trans. 2*, 1982, **78**, 1985.
23. A. Klemm, in ref. 10, p. 566–572.
24. A. N. Campbell and M. K. Nagarajan, *Can. J. Chem.*, 1964, **42**, 1616.
25. K. K. Kelly, *US Bur. Mines Bull.*, 1934, **371**.
26. L. Denielou, J. -P. Petitet and C. Tequi, *J. Phys.*, 1976, **37**, 1017.
27. Y. Iwadate, I. Okada and K. Kawamura, *J. Chem. Eng. Data*, 1982, **27**, 288.
28. A. N. Campbell and E. T. van der Kouwe, *Can. J. Chem.*, 1968, **46**, 1287.
29. R. P. Tye, A. O. Desjarlais and J. G. Bourne, *Proc. 7th Symp. Thermophys. Prop.*, 1977, **7**, 189.
30. J. L. Holm, B. J. Holm, B. Rinnan and F. Grønvold, *J. Chem. Thermodyn.*, 1973, **5**, 97.
31. C. E. Kaylor, G. E. Walden and D. F. Smith, *J. Phys. Chem.*, 1960, **64**, 276.
32. M. B. S. Osman, A. Z. Dakroury, M. T. Dessouky, M. A. Kenawy and A. A. El-Sharkawi, *J. Therm. Anal.*, 1996, **46**, 1697.
33. G. Feick, *J. Am. Chem. Soc*, 1954, **76**, 5858.
34. J. P. Coughlin, *J. Am. Chem. Soc.*, 1951, **73**, 5314.
35. L. Rycerz and M. Gaune-Escard, *Z. Naturforsch. A*, 2002, **57a**(79), 215.

36. L. Rycerz and M. Gaune-Escard, *J Chem. Eng. Data.*, 2004, **49**, 1078.
37. S. S. Todd and J. P. Coughlin, *J. Am. Chem. Soc.*, 1951, **73**, 4184.
38. H. Russell Jr, R. E. Rundel and D. M. Yost, *J. Am. Chem. Soc.*, 1941, **63**, 2825.
39. L. E. Topol and L. D. Ransom, *J. Phys. Chem.*, 1960, **64**, 1339.
40. M. Bizouard and F. Pautt, *Compt. Rend. C*, 1961, **252**, 514.
41. G. Hatem, K. M. Eriksen and R. Fehrmann, *J. Therm. Anal. Calorim.*, 2002, **68**, 25.
42. G. Hatem, F. Abdoun, M. Gaune-Escard, K. M. Eriksen and R. Fehrmann, *Thermochim. Acta*, 1998, **319**, 33.
43. L. Rycerz, S. Gadzuric, E. Ingier-Stocka, R. W. Berg and M. Gaune-Escard, *J. Nucl. Mater.*, 2005, **344**, 115.
44. V. I. Minchenko, M. V. Smirnov and V. P. Stepanov, *Russ. J. Phys, Chem.*, 1975, **49**, 777.
45. I. G. Murgulescu and Cr. Telea, *Rev. Roum. Chim.*, 1977, **22**, 653.
46. P. Ferloni, M. Sanesi and P. Franzosini, *Z. Naturforsch. A*, 1975, **30A**, 1447.
47. H. Eding and D. Cubicciotti, *J. Chem. Eng. Data.*, 1964, **9**, 524.
48. M. Rolla, P. Franzosini and R. Riccardi, *Disc. Faraday Soc.*, 1961, **32**, 84.
49. W. Fuchs and J. Richter, *Z. Naturforsch. A*, 1984, **39A**, 1279.
50. N. Araki, M. Matsuura, A. Makino, T. Hirata and Y. Kato, *Intl. J. Thermophys.*, 1988, **9**, 1071.
51. L. Rycerz, M. Szymanska-Kolodziej, P. Kolodziej and M. Gaune-Escard, *J. Chem. Eng. Data*, 2008, **53**, 1116.
52. H. E. G. Knape and L. M. Torell, *J. Chem. Phys.*, 1975, **62**, 4111.
53. L. Rycerz, E. Ingier-Stocka, B. Ziolek, S. Gadzuric and M. Gaune-Escard, *Z. Naturforsch. A*, 2004, **59A**, 825.
54. L. Rycerz and M. Gaune-Escard, *J. Therm. Anal. Calorim.*, 2002, **68**, 973.

Subject Index

Note: Figures are indicated by *italic page numbers*, and Tables by **enboldened numbers**